Progress in Systems and Control Theory
Volume 26

Series Editor
Christopher I. Byrnes, Washington University

Nonlinear Model Predictive Control

Frank Allgöwer
Alex Zheng
Editors

Springer Basel AG

Editors:

Frank Allgöwer
Institut für Systemtheorie technischer Prozesse
Universität Stuttgart
Pfaffenwaldring 9
70550 Stuttgart
Germany

Alex Zheng
Department of Chemical Engineering
University of Massachusetts at Amherst
159 Goessmann Lab
Amherst, MA 01003-3110
USA

1991 Mathematics Subject Classification 93-06, 49-06, 34-06; 34H05, 34K35

A CIP catalogue record for this book is available from the Library of Congress, Washington D.C., USA

Deutsche Bibliothek Cataloging-in-Publication Data

Nonlinear model predictive control / Frank Allgöwer ; Alex Zheng, ed.. – Basel ; Boston ; Berlin ; Birkhäuser, 2000
(Progress in systems and control theory ; Vol. 26)
ISBN 978-3-0348-9554-5 ISBN 978-3-0348-8407-5 (eBook)
DOI 10.1007/978-3-0348-8407-5

ISBN 978-3-0348-9554-5

Originally published by Birkhäuser Verlag in 2000
Softcover reprint of the hardcover 1st edition 2000
Printed on acid-free paper produced of chlorine-free pulp. TCF ∞

ISBN 978-3-0348-9554-5

9 8 7 6 5 4 3 2 1

Contents

Preface ix

Part I
Theoretical Issues in Nonlinear Predictive Control

G. De Nicolao, L. Magi and R. Scattolini
Stability and Robustness of Nonlinear Receding Horizon Control ... 3

David Mayne
Nonlinear Model Predictive Control:
Challenges and Opportunities 23

Christopher V. Rao and James B. Rawlings
Nonlinear Horizon State Estimation 45

Alberto Bemporad, Manfred Morari
Predictive Control of Constrained Hybrid Systems 71

Basil Kouvaritakis, Mark Cannon and J. Anthony Rossiter
Stability, Feasibility, Optimality and the Degrees of Freedom
in Constrained Predictive Control 99

David Angeli, Alessandro Casavola and Edoardo Mosca
A Predictive Command Governor for Nonlinear Systems
under Constraints 115

Alex Zheng
Some Practical Issues and Possible Solutions for
Nonlinear Model Predictive Control 129

Rolf Findeisen and Frank Allgöwer
Nonlinear Model Predictive Control for
Index–one DAE Systems 145

Masoud Soroush and Kenneth R. Muske
Analytical Predictive Control 163

J.M. Lemos, L.M. Rato and E. Mosca
Integrating Predictive and Switching Control:
Basic Concepts and an Experimental Case Study 181

J.O. Trierweiler and A.R. Secchi
Exploring the Potentiality of Using Multiple Model Approach in Nonlinear Model Predictive Control 191

Mark Cannon and Basil Kouvaritakis
Continuous-time Predictive Control of Constrained Nonlinear Systems 205

Part II
Modelling and Computational Aspects in Nonlinear Predictive Control

Lorenz T. Biegler
Efficient Solution of Dynamic Optimization and NMPC Problems 219

H.G. Bock, M.M. Diehl, D.B. Leineweber, J.P. Schlöder
A Direct Multiple Shooting Method for Real-time Optimization of Nonlinear DAE Processes 245

Jay H. Lee
Modeling and Identification for Nonlinear Model Predictive Control: Requirements, Current Status and Future Research Needs 269

Achim Helbig, Olaf Abel, Wolfgang Marquardt
Structural Concepts for Optimization Based Control of Transient Processes 295

Michael Nikolaou and Diwakar Mantha
Efficient Nonlinear Modeling Using Wavelet Compression 313

Luigi Chisci, J. Anthony Rossiter and Giovanni Zappa
Iterative Active-set Method for Efficient On-line MPC Design 335

Robert Haber, Ruth Bars and Orsolya Lengyel
Nonlinear Predictive Control Algorithms with Different Input Sequence Parametrizations Applied for the Quadratic Hammerstein and Volterra Models 347

Matthias Wellers, Heinrich Rake
Nonlinear Model Predictive Control Based on Stable Wiener and Hammerstein Models 357

Part III
Applications of Nonlinear Predictive Control

S. Joe Qin and Thomas A. Badgwell
An Overview of Nonlinear Model Predictive
Control Applications 369

Joseph Z. Lu
Multi-zone Control under Enterprise Optimization:
Needs, Challenges and Requirements 393

Matthias Schley, Vinay Prasad, Louis P. Russo and B. Wayne Bequette
Nonlinear Model Predictive Control of a
Styrene Polymerization Reactor 403

Francis J. Doyle III, Philip A. Wisnewski
Nonlinear Multi-rate MPC with Large Scale Fundamental Models:
Application to a Continuous Kamyr Digester 419

V. Wertz, L. Magni and G. Bastin
Multivariable Control of Cement Mills 433

G. De Nicolao, L. Magni and R. Scattolini
Nonlinear Receding Horizon Control of
Internal Combustion Engines 449

Gordon J. Sutton and Robert R. Bitmead
Performance and Computational Implementation of
Nonlinear Model Predictive Control on a Submarine 461

Part I

Theoretical Issues in Nonlinear Predictive Control

Progress in Systems and Control Theory, Vol. 26

Stability and Robustness of Nonlinear Receding Horizon Control

G. De Nicolao, L. Magni and R. Scattolini

Abstract. The main design strategies for ensuring stability and robustness of nonlinear *RH* (Receding-Horizon) control systems are critically surveyed. In particular, the following algorithms with guaranteed closed-loop stability of the equilibrium are considered: the zero-state terminal constraint, the dual-mode *RH* controller, the infinite-horizon closed-loop costing, the quasi-infinite method, and the contractive constraint. For each algorithm, we analyse and compare feasibility, performance, and implementation issues. For what concerns robustness analysis and design, we consider: monotonicity-based robustness, inverse optimality robustness margins, nonlinear H_∞ *RH* design, and a new nonlinear *RH* design with local H_∞ recovery.

1. Introduction

Receding-Horizon (*RH*) control, also known as model predictive control, owes its success to the simplicity of its rationale which is well suited to deal with nonlinearities and constraints. In discrete-time, the basic idea is to determine the current control $u(t)$ by solving a finite-horizon optimization problem over the interval $[t, t+N]$. At the next time instant $t+1$, the new control $u(t+1)$ is found by translating the optimization horizon and solving a new problem over $[t+1, t+N+1]$. Being optimization-based, the *RH* scheme can allow for nonlinearities and constraints much more straightforwardly than other methods. In view of the increased efficiency of the hardware, there are more and more plants on which *RH* control can be implemented by solving the finite-horizon optimization on-line.

An important caveat of *RH* control is that closed-loop stability is not guaranteed for a generic finite-horizon cost function, as it was well illustrated by Bitmead, Gevers and Wertz [3]. In the linear case, the first stability result was obtained by complementing the cost function with a terminal zero-state constraint, see e.g. the work by Kwon and Pearson [22]. Rawlings and Muske removed, at least for the stable modes, the need of equality constraints, by introducing a terminal penalty equal to the infinite-horizon cost due to zero control [34]. It is interesting to note that a fairly general stability theory for linear *RH* control can be developed by referring to the monotonicity properties of a suitable difference Riccati equation initialized with the terminal penalty matrix [3].

Although there is a well-established theory in the linear case, its importance is diminished by the fact that, in absence of nonlinearities and constraints, RH control is hardly better than infinite-horizon (IH) linear quadratic (LQ) control. This does not happen in the nonlinear case, because IH optimization becomes computationally intractable.

An important exception is given by LQ RH control with constraints. In such a context, Sznaier and Damborg [40] showed that the IH optimal control law can be found as the solution of a finite-horizon problem where the terminal penalty is equal to the IH cost of the unconstrained LQ problem. Other contributions along this line are due to Scokaert and Rawlings [35] and Chmielewski and Manousiouthakis [11].

For nonlinear systems Chen and Shaw [4], Keerthi and Gilbert [21], and Mayne and Michalska [27] showed for discrete and continuous systems that the terminal zero-state constraint guarantees closed-loop stability also in the nonlinear case. However, the presence of the terminal equality constraint places an heavy requirement on the on-line optimizing controller. This motivated the development by Michalska and Mayne of the dual-mode RH controller, which replaces the equality constraint with an inequality one, namely that $x(t+N)$ belongs to a suitable neighbourhood of the origin where the nonlinear system is stabilized by a linear control law. The scheme is called "dual-mode" because, when such a neighbourhood is eventually reached, the RH controller switches to the linear one [29]. Another stabilization method worked out by Yang and Polak is based on a terminal contractive constraint requiring that the norm of the terminal state $x(t+N)$ is sufficiently smaller than the norm of $x(t)$ [43].

More recently, schemes have been proposed that combine a terminal penalty with a terminal inequality constraint. In particular, Parisini and Zoppoli, and Chen and Allgöwer have shown that stabilization can be enforced by a suitable quadratic terminal penalty [32], [5]. On the other hand, De Nicolao, Magni and Scattolini established closed-loop stability of the equilibrium using a (nonquadratic) terminal penalty equal to the infinite cost due to a locally stabilizing linear control law [17]. The most recent developments concern the design of RH controllers of H_∞ type that are capable of achieving guaranteed robustness margins [7], [25].

The main purpose of the present contribution is to provide a critical survey of the existing literature on stability and robustness of (state-feedback) RH control schemes. The emphasis is on the key ideas rather than on mathematical technicalities. This justifies some simplifications, the use of conservative assumptions, and the lack of formal proofs.

The paper is organized as follows. In Section 2, some preliminary definitions and notions are introduced. Section 3 is devoted to the analysis of the alternative RH stabilization schemes. The results concerning robustness analysis and synthesis are reported in Section 4. Some concluding remarks end the paper.

2. Preliminaries

Consider the time-invariant nonlinear discrete-time system

$$x(k+1) = f(x(k), u(k)), \quad x(t) = \bar{x}, \quad k \geq t \tag{1}$$

where $x(k) \in R^n$ is the state, $u(k) \in R^m$ is the input, $f(\cdot,\cdot)$ is a C^2 function of its arguments, and $f(0,0) = 0$ (the origin is an equilibrium point). The state and input vectors are subject to the constraints

$$x(k) \in X, \quad u(k) \in U, \quad k \geq t \tag{2}$$

where X and U are compact sets of R^n and R^m respectively, both containing the origin as an interior point.

Consider now a control law $u = \kappa(x)$ with $\kappa(0) = 0$. The associated closed-loop system is

$$x(k+1) = F(x(k)) \quad x(t) = \bar{x}, \quad k \geq t \tag{3}$$

where $F(x) := f(x, \kappa(x))$. In the following, $\kappa(\cdot)$ will be said to be *stabilizing* if the origin of (3) is an asymptotically stable equilibrium point (for every $\varepsilon > 0$, there exists δ_ε such that, if $\|\bar{x}\| < \delta_\varepsilon$, then $\|x(k)\| < \varepsilon$ and $\lim_{k\to\infty} \|x(k)\| = 0$, where $x(k)$ is the solution of (3)). Hereafter, with some abuse of terminology, the term "stable" will be used for short in place of the more correct "asymptotically stable". The origin is said to be an exponentially stable equilibrium of (3) if there exist constants r, a, $b > 0$ such that, whenever $\|\bar{x}\| < r$, it results that $\|x(k)\| \leq ae^{-b(k-t)} \|\bar{x}\|$, $\forall k \geq t$.

A sufficient condition for the origin to be exponentially stable is that the linearized matrix $dF/dx|_{x=0}$ has all its eigenvalues inside the unit circle. In the following, we will not discuss exponential stability properties explicitly, but this property will implicitly follow whenever the linearized closed-loop is found to be stable.

Throughout this chapter, our aim is to design a state-feedback controller that stabilizes (1) around the origin complying with the constraints (2).

2.1. Linear quadratic control (LQ)

The easiest (and most used) way to stabilize (1) is by means of a linear control law designed on the base of the linearization of (1) around the equilibrium. To this purpose, let

$$A = \left.\frac{\partial f}{\partial x}\right|_{x=0,u=0} \qquad B = \left.\frac{\partial f}{\partial u}\right|_{x=0,u=0}$$

Assumption: In order to design the linear controller, it is assumed that the pair (A, B) is stabilizable. □

The idea is to find some K such that $A + BK$ is stable and then apply the control law $u = Kx$ to the nonlinear system (1). In particular, K can be found by minimizing the infinite-horizon quadratic cost function

$$J_{IH}(\bar{x}, u(\cdot)) = \sum_{k=t}^{\infty} \{x(k)'Qx(k) + u(k)'Ru(k)\} \tag{4}$$

where $Q > 0$, $R > 0$, subject to the linearized dynamics

$$x(k+1) = Ax(k) + Bu(k), \quad x(t) = \bar{x}, \quad k \geq t \tag{5}$$

As is well known, the solution to the above LQ problem is given by the linear state-feedback

$$u(k) = K^{LQ}x(k), \quad k \geq t \tag{6}$$

where

$$K^{LQ} = -(R + B'P_{\infty}B)^{-1}B'P_{\infty}A$$

and P_{∞} is the (unique) nonnegative definite solution of the ARE (Algebraic Riccati Equation)

$$P = A'PA + Q - A'PB\,(R + B'PB)^{-1}\,B'PA \tag{7}$$

When (6) is applied to (1), we obtain the closed-loop system

$$x(k+1) = f(x(k), K^{LQ}x(k)) \quad x(t) = \bar{x}, \quad k \geq t \tag{8}$$

Proof of stability: As is well known, the stability of the origin of (8) is proven by using $V(x) = x'P_{\infty}x$ as a Lyapunov function [42].

Output admissible set: Due to the stabilizability assumption, the origin is a stable equilibrium point of (8) with a nonzero-measure domain of attraction. However, the controller design completely ignores the constraints (2). Therefore, for a given $\bar{x}$, there is no guarantee that $x(k)$ and $u(k)$ defined by (8), (6) will satisfy (2). Hereafter, the term *output admissible set* [19], referred to the closed-loop formed by (1) joined with the state feedback

$$u(k) = \kappa(x(k)) \tag{9}$$

will denote an invariant set $\tilde{X}$ which is a domain of attraction of the origin and such that $\bar{x} \in \tilde{X}$ implies that $x(k)$, $u(k)$, $k \geq t$ defined by (1), (9) satisfy (2). Moreover $u(k)$, $k \geq t$, is called a *feasible control sequence.* In particular, the maximal output admissible set of the controller (6) will be indicated by $X(K^{LQ})$. Needless to say, $X(K^{LQ})$ may well be unsatisfactorily small, since both the nonlinear nature of the system and the presence of constraints have been neglected.

Under a computational viewpoint, the evaluation of $X(K^{LQ})$ for a nonlinear system may be difficult (or even impossible). It is however possible to obtain an inner bound by computing an output admissible set given by a suitable level set of a local Lyapunov function for the linearized system, see [29] and also [6], [5].

Performance: It is apparent that the linear design optimizes performance only close to the equilibrium point. The acceptability of the performance clearly depends on how well the system (1) can be approximated by its linearization (5) in the considered region.

Implementation: The calculation of the gain K^{LQ} requires only the (off line) solution of the ARE (7), which is a standard task.

2.2. Infinite Horizon Nonlinear Controller (IH)

The limitations of the LQ controller motivate the search for more sophisticated approaches capable of improving both the output admissible set and the performance. The most direct way is to take into account explicitly both nonlinearity and constraints in the minimization of (4). Along this direction, the IH (Infinite Horizon) nonlinear optimal controller is obtained by minimizing (4) subject to (1) and (2).

Assumption: As is the case of the LQ controller, we assume that (A, B) is stabilizable. □

The solution of the $IHNOCP$ (IH Nonlinear Optimal Control Problem) is given by a nonlinear state-feedback

$$u(k) = \kappa^{IH}(x(k)) \tag{10}$$

which stabilizes the origin, see e.g. [21]. In the following $\varphi^{IH}(k, \bar{x}) := x(k)$ where $x(k)$ is the solution of (1) subject to (10).

Proof of stability: Closed-loop stability of the origin is proven by using $V(x) = J^o_{IH}(x)$ as a Lyapunov function where

$$J^o_{IH}(x) = J_{IH}(x, \kappa^{IH}(\varphi^{IH}(\cdot, x)))$$

denotes the minimal value of the cost function, see e.g. [21].

Output admissible set: When $\bar{x} \in X(K^{LQ})$ the control sequence $u(k) = K^{LQ}x(k)$, $k \geq t$, is always an admissible solution for the $IHNOCP$. Consequently, the output admissible set of the $IHNOCP$, hereafter denoted by X^{IH}, is nonempty and such that $X^{IH} \supseteq X(K^{LQ})$. In the present context, we can regard X^{IH} as the largest achievable output admissible set.

Linearization: Rather expectedly, it can be shown that the linearization of $\kappa^{IH}(x)$ coincides with K^{LQ}, i.e.

$$\left.\frac{d\kappa^{IH}(x)}{dx}\right|_{x=0} = K^{LQ}.$$

This observation can help in the tuning of the design matrices Q and R . In fact, a reasonable procedure is to adjust Q and R with reference to the LQ problem (4), (5) by means of well-established methods until the (linearized) closed-loop performance becomes satisfactory. This is clearly easier than tuning Q and R by extensive trials on the nonlinear model (1).

Performance: Assuming that the weights Q and R have been properly chosen, the performance of the IH controller is "optimal" by definition.

Implementation: This is the main problem with IH control. For a generic nonlinear system, analytic solutions of the $IHNOCP$ do not exist and the attempt to approximate the IH cost functional by means of a finite-horizon one (by truncating the series in (4)) leads to a hard optimization problem in a large dimensional space.

3. Nonlinear RH control

The difficulties inherent in the implementation of the IH controller motivate the development of control strategies based on FH (Finite-Horizon) optimization. In particular, letting $u_{t,t+N-1} := [u(t),\ u(t+1), \dots u(t+N-1)]$, we will consider cost functions of the type

$$J_{FH}(x, u_{t,t+N-1}, N) = \sum_{k=t}^{t+N-1} \{x(k)'Qx(k) + u(k)'Ru(k)\} + V_f(x(t+N)) \quad (11)$$

to be minimized with respect to $u_{t,t+N-1}$, subject to (1), (2) as well as to the terminal constraint

$$x(t+N) \in X_f \subset R^n \quad (12)$$

As discussed below, the different algorithms are characterized by the choices of the terminal penalty function $V_f(x)$ and the terminal region X_f. For computational reasons the optimization horizon N should be as short as possible, compatible with the desired performance and output admissible set.

Associated with (11), (12) it is possible to define an RH (Receding Horizon) control strategy in the usual way: at every time instant t, define $\bar{x} = x(t)$ and compute the optimal solution $u^o_{t,t+N-1}$ for the FH problem (11) subject to (1), (2), (12); then apply the control $u(t) = u^o(\bar{x})$ where $u^o(\bar{x})$ is the first column of $u^o_{t,t+N-1}$. Although the FH minimization of (11) has to be performed at each time instant, this is much more viable than solving an IH problem.

The main challenge is to guarantee closed-loop stability and performance also for small values of N. Along this direction, we can take advantage of the experience gained in the RH control of linear systems [22], [2], [3], [34], [11]. In particular, it is well known that, if $V_f(x) \equiv 0$ and $X_f = R^n$, for a given N it may well happen that the RH controller yields an unstable closed-loop system. Nevertheless, by a proper design it is possible to ensure closed-loop stability with a finite horizon N.

In the LQ case a fairly complete stability theory is available which is based on the so-called Fake Riccati analysis. The main point is to choose $V_f(x)$ and X_f so as to force the monotonicity of the solution of a relevant difference Riccati equation. Once monotonicity is established, it follows that

$$J^o_{FH}(x, N) = \min_{u_{t,t+N-1}} J_{FH}(x, u_{t,t+N-1}, N)$$

is a Lyapunov function for the closed-loop.

An analogous rationale can be extended to the nonlinear case. In fact, the closed-loop stability of most RH schemes is proven by showing that $J^o_{FH}(x, N)$ is a Lyapunov function. For this purpose, the main point is to demonstrate that

$$J^o_{FH}(f(x, u^o(x)), N) < J^o_{FH}(x, N) \quad (13)$$

which can be done in two steps. First, by optimality it always holds that

$$\begin{aligned} & J^o_{FH}(f(x,u^o(x)),N-1) \\ = \; & J^o_{FH}(x,N) - x'Qx - u^o(x)'Ru^o(x) < J^o_{FH}(x,N) \end{aligned} \tag{14}$$

The second step is to show that

$$J^o_{FH}(\xi,N) \leq J^o_{FH}(\xi,N-1), \quad \forall \xi \in \Xi \tag{15}$$

where Ξ is a neighbourhood of the origin in R^n. From (15), letting $\xi = f(x,u^o(x))$ and recalling (14), one sees that (13) follows. Now, the inequality (15), which will be hereafter termed as *monotonicity property*, is fulfilled only for a suitable choice of $V_f(x)$ and X_f. In the remaining part of this section we will examine different choices of $V_f(x)$ and X_f that enforce stability.

3.1. The zero-state terminal constraint (ZS)

The idea is to let $V_f(x) \equiv 0$ and $X_f = \{0\}$. In this way, the FH minimization problem becomes equivalent to the minimization of (11) subject to (1), (2) and the $ZSTC$ (*Zero-State Terminal Constraint*) $x(t+N) = 0$. The corresponding RH control law will be indicated as $u(k) = \kappa^{ZS}(x(k))$. Historically, this was the first RH method with guaranteed stability, see [22] for linear systems and [4], [21], [27], [28], [36] for nonlinear systems.

Assumption: For a given integer $N > 0$, there exists a nonempty neighbourhood $X^c(N)$ of the origin such that $\forall \bar{x} \in X^c(N)$ one can find a control sequence $u(k)$, $t \leq k \leq t+N-1$ driving state of (1) to the origin in N steps, i.e. $x(t+N) = 0$, and such that the constraints (2) are satisfied $\forall k \in [t, t+N-1]$.

Proof of stability: The keystone of the proof is establishing the monotonicity property (15). To this purpose, let $u^o_{t,t+N-2}$ be the optimal solution minimizing $J_{FH}(\bar{x}, u_{t,t+N-2}, N-1)$ subject to (1), (2) and $x(t+N-1) = 0$. Consider now the problem of minimizing $J_{FH}(\bar{x}, u_{t,t+N-1}, N)$ subject to (1), (2) and $x(t+N) = 0$. It is clear that $\tilde{u}_{t,t+N-1} = [u^o_{t,t+N-2}, \;\; 0]$ is an admissible solution for the new problem and moreover (recalling that $x(t+N-1) = 0$) we have $J_{FH}(\bar{x}, \tilde{u}_{t,t+N-1}, N) = J_{FH}(\bar{x}, u^o_{t,t+N-1}, N-1) = J^o_{FH}(\bar{x}, N-1)$. Since optimality implies that $J^o_{FH}(\bar{x}, N) \leq J_{FH}(\bar{x}, \tilde{u}_{t,t+N-1}, N)$, the monotonicity property $J^o_{FH}(\bar{x}, N) \leq J^o_{FH}(\bar{x}, N-1)$ follows.

Output admissible set: The output admissible set coincides with the (constrained) controllability region $X^c(N)$. Note that $X^c(N)$ may be "small". In particular, there is no guarantee that $X^c(N)$ is larger than the output admissible set $X(K_{LQ})$ of the LQ controller. Of course, $X^c(N)$ grows with N, but increasing the optimization horizon has computational drawbacks.

Linearization: The linearization

$$K^{ZS} = \left.\frac{d\kappa^{ZS}(x)}{dx}\right|_{x=0}$$

coincides with the gain matrix of the linear RH controller associated with the minimization of the FH cost function (11) (with $V_f(x) \equiv 0$) subject to (5) (the linearized dynamics) and $x(t+N)=0$. Formulas for the computation of K^{ZS} when A is nonsingular can be found in [22], [3] . If A is singular, K^{ZS} can be computed by dualizing the filtering results reported in [12]. It is interesting to note that K^{ZS} can be seen as the optimal LQ gain associated with a "fake" cost function

$$J_{IH}^{F}=\sum_{k=t}^{\infty}\left\{x(k)'\left(Q+\tilde{Q}_N\right)x(k)+u(k)'Ru(k)\right\} \tag{16}$$

where $\tilde{Q}_N \geq 0$ is a suitable matrix (which tends to 0 as N goes to ∞) [3].

Performance: In view of the extra term $\tilde{Q}$ in (16), the linearization of $\kappa^{ZS}(x)$ is likely to require an excessive control effort with respect to $\kappa^{IH}(x)$. This could be compensated by artificially decreasing Q in the design phase but this may not suffice, especially for small values of N. Then, one should increase N with the obvious computational drawbacks.

Implementation: The ZS algorithm enjoys the so-called *feasibility property*: once the FH optimization problem has been solved at time t, then at time $t+1$ a feasible control sequence is immediately obtained by appending zeros to the entries of the previous optimal control sequence that have not been applied yet. Another interesting property is the *improvement property*. By recalling the proof of stability, it is seen that, in order to preserve stability, it suffices that at each step there is a decrease of the FH cost function (which plays the role of a Lyapunov function) without necessarily finding the minimum. In spite of these favourable properties, the actual implementation has to face the problem of solving a nonlinear optimization problem with equality constraints which, as pointed out in [29], can be computationally prohibitive.

3.2. The dual-mode controller (DM)

As seen in the previous subsection, the terminal equality constraint may be unsatisfactory for both performance and implementation issues. Hence, the idea of replacing the equality constraint with an inequality one, which is much easier to handle computationally.

Assumption: The pair (A,B) is stabilizable. □

Consider the LQ controller $u(k)=K^{LQ}x(k)$ and the associated output admissible set $X(K^{LQ})$. Let $W_\alpha \subset X(K^{LQ})$ denote a (smaller) output admissible set for the LQ controller where α is a scalar parameter such that $W_{\alpha''} \supset W_{\alpha'}$ if $\alpha'' > \alpha'$, and $\lim_{\alpha\to 0} W_\alpha = \{0\}$. The dual-mode controller is based on the FH problem (11), (12) with $X_f = W_\alpha$. In nominal conditions the dual-mode control scheme works as follows. If $x(t) \notin W_\alpha$, one lets $\bar{x}=x(t)$ and solves the FH problem. Then, the optimal control sequence $u^o_{t,t+N-1}$ is applied *entirely* so that $x(t+N) \in W_\alpha$. At time $t+N$ the controller switches to the linear state feedback $u(k)=K^{LQ}x(k)$, $k \geq t+N$. When conditions are not nominal, model uncertainties and disturbances are coped with by resorting to an RH implementation whose details can be found in [29], where the continuous-time case is considered. In particular, in the RH implementation a varying horizon N is used.

Proof of stability: Closed-loop stability is straightforward because W_α is an output admissible set for the LQ controller.

Output admissible set: The output admissible set $X^{DM}(N, \alpha)$ grows with N and α. In particular, if $\alpha = 0$, then $X^{DM}(N, 0) = X^c(N)$. If it were possible to take $X_f = X(K^{LQ})$, the output admissible set of the dual-mode controller would include $X(K^{LQ})$, at the cost of using the suboptimal state feedback $u(k) = K^{LQ}x(k)$ whenever $x(k) \in X(K^{LQ})$.

Linearization: The linearization around the origin obviously coincides with the LQ control law.

Performance: There is a trade-off between the size of the output admissible set and performance. To improve the former, α should be as large as possible but this implies the use of the suboptimal LQ control law in a vast region. The switching is somewhat artificial and introduces some kind of discontinuity.

Implementation: The dual-mode controller enjoys the feasibility property and avoids the use of equality constraints. In its RH implementation [29], it enjoys also the improvement property. The numerical difficulties approach those of the $ZSTC$ controller only when $\alpha \to 0$.

3.3. Infinite-horizon closed-loop costing (CL)

In analogy with the dual-mode scheme we let $X_f = X(K^{LQ})$. However, the cost function (11) includes a (generally nonquadratic) terminal penalty $V_f(\cdot)$ defined as

$$V_f(\bar{x}) = \sum_{k=t}^{\infty} \{x(k)'Qx(k) + x(k)'K^{LQ'}RK^{LQ}x(k)\} \tag{17}$$

subject to (1) and $u(k) = K^{LQ}x(k)$. In other words, $V_f(\bar{x})$ in (17) is the IH cost that is incurred by applying the LQ control law for $k \geq t$. This RH scheme has been recently proposed in [15], [16], [17], [23], [44], [45] and has some analogies with ideas originally developed in the context of LQ control with constraints, [39], [40], [41], [34], [35]. One should also mention [1], where infinite as well large, but finite, prediction horizons are used to enforce closed-loop stability of the nonlinear RH controller. Finally, the method proposed in [9], [10] is a kind of hybrid between the DM and CL schemes.

Assumption: The pair (A, B) is stabilizable.

Proof of stability: The proof is exactly the same as for the $ZSTC$ controller provided that one lets $\tilde{u}_{t,t+N-1} = [u^o_{t,t+N-2}, \quad K^{LQ}x(t + N - 1)]$.

Output admissible set: Letting $X_o(N, K^{LQ})$ denote the output admissible set of the CL RH scheme, it is possible to show that

$$X(K^{LQ}) \subseteq X_o(1, K^{LQ}) \subseteq \ldots \subseteq X_o(N, K^{LQ}) \subseteq X_o(N+1, K^{LQ}) \subseteq X_{IH}$$

Moreover, $\lim_{N\to\infty} X_o(N, K^{LQ}) = X_{IH}$. Note that the output admissible set of the CL scheme includes $X(K^{LQ})$ even for $N = 1$.

Linearization: In order to compute the linearization of the CL control law $\kappa^{CL}(x)$, it is useful to evaluate the quadratic approximation of $V_f(x)$ when x is close to the origin. By recalling the definition of $V_f(x)$, when $x \to 0$, $V_f(x)$ tends to $x'\Pi x$ where Π is the solution of the algebraic Lyapunov equation

$$\Pi = \left(A + BK^{LQ}\right)' \Pi \left(A + BK^{LQ}\right) + Q + K^{LQ'}RK^{LQ} \tag{18}$$

Note that the solution of (18) is unique due to the stability $A + BK^{LQ}$. In view of the definition of K^{LQ}, it is not difficult to verify that P_∞ (solution of (7)) satisfies (18). Hence, for small values of $\bar{x}$, the CL scheme reduces to the minimization of (11) subject to (5) with $V_f(x) = x'P_\infty x$, whose solution is $u(k) = K^{LQ}x(k)$. Hence,

$$\left.\frac{d\kappa^{CL}(x)}{dx}\right|_{x=0} = K^{LQ} \tag{19}$$

The fact that the linearization coincides with the LQ control law can ease the adoption of such a scheme. In many cases the nonlinear RH controller may represent a more sophisticated alternative to an existing LQ regulator designed on the linearized plant. In view of (19), one can widen the output admissible set and improve performances without affecting the local dynamics of the existing LQ control system. An example of this type is provided by the cement mill plant dealt with in [24].

Performance: In view of its linearization, the CL RH controller is locally optimal. Far from the origin (where feasibility is presumably more important than performance) it may provide reasonable approximation of $\kappa^{IH}(x)$ in so far $V_f(x)$ is close to $J^o_{IH}(x)$. This is more likely to happen when N increases.

Notably, if the controlled plant (1) is linear, then $\kappa^{CL}(x) = \kappa^{IH}(x)$ [39], [40], [41], [35]. This happens because, if $x \in X(K^{LQ})$, then $V_f(x)$ defined by (17) coincides with $J^o_{IH}(x)$ (recall that the constraints (2) are never active within $X(K^{LQ})$).

Implementation: The CL RH scheme enjoys both the feasibility and improvement properties as defined in Section 3.1. Due to the difficulty of computing the maximal output admissible set $X(K^{LQ})$ for a nonlinear system, the possibility of imposing that the terminal state $x(t+N)$ belongs to the region $X_f = X(K^{LQ})$ is only conceptual. Nevertheless, the algorithm is practically implementable by letting $X_f = \tilde{X}_f$ where $\tilde{X}_f$ is any output admissible set (not necessarily maximal) associated with (8)-(2). In particular, one can use the sets computed in [29], [6], [5].

As for the evaluation of $V_f(x)$ defined in (17), two procedures are possible. In the first one, $V_f(x)$ is approximated by integrating (8) for a sufficiently large number of steps. More precisely, one can use the approximation

$$\begin{aligned} V_f(x(t+N)) \;\simeq\; & \sum_{k=t+N}^{t+M-1} \left\{x(k)'Qx(k) + x(k)'K^{LQ'}RK^{LQ}x(k)\right\} + \\ & + x(t+M)'P_\infty x(t+M) \end{aligned} \tag{20}$$

where $M > N$ is "large" compared to the time constants of $A + BK^{LQ}$. The idea is that $x(t+M)$ is sufficiently close to the origin so that, by standards results of LQ control theory,

$$\sum_{k=t+M}^{\infty} \{x(k)'Qx(k) + x(k)'K^{LQ'}RK^{LQ}x(k)\} \simeq x(t+M)'P_\infty x(t+M).$$

In the second approach, let define

$$\tilde{V}_f(x(t+N), M) = \sum_{k=t+N}^{t+M-1} \{x(k)'Qx(k) + x(k)'K^{LQ'}RK^{LQ}x(k)\}$$

$M > N$, $x(t+N) \in \tilde{X}_f$. Due to the exponential stability of the origin for system (8), it is not difficult to obtain an upper bound for

$$\varepsilon_V(M) = \left|\tilde{V}_f(x(t+N), M) - V_f(x(t+N))\right|.$$

Then, it suffices to replace V_f with $\tilde{V}_f$ in the cost function (11) provided that M is chosen so as to ensure that $\varepsilon_V(M) < x(t-1)'Qx(t-1) + \kappa^{CL}(x(t-1))'R\kappa^{CL}(x(t-1))$ which preserves the improvement property.

3.4. The quasi-infinite (QI)

The idea is to use a quadratic terminal penalty of the type $V_f(x) = x'P_f x$, where $P_f = P_f' > 0$ is a suitable matrix. In [32], [33], [5], [6] and [8] it has been shown that there exist choices of P_f and X_f that guarantee closed-loop stability of the associated RH scheme. Without going into details, P_f is chosen in such a way that $x'P_f x$ is a Lyapunov function for the nonlinear system (1), subject to the linear control law $u(k) = K^{LQ}x(k)$. More precisely, X_f is an output admissible set for the LQ controller and is such that

$$f(x, K^{LQ}x)'P_f f(x, K^{LQ}x) < x'P_f x - x'(Q + K^{LQ'}RK^{LQ})x, \forall x \in X_f \qquad (21)$$

In other words, within X_f it is required that $x'P_f x$ is a Lyapunov function with a sufficiently large rate of decrease. A different rational, which we will not discuss in detail is employed for continuous-time systems in [30], where a FH cost function includes (as an additive term) also the length N of the finite-horizon, and the optimization is carried out with respect to both N and $u_{t,t+N-1}$. Then, it is shown that, by a proper choice of the quadratic terminal penalty the RH control law is a dead-beat stabilizing controller.

Assumption: The pair (A, B) is stabilizable.

Proof of stability: As usual, the proof is based on the monotonicity of the optimal value of the FH cost function $J^o_{FH}(x, N)$. The proof is exactly the same as for the CL scheme and is based on the inequality (21). Two different procedures for finding P_f and X_f are developed in [32], [33], [8], [5] and [6]. In all these works, $P_f \neq P_\infty$.

Output admissible set: For a finite value of the horizon N it seems difficult to show that the output admissible set of the QI method includes the output admissible set $X(K^{LQ})$ of the LQ controller. However, a possible way to enlarge the output admissible set is by modifying the cost function $V_f(x(t+N))$ and the terminal inequality (12) as follows:

$$\begin{aligned} V_f(x(t+N)) &= \sum_{k=t+N}^{t+M-1} \{x(k)'Qx(k) + x(k)'K^{LQ\prime}RK^{LQ}x(k)\} \\ &+ x(t+M)'P_f x(t+M) \qquad (22) \\ x(t+M) &\in X_f \qquad (23) \end{aligned}$$

where P_f and X_f are such that (21) holds (note the similarity of (22) and (20)). It is easy to see that stability of the origin is guaranteed $\forall M \geq N$. Moreover, the output admissible set of the RH scheme tends to $X_o(N, K^{LQ})$ of the CL RH method as $M \to \infty$.

Linearization: When $\bar{x}$ is close to the origin the nonlinear FH problem tends to the FH LQ problem consisting of the minimization of (11) subject to (5) (the linearized dynamics), with $V_f(x) = x'P_f x$. The corresponding RH control law is $u(k) = K^{QI}x(k)$, where

$$K^{QI} = -(R + BP(N-1)B')^{-1}B'P(N-1)A \qquad (24)$$

where $P(N-1)$ is obtained from the difference Riccati equation

$$P(k+1) = A'P(k)A + Q - A'P(k)B\left(R + BP(k)B'\right)^{-1}B'P(k)A \qquad (25)$$

with initial condition $P(0) = P_f$. In general, $P(N-1) \neq P_\infty$ but $P(N-1) \to P_\infty$ as $N \to \infty$ and/or $P_f \to P_\infty$. In conclusion

$$\left.\frac{d\kappa^{QI}(x)}{dx}\right|_{x=0} = K^{QI},$$

with $K^{QI} \neq K^{LQ}$. Conversely, if the alternative scheme (22), (23) is used, then the linearization is still given by (24) but the Riccati equation (25) has to be integrated with initial condition $P(0) = \Sigma(M-N)$ where $\Sigma(M-N)$ is obtained from the difference Lyapunov equation

$$\Sigma(k+1) = \left(A + BK^{LQ}\right)'\Sigma(k)\left(A + BK^{LQ}\right) + Q + K^{LQ\prime}RK^{LQ}$$

with initial condition $\Sigma(0) = P_f$. It can be seen that $P(N-1) \to P_\infty$ as $M \to \infty$ also for a short optimization horizon N.

Performance: In view of the linearization analysis, the QI scheme does not recover the performance of the optimal LQ controller when $\|x\| \to 0$, unless $P_f = P_\infty$. Of course, it is still possible to perform an iterative adjustment of Q and R, computing $P(N-1)$ and K^{QI} at each step, until the performances of the linearized closed-loop $A + BK^{QI}$ become satisfactory.

Implementation: The QI controller enjoys both the feasibility and the improvement properties (this can be seen by examining the details of the stability proof). Due to the use of a simple quadratic terminal penalty it is perhaps the easiest to implement.

3.5. The contractive constraint (CC)

In this approach the terminal constraint in the FH cost function (11) is defined as a *contractive constraint* of the type

$$X_f = \{x(t+N) | \, \|x(t+N)\| \leq \alpha \|\bar{x}\|\}, \quad \alpha \in [0,1) \tag{26}$$

where α is a design parameter and $\bar{x} = x(t)$ as in (1). A further constraint is also introduced to be satisfied at each time within the optimization horizon

$$\|x(t+i)\| \leq \beta \|\bar{x}\|, \quad 1 \leq i \leq N, \quad \beta \in (1, \infty] \tag{27}$$

β being another design parameter. As it will be seen later, the terminal penalty function V_f does not affect stability. Here, we assume that $V_f(x) = x'Px$, where $P \geq 0$ is a suitable matrix. In the RH implementation, at some given time t the FH problem (11), subject to (1), (2), (12), (26), (27), is solved and the optimal control $u^o_{t,t+N-1}$ is applied entirely, i.e. $u(k) = u^o(k)$, $t \leq k \leq t+N-1$. At time $t+N$, the FH problem with initial state $x(t+N)$ is solved on the interval $[t+N, t+2N-1]$ and the corresponding optimal control applied. Note that differently from the usual RH scheme in which the optimization horizon is moved stepwise, this scheme moves the optimization horizon intervalwise, i.e. once every N steps. Letting $u^{cc}(k)$ be the control applied by the RH controller, in nominal conditions it results that $u^{cc}(k) = \kappa^{cc}(x(k), k)$, with $\kappa^{cc}(x, k+N) = \kappa^{cc}(x, k)$, i.e. the RH scheme leads to a periodically time-varying control law.

The scheme described above is an adaptation to the discrete-time case of the nonlinear RH controller developed in [43] for continuous-time systems. The contractive constraint method has also been studied in a purely RH setting in [18], [31] where continuous systems subject to sampled-data control are considered.

Assumption: It is assumed that there exists a nonempty neighbourhood $X^{cc}(N, \alpha, \beta)$ of the origin such that $\forall \bar{x} \in X^{cc}(N, \alpha, \beta)$ the FH optimal control problem (11) subject to (1), (2), (12), (26), (27) has a solution, see Assumption 2.3 in [43].

Proof of stability: The proof easily follows from the constraints (26) and (27). In fact, let $x^{cc}(k)$ the solution of (1) subject to $u(k) = \kappa^{cc}(x(k), k)$, $k \geq t$. Then, (26) implies $\|x^{cc}(t+jN)\| \leq \alpha^j \|\bar{x}\|$. Moreover, in view of (27), it follows that $\|x^{cc}(t+jN+i)\| \leq \beta \|x^{cc}(t+jN)\| \leq \beta\alpha^j \|\bar{x}\|$. Finally, since $\beta\alpha^j \to 0$ as $j \to \infty$, it follows that $x^{cc}(k) \to 0$ as $k \to \infty$.

Output admissible set: It is clear that there could be triplets (N, α, β) such that $X^{cc}(N, \alpha, \beta)$ is empty. In particular this happens when β is too small in a system with overshoot. In general, it is possible to enlarge the output admissible set by letting $\alpha \to 1$, $\beta \to \infty$. Even in this case, however, there may be a minimal value of N for $X^{cc}(N, \alpha, \beta)$ to be nonempty. If β is reasonably large, a possible way to overcome problems related to the size of the output admissible set is by resorting to a variable horizon N [43].

Linearization: Due to the constraints (26) and (27), it is difficult to obtain the linearization of the control law around the origin. However, if such constraints are not active, the linearization is a periodic linear control law that can be derived using linear periodic RH algorithms [13].

Performance: A possible strategy in order to obtain satisfactory performances, at least in a neighborhood of the origin, is as follows. First, constraints (26) and (27) are neglected and a linear periodic RH controller is designed on the basis of the linearized dynamics (5). This means that the parameters P and N are tuned until satisfactory performances are derived (in view of periodicity this is less standard then tuning a linear time invariant LQ). Then, α and β are selected such that the constraints (26) and (27) are not active in a neighborhood of the origin (obviously this is possible only if the linear periodic RH control law is stabilizing). In this way, the constraints (26) and (27) do not affect the local performance but take care of ensuring convergence when the state is "far" from the equilibrium.

Implementation: The intervalwise schemes have the advantage that the FH problem has to be solved only every N steps so reducing computations. However, at the end of the N-step horizon, the optimization has to be started all over again. Although (27) introduces $N-1$ inequality constraints, if β is large enough, it is likely that they will not be active most of the time.

4. Robustness analysis and synthesis

Robustness analysis and synthesis for nonlinear systems is considerably more complex than in the linear case, especially in presence of state and input constraints. This explains why the available results for nonlinear RH control are far to form a complete theory. In the present section, four aspects will be considered: monotonicity-based robustness, the use of inverse optimality to establish robustness margins, nonlinear H_∞ RH control, and local H_∞ design. In most of the following treatment we will deliberately ignore state and input constraints pointing out the few results that allow for the presence of constraints.

By the way, we mention the fact that, whenever the equilibrium is exponentially stable, it remains an asymptotically stable fixed point also in the presence of disturbances asymptotically decaying to zero [36]. This result is of particular importance for establishing closed-loop stability in the output feedback case.

4.1. Monotonicity-based robustness

The methods whose stability proof is based on monotonicity enjoy some sort of embedded robustness also in the presence of state and input constraints. In fact, the keystone of the stability argument is the decrease of the optimal value of the FH cost, i.e. $J^o_{FH}(f(\bar{x},\kappa^{RH}(\bar{x})),N) < J^o_{FH}(\bar{x},N)$. Now, assume that, due to modelling errors, the real system is

$$x(k+1) = f_r(x(k),u(k)), \quad x(t)=\bar{x}, \quad k \geq t \tag{28}$$

with $f_r(0,0)=0$ and $f_r(\cdot,\cdot) \neq f(\cdot,\cdot)$. In other words, the nominal model (1) used to compute $\kappa^{RH}(x)$ through the solution of the FH problem (11), (12) is only an approximate representation of the real system. In view of this discrepancy, there is no guarantee that

$$J^o_{FH}(f_r(\bar{x}, \kappa^{RH}(\bar{x})), N) < J^o_{FH}(\bar{x}, N) \tag{29}$$

and closed-loop stability may be lost. However, recalling (14), (15) if the *RH* controller at hand enjoys the monotonicity property, it holds that

$$J^o_{FH}(f(\bar{x}, \kappa^{RH}(\bar{x})), N) - J^o_{FH}(\bar{x}, N) \leq -\bar{x}'Q\bar{x} - \kappa^{RH}(\bar{x})'R\kappa^{RH}(\bar{x})$$

Hence, there is some margin and, provided that $\|f(x,u) - f_r(x,u)\|$ is not "too large", compared to $\|(x,u)\|$, (29) will hold so that stability is preserved in a suitable region also in the perturbed case. Note, however, that a thorough robustness analysis should take into account the possibility that the model error causes the state to drift into regions where the optimal control problem is infeasible.

Although the proof is based on a different rationale, also the dual-mode *RH* controller is robust with respect to generic perturbations of the type (28) [29]. Furthermore, with a suitable modification, the *DM RH* scheme has been shown to achieve robust stability also in the presence of state and input constraints.

4.2. Robustness based on inverse optimality

A whole class of stability margins can be derived by showing that the *RH* controller coincides with the optimal control law associated with a suitable defined *IH* nonlinear optimization problem. To this purpose, consider an *RH* control law $u = \kappa^{RH}(x)$ that enjoys the monotonicity property and let

$$l(x) := x'Qx + J^o_{FH}(x, N-1) - J^o_{FH}(x, N)$$

In view of monotonicity, $l(x) > 0$ for all $x \neq 0$ belonging to the output admissible set. Then, consider the "fake" *IH* cost function

$$J^F_{IH}(\bar{x}, u(\cdot)) = \sum_{k=t}^{\infty} \{l(x(k)) + u(k)'Ru(k)\} \tag{30}$$

Under regularity assumptions, it is possible to show that $\kappa^{RH}(x)$ coincides with the optimal *IH* control law solving (30), see e.g. [26] where the continuous-time case is considered. As such, $\kappa^{RH}(x)$, enjoys all the stability margins of *IH* regulators. In the continuous-time case, for instance, if the plant is affine in u (i.e. $\dot{x} = a(x) + b(x)u$) one can guarantee the celebrated "infinite gain margin-50% gain reduction tolerance" as well tolerances to suitable classes of dynamic input uncertainties [20], [37]. As for the robustness margins of discrete-time *RH* controllers, see [14].

4.3. Nonlinear H_∞ RH control

In order to derive an H_∞ *RH* controller, the main idea is to compute the control by solving an *FH* minmax problem, where the designer plays against the nature so as to account for the worst-case disturbances. A control scheme of this type has been proposed in continuous-time [7], showing that it guarantees robust stability in the face of unstructured and parametric uncertainties also in the presence of constraints on the control inputs. However, as recognized by the authors, the proposed scheme is still far from a practical solution especially for the high computational demand involved by the solution of the minmax problem. A general analysis of H_∞ RH control schemes is carried out in [25], where it is shown that, if the FH optimal cost function satisfies a monotonicity condition, the RH control law can be interpreted as the solution of a suitable IH H_∞ control problem. In this way, it is easy to show that the closed-loop attains a prescribed disturbance attenuation level.

4.4. Local H_∞ design

In view of the difficulty of implementing an H_∞ RH controller, it may be interesting to design a nonlinear RH controller whose linearization in the equilibrium coincides with a prescribed linear H_∞ control law designed on the linearized plant. In this way, H_∞ robustness is recovered locally whereas the nonlinear RH controller takes care of enlarging the output admissible set.

In order to design the local H_∞ state feedback, consider the system

$$\begin{aligned} x(k+1) &= Ax(k) + Bu(k) + Lw(k) \\ y(k) &= Cx(k) + Du(k) \end{aligned}$$

where $C'D = 0$, $D'D = R > 0$, and $w(k)$ is a disturbance. The linear H_∞ design aims at finding a state feedback $u(k) = K^H x(k)$ such that the H_∞-norm from $w(k)$ to $y(k)$ is less than some prescribed positive scalar γ. For the sake of simplicity, it is assumed that $C'C > 0$ and that the H_∞ ARE

$$P = A' \left(P^{-1} + BR^{-1}B' - \frac{LL'}{\gamma^2} \right)^{-1} A + C'C$$

admits a positive definite solution P^H such that $\left(P^H\right)^{-1} + BR^{-1}B' - \frac{LL'}{\gamma^2} > 0$ and $\left(I + P^H \left(BR^{-1}B' - \frac{LL'}{\gamma^2}\right)\right)^{-1} A$ is stable. Then, the prescribed attenuation level γ is guaranteed by the state-feedback [38]

$$\begin{aligned} u(k) &= K^H x(k) \\ K^H &= -\left(R + B'P^H B\right)^{-1} B'P^H A \end{aligned} \tag{31}$$

Now, define

$$\tilde{Q} := A' \left(P^{-1} + BR^{-1}B' - \frac{LL'}{\gamma^2} \right)^{-1} A - A' \left(P^{-1} + BR^{-1}B'\right)^{-1} A$$

It easy to see that $\tilde{Q} \geq 0$. Then, P^H satisfies the equation

$$P = A' \left(P^{-1} + BR^{-1}B'\right)^{-1} A + C'C + \tilde{Q}$$

which, letting

$$Q := C'C + \tilde{Q} \tag{32}$$

can be rewritten, after some manipulations, as a standard LQ ARE:

$$P = A'PA - A'PB(R + B'PB)^{-1}B'PA + Q$$

In other words, provided that Q is defined according to (32), $P^H = P_\infty$ and the H_∞ gain K^H coincides with the optimal LQ gain K^{LQ} solving (4) subject to (5). Hence, if one adopts the CL RH scheme (Section 3.3) with Q given by (32), the resulting nonlinear RH control law will be tangent in the origin to the linear H_∞ state feedback (31). It is however clear that H_∞ robustness is recovered only locally as $\|x\| \to 0$.

5. Concluding remarks

At this stage, it is convenient to draw some conclusions on the relative merits of the various stabilization methods with respect to the main issues of output admissible set, performance and implementation. The size of the output admissible set has to do with the achievement of the largest possible region in the state-space inside which constraints satisfaction and convergence to the equilibrium is guaranteed. In this respect, the best methods seem to be the dual-mode (DM) RH controller and the infinite-horizon closed-loop costing (CL). Nevertheless, by a proper adjustment of the parameters, it should be possible to obtain good results also with the quadratic terminal penalty (QI) (provided that the modification (22), (23) is adopted) and the contractive constraint (CC).

If we consider performance, the CL method provides a good approximation to the infinite-horizon optimal controller. The QI method with the modification (22), (23) can also give comparable performances, provided that the design parameter M is large enough. The DM and CC methods can be tuned so as to be optimal, at least locally, by imposing that their linearization around the equilibrium is equal to the linear LQ control law.

Coming to the implementation issue, the DM, QI, and CC methods seem to be about equally efficient, since they all require the solution of a finite-horizon optimization problem with inequality constraints and (possibly) a quadratic terminal penalty. The CL method is more expensive computationally because the evaluation of the nonquadratic terminal penalty requires the integration of the nonlinear system subject to a linear control law over a long horizon (theoretically infinite), although cheaper approximations are possible. On the other hand, the CL method works (maintaining its remarkable properties) also with very short optimization horizons and this may compensate the cost of evaluating the terminal penalty.

The robustness results are more preliminary. The issue of robust design is involved although H_∞ RH schemes appear to be promising.

References

[1] M. Alamir and G. Bornard. Stability of a truncated infinite constrained receding horizon scheme: the general discrete nonlinear case. *Automatica*, 31(9):1353–1356, 1995.

[2] R. Bitmead, M. Gevers, I. Petersen, and R.J.Kaye. Monotonicity and stabilizability properties of solutions of the Riccati difference equation: Propositions, lemmas, theorems, fallacious conjectures and counterexamples. *System & Control Letters*, 5:309–315, 1985.

[3] R. Bitmead, M. Gevers, and V. Wertz. *Adaptive Optimal Control: The Thinking Man's GPC.* Prentice Hall, 1990.

[4] C. Chen and L. Shaw. On receding horizon feedback control. *Automatica*, 18(3):349–352, 1982.

[5] H. Chen and F. Algöwer. A quasi-infinite horizon nonlinear model predictive control scheme with guaranteed stability. *Automatica*, 1998.

[6] H. Chen and F. Allgöwer. A quasi-infinite horizon nonlinear predictive control. In *European Control Conference '97*, 1997.

[7] H. Chen, C. Scherer, and F. Allgöwer. A game theoretical approach to nonlinear robust receding horizon control of constrained systems. In *American Control Conference '97*, 1997.

[8] M. Chen. *Stability and Robustness Considerations in Nonlinear Model Predictive Control.* PhD thesis, Stuttgarg, 1997.

[9] L. Chisci and A. Lombardi. Dual receding horizon control with deadzone. In *Proceeding of 3rd European Control Conference*, 1995.

[10] L. Chisci, A. Lombardi, and E. Mosca. Dual receding-horizon control of constrained discrete-time systems. *European Journal of Control*, 2:278–285, 1996.

[11] D. Chmielewski and V. Manousiouthakis.On constrained infinite-time linear quadratic optimal control. *System & Control Letters*, 29:121-129, 1996.

[12] L. Danyang and L. Xuanhuang. Optimal state estimation without the requirement of a priori statistics information on the initial state. *IEEE Trans. on Automatic Control*, 39:2087–2091, 1994.

[13] G. De Nicolao. Cyclomonotonicity and stabilizability properties of solutions of the difference periodic Riccati equation. *IEEE Trans. on Automatic Control*, 37:1405–1410, 1992.

[14] G. De Nicolao, L. Magni, and R. Scattolini. On the robustness of receding-horizon control with terminal constraints. *IEEE Trans. Automatic Control*, 41:451–453, 1996.

[15] G. De Nicolao, L. Magni, and R. Scattolini. Stabilizing nonlinear receding horizon control via a nonquadratic terminal penalty. In *IMACS Multiconference CESA '96*, pages Vol. 1, 185–187, 1996.

[16] G. De Nicolao, L. Magni, and R. Scattolini. Stabilizing predictive control of nonlinear *ARX* models. *Automatica*, 33:1691–1697, 1997.

[17] G. De Nicolao, L. Magni, and R. Scattolini. Stabilizing receding-horizon control of nonlinear time-varying systems. *IEEE Trans. on Automatic Control*, 43:1030-1036, 1998.

[18] S. De Oliveira. *Model Predictive Control (MPC) for Constrained Nonlinear Systems.* PhD thesis, Institute of Technology, Pasadena , California, 1996.

[19] E. G. Gilbert and K. T. Tan. Linear systems with state and control constraints: the theory and application of maximal output admissible sets. *IEEE Trans. on Automatic Control*, 36:1008-1020, 1991.

[20] S. Glad. Robustness of nonlinear state feedback – a survey. *Automatica*, 23:425–435, 1987.

[21] S. Keerthi and E. Gilbert. Optimal, infinite-horizon feedback laws for a general class of constrained discrete-time systems. *J. Optimiz. Th. Appl.*, 57:265–293, 1988.

[22] W. Kwon and A. Pearson. On feedback stabilization of time-varying discrete-linear systems. *IEEE Trans. on Automatic Control*, 23:479–481, 1978.

[23] L. Magni. *Nonlinear Receding Horizon Control: Theory and Application.* PhD thesis, Dipartimento di Informatica e Sistemistica, Università degli Studi di Pavia, 1994-1997.

[24] L. Magni, G. Bastin, and V. Wertz. Multivariable nonlinear predictive control of cement mills. *IEEE Trans. on Control and Systems Technology*, 1998. (to appear).

[25] L. Magni, H. Nijmaijer, and A. van der Schaft. A receding-horizon approach to the nonlinear H_∞ control problem. *Submitted to Automatica.*

[26] L. Magni and R. Sepulchre. Stability margins of nonlinear receding horizon control via inverse optimality. *System & Control Letters*, 32:241–245, 1997.

[27] D. Mayne and H. Michalska. Receding horizon control of nonlinear systems. *IEEE Trans. on automatic Control*, 35:814–824, 1990.

[28] E. Meadows, M. Henson, J. Eaton, and J. Rawlings. Receding horizon control and discontinuous state feedback stabilization. *Int. J. Control*, 62:1217–1229, 1995.

[29] H. Michalska and D. Mayne. Robust receding horizon control of constrained nonlinear systems. *IEEE Trans on Automatic Control*, 38:1512–1516, 1993.

[30] H. Michalska. A new formulation of receding horizon stabilizing control without terminal constraint on the state. *European Journal of Control*, 3:2–14, 1997.

[31] M. Morari and S.L. de Oliveira. Contractive model predictive control for constrained nonlinear systems. *IEEE Trans on Automatic Control*, 1998, in press.

[32] T. Parisini and R. Zoppoli. A receding-horizon regulator for nonlinear systems and a neural approximation. *Automatica*, 31:1443–1451, 1995.

[33] T. Parisini, M. Sanguineti, and R. Zoppoli. Nonlinear stabilization by receding-horizon neural regulators. In *Int. J. Control*, 70:341–362, 1998.

[34] J. Rawlings and K. Muske. The stability of constrained receding horizon control. *IEEE Trans on Automatic Control*, 38:1512–1516, 1993.

[35] P. Scokaert and J. Rawlings. Infinite horizon linear quadratic control with constraints. In *IFAC 13th Triennal World Congress*, pages vol.M 109–114, 1996.

[36] P. Scokaert, J. Rawlings, and E. Meadows. Discrete-time stability with perturbations: Application to model predictive control. *Automatica*, 33:463–470, 1997.

[37] R. Sepulchre, M. Jankovic, and P. Kokotovic. *Constructive Nonlinear Control.* Springer-Verlag, 1996.

[38] A. Stoorvogel and A. Weeren. The discrete-time Riccati equation related to the H_∞ control problem. *IEEE Trans. on Automatic Control*, 39:686–691, 1994.

[39] M. Sznaier and M. Damborg. Suboptimal control of linear systems with state and control inequality constraints. In *Proceedings of the 26th Conference on Decision and Control*, pages 761–762, 1987.

[40] M. Sznaier and M. Damborg. Control of constrained discrete time linear systems using quantized controls. *Automatica*, 25:623–628, 1989.

[41] M. Sznaier and M. Damborg. Heuristically enhanced feedback control of constrained discrete-time linear systems. *Automatica*, 26:521–532, 1990.

[42] M. Vidyasagar. *Nonlinear System Theory*. Englewood Cliffs, NJ, Prentice-Hall, 2nd edition, 1993.

[43] T. Yang and E. Polak. Moving horizon control of nonlinear systems with input saturation, disturbances and plant uncertainty. *Int. J. Control*, 58:875–903, 1993.

[44] A. Zheng. Model predictive control: Is QP necessary? In *AIChE Annual Meeting*, 1996.

[45] A. Zheng. A computationally efficient nonlinear model predictive control algorithm. In *Proceedings of American Control Conference*, 1997.

Dipartimento di Informatica e Sistemistica
Università degli Studi di Pavia
Via Ferrata 1, 27100 Pavia, Italy
tel +39-382-505484 fax: +39-382-505373/525638
E-mail address: {denicolao,magni,scatto}@conpro.unipv.it

Progress in Systems and Control Theory, Vol. 26

Nonlinear Model Predictive Control: Challenges and Opportunities

David Mayne

Abstract. Linear Model Predictive Control has achieved a state of considerable maturity. With over two thousand applications, linear model predictive control is the preferred method of control in several application areas; a major motivation for its use is the presence of hard constraints on controls and states that are difficult to handle by other methods. Theoretical understanding has progressed and, while there remain open problems, notably in estimation, adaptation and robustness, success in implementation and maturity in underlying theory ensure the continuing development and application of linear model predictive control. Research on Nonlinear Model Predictive Control is at an interesting point. While applications are few, they appear to be rapidly increasing. There is a noticeable convergence in the methodologies proposed in the literature. Open problems include, as in linear model predictive control, estimation, adaptation and robustness but, in addition, there is a daunting challenge: the solution, online, of non-convex optimal control problems. This paper reviews the convergence of methodologies for Nonlinear Model Predictive Control, assesses the challenges arising from uncertainty and from non-convexity of the optimal control problem, and attempts to define fruitful avenues for future research.

1. Introduction

Model predictive control has a rich and unusual history, with strands that include optimal control theory, which provides theoretical foundations, the 'process control literature' that is responsible for its wide-scale adoption by industry, and 'modern' control that addresses such problems as robustness. The main reason for its wide-scale adoption is its ability to handle hard constraints on controls and states that arise in most applications. These constraints are particularly important in the petro-chemical industry where optimization of set points results in steady-state operation on, or close to, the boundary of the set of permissible states. Satisfactory operation at a set point, or during the transition from one set point to another, requires control that respects hard constraints. Model Predictive Control (MPC) is one of very few methods available for handling such constraints, and it is precisely this fact that makes it so useful for the control engineer, particularly in the process industries where plants being controlled are sufficiently 'slow' to permit its implementation. When the plant is linear and the cost quadratic, the optimal control problem solved online reduces to a (convex) quadratic program for which efficient algorithms exist. Over two thousand applications of linear model predictive control have been reported, making it the industry standard in certain areas. Nonlinear model predictive control has progressed more slowly but its use is accelerating; the number of reported applications has increased from one to eighty six in the last three years [14].

Theoretical analysis has been mainly directed towards establishing stability of the nominal system. Whereas in the early 'industrial' literature attention was focussed on achieving stability by 'playing games' [1] with tuning parameters (such as cost coefficients and control and prediction horizons), many methods for enforcing stability *a-priori* (by the incorporation of appropriate 'ingredients' in the optimal control problem) have since been proposed. It has recently been observed that these methods have much in common; this emerging consensus is discussed in §2 and provides a basis for discussing current challenges. Many important aspects of control, such as performance, state estimation and its use in model predictive control, adaptation and robustness have received relatively little attention although there is encouraging recent progress in some areas such as state estimation [15] and robustness [8].

What, then, are the major challenges facing nonlinear model predictive control? The first, arising directly from nonlinearity of the plant, is the online solution of non-convex optimal control problems; failure to respond to this challenge will severely inhibit the use on nonlinear model predictive control. The second challenge is the need to develop robust versions of model predictive control to cope with inevitable uncertainties, due either to model error or unmeasured disturbances. Observer-based and adaptive nonlinear model predictive control, for example, both require robustness of the nominal system; it is interesting to note in [9] that the extension of the modular approach in linear adaptive control (in which the estimation and control modules are independently designed) to the nonlinear case requires the state-feedback controller to be robustly stable. Hence robustness appears to be the next most pressing challenge; a successful response would be useful for linear, as well as non-linear, model predictive control. We address the challenges posed by plant nonlinearity and robustness in, respectively, §3 and §4.

2. Emerging Consensus

2.1. Preliminaries

Assume the plant (the system being controlled) is described by:

$$x(k+1) = f(x(k), u(k)) \tag{2.1}$$

The objective is to choose a control law $h(\cdot)$ such that the origin is asymptotically stable (for the system $x(k+1) = f(x(k), h(x(k)))$ and the region of attraction is large. Let $\mathbf{u}$ denote the control (sequence) $\{u(0), u(1), \ldots, u(N-1)\}$ where N is the horizon. The cost at *event* (x, j) (state x, time j) is defined by

$$V_N(x, \mathbf{u}) := \sum_{k=0}^{N-1} \ell(x(k), u(k)) + F(x(N)) \tag{2.2}$$

where $x(k) := x^{\mathbf{u}}(k; x, 0)$, the solution at time k of (2.1) satisfying the initial condition $x(0) = x$; $\ell(\cdot)$ is the *stage* cost and $F(\cdot)$ the *terminal* cost. The cost is time-invariant (independent of j) because the functions $f(\cdot)$, $\ell(\cdot)$ and $F(\cdot)$ are time-invariant. The plant is required to satisfy the hard control and state constraints

$$u(k) \in \mathbb{U} \tag{2.3}$$

$$x(k) \in \mathbb{X} \tag{2.4}$$

for all k. The model predictive control at event (x, j) is determined by solving the finite horizon optimal control problem

$$\mathcal{P}_N(x): \quad V_N^0(x) = \min_{\mathbf{u}}\{V_N(x, \mathbf{u}) \mid \mathbf{u} \in \mathcal{U}_N(x)\} \tag{2.5}$$

where $\mathcal{U}_N(x)$ is the set of controls $\mathbf{u} = \{u(0), \ldots, u(N-1)\}$ satisfying

$$u(k) \in \mathbb{U}, \quad \forall k \in \{0, \ldots, N-1\} \tag{2.6}$$

$$x(k) \in \mathbb{X}, \quad \forall k \in \{0, \ldots, N\} \tag{2.7}$$

$$x(N) \in W \tag{2.8}$$

where, as before, $x(k) := x^{\mathbf{u}}(k; x, 0)$. Hence, in the optimal control problem $\mathcal{P}_N(x)$, the control and state are required to satisfy the same control and state constraints under which the plant operates and, in addition, the *stability* terminal constraint (2.8). Let $X_N := \{x \mid \mathcal{U}_N(x) \neq \emptyset\}$, the set of states that can be steered to W in time N along trajectories satisfying the control and state constraints; clearly $X_N \subset \mathbb{X}$. For given x, a solution to $\mathcal{P}_N(x)$ exists if

- $\mathbf{u} \mapsto V_N(x, \mathbf{u})$ is continuous.
- $\mathcal{U}_N(x)$ is compact.
- $x \in X_N$ $(\mathcal{U}_N(x) \neq \emptyset)$.

These conditions are satisfied if $f(\cdot)$, $\ell(\cdot)$ and $F(\cdot)$ are continuous, $\mathbb{U}$ is compact, and $\mathbb{X}$ and W are closed. We assume, therefore, that the following assumptions are satisfied in the sequel :

A1: $f(\cdot)$, $\ell(\cdot)$ and $F(\cdot)$ are continuous,

A2: $f(0, 0) = 0$, $\ell(0, 0) = 0$ and $F(0) = 0$,

A3: $\mathbb{U}$ is compact and contains the origin in its interior,

A4: $\mathbb{X}$ is closed and contains the origin in its interior.

These assumptions do not hold if the operating point (assumed here to be the origin) lies on the boundary of the feasible set, a case of practical interest. For any $x \in X_N$, the solution to $\mathcal{P}_N(x)$ is the (optimal) control sequence

$$\mathbf{u}^0(x) := \arg\min_{\mathbf{u}}\{V_N(x, \mathbf{u}) \mid \mathbf{u} \in \mathcal{U}_N(x)\} \tag{2.9}$$

where

$$\mathbf{u}^0(x) = \{u_x^0(0), u_x^0(1), \dots, u_x^0(N-1)\} \tag{2.10}$$

The model predictive control applied to the plant at state x is $u_x^0(0)$, the first element of the sequence $\mathbf{u}^0(x)$. Hence model predictive control implicitly defines a control law $h(\cdot)$ defined by

$$h(x) := u_x^0(0) \tag{2.11}$$

2.2. Common ingredients

A variety of techniques (e.g. [7, 13, 18, 4, 2, 19, 5]) have been employed to enforce asymptotic stability. All the techniques require some modification to the optimal control problem $\mathcal{P}_N(x)$ solved online. Despite their apparent diversity, all versions of the optimal control problem possess, implicitly or explicitly, the following ingredients:

(i): A positive definite terminal cost function $F(\cdot)$.
(ii): A terminal (stability) constraint $x(N) \in W$.
(iii): A continuous local control law $h_W(\cdot)$.

The ingredients satisfy the following assumptions:

A5: $W \subset X_N \subset \mathbb{X}$, is compact, and $0 \in \overset{\circ}{W}$ (unless $W = \{0\}$),

A6: $h_W(W) \subset \mathbb{U}$,

A7: $f_W(W) \subset W$ $(f_W(x) := f(x, h_W(x)))$,

A8: $F(x) \geq \ell(x, h_W(x)) + F(f(x, h_W(x))$ for all $x \in W$.

where $\overset{\circ}{W}$ denotes the interior of W. One consequence of these assumptions is $h_W(0) = 0$. The closed loop system, under the local control law $h_W(\cdot)$, is $x(k+1) = f_W(x(k))$. A5 and A6 ensure, respectively, satisfaction by the closed loop system of the state and control constraints everywhere in W. A7 ensures W is *positively invariant* for the closed loop system $x(k+1) = f_W(x(k))$ (once the state enters W, the controller $h_W(\cdot)$ ensures the state remains in W). Assumption A8 ensures that $F(\cdot)$ decreases along trajectories of the closed loop system $x(k+1) = f_W(x(k))$ in W and that (as will be shown later) the optimal value function $V_N^0(\cdot)$ is a Lyapunov function for the (model predictive) controlled system $x(k+1) = f_h(x(k))$ where

$$f_h(x) := f(x, h(x)) \tag{2.12}$$

If $\ell(\cdot)$ and $F(\cdot)$ are chosen appropriately, A8 implies A7. Satisfaction of A5–A8 impose some implicit assumptions on the system, such as differentiability of $f(\cdot)$ and stabilizability of the linearized system at the origin, that are not necessary for satisfactory model predictive control [12].

A second feature, common to most versions of model predictive control, is the construction, from the solution $\mathbf{u}^0(x)$ to $\mathcal{P}_N(x)$, of a feasible control sequence $\tilde{\mathbf{u}}(x)$ for the successor problem $\mathcal{P}_N(x')$, where $x' := f_h(x)$ is the next state of the controlled system. The control $\tilde{\mathbf{u}}(\cdot)$ is used theoretically (to establish stability) and computationally (to provide a feasible solution to the current optimal control problem). The sequence $\tilde{\mathbf{u}}(x)$ is obtained from $\mathbf{u}^0(x)$ as follows:

$$\mathbf{u}^0(x) = \{u_x^0(0), u_x^0(1), \dots, u_x^0(N-1)\} \tag{2.13}$$

$$\tilde{\mathbf{u}}(x) = \{u_x^0(1)), \dots, u_x^0(N-1)), h_W(\bar{x})\} \tag{2.14}$$

where $\bar{x} := x^{\mathbf{u}^0(x)}(N; x, 0)$, the state at time N resulting from an initial state x (at time 0) and control $\mathbf{u}^0(x)$.

Proposition 2.1. *Suppose Assumptions A1–A8 are satisfied. Let $x \in X_N$ and let $x' := f_h(x)$ denote the successor state (under model predictive control) to x. Then $\tilde{\mathbf{u}}(x)$, defined in (2.14), is feasible for $\mathcal{P}_N(f_h(x))$ and X_N is positively invariant (for the system $x(k+1) = f_h(x(k))$.*

Proof. Since $x \in X_N$, there exists a solution $\mathbf{u}^0(x)$ to $\mathcal{P}_N(x)$. Because $\mathbf{u}^0(x) \in \mathcal{U}_N(x)$, the corresponding terminal state $\bar{x} := x^{\mathbf{u}^0(x)}(N; x, 0)$ lies in W. Let $\tilde{\mathbf{u}}(x)$ be defined by (2.14). We claim $\tilde{\mathbf{u}}(x)$ is feasible for $\mathcal{P}_N(x')$, $x' := f_h(x)$. Firstly, the controls $u_x^0(1), \dots, u_x^0(N-1)$, being elements of $\mathbf{u}^0(x) \in \mathcal{U}_N(x)$, all lie in $\mathbb{U}$, and it follows from P2 (since $\bar{x} \in W$) that $h_W(\bar{x}) \in \mathbb{U}$, i.e. every element of the sequence $\tilde{\mathbf{u}}(x)$ satisfies the control constraint. Next consider the state constraint. Let $\mathbf{x}^0 := \{x^0(0), x^0(1), \dots, x^0(N)\}$ denote the state sequence resulting from initial state x and control sequence $\mathbf{u}^0(x)$, i.e.

$$x^0(j) := x^{\mathbf{u}^0(x)}(j; x, 0), \quad j = 0, 1, \dots, N \tag{2.15}$$

so that $x^0(0) = x$ and $x^0(N) = \bar{x}$. It follows that the state sequence resulting from initial state x' and control sequence $\tilde{\mathbf{u}}(x)$ is $\tilde{\mathbf{x}} := \{\tilde{x}(0), \tilde{x}(1), \dots, \tilde{x}(N)\}$ where

$$\tilde{x}(j) = x^0(j+1), \quad j = 0, \dots, N-1 \tag{2.16}$$

$$\tilde{x}(N) = f_W(\bar{x}) \tag{2.17}$$

and $\tilde{x}(0) = x^0(1) = f(x, u_x^0(0)) = f_h(x)$. The states $x^0(1), \dots, x^0(N)$ clearly satisfy the state constraint. Since $\bar{x} \in W$, assumptions A5 and A7 imply that $f_W(\bar{x}) \in W \subset \mathbb{X}$, so that every element of the state sequence $\{\tilde{x}(0), \tilde{x}(1), \tilde{x}(N)\}$ satisfies the state constraint, and the new terminal state $\tilde{x}(N) = f_W(\bar{x})$ satisfies the stability constraint. Hence $\tilde{\mathbf{u}}(x)$ is feasible for $\mathcal{P}_N(f_h(x))$ and $f_h(x) \in X_N$. □

A second important consequence of A1–A8 is:

Proposition 2.2. *Suppose Assumptions A1–A8 are satisfied. Then*

$$V_N^0(f_h(x)) - V_N^0(x) \le -\ell(x, h(x)) \tag{2.18}$$

for every $x \in X_N$.

Proof. Since the sequence pairs $(\mathbf{u}^0(x), \tilde{\mathbf{u}}(x))$ and $(\mathbf{x}^0, \tilde{\mathbf{x}})$ (defined in the proof of Proposition 2.2) have common elements (see Figure 1), it follows that:

$$\begin{aligned} V_N^0(f_h(x)) - V_N^0(x) &\le V_N(x, \tilde{\mathbf{u}}(x)) - V_N(x, \mathbf{u}^0(x)) \\ &= -\ell(x, h(x)) + \ell(\bar{x}, h_W(\bar{x})) - F(\bar{x}) + F(f_W(\bar{x})) \end{aligned} \tag{2.19}$$

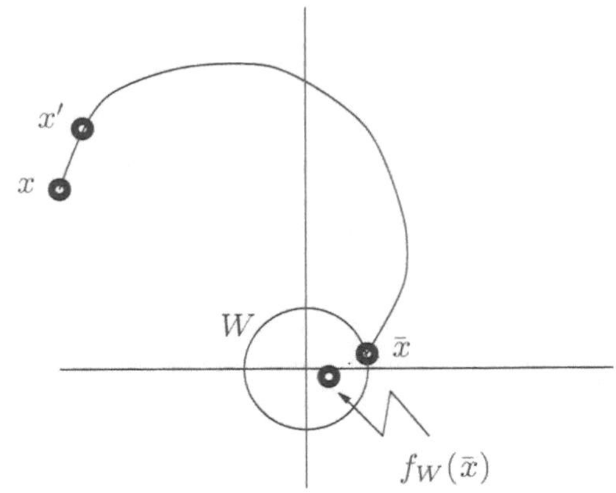

FIGURE 1. $\mathbf{u}_x^0$ and $\tilde{\mathbf{u}}_x$
(a) $\mathbf{u}^0(x)$ steers x to $\bar{x}$
(b) $\tilde{\mathbf{u}}(x)$ steers $x' = f_h(x)$ to $f_W(\bar{x})$

But, since $\bar{x} \in W$, Assumption A8 implies:

$$-F(\bar{x}) + F(f_W(\bar{x})) \leq -\ell(\bar{x}, h_W(\bar{x})) \tag{2.20}$$

Hence (2.18) is satisfied for every $x \in X_N$. □

The controller $h_W(\cdot)$ is not necessarily employed. Its primary purpose is to ensure satisfaction of (2.18). However $h_W(\cdot)$ may be employed in place of $h(\cdot)$ in W (dual-mode control). Under some mild additional assumptions, every model predictive controller possessing the ingredients above is asymptotically stabilizing with a region of attraction X_N.

D1: A function $\alpha : \mathbb{R}_+ \to \mathbb{R}_+$ belongs to class $\mathcal{K}$ if it is continuous, positive definite, and strictly increasing; it belongs to class $\mathcal{K}_\infty$ if, in addition, $\alpha(s) \to \infty$ as $s \to \infty$.

We add to our previous assumptions:

A9: $\ell(x, u) = (1/2)x^T Q x + (1/2)u^T R u$ where Q and R are positive definite,

Assumption A9 is unnecessarily strong; it is made to simplify subsequent analysis. The functions $x \mapsto (1/2)x^T Q$ and $u \mapsto (1/2)u^T R u$ may be replaced by $x \mapsto (1/2)h(x)^T h(x)$ and $u \mapsto \phi(u)$ respectively where the system (2.1) with output $y = h(x)$ satisfies a suitable detectability property and $\phi(\cdot)$ is positive definite. For all $r > 0$, let $B_r := \{x \mid \|x\| \leq r\}$.

Theorem 2.3. *Suppose A1–A9 are satisfied. Then the system*

$$x(k+1) = f_h(x(k))$$

is asymptotically stable (at the origin) with a region of attraction X_N.

Proof. (i) (Stability) It follows from our assumptions that $V_N^0(0) = 0$, $V_N^0(\cdot)$ is continuous at the origin, and that the origin lies in the interior of X_N. Hence there exists a $\mathcal{K}$ function $\alpha(\cdot)$ and a $r > 0$ such that $B_r \subset X_n$ and $V_N^0(x) \leq \alpha(\|x\|)$ for all $x \in B_r$. Since Q is positive definite, there exists a $\mathcal{K}$ function $\beta(\cdot)$ such that $(1/2)x^T Qx \geq \beta(\|x\|)$ for all $x \in \mathbb{R}^n$. Hence

$$\alpha(\|x\|) \geq V_N^0(x) \geq \beta(\|x\|)$$

for all $x \in B_r$. By Proposition 2.2, $\{V_N^0(x(k))\}$ is a decreasing sequence along any solution of $x(k+1) = f_h(x(k))$. Stability of the origin follows by the usual arguments.
(ii) (Convergence) By Proposition 2.2, for any $x \in X_N$, $\{V_N^0(x(k)\}$ is a decreasing sequence, bounded from below by zero. Hence $\{V_N^0(x(k)\}$ converges which implies (since Q is positive definite) that $x(k) \to 0$ as $k \to \infty$. Asymptotic stability with a region of attraction X_N follows. □

A range of model predictive controllers have the structure assumed in this section. In the discussion below it is assumed that the model predictive control $h(\cdot)$ is used everywhere in X_N; alternatively, the local control law $h_W(\cdot)$ may be used in W (if it is stabilizing in W, which is always the case); this is dual-mode control [13].

2.3. 'Classical' MPC

The ingredients for this form of model predictive control [7] are:

- $F(x) \equiv 0$.
- $W = \{0\}$.
- $h_W(x) \equiv 0$.

It is easily verified that (F, W, h_W) satisfy A5–A8. This form of model predictive control employs a stability equality constraint $x(N) = 0$ ($W = \{0\}$) so that terminal cost is irrelevant ($F(x) \equiv 0$). The local control law $h_W(\cdot)$ ($h_W(x) \equiv 0$) maintains the state at the origin, as required.

2.4. Infinite horizon MPC for constrained linear systems

The optimal control problem for this version has a finite horizon N; the term 'infinite horizon' in its title refers to the fact that the finite horizon optimal control problem is equivalent (by suitable choice of its ingredients) to an infinite horizon problem. The ingredients for this form of model predictive control [18, 4, 19] are:

- $F(\cdot)$ is the optimal value function for the infinite-horizon, unconstrained, optimal control problem.
- $W(\cdot)$ is the output admissible set [6] for the linear system $x(k+1) = f_W(x(k))$.
- $h_W(x) = Kx$ is the optimal controller for the infinite-horizon, unconstrained, optimal control problem.

Again, (F, W, h_W) satisfy A5–A8 so that asymptotic stability follows from Theorem 2.3 if the horizon N is fixed. The algorithm proposed in the literature differs as follows. The stability constraint $x(N) \in W$ is not included in the optimal control problem $\mathcal{P}_N(\cdot)$ but, at each k, the horizon N is increased, if necessary, until the constraint $x(N) \in W$ is satisfied. If N is chosen in this way, $\mathcal{P}_N(\cdot)$ is equivalent to the infinite-horizon, constrained, optimal control problem, so that the trajectory followed by the controlled system is identical to that predicted by the solutions to $\mathcal{P}_N(\cdot)$; this facilitates tuning.

2.5. Infinite horizon MPC for constrained nonlinear systems

A *conceptual* version of the above version of MPC is easily stated:

- $F(\cdot)$ is the optimal value function for the infinite-horizon, unconstrained, nonlinear optimal control problem.
- W is the output admissible set for the nonlinear system $x(k+1) = f_W(x(k)$.
- $h_W(x)$ is the optimal controller for the infinite-horizon, unconstrained, nonlinear optimal control problem.

Here the output admissible set W is the obvious extension to nonlinear systems of the definition for linear systems [7]; W is the largest set such that every trajectory of $x(k+1) = f_W(x(k))$ commencing in W remains in W, converges to the origin, and satisfies the state and control constraints. While (F, W, h_W) satisfy A5–A8, determination of these quantities is practically impossible (except in low dimensional problems). To alleviate this difficulty, one proposal [5] is to use a locally stabilizing, linear control law $h_W(x) = Kx$ (which is easily obtained from the linearization of $f(\cdot)$ at the origin) but this leaves the problem of determining $F(\cdot)$ (the (non-optimal) value function associated with controller h_W) and W (the output admissible set associated with $h_W(\cdot)$); the authors propose approximating $F(\cdot)$ by computing the sum of a finite number of terms in its definition.

2.6. Quasi-infinite horizon MPC for constrained nonlinear systems

The ingredients here [2] (slightly modified) are:

- $F(\cdot)$ is the (optimal) value function for the infinite horizon unconstrained linear optimal control problem in which the system is the linearization of (2.1) at the origin, and the stage cost is $\ell'(x,u) = x^TQx + (1/2)u^TRu \geq \ell(x,u)$.
- W is a level set of $F(\cdot)$ chosen sufficiently small (as in [13]) to ensure satisfaction of A5–A8.
- $h_W(\cdot)$ is the linear optimal control law (for the linear, unconstrained system) associated with $F(\cdot)$.

The underlying idea is that $F(\cdot)$ satisfies (by definition)

$$F(x) = \ell'(x, h_W(x)) + F(\hat{f}(x, h_W(x))$$

where $\hat{f}(x,u) = f_x(0,0)x + f_u(0,0)u$ is the linearization of $f(\cdot)$ at the origin. Because $\ell'(x,u) = \ell(x,u) + (1/2)x^TQx$, under suitable differentiability assumptions:

$$F(x) \geq \ell(x, h_W(x)) + F(f(x, h_W(x))$$

for all $x \in W$ if W is sufficiently small, thus satisfying A8. Because W is a level set of $F(\cdot)$, A7 follows. Conditions A5 and A6 are satisfied by an appropriate choice of W.

2.7. Dual-mode MPC, for linear constrained systems

In dual-model MPC, the control law $h(\cdot)$ is employed outside W and $h_W(\cdot)$ inside. We discuss here fixed horizon versions of the dual-mode model predictive control proposed in [13]; see also [3, 17]. The ingredients of the first version are:

- $F(\cdot) = \psi^0(\cdot)$, the (optimal) value function for the infinite horizon unconstrained optimal control problem.
- W is a sufficiently small level set of $F(\cdot)$.
- $h_W(\cdot)$ is the optimal linear control law associated with $F(\cdot) = \psi^0(\cdot)$.

This version is identical in behavior to infinite horizon MPC for linear constrained systems (if N is fixed) since $h(\cdot) = h_W(\cdot)$ in W. The ingredients of the second version are

- $F(x) \equiv 0$.
- W is a sufficiently small level set of the value function $\psi^0(\cdot)F(\cdot)$ for the infinite horizon unconstrained optimal control problem.
- $h_W(\cdot)$ is the optimal linear control law associated with $\psi^0(\cdot)$.

In this version we choose $\ell(x,u) = \alpha(x)(1/2)(x^TQx + u^TRu)$ where $\alpha(x) = 1$ outside W and $\alpha(x) = 0$ in W (so that both $\ell(x,u)$ and $F(x)$ are zero in W, satisfying A8 trivially). A variable horizon version of dual-mode model predictive control for nonlinear constrained systems is presented in [13].

3. The challenge of nonlinearity

An important problem facing model predictive control is nonlinearity. This does not affect, in any material way, the underlying theory; the consensus, discussed in the preceding section, reveals common ingredients in most model predictive controllers, linear or nonlinear. However, nonlinearity does affect implementation, which commonly requires at each time k, solution of an optimal control problem. When this problem is formulated in discrete-time, it becomes a conventional (finite-dimensional) mathematical program. The big divide in mathematical programming is not between linear and nonlinear, but between convex and nonconvex. In convex programs, any local minimizer is a global minimizer, and computation of the global minimum is possible. Thus, model predictive control of linear systems with convex constraints results in a convex optimal control problem; if the cost is quadratic and the constraints linear, the optimal control problem is a quadratic program for which efficient algorithms (for finding global minima) exist. However, if $f(\cdot)$ is nonlinear, the optimal control problem $\mathcal{P}_N(x)$ is, in general, non-convex, with the result that only local, rather than global, minima can be computed (global optimization is not, in general, practical, especially for online use). This presents a dilemma; stability theory for model predictive control, summarized above, generally requires global optimality of the optimal control problem solved online.

Model predictive control requires the solution of many optimal control problems. However, each problem is closely related to the previous problem (since the change in the data x is relatively small). To escape the dilemma above, we may enquire whether, given a globally optimal solution to the intial optimal control problem $\mathcal{P}_N(x(0))$, if it is possible to 'track' the globally solutions of $\mathcal{P}_N(x(k))$ as k progresses. The answer is, unfortunately negative, as is illustrated in Figure 2, which shows how a global minimum may be transformed, successively, into a local minimum, a stationary point, and, finally, a global maximum as the state x evolves. A descent algorithm, which determines stationary points, may indeed yield these solutions; an algorithm that yields local minima, will (in this example) determine the local minimum and the stationary point, but, at the last value of x, will determine either the local minimum or the global minimum rather than the global maximum. In either case, global optimality is lost. Since obtaining a global

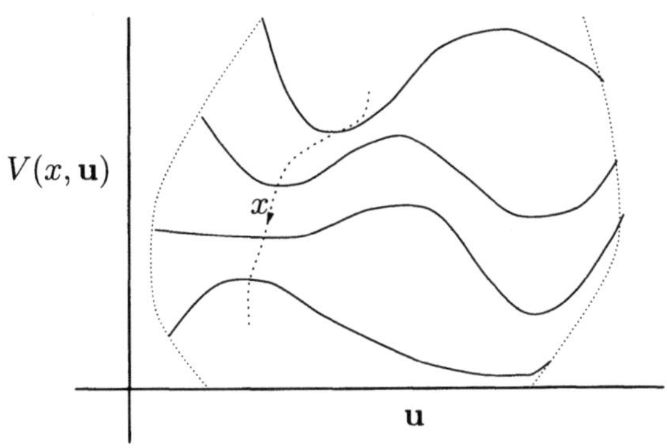

FIGURE 2. Transformation of Global Minimum

solution to the optimal control problem is not generally practical, the requirement for global optimality must be jettisoned. Can stability, nontheless, be guaranteed? A positive answer is provided in [17]. Suppose that a suboptimal model predictive controller generates state and control sequences $\{x(k)\}$ and $\{\mathbf{u}(k)\}$ where, for each k, $\{\mathbf{u}(k)\}$ is a sequence with N elements:

$$\mathbf{u}(k) = \{u_k(0), u_k(1), \dots, u_k(N-1)\} \tag{3.1}$$

The sequences are not necessarily unique (since $\mathbf{u}(k)$ is not necessarily a global minimizer for $\mathcal{P}_N(x(k))$). The control applied to the plant at time k is the first element in $\mathbf{u}(k)$, i.e.

$$u(k) = u_k(0) \tag{3.2}$$

for all k. Let $V_N(\cdot)$ be defined as in (2.2). In view of our assumptions, there exists a $\mathcal{K}$ function $\beta(\cdot)$ such that

$$V(x,\mathbf{u}) \geq \beta(\|x\|) \tag{3.3}$$

for all $x \in X_N$, all $\mathbf{u} \in \mathcal{U}_N(x)$. Given a $\mathcal{K}$ function $\sigma(\cdot)$, and a $r > 0$ such that $B_r \subset W$, let $\hat{\mathcal{U}}(x) \subset \mathcal{U}_N(x)$ be defined by

$$\hat{\mathcal{U}}(x) := \begin{cases} \mathcal{U}_N(x) & \text{if } x \notin B_r, \\ \{\mathbf{u} \in \mathcal{U}_N(x) \mid \|\mathbf{u}\| \leq \sigma(\|x\|)\} & \text{if } x \in B_r. \end{cases} \tag{3.4}$$

The constraint $\|\mathbf{u}\| \leq \sigma(\|x\|)\}$ is introduced in order to force $\mathbf{u}(k)$ to converge to zero as $k \to \infty$; in the absence of optimality, this does not necessarily occur. The class of stabilizing sub-optimal model predictive controllers is large; a model predictive controller is said to be sub-optimal if, at each time k, the control $\mathbf{u}(k)$ is chosen to be *any* u in $\hat{\mathcal{U}}(x(k))$ that satisfies

$$V(x(k),\mathbf{u}(k)) \leq V(x(k-1),\mathbf{u}(k-1)) - \gamma(\|(x(k-1),u(k-1))\|) \tag{3.5}$$

where $\gamma(\cdot)$ is a $\mathcal{K}$ function. The conditions under which such a control is stabilizing are Hypotheses 1, 2 and 3 of the following result:

Theorem 3.1. *Let there exist*

1. *A function $V(\cdot,\cdot)$, continuous at the origin, $V(0,0) = 0$ and $\mathcal{K}$-function $\alpha(\cdot)$ such that,*
$$V(x,\mathbf{u}) \geq \alpha(\|x\|),$$
for all $x \in X_N$, all $u \in \mathcal{U}_N(x)$,
2. *A $\mathcal{K}$-function $\gamma(\cdot)$ such that every realization, $\{x(k),\mathbf{u}(k)\}$, of the controlled system, with $x_0 \in X_N$, satisfies $\{x(k)\} \subset X_N$ and*
$$V(x(k+1)),\mathbf{u}(k+1)) - V(x(k),\mathbf{u}(k)) \leq -\gamma(\|(x(k),u(k))\|)$$
for all k ($u(k)$ denotes the first element of $\mathbf{u}(k)$),
3. *A constant, $r > 0$, and a K-function, $\sigma(\cdot)$, such that every realization $\{x(k),\mathbf{u}(k)\}$ of the controlled system satisfies*
$$\|\mathbf{u}(k)\| \leq \sigma(\|x(k)\|).$$
for all k such that $x(k) \in B_r$.

Then, the controlled system is asymptotically stable with a region of attraction X_N.

Proof. See [17]. Stability follows from Hypotheses 1 and 3, and convergence to the origin from Hypothesis 2. □

Hypotheses 1–3 in this result are independent of A1–A8. It has been shown in the literature [3, 17] that these hypotheses, with $V(\cdot) := V_N(\cdot)$, are relatively easily satisfied if assumptions (similar to) A1–A9 are assumed to hold. However, to show that these hypotheses are satisfied in a sub-optimal model predictive controller with $V(\cdot) := V_N(\cdot)$ usually requires that assumptions similar to A1 to A9 are satisfied. We assume satisfaction of A1–A9 in the sequel.

We next show that the suboptimal model predictive controller, which satisfies the hypotheses of Theorem 3.1 with $V(\cdot) := V_N(\cdot)$, is well defined, i.e. that, at each k, $x(k) \in X_N$ and $\hat{\mathcal{U}}(x(k)) \neq \emptyset$. For each k, let $\tilde{\mathbf{u}}(k)$ be constructed from $\mathbf{u}(k) \in \hat{\mathcal{U}}(x(k))$ as follows

$$\mathbf{u}(k) = \{u_k(0), u_k(1), \dots, u_k(N-1)\} \tag{3.6}$$

$$\begin{aligned}\tilde{\mathbf{u}}(k) &= \{\tilde{u}_k(0), \tilde{u}_k(1), \dots, \tilde{u}_k(N-1)\} \\ &:= \{u_k(1), \dots, u_k(N-1), h_W(x_k(N)\}\end{aligned} \tag{3.7}$$

where $x_k(j) := x^{\mathbf{u}(k)}(j; x(k), 0)$ (the state at time j due to control $\mathbf{u}(k)$ and initial state $x(k)$, $j = 0, 1, \dots, N$). For future reference let $\mathbf{x}(k)$ be defined by

$$\mathbf{x}(k) := \{x_k(0), x_k(1), \dots, x_k(N)\} \tag{3.8}$$

Proposition 3.2. *Suppose A1–A9 are satisfied. For each k, if $x(k) \in X_N$ and $\mathbf{u}(k) \in \mathcal{U}_N(x(k))$, then*
(i) $\tilde{\mathbf{u}}(k) \in \mathcal{U}_N(x(k+1))$ where $x(k+1) = f(x(k), u(k)$ with $u(k)$ the first element of $\mathbf{u}(k)$, and
(ii) $V(x(k+1, \tilde{\mathbf{u}}(k)) \leq V(x(k), \mathbf{u}(k)) - \ell(x(k), u(k))$

Proof. The proofs of (i) and (ii) are similar to those of Propositions 2.1 and 2.2, respectively. □

Proposition 3.2 shows the algorithm is well defined when $x(k) \notin B_r$; the control $\tilde{\mathbf{u}}(k-1)$, simply obtained from $\mathbf{u}(k-1)$, may be employed, at least as an initial candidate, for $\mathbf{u}(k)$. Satisfaction of Hypothesis 3 (necessary when $x(k) \in B_r$) requires an additional mechanism. For each $x \in W$, let $\mathbf{u}_W(x) = \{u_W(k)\}_0^{N-1} := \{u_W(0), u_W(1), \dots, u_W(N-1)\}$ be the sequence satisfying $u_W(k) = h_W(x_W(k))$, $k = 0, \dots, N-1$, where $\{x_W(k)\}$ is the solution at time k of:

$$x_W(0) = x \tag{3.9}$$

$$x_W(k+1) = f_W(x_W(k)) \tag{3.10}$$

If A1–A9 are satisfied and W, $F(\cdot)$ and $h_W(\cdot)$ are defined in the usual way ($F(x) = (1/2)x^T Px$, $h_W(x) = Kx$), there exists a $\mathcal{K}$ function $\sigma(\cdot)$ such that

$$\|\mathbf{u}_W(x)\| \leq \sigma(\|x\|)$$

for all $x \in W$. One way of ensuring that Hypothesis 3 can be satisfied when $x(k) \in B_r$, is to require that $\mathbf{u}(k) = \{u_k(0), u_k(1), \dots, u_k(N-1)\}\}$ satisfies:

$$u_k(j) = \tilde{u}_k(j) \quad \text{if} \quad x_k(j) \in W \tag{3.11}$$

With this mechanism, which reduces computation as in [20] (by restricting the number of variables to be 'optimized) $\mathbf{u}(k) = \mathbf{u}_W(x(k))$ for all k sufficiently large; Hypothesis 3 of Theorem 3.1 is then satisfied. Alternative mechanisms are easily obtained. We conclude it is relatively easy to ensure asymptotic stability by requiring feasible, rather than global optimal, solutions of $\mathcal{P}_N(x)$. Moreover, given an initial feasible solution, subsequent feasible solutions with the required properties are easily obtained (Proposition 3.2). Of course, an initial feasible control is required, but computation of a feasible solution is enormously less complex than computation of a globally optimal solution. This is illustrated in Figure 3. Indeed, merely to confirm that a point is globally optimal requires, in the absence

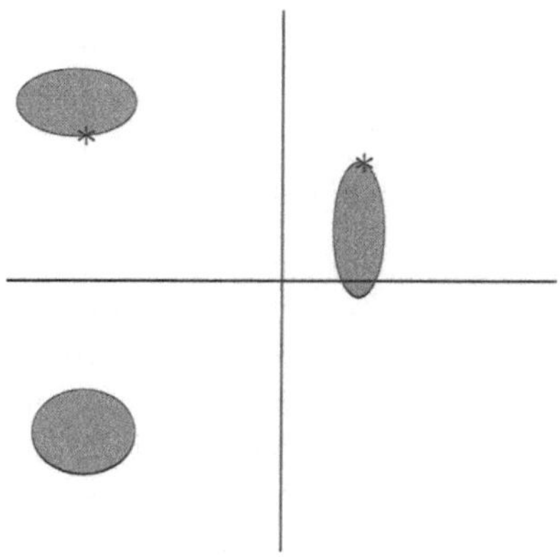

FIGURE 3. Feasible controls
(a) The ellipses are feasible regions
(b) $*$ is a globally optimal point

of global information, dense sampling of the decision space, whereas it is easily verified whether or not a point is feasible.

These results show that global optimality is *not* required for stability; unrestrictive conditions (mainly feasibility) suffice. It is therefore possible that many variants of model predictive control, perhaps with minor modifications, are stabilizing even if local optimality, rather than global optimality, is achieved at each iteration. The results provide an excellent opportunity for the development of new variants of model predictive control with desirable characteristics, such as reduced computational complexity. Other examples of suboptimal model predictive controllers are given in [3, 17].

4. The challenge of uncertainty

4.1. Introduction

An important feature of model predictive control of deterministic systems (no model error and no disturbances) is the fact that $h(\cdot) = h^0(\cdot)$ where $h(\cdot)$ is the optimal model predictive controller (defined via the solution of an optimal *open-loop* control problem) and $h^0(\cdot)$ is the optimal feedback controller (obtained through solving the Hamilton-Jacobi difference equation); feedback is not necessary in the deterministic case. An important consequence of determinism is the positive invariance of X_N under $f_h(\cdot)$.

Uncertain systems may be modelled by the inclusion:

$$x(k+1) \in \mathcal{F}(x(k), u(k)) \tag{4.1}$$

where the set valued function $\mathcal{F}(\cdot)$ may conveniently be expressed as

$$\mathcal{F}(x,u) = f(x,u) + D(x,u) \tag{4.2}$$

Equivalently, the system is described by

$$x(k+1) = f(x(k), u(k)) + d(k) \tag{4.3}$$

where the disturbance $d(k)$ takes arbitrary values in the set $D(x(k), u(k))$. This description models most uncertainties such as bounded disturbance (in which case $D(x,u)$ is the constant set $D \subset I\!R^n$) or model error in which case it is often assumed that $d \in \mathcal{D}(x,u)$ implies $\|d\| \leq \delta\|(x,u)\|$. In the case of bounded disturbance, it is impossible to control to the origin; the most that can be hoped for is to control the state to a set W that is robustly positively invariant under some control law $h_W(\cdot)$, i.e.

$$f_W(W) + D \subset W \tag{4.4}$$

where $f_W(x) := f(x, h_W(x))$. We do not consider this case further (except in §4.5.2). Instead we assume that the uncertainty set $D(x,u)$ converges to $\{0\}$) as (x,u) converges to zero (i.e. $\|D(x,u)\| := \max\{\|z\| \mid z \in D(x,u)\}$ converges to 0 as (x,u) converges to zero). There are several possible strategies for handling uncertainty.

4.2. Nominal model predictive control

Let $h(\cdot)$ denote the nominal model predictive control law, obtained above *ignoring* uncertainty, and let $h^0(\cdot)$ denote the optimal control law for the *uncertain* control problem. It is no longer true that

$$h(x) \equiv h^0(x) \tag{4.5}$$

This is a consequence of the fact that feedback is necessary in the uncertain case. Moreover the set X_N is *not* positively invariant for the controlled system:

$$x(k+1) \in \mathcal{F}(x(k), h(x(k))) \tag{4.6}$$

Hence the state $x(k)$ may drift out of X_N causing the optimal control problem $\mathcal{P}_N(x)$ to become insoluble. Thus, using $h(\cdot)$ (i.e. ignoring uncertainty) is a dangerous strategy, although it may be possible to determine a subset X_u of X_N in which the system (4.6) is positively invariant (equivalently, the system $x(k+1) = f(x((k), u(k))$ is *robustly* positively invariant). The problem that $x(k)$ may drift out of X_N appears to be largely ignored in the literature that concentrates on the fact that, if the disturbance $d(k)$ is sufficiently small, the deterministic control law $h(\cdot)$ still decreases the Lyapunov function $V_N^0(x(k))$; this does not guarantee that the state remains in X_N.

4.3. Min-max model predictive control (open-loop)

The uncertain system is described by

$$x(k+1) = f(x(k), u(k)) + d(k) \tag{4.7}$$

where

$$d(k) \in D(x(k), u(k)) \tag{4.8}$$

Let $x^{(\mathbf{u},\mathbf{d})}(k, x, 0)$ denote the solution, at time k, of (4.8) due to control sequence $\mathbf{u} = \{u(0), u(1), \ldots, u(N-1)\}$ and disturbance sequence $\mathbf{d} = \{d(0), d(1), \ldots, d(N-1)\}$. The (open-loop) min-max optimal control problem is:

$$\mathcal{P}_{\text{mm}}(x): \quad V^0_{\text{mm}}(x) = \min_{\mathbf{u}} \max_{\mathbf{d}} \{V_N(x, \mathbf{u}, \mathbf{d}) \mid \mathbf{u} \in \mathcal{U}_{\text{mm}}(x), \mathbf{d} \in \mathcal{D}(x, \mathbf{u})\} \tag{4.9}$$

where (with $x(k) := x^{(\mathbf{u},\mathbf{d})}(k, x, 0)$)

$$V^0_{\text{mm}}(x) := \sum_{k=0}^{N-1} \ell(x(k), u(k)) + F(x(N)), \tag{4.10}$$

$\mathcal{U}_{\text{mm}}(x)$ is the set of control sequences $\mathbf{u}$ satisfying, for all disturbance sequences $\mathbf{d} \in \mathcal{D}(x, \mathbf{u})$, the constraints:

$$u(k) \in \mathbb{U}, \quad \forall k \in \{0, 1, \ldots, N-1\} \tag{4.11}$$

$$x(k) \in \mathbb{X}, \quad \forall k \in \{0, 1, \ldots, N\} \tag{4.12}$$

$$x(N) \in W \tag{4.13}$$

and $\mathcal{D}(x, \mathbf{u})$ is the set of admissible disturbances, i.e. the set of disturbance sequences $\mathbf{d}$ satisfying

$$d(k) \in D(x(k), u(k)), \quad \forall k \in \{0, 1, \ldots, N-1\} \tag{4.14}$$

Let $(\mathbf{u}^0(x), \mathbf{d}^0(x))$ denote the solution to $\mathcal{P}_{mm}(x)$, i.e.

$$(\mathbf{u}^0(x), \mathbf{d}^0(x)) := \arg \min_{\mathbf{u}} \max_{\mathbf{d}} \{V_N(x, \mathbf{u}, \mathbf{d}) \mid \mathbf{u} \in \mathcal{U}_{\text{mm}}(x), \mathbf{d} \in \mathcal{D}(x, \mathbf{u})\} \tag{4.15}$$

where

$$\mathbf{u}^0(x) = \{u^0_x(0), u^0_x(1), \ldots, u^0_x(N-1)\} \tag{4.16}$$

the (open-loop) min-max model predictive control law $h_{\text{mm}}(\cdot)$ is defined by:

$$h_{\text{mm}}(x) := u^0_x(0) \tag{4.17}$$

This version of model predictive control possesses the same ingredients as before, namely a positive definite terminal cost function $F(\cdot)$, a terminal (stability) constraint ($x(N) \in W$), and a continuous local control law $h_W(\cdot)$. The conditions these ingredients must satisfy differ slightly from those imposed above. Assumptions **A5** and **A6** are unaltered, but **A7** and **A8** must be modified. Let $f_W(\cdot)$ be defined, as before, by $f_W(x) := f(x, h_W(x))$ and let $\mathcal{D}_W(x) := \mathcal{D}(x, h_W(x))$. Our modified conditions are:

A7a: $f_W(x) + \mathcal{D}_W(x) \subset W$ for all $x \in W$.

A8a: $F(x) \geq \ell(x, h_W(x)) + F(f(x, h_W(x) + d)$ for all $(x, d) \in W \times \mathcal{D}_W(x)$.

Assumption A7a states that W is robustly positively invariant under the control $h_W(\cdot)$, i.e. $h_W(\cdot)$ maintains the state in W, despite the disturbance, if the initial state is in W. Assumption A8a ensures that $F(\cdot)$ is a robust Lyapunov function (that decreases, depite d, under $h_W(\cdot)$) in W. Let X_{mm} denote the set of states x for which there exists a $\mathbf{u} \in \mathcal{U}_{\mathrm{mm}}(x)$, i.e. $X_{\mathrm{mm}} := \{x \mid \mathcal{U}_{\mathrm{mm}}(x) \neq \emptyset\}$. The proofs of the following results are similar to their earlier counterparts.

Proposition 4.1. *Suppose that Assumptions A1–A6, A7a and A8a are satisfied and that $x \in X_{mm}$. Then $\tilde{\mathbf{u}}(x)$ (defined in (2.14)) is feasible for any $x' \in \mathcal{F}(x, h(x))$ and X_{mm} is positively invariant for the system $x(k+1) \in \mathcal{F}(x(k), h(x(k)))$ (is robustly positively invariant for the system $x(k+1) = f_h(x(k))$).*

Proposition 4.2. *Suppose Assumptions A1–A6, A7a and A8a are satisfied. Then, for all $x \in X_{mm}$*

$$V^0_{mm}(x') - V^0_{mm}(x) \leq -\ell(x, h_{mm}(x)) \tag{4.18}$$

for all $x' \in \mathcal{F}(x, h_{mm}(x)$.

Theorem 4.3. *Suppose Assumptions A1–A6, A7a, A8a, and A9 are satisfied. Then the system*

$$x(k+1) \in \mathcal{F}(x(k), h_{mm}(x(k))$$

is asymptotically stable at the origin with a region of attraction X_{mm} (the system $x(k+1) = f_(x(k), h_{mm}(x(k)$ is robustly asymptotically stable at the origin with a a region of attraction X_{mm}).

With these desirable properties, (open-loop) min-max model predictive control would appear to provide a good solution to the robustness problem. However, the solution is very conservative [10, 11]. Because feedback control is not permitted in $\mathcal{P}_{\mathrm{mm}}(\cdot)$, the scenarios (state trajectories) generated when solving $\mathcal{P}_{\mathrm{mm}}(x)$ are wildly unrealistic. For example, when the nominal system is unstable, the set of states $\{x^{(\mathbf{u},\mathbf{d})}(N; x, 0) \mid d \in \mathcal{D}(x, \mathbf{u})\}$ at time N expands, without bound, as $N \to \infty$. The inevitable consequence is that X_{mm} is unduly small or, even, empty, in which case $h_{\mathrm{mm}}(\cdot)$ does not exist. Open-loop min-max model predictive control does not, in general, provide a good solution to the robustness problem, although it may provide a conservative solution when the plant being controlled is robustly stable. Feedback min-max model predictive control is required for a non-conservative solution to the robust control problem.

4.4. Feedback min-max model predictive control

In feedback min-max model predictive control, feedback is explicitly allowed in solving the optimal control control problem. This is done by redefining the decision variable $\mathbf{u}$ to permit feedback:

$$\mathbf{u} := \{u(0), h_1(\cdot), h_2(\cdot), \ldots, h_{N-1}(\cdot)\} \tag{4.19}$$

where $u(0) \in \mathbb{U}$ is a control action, and, for each $i \in \{1, \ldots, N-1\}$, $h_i : X_{\text{fb}} \to \mathbb{U}$ is a control law (X_{fb} is defined below). We refer, somewhat imprecisely, to $\mathbf{u}$ as a policy. The decision variable $\mathbf{d}$ is defined as before to be a sequence of disturbance actions. The notation $x^{(\mathbf{u},\mathbf{d})}(k; x, 0)$ now denotes the solution, at time k, of the following system:

$$\begin{aligned} x(0) &= 0 \\ x(1) &= f(x(0), u(0)) + d(0) \\ x(k+1) &= f(x(k), h_k(x(k)) + d(k), \quad k \in \{1, 2, \ldots, N-1\} \end{aligned} \tag{4.20}$$

The set $\mathcal{D}_{\text{fb}}(x, \mathbf{u})$ of admissible disturbance sequences is defined to be the set of disturbance sequences $\mathbf{d}$ satisfying:

$$d(0) \in D(x, u(0)) \tag{4.21}$$

$$d(k) \in D(x^{(\mathbf{u},\mathbf{d})}(k; x, 0), u(k)) \quad k \in \{1, \ldots, N-1\} \tag{4.22}$$

(the definition is not circular because the constraint on $d(k)$ in (4.22) depends (via $x^{(\mathbf{u},\mathbf{d})}(k; x, 0)$) on $\{d(0), \ldots, d(k-1)\}$ and not the entire sequence $\mathbf{d}$). Let $\mathcal{U}_{\text{fb}}(x)$ denote the set of *policies* $\mathbf{u} := \{u(0), h_1(\cdot), h_2(\cdot), \ldots, h_{N-1}(\cdot)\}$ satisfying, for *all* $d \in \mathcal{D}_{\text{fb}}(x, \mathbf{u})$ and all $k \in \{1, \ldots, N-1\}$, the constraints

$$\begin{aligned} u(0) &\in \mathbb{U} \\ h_k(x^{(\mathbf{u},\mathbf{d})}(k; x, 0)) &\in \mathbb{U} \\ x^{(\mathbf{u},\mathbf{d})}(N; x, 0)) &\in W \end{aligned} \tag{4.23}$$

The set X_{fb} is the set of states x that may be steered, in time N, to W by a policy $\mathbf{u} \in \mathcal{U}_{\text{fb}}(x)$:

$$X_{\text{fb}} := \{x \mid \mathcal{U}_{\text{fb}}(x) \neq \emptyset\} \tag{4.24}$$

The *feedback* min-max optimal control problem is

$$\mathcal{P}_{\text{fb}}(x): \quad V_{\text{fb}}^0(x) = \min_{\mathbf{u}} \max_{\mathbf{d}} \{V_N(x, \mathbf{u}, \mathbf{d}) \mid \mathbf{u} \in \mathcal{U}_{\text{fb}}(x), \mathbf{d} \in \mathcal{D}_{\text{fb}}(x, \mathbf{u})\} \tag{4.25}$$

Let $(\mathbf{u}^0(x), \mathbf{d}^0(x))$, where $\mathbf{u}^0(x) = \{u_x^0(0), h_1^0(x)(\cdot), \ldots, h_{N-1}^0(x)(\cdot)\}$ denote the solution to this problem (when it exists). The feedback min-max model predictive control law is $h_{\text{fb}}(\cdot)$, defined by

$$h_{\text{fb}}(x) := u_x^0(0) \tag{4.26}$$

Let $\tilde{\mathbf{u}}(x, x')$, the candidate solution for the successor problem $\mathcal{P}_{\text{fb}}(x')$ where $x' \in \mathcal{F}(x, h_{fb}(x, h_{\text{fb}}(x))$, is defined by:

$$\tilde{\mathbf{u}}(x, x') := \{h_1^0(x)(x'), h_2^0(x)(\cdot), \ldots, h_{N-1}^0(x)(\cdot), h_W(\cdot)\} \tag{4.27}$$

($\tilde{\mathbf{u}}$ has two arguments because x' is not uniquely defined by x and h_{fb}). Results, similar to those for open-loop min-max model predictive control, are easily established.

Proposition 4.4. *Suppose that Assumptions A1–A6, A7a and A8a are satisfied and that $x \in X_{fb}$. Then $\tilde{\mathbf{u}}(x, x')$ (defined in (4.27)) is feasible for any $x' \in \mathcal{F}(x, h_{fb}(x))$ and X_{fb} is positively invariant for the system $x(k+1) \in \mathcal{F}(x(k), h_{fb}(x(k)))$ (is robustly positively invariant for the system $x(k+1) = f(x(k), h_{fb}(x(k)))$).*

Proposition 4.5. *Suppose Assumptions A1–A6, A7a and A8a are satisfied. Then, for all $x \in X_{fb}$*

$$V^0_{fb}(x') - V^0_{fb}(x) \leq -\ell(x, h_{fb}(x)) \tag{4.28}$$

for all $x' \in \mathcal{F}(x, h_{fb}(x))$.

Theorem 4.6. *Suppose Assumptions A1–A6, A7a, A8a, and A9 are satisfied. Then the system*

$$x(k+1) \in \mathcal{F}(x(k), h_{fb}(x(k)))$$

is asymptotically stable at the origin with a region of attraction X_{fb} (the system $x(k+1) = f(x(k), h_{fb}(x(k)$ is robustly asymptotically stable at the origin with a a region of attraction X_{fb}).

Because feedback is permitted in the definition of the optimal control problem, the scenarios are realistic and $X_{\text{fb}} \supset\supset X_{\text{mm}}$, i.e. feedback min-max model predictive control is far less conservative that open-loop min-max model predictive control. It appears to provide a good solution to the robustness problem, provided it is implementable. The expansion of the decision space complicates significantly the complexity of the optimal control problem. However, it is encouraging to note that a few versions of feedback model predictive control have already appeared in the literature.

4.5. Examples of feedback model predictive control

4.5.1. Example 1 Our first example is the interesting proposal recently made in [8]. The system considered is

$$x(k+1) = Ax(k) + Bu(k), \quad y(k) = Cx(k)$$

Uncertainty enters via the parameter $\theta := (A, B)$ that is known only to the extent that it lies in the set Θ defined by

$$\Theta := (A, B) \in \Theta := \text{co}\{(A_i, B_i) \mid i \in I\}$$

where "co" denotes "convex hull". The cost in the optimal control problem (to be defined) is

$$V(x, P) := x^T P x$$

The control and state constraints are, respectively,

$$u(k) \in \mathbb{U} := \{u \mid \|u\| \leq c_u\}$$

and

$$x(k) \in \mathbb{X} := \{x \mid \|Cx\| \leq c_x\}$$

The control in the optimal control problem is defined to be a linear *feedback* control $u = Kx$; this is equivalent to defining the decision variable $=\{u(0), h_K(\cdot), \ldots, h_K(\cdot)\}$ where $h_K(x) := Kx$. Since K is a decision variable, the model predictive control law is $h(x) = K^0(x)x$ where $K^0(x)$ is obtained from the solution to the optimal control problem, i.e. the 'linear' feedback actually applied to the plant is state dependent. Let $A_K := A + BK$. The decision variable for the optimal control problem $\mathcal{P}(x)$ is (γ, P, K), and the optimal control problem is defined by

$$\mathcal{P}(x): \quad \min_{(\gamma,P,K)} \{\gamma \mid (\gamma, P, K) \in Z(x)\}$$

where $Z(x)$ is set of (γ, P, K) satisfying the constraints

(i) $P > 0$
(ii) $V(x, P) \leq \gamma$
(iii) $P \geq A_K^T P A_K + Q + K^T R K \quad \forall (A, B) \in \Theta$
(iv) $\sqrt{\gamma} \| K P^{-(1/2)} \| \leq c_x$
(v) $\sqrt{\gamma} \| K P^{-(1/2)} \| \leq c_u$

The first constraint ensures the cost is positive definite. The second is merely a device to ensure minimizing γ is equivalent to minimizing the cost. The third ensures that

$$V(x, P) \geq \ell(x, h_K(x)) + V(Ax + Bh_K(x)) \ \ \forall x \in W(\gamma, P)$$

which is identical to A8 with $F(x) = V(x, P)$, $h_W(x) = h_K(x)$ and $W = W(\gamma, P) := \{x \mid V(x, P) \leq \gamma\}$. The fourth and fifth constraints ensure, respectively, satisfaction of A5 ($W(\gamma, P) \subset \mathbb{X}$) and A6 ($h_K(W(\gamma, P)) \subset \mathbb{U}$). Problem $\mathcal{P}(x)$ may be recast, as shown in [8] as a linear matrix inequality for which good software exists. If $(\gamma, P, K) \in Z(x)$, then

(i) $x \in W(\gamma, P) := \{x \mid V(x, P) \leq \gamma\}$
(ii) $W(\gamma, P)$ is positively invariant under the control law $u = Kx$.

Consequence (ii) implies satisfaction of A5. *If* a solution to the initial problem $\mathcal{P}(x)(0))$ exists, the model predictive controller generates a sequence $\{(x(k), \gamma(k), P(k), K(k))\}$ satisfying, for all k and all admissible parameter sequences, $x(k) \in W(\gamma(k), P(k))$ and $\gamma(k+1) \leq \gamma(k) - x(k)^T Q x(k)$ (an admissible parameter sequence $\{\theta(k)\}$ satisfies $\theta(k) \in \Theta$ for all k). It readily follows that $x(k) \to 0$ as $k \to \infty$. This is an interesting example of robust feedback model predictive control although it is conservative in that the sets $W(\gamma(k), P(k))$, being ellipsoids, are conservative inner approximations to X_∞, the set of states steerable to the origin by admissible controls.

4.5.2. Example 2 A second example of feedback model predictive control appears in [11, 16] where the uncertainty is due to a bounded disturbance. The system considered is

$$x(k+1) = Ax(k) + Bu(k) + d(k)$$

and is subject to the same control and state constraints as previously considered. The disturbance is bounded, satisfying

$$d(k) \in D, \ D := \text{co}\{d_i \mid i \in I\}$$

where the set I has finite cardinality. Because the disturbance is merely bounded, the controller steers the state to a compact set (that is robustly positively invariant) rather than to the origin. A convex optimal control problem is defined, with decision variable

$$\mathbf{u} = \{u(0), h_1(\cdot), h_2(\cdot), \dots, h_{N-1}(\cdot)\}$$

as is appropriate for feedback model predictive control. The complexity of the optimal control problem is substantially reduced (at some loss to optimality) by finitely parameterizing $\mathbf{u}$. This is done by computing the values of $h_k(\cdot)$ *only* at states corresponding to vertices of D, $k = 1, \dots, N-1$. This is equivalent, in the optimal control problem, to considering only those disturbance sequences $\mathbf{d}$ whose elements lie in the set of vertices of D (rather than in D itself). Convexity ensures that the resultant model predictive control law is robust to disturbance sequences whose elements take values anywhere in D.

4.6. Summary

Feedback model predictive control is generally required to cope satisfactorily with uncertainty. The results above show that this form of model predictive control is possible if the system is linear, even if it is constrained. The possibility of achieving feedback model predictive control for nonlinear systems remains an important, but challenging, open question.

5. Conclusion

Open-loop model predictive control has been shown, both theoretically and practically, to provide an excellent solution to the problem of controlling constrained, linear, deterministic systems. The main challenges facing the development of this form of control are due to nonlinearity and uncertainty. The challenge due to nonlinearity is the intractable nature of global optimization when the optimal control problem is non-convex. The challenge due to uncertainty arises from the lack of feedback in conventional (open-loop) model predictive control.

There exist excellent opportunities for responding to these challenges. The difficulties due to nonlinearity may be alleviated significantly by exploiting the fact that feasibility, rather than optimality, suffices for stability. The consequence is that forms of model predictive control that rely on local, rather than global optimality, can be developed. Moreover, it is possible to devise versions of model predictive control that reduce significantly the amount of computation required without jeopardizing stability. The presence of uncertainty requires a substantial modification to existing forms of model predictive control because of the lack of feedback in the optimal control problem that defines the controller. This can be done by redefining the decision variable to permit feedback. While this increases considerably the complexity of the optimal control problem, examples exist, at least when the system is linear, of feedback versions of model predictive control. The development of feedback model predictive control for uncertain nonlinear systems remains an important, but challenging, open question. The fact that feasibility, rather than optimality, is sufficient for stability may be exploited again in order to reduce to a satisfactory level the considerable complexity of the feedback optimal control problem.

References

[1] R. R. Bitmead, M. Gevers, and V. Wertz. *Adaptive Optimal Control – The Thinking Man's GPC.* Prentice Hall Int., 1990.

[2] H. Chen and F. Allgöwer. A quasi-infinite predictive control scheme for constrained nonlinear systems. *Automatica*, 1997. In press.

[3] L. Chisci, A. Lombardi, and E. Mosca. Dual receding horizon control of constrained discrete-time systems. *European Journal of Control*, 2:278–285, 1996.

[4] D. Chmielewski and V. Manousiouthakis. On constrained infinite-time linear quadratic optimal control. *Systems & Control Letters*, 29:121–129, 1996.

[5] G. de Nicolao, L. Magni, and R. Scattolini. Stability and robustness of nonlinear model predictive control. In *Preprints of International Symposium on Nonlinear Model Predictive Control: Assessment and Future Directions*, pages 77–90, Monte Verità, Ascona, Switzerland, 1998.

[6] E. G. Gilbert and K. T. Tan. Linear systems with state and control constraints: the theory and application of maximal output admissible sets. *IEEE Transactions on Automatic Control*, AC-36:1008–1020, 1991.

[7] S. S. Keerthi and E. G. Gilbert. Optimal, infinite horizon feedback laws for a general class of constrained discrete time systems: Stability and moving-horizon approximations. *Journal of Optimization Theory and Applications*, 57:265–293, 1988.

[8] M. V. Kothare, V. Balakrishnan, and M. Morari. Robust constrained model predictive control using linear matrix inequalities. *Automatica*, 32(10):1361–1379, 1996.

[9] M. Krstič, I. Kanellakopoulis, and P. V. Kokotovič. *Nonlinear and Adaptive Control Design.* John Wiley and Sons, New York, 1995.

[10] D. Q. Mayne. Optimization in model predictive control. In R. Berber, editor, *Methods of Model Based Process Control*, Nato Advanced Study Institute series: E Applied Sciences 293, pages 367–398. Kluwer Academic Publishers, Dordrecht, The Netherlands, 1995.

[11] D. Q. Mayne. Nonlinear model predictive control: an assessment. In Jeffrey C. Kantor, Carlos E. García, and Brice Carnahan, editors, *Fifth International Conference on Chemical Process Control*, pages 217–231. CACHE, AIChE, 1997.

[12] E. S. Meadows, M. A Henson, J. W. Eaton and J. B. Rawlings. Receding Horizon Control and Discontinuous State Feedback Stabilization. *International Journal of control*, 62:1217–1229,1995.

[13] H. Michalska and D. Q. Mayne. Robust receding horizon control of constrained nonlinear systems. *IEEE Transactions on Automatic Control*, 38:1623–1632, 1993.

[14] S. J. Qin and T. A. Badgwell. An overview of nonlinear model predictive control applications. In *Preprints of International Symposium on Nonlinear Model Predictive Control: Assessment and Future Directions*, pages 128–145, Monte Verità, Ascona, Switzerland, 1998.

[15] C. V. Rao and J. B. Rawlings. Nonlinear moving horizon estimation. In *Preprints of International Symposium on Nonlinear Model Predictive Control: Assessment and Future Directions*, pages 146–163, Monte Verità, Ascona, Switzerland, 1998.

[16] P. O. M. Scokaert and D. Q. Mayne. Min-max feedback control for constrained linear systems. *IEEE Transactions on Automatic Control*, 43:1136–1142, 1998.

[17] P. O. M. Scokaert, D. Q. Mayne, and J. B. Rawlings. Suboptimal model predictive control. *IEEE Transactions on Automatic Control*, 1998. In press.

[18] P. O. M. Scokaert and J. B. Rawlings. Infinite horizon linear quadratic control with constraints. In *Proceedings of the 13th IFAC triennial world congress*, volume M, pages 109–114, San Francisico, June 1996.

[19] P. O. M. Scokaert and J. B. Rawlings. Constrained linear quadratic regulation. *IEEE Transactions on Automatic Control*, 43:1163–1169, 1998.

[20] A. Zheng. A computationally efficient nonlinear MPC algorithm. In *Proceedings of the 1997 American Control Conference*, pages 1623–1627, Albuquerque, New Mexico, June 1997.

Department of Electrical and Electronic Engineering,
Imperial College of Science, Technology and Medicine,
London SW7 2BT,
England, U.K.
E-mail address: d.mayne@ic.ac.uk

Progress in Systems and Control Theory, Vol. 26

Nonlinear Moving Horizon State Estimation

Christopher V. Rao and James B. Rawlings

Abstract. MHE is an optimization based strategy for state estimation that explicitly allows for nonlinear models and inequality constraints. In this work we investigate strategies to guarantee the stability of moving horizon estimation (MHE). We begin our discussion by analyzing the stability of the abstract MHE problem. Sufficient conditions for asymptotic stability are established. Using forward dynamic programming to analyze MHE, we introduce the fundamental concept of arrival cost to bound the size of the estimation problem. The key result is if one can construct a global lower bound for the arrival cost, then asymptotic stability is guaranteed. In the second part of our discussion, we propose a strategy that circumvents the need for a global lower bound for the arrival cost. In particular, by including bounds to guarantee the existence of a decreasing nominal sequence, we circumvent the need for a global lower bound for the arrival cost. This result is significant, because, in general, we cannot calculate nor bound the arrival cost. Notable exceptions are linear systems, where the Kalman filter covariance generates a global lower bound for the arrival cost.

1. Introduction

Many control and monitoring systems are based on state space models. The state is a natural construct in the fundamental modeling of chemical and biological processes, because it compactly summarizes the past information needed to understand the future behavior of the process. For example, temperature, pressure, and concentrations comprise the state of a single phase chemically reactive system. Whether full spatial or simple functional representations such as lumping are employed depends on the accuracy required. However, the state is rarely available directly from process measurements, and the state typically needs to be inferred from secondary process measurements or a measurable subset of the state. For example, the average molecular weight of many polymer systems is inferred from viscosity measurements. Also, the concentration in a simple chemically reactive system may be inferred from the reactor temperature, a more easily measured state variable.

Even when the state is directly accessible from the process measurements, one still needs to address measurement noise. If the state is static, then the standard approach is to invoke a weighted averaging technique. The theory of point estimation is well developed and its application is ubiquitous in scientific inquiry. However, for many process systems, the state evolves with time. The averaging technique needs to account for the dynamic evolution of the state. Furthermore, the dynamic evolution of the state intimately connects the estimation process with the model of the system. In addition to model infidelity, we seldom have accurate knowledge of the system parameters and inputs due to noisy measurements and process variations. If the uncertainty is significant or accurate estimates are required, then the state estimator must account for uncertain parameters or process variations.

One general approach for constructing a state estimator is to model the evolving state as a stochastic process. Probability theory is a natural framework in which to make decisions in uncertain environments. Quoting Laplace, *probability theory is nothing but common sense reduced to calculation.* The optimal estimate of the state at time k given process measurements $\{y_k\}_{k=0}^{N-1}$ is a function F of the conditional probability density of the state given the process measurements. We represent the optimal estimate as

$$\{\hat{x}_{k|N-1}\}_{k=0}^{N} = F(p(x_0, \dots, x_N \mid y_0, \dots, y_{N-1})),$$

where x_k denotes the state and y_k denotes the measured output at time j. There exist many criteria for specifying what optimality is. A discussion of some of the various strategies can be found in [8]. In this work, we define the optimal state estimate as the maximum a posteriori Bayesian estimate

$$\{\hat{x}_{k|N-1}\}_{k=0}^{N} \in \arg \max_{\{x_k\}_{k=0}^{N}} p(\{x_k\}_{k=0}^{N} \mid y_0, \dots, y_{N-1}). \tag{1}$$

Our choice of the Bayesian criterion is motivated by its simplicity and generality; without further information regarding the intended application of the state estimate, the natural choice is the most probable state given the data.

In order to evaluate (1) we need an expression for the conditional probability. Following the developments in [4] and [8], we determine the conditional probability as follows. If we model the system as a Markov process, we can express the joint probability of the state as

$$p(x_0, \dots, x_N) = p(x_0) \prod_{k=0}^{N-1} p(x_{k+1}|x_k),$$

where $p(x_0)$ denotes the prior information concerning the initial state of the system. If we assume the measurement noise is additive, i.e $y_k = h(x_k) + v_k$, and independent, we have the relationship

$$p(y_0, \dots, y_{N-1} \mid x_0, \dots, x_{N-1}) = \prod_{k=0}^{N-1} p_{v_k}(y_k - h(x_k)).$$

Using Bayes' rule, we obtain

$$p(x_0,\dots,x_N \mid y_0,\dots,y_{N-1}) \propto p(x_0)\prod_{k=0}^{N-1} p_{v_k}(y_k - h(x_k))p(x_{k+1}|x_k).$$

Taking logarithms, it follows immediately that the estimation problem (1) is equivalent to solving an optimization of the form

$$\{\hat{x}_{k|N-1}\}_{k=0}^{N} \in \arg \min_{x_0,\{w_k\}_{k=0}^{N-1}} \sum_{k=0}^{N-1} L(v_k, w_k) + \Gamma(x_0), \tag{2}$$

where the vector w_k models the randomness attributed to the state transition from x_k to x_{k+1}, $L(v_j, w_j) = -\log p_{v_k}(v_k) - \log p(x_{k+1}|x_k)$, and $\Gamma(x_0) = -\log p(x_0)$. We may interpret the vector w_k as representing state disturbances and the vector v_k as modeling measurement error. We refer to (2) as the full information estimation problem, because we consider all of the available data.

The size of the estimation problem (2) increases with time as more data is considered. To bound the size of the optimization problem, we require the existence of an equivalent statistic z_k of fixed and finite dimension such that

$$p(x_N \mid y_0,\dots,y_{N-1}) = p(x_N \mid z_{N-1}(y_0,\dots,y_{N-1})).$$

The role of the equivalent statistic is to summarize the past information such that the reduced estimation problem with the statistic is equivalent to the full estimation problem. In particular, we have

$$\begin{aligned} &p(x_{N-M},\dots,x_N|y_0,\dots,y_{N-1}) = \\ &\quad p(x_{N-M+1},\dots,x_N|x_{N-M},y_{N-M},\dots,y_{N-1}) \cdot \\ &\quad p(x_{N-M}|z_{N-M-1}(y_0,\dots,y_{N-M-1})). \end{aligned}$$

With the exception of linear Gaussian systems, equivalent statistics of fixed and finite dimension are generally unavailable. For linear Gaussian systems

$$p(x_N \mid y_0,\dots,y_{N-1}) \sim N(\hat{x}_N, \Pi_{N|N-1}),$$

where $\Pi_{N|N-1}$ is the covariance of the state estimate at time index N given observations $\{y_k\}_{k=0}^{N-1}$. A discussion of equivalent statistics may be found in [26].

One strategy to bound the estimation problem is to consider only a fixed amount of data. The resulting estimation problem is finite dimensional, so the implementation barrier is only the ability to solve the nonlinear program in real time. We refer to this strategy as moving horizon estimation (MHE). The basic strategy is to estimate the state using a moving and fixed size window of data. When we estimate the state at the next time index, we add the new measurement to the data window and remove the oldest measurement. By considering a fixed amount of data, we can bound the problem size. However, the approach is suboptimal, because we may incorrectly treat the influence of the past data. More importantly, incorrect treatment of the past data may lead to instability [18].

Although equivalent statistics are generally unavailable for summarizing the past data of nonlinear systems, we expect that linearized statistics yield satisfactory performance in conjunction with the moving horizon strategy. In order for the approximation to be stable, we need to lower bound the arrival cost (the analogue to the cost to go in forward dynamic programming). However, unlike receding horizon control, the classical dual-mode arguments for constructing bounds fail, because there is no fixed origin about which to construct a local invariant region. The lack of locality complicates the stability analysis of the estimator, because the linear statistics need to provide a *global* lower bound for the arrival cost.

The common practice for constructing a state estimator for nonlinear systems is to transform the nonlinear system to a linear system, and then apply a linear state estimator. The linear transformations is usually achieved either by a coordinate transformation or a first order Taylor series approximation. Because general linearizing coordinate transformations require the solution of a set of partial differential equations, solutions may not exist for all systems. For systems that display strong nonlinearities, approximate linearizations can yield poor performance. Furthermore, precise statements other than local stability are generally unavailable. For a current survey of nonlinear estimation strategies, see [17]. In addition to nonlinear state estimation, there is the related field of nonlinear observers. An excellent, though slightly dated, comparative study of nonlinear observers is given in [28].

In this work we formulate a general theory for MHE. The paper is organized as follows. We begin our exposition by deriving sufficient conditions for the stability of the abstract estimation problem (2) and its moving horizon approximation. We proceed by applying the abstract concepts to the construction of implementable strategy for nonlinear MHE. We conclude by discussing duality.

2. Notation, definitions, and basic assumptions

The Cartesian product $\times_{k=1}^{N}\mathbb{A}$ of a set $\mathbb{A}$ is denoted by $\mathbb{A}^N$. The nonnegative elements of a set $\mathbb{S}$ are denoted by $\mathbb{S}_+$. Let $\underline{\mathcal{C}}(\mathbb{R}^n)$ denote the space of lower semicontinuous functions that map from $\mathbb{R}^n$ to $\mathbb{R}$ and $\mathcal{C}^p(\mathbb{R}^n, \mathbb{R}^m)$ denote the space of functions with p continuous derivatives that map from $\mathbb{R}^n$ to $\mathbb{R}^m$.

Definition 2.1. *A function* $\alpha : \mathbb{R}_+ \to \mathbb{R}_+ : x \mapsto \alpha(x)$ *is a* **$\mathbf{K}_+$-function** *if it is continuous, strictly monotone increasing,* $\alpha(x) > 0$ *for* $x \neq 0$*, and* $\alpha(0) = 0$.

We assume the system Σ generating the data sequence $\{y_k\}$ is modeled by the following nonlinear difference equation

$$\hat{\Sigma}: \quad \begin{aligned} x_{k+1} &= f(x_k) + g(x_k, w_k), \\ y_k &= h(x_k) + v_k, \end{aligned} \tag{3}$$

where the functions $f \in \mathcal{C}^1(\mathbb{R}^n, \mathbb{R}^n)$, $g \in \mathcal{C}^1(\mathbb{R}^n \times \mathbb{R}^m, \mathbb{R}^n)$, and $h \in \mathcal{C}^1(\mathbb{R}^n, \mathbb{R}^p)$. We also assume that $g(\cdot, 0) = 0$. We wish to distinguish between the system Σ and the model $\hat{\Sigma}$. The need for this separation arises when we consider the deterministic estimation (observer) problem. More general model structures are possible. However, for mathematical convenience, we restrict ourselves to the model structure (3). Let

$$x_{\hat{\Sigma}}(N, \bar{x}, \{w_k\}_{k=0}^{N-1})$$

denote the solution to the difference equation (3), or the N-step state transition map for $\hat{\Sigma}$, with the initial condition $x_0 = \bar{x}$ and disturbance sequence $\{w_k\}_{k=0}^{N-1}$. When we consider the disturbance free response of the system, i.e. $w_k = 0$, we use the following simplification,

$$x_{\hat{\Sigma}}(N, \bar{x}),$$

for the solution to the difference equation (3) with the initial condition $x_0 = \bar{x}$. We define the N-step output response map of $\hat{\Sigma}$ given the initial condition $x_0 = \bar{x}$ and disturbance sequence $\{w_k\}_{k=0}^{N-1}$ as

$$y_{\hat{\Sigma}}(N, \bar{x}, \{w_k\}_{k=0}^{N-1}) \left\{ \triangleq h\left(x_{\hat{\Sigma}}(N, \bar{x}, \{w_k\}_{k=0}^{N-1})\right) \right\},$$

with the similar simplification,

$$y_{\hat{\Sigma}}(N, \bar{x}) \left\{ \triangleq h\left(x_{\hat{\Sigma}}(N, \bar{x})\right) \right\},$$

for the disturbance free response of the system. Note the difference between y_k and $y_{\hat{\Sigma}}(k, \cdot)$. The vector y_k denotes the observation of the output at time index k and the vector $y_{\hat{\Sigma}}(k, \cdot, \cdot)$ denotes the predicted output at time index k.

Unlike linear system theory, there does not exist a widely accepted definition of observability for nonlinear systems. While the definition of observability corresponds to the ability to infer the state of an idealized system from the data, in the development of the estimator we use the technical condition of incremental observability instead. As discussed in [9], the classical notion of observability arose purely as a technical condition in the proof of the stability of the Kalman filter, and the conceptual view of observability was incidental. The main justification for the observability condition lies in the theorems that can be proved with it.

Definition 2.2. *A system $\hat{\Sigma}$ is* **incrementally observable** *if there exists a positive integer N_o and a K_+-function φ such that for any two states x_1 and x_2,*

$$\varphi(\|x_1 - x_2\|) \leq \sum_{j=0}^{k-1} \|y_{\hat{\Sigma}}(j, x_1) - y_{\hat{\Sigma}}(j, x_2)\|,$$

for all $k \geq N_o$.

The observability condition states that if the prediction residuals are small for sufficient time, then the estimation error is small.

The goal of the full information estimation problem at time index N is to determine the state sequence $\{x_k\}_{k=0}^{N}$ and disturbance sequence $\{w_k\}_{k=0}^{N-1}$, subject to the model $\hat{\Sigma}$, that fits the data sequence $\{y_k\}_{k=0}^{N-1}$ in some optimal sense. Since the full information estimator considers all of the available data, the problem grows without bound as we process more data. As discussed in the introduction, we employ a moving horizon approximation to bound the size of the optimization problem. The goal of the MHE problem at time N is to determine the state sequence $\{x_k\}_{k=N-M}^{N}$ and disturbance sequence $\{w_k\}_{k=N-M}^{N-1}$, subject to the model $\hat{\Sigma}$, that fits the fixed horizon of data $\{y_k\}_{k=N-M}^{N-1}$ in some optimal sense. Because the MHE considers only a subset of the data, the *goodness of fit* is adjusted to account for prior knowledge of past data.

We formulate the full information estimation problem as the solution to the following optimal control problem with Bolza objectives

$$(P1) \qquad \Phi_N^* = \min_{x_0, \{w_k\}_{k=0}^{N-1}} \Phi_N(x_0, \{w_k\}_{k=0}^{N-1})$$

subject to the constraints

$$\{x_k\}_{k=0}^{N} \in \mathbb{X}^{N+1}, \qquad \{w_k\}_{k=0}^{N-1} \in \mathbb{W}^N,$$

where the objective function is of the form

$$\Phi_N(x_0, \{w_k\}_{k=0}^{N-1}) \triangleq \sum_{k=0}^{N-1} L(w_k, v_k) + \Gamma(x_0).$$

The dynamics are specified by the model $\hat{\Sigma}$. The sets $\mathbb{W}$ and $\mathbb{X}$ represent constraints on the disturbance and state sequences. We assume the stage cost function $L \in \underline{\mathcal{C}}(\mathbb{R}^n \times \mathbb{R}^m)$ and the initial penalty $\Gamma \in \underline{\mathcal{C}}(\mathbb{R}^n)$. The solution to $(P1)$ at time N is the state $\hat{x}_{0|N-1}$ and disturbance sequence $\{\hat{w}_{k|N-1}\}_{k=0}^{N-1}$. Let $\{\hat{x}_{k|N-1}\}_{k=1}^{N}$ denote the state sequence generated by the initial condition $\hat{x}_{0|N-1}$ and the disturbance sequence $\{\hat{w}_{k|N-1}\}_{k=0}^{N-1}$. Recall $v_k = y_k - h(x_k)$, so the decision variable x_k is implicit in the objective function through the definition of $\hat{\Sigma}$. The arguments may be extended with relative ease, mostly index bookkeeping, to encompass time-varying stage cost functions. We assume the set $\mathbb{W} \times \mathbb{X} \subseteq \mathbb{R}^m \times \mathbb{R}^n$ is closed. We also assume the set $\mathbb{W}$ contains the origin.

Unlike the full information estimator, the MHE does not determine the full state sequence $\{x_k\}_{k=0}^{N}$. Rather, the MHE determines the truncated sequence $\{x_k\}_{k=N-M}^{N}$. We formulate the moving horizon approximation to the full information estimation problem as the following optimal control problem with Bolza objectives

$$(P2) \qquad \Theta_N^* = \min_{x_{N-M}, \{w_k\}_{k=N-M}^{N-1}} \Theta_N(x_{N-M}, \{w_k\}_{k=N-M}^{N-1})$$

subject to the constraints

$$\{x_k\}_{k=N-M}^{N} \in \mathbb{X}^{M+1}, \qquad \{w_k\}_{k=N-M}^{N-1} \in \mathbb{W}^M,$$

where the objective function is of the form

$$\Theta_N(x_{N-M}, \{w_k\}_{k=N-M}^{N-1}) \triangleq \sum_{k=N-M}^{N-1} L(w_k, v_k) + \Gamma_{N-M}(x_{N-M}),$$

and $\Gamma_{N-M} \in \underline{\mathcal{C}}(\mathbb{R}^n)$ for all $N \in \mathbb{I}_+$. We define the stage cost functions L and the sets $\mathbb{X}$ and $\mathbb{W}$ as equivalent to their full information counterpart. The solution to $(P2)$ at time N is the state $\hat{x}^{\text{mh}}_{N-M|N-1}$ and disturbance sequence $\{\hat{w}^{\text{mh}}_{k|N-1}\}_{k=N-M}^{N-1}$. Let $\{\hat{x}^{\text{mh}}_{k|N-1}\}_{k=N-M+1}^{N}$ denote the state sequence generated by the initial condition $\hat{x}^{\text{mh}}_{N-M|N-1}$ and disturbance sequence $\{\hat{w}^{\text{mh}}_{k|N-1}\}_{k=N-M}^{N-1}$. We drop the superscript when it is obvious from context that we are referring to the solution of $(P2)$. The initial penalty Γ_{N-M} attempts to summarize the past data not considered in the fixed horizon. To maintain the continuity between the estimates conditioned on disjoint data horizons, we condition the moving horizon estimate at time N with the optimal estimate at time $N-M$ by reinitializing the system with the optimal estimate from time $N-M$. In this regard we may interpret Γ_{N-M} as representing (or approximating) the equivalent statistic. The initial penalty serves as an equivalent statistic by penalizing deviation of x_{N-M} away from $\hat{x}^{\text{mh}}_{N-M|N-M-1}$. If we have high confidence in our estimate $\hat{x}^{\text{mh}}_{N-M|N-M-1}$, then the penalty for deviation should be large. Otherwise, if we have little confidence in the estimate $\hat{x}^{\text{mh}}_{N-M|N-M-1}$, then the penalty for deviation should be small.

Suppose the data sequence $\{y_0, \cdots, y_{N-1}\}$ is generated by the system Σ. We construct the full information estimator, or the **$\hat{\Sigma}$:estimator** for the system Σ, as the dynamical system that generates the sequence $\{\hat{x}_{k|k-1}\}$. Our reason for the prefix is we wish to differentiate the model used by the estimator and the actual system. We construct the MHE, or the **$\hat{\Sigma}$:mh-estimator** for the system Σ, as the dynamical system that generates the sequence $\{\hat{x}^{\text{mh}}_{k|k-1}\}$. We define the N-step transition matrix for the estimator, full information or moving horizon, with initial condition (prior information) $\bar{x}$ as

$$\hat{x}_{\hat{\Sigma}}(N, \bar{x}) \left\{ \triangleq \hat{x}_{N|N-1} \right\}.$$

For notational simplicity, we make that following definition: $\hat{(\cdot)}_k \triangleq \hat{(\cdot)}_{k|k-1}$.

In order to guarantee the problem is well posed, we impose the following conditions on the stage cost function L and the initial penalty Γ.

A1: $L(0,0) = 0$ and $\Gamma(0) = 0$.

A2: There exist K_+-functions η and γ such that

$$\begin{aligned} \eta(\|(w,v)\|) < L(w,v) &< \gamma(\|(w,v)\|), \\ \eta(\|x-\tilde{x}\|) &< \Gamma(x) - \Gamma(\tilde{x}) < \gamma(\|x-\tilde{x}\|), \end{aligned}$$

for all $(w,v) \in (\mathbb{W} \times \mathbb{R}^p) \setminus (0 \times 0)$ and $x \in \mathbb{X}$. The vector $\tilde{x}$ denotes the reference state or initial guess for the state. Often we take $\Gamma(\tilde{x}) = 0$.

We also impose the following analogous conditions on the MHE initial penalty Γ_N for all $N \in \mathbb{I}_+$ to guarantee the problem is well posed

A1′: $\Gamma_N(0) \geq 0$,
A2′: There exist K_+-functions $\bar{\eta}$ and $\bar{\gamma}$ and a constant $\alpha \in \mathbb{R}_+$ such that

$$\alpha\bar{\eta}(\|x - \hat{x}_N\|) \leq \Gamma_N(x) - \Gamma_N(\hat{x}_N) < \bar{\gamma}(\|x - \hat{x}_N\|)$$

for all $x \in \mathbb{X} \setminus \hat{x}_N$.

The addition of constraints presents the potential for infeasibility. Particular choices of the feasible region $\mathbb{W} \times \mathbb{X}$ may preclude all possible solutions to either $(P1)$ or $(P2)$. To circumvent infeasibility and guarantee existence, we assume the feasible region is non-empty by imposing the following condition:

A3: There exists $\bar{x}$ such that $x_{\hat{\Sigma}}(k, \bar{x}) \in \mathbb{X}$ for all $k \geq 0$.

Condition $(A3)$ guarantees the existence of an initial condition such that the resulting evolution of the system $\hat{\Sigma}$ is feasible.

Consider again the full information problem $(P1)$ for $N > M$. In an analogous manner to the introduction of an equivalent statistic, we can reformulate the objective function by breaking the time interval into two pieces as follows.

$$\begin{aligned}\Phi_N(x_0, \{w_k\}_{k=0}^{N-1}) &= \sum_{k=0}^{N-1} L(w_k, v_k) + \Gamma(x_0), \\ &= \sum_{k=N-M}^{N-1} L(w_k, v_k) + \sum_{k=0}^{N-M-1} L(w_k, v_k) + \Gamma(x_0).\end{aligned}$$

By the Markov property, the quantity

$$\sum_{k=N-M}^{N-1} L(w_k, v_k)$$

depends only on the state x_{N-M} and the decision variables w_k and v_k for $k \in [(N-M), \ldots, (N-1)]$. Hence, we can reformulate $(P1)$ as the following equivalent optimal control problem

$$(P1') \qquad \min \sum_{k=N-M}^{N-1} L(w_k, v_k) + \bar{\Xi}_{N-M}(x_{N-M})$$

where

$$\bar{\Xi}_{N-M}(\bar{x}) = \min_{x_0, \{w_k\}_{k=0}^{N-M-1}} \left\{\Phi_{N-M}(x_0, \{w_k\}_{k=0}^{N-M-1}) \ : \ x_{N-M} = \bar{x}\right\},$$

subject to the constraints

$$\{x_k\}_{k=0}^{N-M} \in \mathbb{X}^{N-M+1}, \qquad \{w_k\}_{k=0}^{N-M-1} \in \mathbb{W}^{N-M}.$$

We refer to the initial penalty $\bar{\Xi}_j$ as the **arrival cost**. Arrival cost is a fundamental concept in MHE, because it allows us to relate the full information $(P1)$ and fixed horizon $(P1')$ problems. To establish the equivalence between the problems $(P1)$

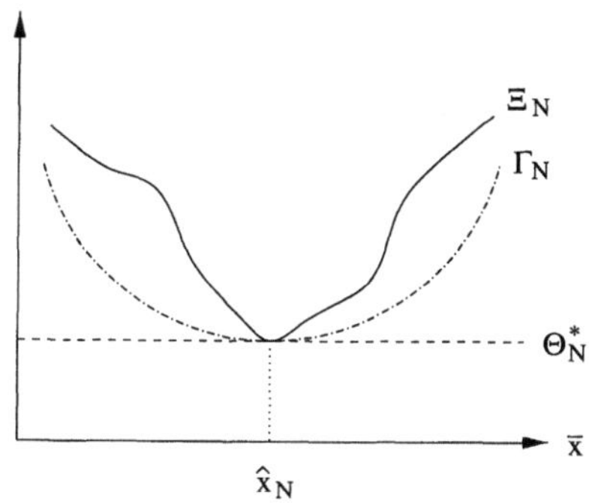

FIGURE 1. A graphical representation of conditions ($A4$) and ($A5$).

and ($P2$), we redefine arrival cost with a fixed horizon of length M as the solution to the following mathematical program

$$\Xi_j(\bar{x}) = \min_{x_{j-M},\{w_k\}_{k=j-M}^{j-1}} \left\{ \Theta_j(x_{j-M}, \{w_k\}_{k=j-M}^{j-1}) \ : \ x_j = \bar{x} \right\},$$

subject to the constraints

$$\{x_k\}_{k=j-M}^{j} \in \mathbb{X}^{M+1}, \qquad \{w_k\}_{k=j-M}^{j-1} \in \mathbb{W}^{M}.$$

The redefinition of the arrival cost is not necessary to establish the equivalence between ($P1$) and ($P2$). However, it is convenient to define the arrival cost in terms of the Θ_j. If $\Gamma_j \equiv \Xi_j$, then

$$\{\hat{x}_{k|N-1}^{\mathrm{mh}}\}_{k=N-M}^{N} = \{\hat{x}_{k|N-1}\}_{N-M}^{N},$$

(i.e. the MHE is equivalent to the full information estimator). We can establish the equivalence of ($P1$) to ($P2$) simply using forward dynamic programming.

If we are able to construct the arrival cost, then the MHE problem is equivalent to the full information problem. With the notable exception of the linear quadratic problem, closed form expressions for the arrival cost are generally unavailable. To ensure stability we require instead that the initial penalty Γ_N is a lower bound to the arrival cost as specified by the following conditions.

A4: $\Gamma_N(x) \geq \Theta_N^*$ for all $x \in \mathbb{X}$;
A5: $\Gamma_N(x) \leq \Xi_N(x)$ for all $x \in \mathbb{X}$.

A graphical representation of conditions ($A4$) and ($A5$) is shown in Figure 1. We may interpret the arrival cost as the dynamic programming analogue to the equivalent statistic.

Remark 2.3. *The arrival cost Ξ_k is proportional to the negative logarithm of the conditional probability density $p(x_k \mid y_0, \dots, y_{k-1})$ The conditional probability density function is precisely the equivalent statistic discussed in Section 1.*

Remark 2.4. *When the system $\hat{\Sigma}$ is modeled with the linear difference equation*

$$x_{k+1} = Ax_k + Gw_k, \tag{4a}$$
$$y_k = Cx_k + v_k, \tag{4b}$$

and $\{w_k\}$ and $\{v_k\}$ are sequences of zero mean Gaussian random variables with covariances Q and R. then we can calculate the arrival cost explicitly when constraints are not present (i.e. $\mathbb{X} = \mathbb{R}^n$ and $\mathbb{W} = \mathbb{R}^m$) using the Kalman filter covariance update recursion

$$\Pi_{N+1|N} = GQG^T$$
$$+A(\Pi_{N|N-1} - \Pi_{N|N-1}C^T(R + C\Pi_{N|N-1}C^T)^{-1}C_N\Pi_{N|N-1})A^T.$$

In particular, we have

$$\Xi_N(x_N) = (x_N - \hat{x}_N)^T \Pi_{N|N-1}^{-1}(x_N - \hat{x}_N) + \Phi_N^*. \tag{5}$$

When constraints are present, we can satisfy (A4) and (A5) with the expression (5).

Remark 2.5. *A simple strategy to satisfy assumptions (A4) and (A5) is to set $\Gamma_N(x) = \Theta_N^*$ for all $x \in \mathbb{X}$. This construction was employed in [18] to generate a stable nonlinear MHE.*

Remark 2.6. *During the initialization phase when $N \leq M$, the full information estimator and the MHE are equivalent. Characterizing or approximating the arrival cost is necessary only when $N > M$. When $N \leq 2M$, the arrival cost is evaluated over a truncated horizon of length $N - M$.*

3. Stability analysis

The following discussion of observer stability is premised on classical Lyapunov stability theory for dynamical systems. The concepts are completely analogous to their classical counterpart. To account for constraints, we have modified the definition of stability in an analogous manner to [10].

Before proceeding with the abstract definition of observer stability, we first introduce the concept of an observer. When the state is not directly available from measurements, it is necessary to infer the current state of the system. The observer addresses the problem of inferring the current state of the following deterministic (i.e. $(w_k, v_k) \equiv 0$) system

$$\Sigma^+ : \quad \begin{aligned} x_{k+1} &= f(x_k), \\ \hat{y}_k &= h(x_k). \end{aligned} \tag{6}$$

When the functions f and h are linear, the Luenberger observer [11, 12] solves the observer problem. The basic strategy of the Luenberger observer is to design a linear output feedback law that stabilizes the error dynamics of the system Σ^+. Note the observer problem is solved if the initial condition x_0 is known exactly, because the model will track the system perfectly. Luenberger emphasized any system is an observer for another system. Whether the observer is effective is a different question. However, many observers are directly available without the need for rational design. This concept is clearly demonstrated in [25], where the author shows many reactor models are stable observers without modification.

One strategy for the observation problem is to solve directly for x_0. If the system Σ^+ is incrementally observable and there is sufficient data, the initial condition can be uniquely determined from the data. This strategy is employed iteratively using moving horizon observers in [7], [29], [16], and [15]. The terminology moving horizon observer is used, because only the state of a deterministic system is reconstructed. This strategy is limited for many reasons. The primary reason is the problem statement represents an idealization, because our knowledge of the true system is not complete. In addition, state and measurement noise prevent a consistent state estimate being recovered from a fixed window of data. By assuming a deterministic model structure, the moving horizon observer strategies may amplify rather than filter the effect of noise on the estimate. Finally, the iterative strategies may break down, because cost decreases are not always possible. Even though the problem statement is an idealization, it is prudent to ask whether the estimation strategy proposed in this work can address this *simple* problem. The ability to reconstruct the state of Σ^+ from the process measurements is what we define as observer stability.

In addition to representing a degree of basic competency for the estimator, observer stability has the additional property that small numerical errors are bounded and asymptotically rejected. This property is important, because the calculations are performed with finite precision arithmetic.

We begin our analysis by making precise the notion of observer stability. The concept of observer stability and asymptotic observer stability is equivalent to stating that the error $\bar{e}\left\{\stackrel{\Delta}{=} x - \hat{x}\right\} = 0$ is an asymptotically stable equilibrium point for the error dynamics of the system Σ^+ combined with the observer.

Definition 3.1. *The $\hat{\Sigma}$:estimator $\left(\hat{\Sigma}\text{:mh-estimator}\right)$ is a* **stable observer** *for the system Σ^+ if for any $\epsilon > 0$ there corresponds a number $\delta(\epsilon) > 0$ and a positive integer $\bar{N}$ such that if $\|\bar{x} - \tilde{x}\| \leq \delta(\epsilon)$ and $\tilde{x} \in \mathbb{X}$, then $\|x_{\Sigma^+}(N, \bar{x}) - \hat{x}_{\hat{\Sigma}}(N, \tilde{x})\| \leq \epsilon$ for all $N \geq \bar{N}$.*

Definition 3.2. *The $\hat{\Sigma}$:estimator $\left(\hat{\Sigma}\text{:mh-estimator}\right)$ is an* **asymptotically stable observer** *for the system Σ^+ if it is stable and there is a fixed number $\vartheta > 0$ such that if $\|\bar{x} - \tilde{x}\| \leq \vartheta$ and $\tilde{x} \in \mathbb{X}$, then for all $\mu > 0$ there exists a positive integer $\tilde{N}(\mu)$ such that $\|x_{\Sigma^+}(N, \bar{x}) - \hat{x}_{\hat{\Sigma}}(N, \tilde{x})\| \leq \mu$ for all $N \geq \tilde{N}$.*

The implications of constraints on the observer are more subtle than for the regulator. In particular, the estimator has no control over the evolution of the state of the system. Therefore, we require the evolution of the system Σ^+ respects the constraints. We satisfy this requirement with the following assumption.

A6: The evolution of the system Σ^+ respects the constraints $\mathbb{X}$. In other words, $x_{\Sigma^+}(k, \bar{x}) \in \mathbb{X}$ for all $k \geq 0$, where $\bar{x}$ is the initial condition for the system Σ^+.

Assumption ($A6$) guarantees we can always solve the observer problem by choosing the true state as our estimate. Without assumption ($A6$), divergence of the estimator is possible. Some examples of the instability of MHE due to constraints are given in [27].

Before proceeding with the stability analysis of the MHE, we first analyze the stability of the full information estimator for pedagogical purposes. A preliminary analysis of the stability of the full information estimator is given in [19] for linear systems and [18] for nonlinear systems. In order to invoke the Lyapunov arguments, we require an existence condition and a technical lemma.

Proposition 3.3. *If we assume* ($A1$)-($A3$), *then a solution exists to* ($P1$) *for all* $N \in \mathbb{N}$.

Proof. By assumption ($A3$), a feasible solution exists for all N. Let $\tilde{\Phi}_N$ denote the finite cost, by ($A2$), associated with $\tilde{x}_0$ and the sequence $\{\tilde{w}_{k|N-1}\}_{k=0}^{N-1}$ and consider the level set

$$\Lambda = \left\{ x_0, \{w_k\}_{k=0}^{N-1} \;:\; \begin{array}{c} \{w_k\}_{k=0}^{N-1} \in \mathbb{W}^N,\; \{x_k\}_{k=0}^{N} \in \mathbb{X}^{N+1} \\ \Phi_N(x_0, \{w_k\}_{k=0}^{N-1}) \leq \tilde{\Phi}_N \end{array} \right\}.$$

By assumption, the function $\Phi_N(x_0, \{w_k\}_{k=0}^{N-1})$ is lower semi-continuous. The set

$$F = \left\{ x_0, \{w_k\}_{k=0}^{N-1} \;:\; \Phi_N(x_0, \{w_k\}_{k=0}^{N-1}) \in [0, \tilde{\Phi}_N] \right\}$$

is closed, because the inverse image of a closed set under a lower semi-continuous function is closed [1]. The set

$$G = \left\{ x_0, \{w_k\}_{k=0}^{N-1} \;:\; \{w_k\}_{k=0}^{N-1} \in \mathbb{W}^N,\; \{x_k\}_{k=0}^{N} \in \mathbb{X}^{N+1} \right\}$$

is closed, because the functions f and g are continuous and the sets $\mathbb{X}$ and $\mathbb{W}$ are closed. The set Λ is closed, because it is the intersection of the sets F and G. By assumption ($A2$), there exists a K_+-function η such that $L(w, v) > \eta(\|w, v\|)$ and $\Gamma(x) > \eta(\|x\|)$. This inequality implies the set F is also bounded, because $\Gamma(x_0) < \tilde{\Phi}_N$ and $L(w_k, v_k) < \tilde{\Phi}_N$ for $k = 0, \ldots, N-1$. The set Λ is bounded, because $\Lambda \subseteq F$. Hence, the set Λ is compact. Existence of a solution follows from the Weierstrass Maximum Theorem. □

When α, specified by ($A2'$), is zero, we can cannot invoke the same arguments for the existence of a solution to MHE. To guarantee a solution exists, we also require the set $\mathbb{X}$ is compact. If this condition is too strict, we can relax the requirement to requiring only that x_{N-M} resides in a compact set. For simplicity we state the following corollary existence condition.

Corollary 3.4. *If we assume* ($A1$)-($A3$) *and* $\mathbb{X}$ *is compact, then a solution exists to* ($P2$) *for all* $N \in \mathbb{N}$.

Proof. Let α be specified by ($A2'$). If $\alpha \neq 0$, then the Corollary follows directly from Proposition 3.3 regardless of whether $\mathbb{X}$ is compact, because condition ($A3$) guarantees a feasible solution. In particular, Γ_N is bounded below by a K_+-function. If $\alpha = 0$, let $\tilde{\Theta}_N$ denote the cost due to the feasible solution resulting from ($A3$). Consider the level set

$$\Lambda = \left\{ x_{N-M}, \{w_k\}_{k=N-M}^{N-1} \;:\; \begin{array}{l} \{w_k\}_{k=N-M}^{N-1} \in \mathbb{W}^M, \\ \{x_k\}_{k=N-M}^{N} \in \mathbb{X}^{M+1}, \\ \Theta_N(x_{N-M}, \{w_k\}_{k=N-M}^{N-1}) \leq \tilde{\Theta}_N \end{array} \right\}.$$

The set Λ is compact, because $\mathbb{X}$ is compact, Γ_N is lower semi-continuous, and the stage cost functions are lower semi-continuous and satisfy assumption ($A1$) and ($A2$). The remaining steps of the proof directly follow those in the proof of Proposition 3.3. □

Remark 3.5. *When Σ^+ generates the data and $\alpha = 0$ as defined by ($A2'$), condition ($A6$) guarantees the existence of a solution to (P2). The optimal value function is precisely zero. This scenario is briefly discussed in* [18].

Remark 3.6. *In light of ($A6$), the requirement that $\mathbb{X}$ is compact is tantamount to requiring Σ^+ is stable in the sense that the state evolution is bounded.*

The following technical lemma states if the residual sequence $\{\hat{w}_{k|k}, \hat{v}_{k|k}\}$ is small, then the estimation error is small. This notion is ancillary to incremental input/output-to-state observability [24]. Rather than state the condition as an assumption, we establish it with the following technical lemma.

Lemma 3.7. *Suppose f, g, and h are Lipschitz continuous with constants c_f, c_g, and c_h, and $\hat{\Sigma}$ is incrementally observable. For all $\epsilon > 0$, there exists $\varrho > 0$ such that if*

$$\max_{j \in \{0,\dots,k-1\}} \{\|\hat{w}_{N+j|N+k-1}\|, \|\hat{v}_{N+j|N+k-1}\|\} \leq \varrho$$

for some $k \geq N_0$ and $N \geq 0$, then $\|x_{N+k} - \hat{x}_{N+k}\| \leq \epsilon$.

Proof. Let $\epsilon > 0$. If we choose $k \geq N_o$, then utilizing the inverse triangle inequality, we obtain the following inequality

$$\sum_{j=0}^{k-1} \|\hat{v}_{N+j|N+k-1}\| = \sum_{j=0}^{k-1} \|y_{N+j} - \hat{y}_{N+j|N+k-1}\|,$$

$$= \sum_{j=0}^{k-1} \|y_{N+j} - y_{\hat{\Sigma}}(j, \hat{x}_{N|N+k-1}) + y_{\hat{\Sigma}}(j, \hat{x}_{N|N+k-1}) - \hat{y}_{N+j|N+k-1}\|,$$

$$\geq \sum_{j=0}^{k-1} \|y_{N+j} - y_{\hat{\Sigma}}(j, \hat{x}_{N|N+k-1})\| - \sum_{j=0}^{k-1} \|y_{\hat{\Sigma}}(j, \hat{x}_{N|N+k-1}) - \hat{y}_{N+j|N+k-1}\|.$$

Recall that $y_{\hat{\Sigma}}(j, \hat{x}_{N|N+k-1})$ denotes the unforced response of the system $\hat{\Sigma}$ given the initial condition $\hat{x}_{N|N+k-1}$.

Let φ be specified by the incremental observability condition. Rearranging the inequality and utilizing the observability condition, we obtain the inequality

$$\begin{aligned}\sum_{j=0}^{k-1} \|\hat{v}_{N+j|N+k-1}\| &+ \sum_{j=0}^{k-1} \|y_{\hat{\Sigma}}(j, \hat{x}_{N|N+k-1}) - \hat{y}_{N+j|N+k-1}\| \\ &\geq \sum_{j=0}^{k-1} \|y_{N+j} - y_{\hat{\Sigma}}(j, \hat{x}_{N|N+k-1})\| \geq \varphi(\|x_N - \hat{x}_{N|N+k-1}\|). \end{aligned} \tag{7}$$

By the Lipschitz continuity of f and g, we have the following inequality

$$\begin{aligned}\|x_{\hat{\Sigma}}(1, \hat{x}_{N|N+k-1}) - \hat{x}_{N+1|N+k-1}\| &\leq \|f(\hat{x}_{N|N+k-1}) - f(\hat{x}_{N|N+k-1})\| + \\ &\quad \|g(\hat{x}_{N|N+k-1}, 0) - g(\hat{x}_{N|N+k-1}, \hat{w}_{N|N+k-1})\|, \\ &\leq c_g \|\hat{w}_{N|N+k-1}\|. \end{aligned}$$

Note by assumption, we have $g(\hat{x}_{N|N+k-1}, 0) = 0$. Proceeding for $N+2$, we obtain the inequality

$$\begin{aligned}\|x_{\hat{\Sigma}}(2, \hat{x}_{N|N+k-1}) - \hat{x}_{N+2|N+k-1}\| &\leq c_f \|x_{\hat{\Sigma}}(\hat{x}_{N|N+k-1}, 1) - \hat{x}_{N+1|N+k-1}\| + \\ &\quad \|g(\hat{x}_{N+1|N+k-1}, 0) - g(\hat{x}_{N+1|N+k-1}, \hat{w}_{N+1|N+k-1})\|, \\ &\leq c_f c_g \|\hat{w}_{N|N+k-1}\| + c_g \|\hat{w}_{N+1|N+k-1}\|. \end{aligned}$$

Likewise, for arbitrary q, we have the following inequality

$$\|x_{\hat{\Sigma}}(q, \hat{x}_{N|N+k-1}) - \hat{x}_{N+q|N+k-1}\| \leq \sum_{j=0}^{q-1} c_f^j c_g \|\hat{w}_{N+j|N+k-1}\|. \tag{8}$$

Using the Lipschitz continuity of h and (8), we obtain the following inequality

$$\|y_{\hat{\Sigma}}(q, \hat{x}_{N|N+k-1}) - \hat{y}_{N+q|N+k-1}\| \leq c_h \sum_{l=0}^{q-1} c_f^l c_g \|\hat{w}_{N+l|N+k-1}\|. \tag{9}$$

The inequalities (7) and (9) yield the following bound

$$\varphi(\|x_N - \hat{x}_{N|N+k-1}\|) \leq \sum_{j=0}^{k-1} \|\hat{v}_{N+j|N+k-1}\| + \sum_{j=0}^{k-1} c_h \sum_{l=0}^{j-1} c_f^l c_g \|\hat{w}_{N+l|N+k-1}\|.$$

In an analogous manner to (8), we can construct the bound

$$\|x_{N+k} - \hat{x}_{N+k}\| \leq c_f^k \|x_N - \hat{x}_{N|N+k-1}\| + \sum_{j=0}^{k-1} c_f^j c_g \|\hat{w}_{N+j|N+k-1}\|.$$

If we choose $\varrho > 0$ such that

$$\varrho \leq \frac{1}{2}\min\left\{ \frac{\varphi\left(\epsilon/c_f^k\right)}{2\max\left\{k, \sum_{j=0}^{k-1} c_h \sum_{l=0}^{j-1} c_f^l c_g\right\}}, \frac{1}{\sum_{j=0}^{k-1} c_f^j c_g} \right\},$$

then the lemma follows. □

Proposition 3.8. *If we assume (A1), (A2), (A6) and $\hat{\Sigma}$ is incrementally observable, then the $\hat{\Sigma}$:estimator is an asymptotically stable observer for the system Σ^+.*

Proof. We assume throughout the proof $N \geq N_o$. We first demonstrate convergence. Let $V^+ \triangleq \Gamma(\bar{x})$, where $\bar{x}$ denotes the initial condition of Σ^+. The existence of V^+ follows from (*A*6). We know an optimal solution exists from Proposition 3.3, because a feasible solution exists for all k. Hence, by optimality, we have that $V^+ \geq \Phi_k^*$ for all k. Writing out the cost function explicitly, we regroup the optimal cost as follows

$$\begin{aligned} \Phi_N^* &= \sum_{k=0}^{N-1} L(\hat{w}_{k|N-1}, \hat{v}_{k|N-1}) + \Gamma(\hat{x}_{0|N-1}), \\ &= L(\hat{w}_{N-1|N-1}, \hat{v}_{N-1|N-1}) + \sum_{k=0}^{N-2} L(\hat{w}_{k|N-1}, \hat{v}_{k|N-1}) + \Gamma(\hat{x}_{0|N-1}). \end{aligned}$$

Because $\{\hat{w}_{k|N-1}\}_{k=0}^{N-2}$ and $\hat{x}_{0|N-1}$ are feasible at time index $k = N-1$, we obtain the inequality

$$\sum_{k=0}^{N-2} L(\hat{w}_{k|N-1}, \hat{v}_{k|N-1}) + \Gamma(\hat{x}_{0|N-1}) \geq \Phi_{N-1}^*$$

by optimality. This inequality implies

$$\Phi_N^* - \Phi_{N-1}^* \geq L(\hat{w}_{N-1|N-1}, \hat{v}_{N-1|N-1}).$$

Since Φ_N^* is nondecreasing and bounded above by V^+, the sequence of optimal costs Φ_N^* converges to $\Phi_\infty^* < \infty$. Convergence implies the partial sum

$$\sum_{j=0}^{k-1} L(\hat{w}_{N+j|N+k-1}, \hat{v}_{N+j|N+k-1}) \to 0,$$

as $N \to \infty$ for $k \geq N_o$. By (*A*2) and Lemma 3.7, the estimation error $\|\hat{x}_N - x_{\Sigma^+}(N, \bar{x})\| \to 0$ as claimed.

To prove stability, let $\epsilon > 0$ and choose $\varrho > 0$ for $N = N_0$ as specified by Lemma 3.7. Let η and γ be specified by (*A*2). Without loss of generality, assume

$\Gamma(\bar{x}) = 0$. If we choose $\delta > 0$ such that $\eta^{-1}(\gamma(\delta)) < \varrho$, then we obtain the following inequality for all $N > N_o$.

$$\begin{aligned}\gamma(\delta) &\geq \Gamma(\tilde{x}) \geq \sum_{k=0}^{N-1} L(\hat{w}_{k|N-1}, \hat{v}_{k|N-1}) + \Gamma(\hat{x}_{0|N-1}), \\ &\geq \max_{k \in \{0,\dots,N-1\}} \{\eta(\|\hat{w}_{k|N-1}\|), \eta(\|\hat{v}_{k|N-1}\|)\}.\end{aligned}$$

The K_+-function η^{-1} exists, because η is homeomorphic and strictly increasing [22]. Hence, if the initial estimation error $\|\bar{x} - \tilde{x}\| \leq \delta$, then the estimation error $\|x_{\Sigma^+}(N, \bar{x}) - \hat{x}_N\| \leq \epsilon$ for all $N > N_o$ as claimed. □

The proof for the stability of the MHE is similar to the arguments used for the full information estimation. The key difference is we need to establish V^+ is a uniform bound for Θ_k. Establishing this condition is where we invoke the lower bound for the arrival cost specified by assumptions ($A4$) and ($A5$).

Proposition 3.9. *If we assume* ($A1$), ($A2$), ($A4$)-($A6$), $\hat{\Sigma}$ *is incrementally observable,* $\mathbb{X}$ *is compact, and* $M > N_o$, *then the* $\hat{\Sigma}$*:mh-estimator is an asymptotically stable observer for the system* Σ^+.

Proof. We first demonstrate convergence. The proof closely follows the arguments used in Proposition 3.8. Let $V^+ \triangleq \Gamma_0(\bar{x})$, where $\bar{x}$ denotes the initial condition of Σ^+. The existence of V^+ follows from ($A6$). By Corollary 3.4 an optimal solution exists, because a feasible solution exists for all k and $\mathbb{X}$ is compact. By definition,

$$\Theta^*_{N+M} - \Theta^*_N \geq \sum_{k=N}^{N+M-1} L(\hat{w}_{k|N+M-1}, \hat{v}_{k|N+M-1}). \tag{10}$$

To demonstrate V^+ is a uniform bound, we proceed using a quasi-induction argument. For $N < N_o \leq M$, by optimality,

$$\Theta^*_N \leq \Xi_N(x_{\Sigma^+}(N, \bar{x})) \leq V^+.$$

By construction and optimality, $\Xi_N(x_{\Sigma^+}(N, \bar{x})) \leq V^+$ for $N \leq M$. Utilizing the optimality principle and assumptions ($A4$) and ($A5$), for all $N \geq M$,

$$\begin{aligned}V^+ &\geq \min_{x_N, \{w_k\}_{k=N}^{N+M-1}} \left\{ \sum_{k=N}^{N+M-1} L(w_k, v_k) + \Xi_N(x_N) \ : \ \begin{matrix} x_{N+M} = \\ x_{\Sigma^+}(N+M, \bar{x}) \end{matrix} \right\}, \\ &\geq \min_{x_N, \{w_k\}_{k=N}^{N+M-1}} \left\{ \sum_{k=N}^{N+M-1} L(w_k, v_k) + \Gamma_N(x_N) \ : \ \begin{matrix} x_{N+M} = \\ x_{\Sigma^+}(N+M, \bar{x}) \end{matrix} \right\}, \\ &= \Xi_{N+M}(x_{\Sigma^+}(N+M, x_0)) \geq \Theta_{N+M},\end{aligned}$$

where both minimizations are subject to the constraints

$$\{w_k\}_{k=N}^{N+M-1} \in \mathbb{W}^M, \qquad \{x_k\}_{k=N}^{N+M} \in \mathbb{X}^{M+1}.$$

Hence, Θ_N^* is a monotone increasing sequence that is bounded above by V^+. Convergence implies the sum

$$\sum_{j=0}^{M-1} L(\hat{w}_{N+j|N+M-1}, \hat{v}_{N+j|N+M-1}) \to 0,$$

as $N \to \infty$, because the summation (10) is nonnegative. By $(A2)$ and Lemma 3.7 the estimation error $\|\hat{x}_N - x_{\Sigma^+}(N,\bar{x})\| \to 0$ as claimed.

To prove stability, let $\epsilon > 0$ and choose $\varrho > 0$ for $N = N_o$ as specified by Lemma 3.7. Let η and $\bar{\gamma}$ be specified by $(A2)$. Without loss of generality, assume $\Gamma_0(\bar{x}) = 0$. If we choose $\delta > 0$ such that $\eta^{-1}(\bar{\gamma}(\delta)) < \varrho$, then we obtain the following inequality for all $N > N_o$.

$$\begin{aligned} \bar{\gamma}(\delta) &\geq \Gamma_0(\tilde{x}) \geq \sum_{k=N-M}^{N-1} L(\hat{w}_{k|N-1}, \hat{v}_{k|N-1}) + \Gamma_{N-M}(\hat{x}_{N-M|N-1}), \\ &\geq \max_{k\in\{N-M,\dots,N-1\}} \left\{\eta(\|\hat{w}_{k|N-1}\|), \eta(\|\hat{v}_{k|N-1}\|)\right\}. \end{aligned}$$

The K_+-function η^{-1} exists, because η is homeomorphic and strictly increasing [22]. Hence, if the initial estimation error $\|\bar{x} - \tilde{x}\| \leq \delta$, then the estimation error $\|x_{\Sigma^+}(N,\bar{x}) - \hat{x}_N\| \leq \epsilon$ for all $N > N_o$ as claimed. □

Remark 3.10. *In actual implementation, we take $\Gamma_N(\hat{x}_N) = 0$, because translation of the origin does not affect the result of the optimization.*

3.1. Constrained Nonlinear Estimation

When the system dynamics are nonlinear, we are unable to calculate the arrival cost or generate a global lower bound with the exception of the obvious choice $\Gamma_N = \Theta_N^*$. Unlike the regulator however, we are not concerned with precluding unstable or undesirable policies. We are concerned rather with maintaining a uniform upper bound to the optimal cost. As demonstrated in the proof of Proposition 3.9, the global lower bound of the arrival cost is sufficient to guarantee V^+ is a uniform upper bound to the optimal cost.

While a global lower bound is ideal, we may circumvent the issue by explicitly ensuring the bound is uniform in nominal application. Suppose we have some strategy for generating a sequence of initial penalties $\{\Gamma_N'\}_{N=0}^{\infty}$ that satisfy assumptions $(A1')$ and $(A2')$. Assumption $(A4)$ follows by translating the origin such that $\Gamma_N'(\hat{x}_N) = \Theta_N^*$. The goal of providing a lower bound to the arrival cost is to ensure the initial penalties are monotone decreasing:

$$\Gamma_j(\hat{x}_j) \geq \Gamma_{j+M}(\hat{x}_{j+M}) \geq \cdots \geq \Gamma_{j+kM}(\hat{x}_{j+kM}),$$

for $j,k \in \mathbb{I}_+$. Uniformity of the upper bound is satisfied by pointwise monotonicity of the initial penalties. Rather than rely on the general structure of Γ_j to provide the uniformity, we can explicitly prescribe the pointwise monotonicity in the algorithm without the need for a global lower bound to the arrival cost. In particular, by scaling the initial penalties $\{\Gamma'_N\}$, we can maintain the uniformity as follows.

Consider the MHE problem where we ignore the state disturbances,

$$\{w_k\}_{k=N-M}^{N-1} = 0,$$

and consider only the decision variable x_{N-M}. We formulate this estimation problem as the following optimal control problem

$$(P3) \qquad \Psi_N^* = \min_{x_{N-M}} \Psi_N(x_{N-M})$$

subject to the constraint

$$\{x_k\}_{k=N-M}^{N} \in \mathbb{X}^{M+1},$$

where the cost function is of the form

$$\Psi_N(w_{N-M}) \triangleq \sum_{k=N-M}^{N-1} L(0, v_k).$$

Let $\bar{x}_{N-M|N-1}$ denote the optimal solution to $(P3)$, and

$$\bar{\Theta}_N(\beta_N, \bar{x}_{N-M|N-1}) \triangleq \Psi_N^* + \beta_N \Gamma'_{N-M}(\bar{x}_{N-M|N-1})$$

denote the cost of the policy with the objective function Θ_N and the scaled initial penalty. The dynamics are specified by the system $\hat{\Sigma}$. The scalar β_N is the scaling factor. By optimality, $\bar{\Theta}_N$ is an upper bound to the optimal cost function Θ_N. Hence, we may view $\bar{\Theta}_N$ as our estimate for the arrival cost $\Gamma_N(\hat{x}_{N|N+M-1})$. If we consider time index $N+M$, then in order to maintain the uniformity of V^+, we require that

$$\begin{aligned}\bar{\Theta}&_{N+M|N-1}(\beta_N, \bar{x}_{N-M|N-1}) \\ &\triangleq \sum_{k=0}^{2M-1} L\left(0, y_{k+N-M} - y_{\hat{\Sigma}}(k, \bar{x}_{N-M|N-1})\right) + \beta_N \Gamma'_{N-M}(\bar{x}_{N-M|N-1}) + \Theta_{N-M}^*\end{aligned}$$

is an upper bound to $\bar{\Theta}_{N+M}(\beta_{N+M}, \bar{x}_{N|N+M-1})$ for some $\beta_{N+M} \in [0,1]$. We extend the prediction horizon, because the bounds may not be strictly binding due to measurement and state noise in actual implementation. Therefore, we require overlap between the prediction horizons. In the abstract case, where we assume the dynamics are specified by the system Σ^+, then

$$\bar{\Theta}_{N+M|N}(\beta_N, \bar{x}_{N-M|N-1}) = \bar{\Theta}_{N|N}(\beta_N, \bar{x}_{N-M|N-1}) = \beta_N \Gamma'_{N-M}(\bar{x}_{N-M|N-1}),$$

because the residuals are zero. We are not always guaranteed there is a $\beta_N > 0$ such that

$$\bar{\Theta}_N(\beta_N, \bar{x}_{N-M|N-1}) \leq \bar{\Theta}_{N|N-M}(\beta_{N-M}, \bar{x}_{N-2M|N-M-1}).$$

To overcome this problem, we can always choose $\beta_N = 0$, which is a global, albeit trivial, lower bound.

We formulate the estimation strategy as the following algorithm.

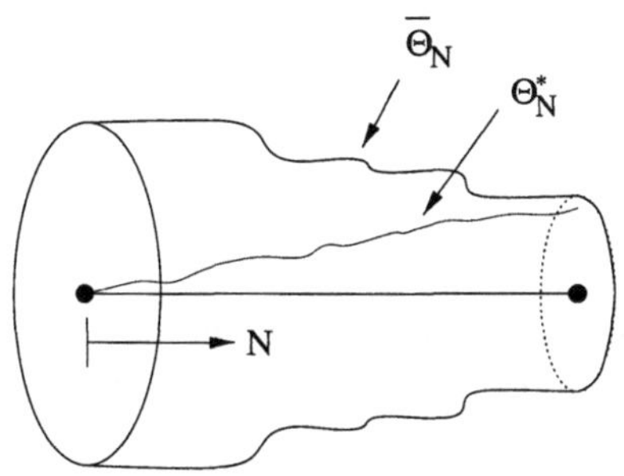

FIGURE 2. Rather than upper bound the monotone increasing sequence Θ^*_N with V^+, in the proposed strategy we bound Θ^*_N with the monotone decreasing sequence $\bar{\Theta}_N$. Convergence of the state estimate results from *pinching* the two sequences.

ESTIMATION STRATEGY For $N \leq M$, the estimation strategy is equivalent to the full information strategy. Suppose $N > M$ and let $\beta_M = 1$ with $\bar{\Theta}_M = \bar{\Theta}_k$ for $k \leq M$, then the estimation strategy is as follows:

1. If

$$\bar{\Theta}_N\left(1, x_{\hat{\Sigma}}(M, \bar{x}_{N-2M|N-M-1})\right) \leq \bar{\Theta}_{N|N-M}(\beta_{N-M}, \bar{x}_{N-2M|N-M-1}),$$

   then set $\Gamma_{N-M} = \Gamma'_{N-M}$ and $\beta_N = 1$;
   else solve $(P3) \Longrightarrow \{\bar{x}_{N-M|N-1}\}$
   (a) Solve the problem

$$\beta_N = \max_{\beta \in [0,1]} \left\{ \beta : \begin{array}{l} \bar{\Theta}_N(\beta, \bar{x}_{N-M|N-1}) \leq \\ \bar{\Theta}_{N|N-M}(\beta_{N-M}, \bar{x}_{N-2M|N-M-1}) \end{array} \right\}.$$

   (b) Set $\Gamma_{N-M} = \beta_N \Gamma'_{N-M}$
2. Solve $(P2)$ with $\Gamma_{N-M} \Longrightarrow \{\hat{x}_N\}$
3. Repeat for $N \leftarrow N + 1$.

Figure 2 displays the key elements of the stability proof. Rather than attempt to bound the sequence Θ^*_N by lower bounding the arrival cost with Γ_N, we generate the bound with the monotonically decreasing sequence $\bar{\Theta}_N$. This strategy explicitly guarantees V^+ is a uniform bound, because $V^+ \geq \bar{\Theta}^*_0$ by construction.

Proposition 3.11. *If we assume (A1), (A2), (A6), Σ is incrementally observable, and $M > N_o$, then the $\hat{\Sigma}$:mh-estimator using the proposed strategy for generating Γ_N is an asymptotically stable observer for the system Σ^+.*

Proof. It suffices to show V^+ is an upper bound to Θ_N for all N. See the proof of Proposition 3.9 for the details of the complete argument. We proceed by induction and consider $N = M+1$. By construction we have $\Gamma_0(\bar{x}_{0|M-1}) = V^+$ and

$$\begin{aligned}\bar{\Theta}_{M+1}(\beta_1, \bar{x}_{1|M}) &\leq \bar{\Theta}_{M+1|M}(1, \bar{x}_{0|M-1}),\\ &= \bar{\Theta}_M(1, \bar{x}_{0|M-1}),\\ &= \Gamma_0(\bar{x}_{0|M-1}).\end{aligned}$$

The equalities follow because Σ^+ generates the data and assumption ($A6$). The existence of β_{M+1} is guaranteed, because $\beta_{M+1} = 0$ implies that

$$\bar{\Theta}_{M+1}(\beta_{M+1}, \bar{x}_{1|M}) = 0,$$

and the inequality directly follows. Hence, by optimality,

$$\Theta^*_{M+1} \leq \bar{\Theta}^*_{M+1} \leq V^+.$$

Now consider arbitrary N and assume that $\bar{\Theta}_N \leq V^+$. By construction we obtain the inequality

$$\begin{aligned}\bar{\Theta}_{N+M}(\beta_{N+M}, \bar{x}_{N-2M|N+M-1}) &\leq \bar{\Theta}_{N+M|N}(\beta_N, \bar{x}_{N-M|N-1}),\\ &\leq V^+.\end{aligned}$$

The existence of β_{N+M} is guaranteed, because $\beta_{N+M} = 0$ is feasible. □

A natural question is how the proposed estimation strategy performs when Γ_N satisfies assumptions ($A4$) and ($A5$). Note the Kalman filter and the constrained linear estimator both satisfy these conditions. In particular, we desire that $\beta_N = 1$ under these circumstance.

Proposition 3.12. *Suppose Γ'_N satisfies assumptions* ($A4$) *and* ($A5$) *for all $N \in \mathbb{I}_+$, Σ is incrementally observable,* ($A1$), ($A2$) *and* ($A6$), *then $\beta_N = 1$ for all $N \in \mathbb{I}_+$.*

Proof. Expanding $\bar{\Theta}_N(1, \bar{x}_{N-M|N-1})$ and using assumptions ($A4$) and ($A5$), we have for $N > M$,

$$\begin{aligned}\bar{\Theta}_N(1, \bar{x}_{N-M|N-1}) &= \Psi^*_N + \Gamma'_{N-M}(\bar{w}_{N-M|N-1})\\ &\leq \Psi^*_N + \Xi_{N-M}(\bar{x}_{N-M|N-M-1})\\ &\leq \Psi^*_N + \sum_{k=0}^{M-1} L\left(0, y_{k+N-2M} - y_{\hat{\Sigma}}(k, \bar{x}_{N-2M|N-M-1})\right) +\\ &\quad\ \Gamma'_{N-2M}(\bar{x}_{N-2M|N-M-1}),\end{aligned}$$

where $\bar{x}_{N-M|N-M-1}$ is chosen such that

$$\bar{x}_{N-M|N-M} = x_{\hat{\Sigma}}(M, \bar{x}_{N-2M|N-M-1}).$$

Hence the proposition follows. □

Remark 3.13. *Note we have made no assumption regarding the system generating the data in Proposition 3.12. In particular, the underlying dynamics need not be specified by the deterministic system Σ^+.*

We still have not specified the particular structure of Γ'_N. One strategy for generating an approximate lower bound for the arrival cost is to use a first order Taylor series approximation of $\hat{\Sigma}$ around the estimated trajectory $\{\hat{x}_k\}_{k=0}^N$. This strategy yields an extended Kalman filter covariance update formula for constructing Γ'_N. We interpret this strategy as a neighboring extremal paths strategy in the context of estimation. Neighboring extremal paths is a technique to generate approximate optimal feedback laws for nonlinear systems by employing an extended linearization [2]. The basic idea is as follows. If the deviation from the optimal path is small and regularity conditions hold (see [6]), then a linear approximation of the optimal path accurately represents the neighboring optimal path.

This concept is used to relax the endpoint penalty in nonlinear receding horizon control [20, 3] using a dual-mode framework [14]. For small estimation errors, we expect the local lower bound is stabilizing. These arguments would follow the local stability results for the extended Kalman filter [23]. However, unlike receding horizon control, the standard dual-mode arguments for *global* stability fail, because there is no fixed origin about which to construct a local invariant region. Nevertheless, we propose the extended linearization for generating approximate lower bounds for the arrival cost, because we expect the extended linearization yields an effective lower bound locally and for weakly nonlinear systems.

Let

$$A_k \triangleq \left[\left.\frac{\partial f(x_k)}{\partial x_k}\right|_{\hat{x}_{k|k}} + \left.\frac{\partial g(x_k, \hat{w}_{k|k})}{\partial x_k}\right|_{\hat{x}_{k|k}} \right],$$

$$G_k \triangleq \left.\frac{\partial g(\hat{x}_{k|k}, w_k)}{\partial w_k}\right|_{\hat{w}_{k|k}}, \qquad C_k \triangleq \left.\frac{\partial h(x_k)}{\partial x_k}\right|_{\hat{x}_{k|k}},$$

denote the linearized dynamics of the system and

$$R_k^{-1} \triangleq \left.\frac{\partial^2 \underline{L}(\hat{w}_{k|k}, v_k)}{\partial v_k^2}\right|_{\hat{x}_{k|k}}, \qquad M_k \triangleq \left.\frac{\partial^2 \underline{L}(w_k, v_k)}{\partial w_k \partial x_k}\right|_{\hat{w}_{k|k},\, \hat{x}_{k|k}},$$

$$Q_k^{-1} \triangleq \left.\frac{\partial^2 \underline{L}(w_k, \hat{v}_{k|k})}{\partial w_k^2}\right|_{\hat{w}_{k|k}},$$

denote the linearized stage penalties, where $\underline{L} \in \mathcal{C}^2(\mathbb{R}^m \times \mathbb{R}^p, \mathbb{R}_+)$ is a lower bound to L and satisfies assumptions $(A1)$ and $(A2)$. If we assume for simplicity $M_k = 0$, then we take $\Gamma'_N(x) = (x - \hat{x}_N)^T \Pi_{N|N-1}^{-1} (x - \hat{x}_N)$, using the recursion,

$$\begin{aligned}\Pi_{N+1|N} &= G_N Q_N G_N^T + \\ &A_{N-1}(\Pi_{N|N-1} - \Pi_{N|N-1} C_N^T (R_N + C_N \Pi_{N|N-1} C_N^T)^{-1} C_N \Pi_{N|N-1}) A_{N-1}^T,\end{aligned}$$

to calculate $\Pi_{N|N-1}$. This result is equivalent to the covariance update formula for the extended Kalman filter. See [8] for further details.

4. Duality

Given the strong duality between the linear quadratic regulator and the Kalman filter, the natural question to pose is whether MHE is the dual to receding horizon control (RHC). Assume the process model for control is given by the following deterministic difference equation

$$x_{k+1} = f_c(x_k, u_k).$$

Proceeding in an informal manner, we formulate RHC, given x_k, as the solution to

$$(P4) \qquad V_N^*(x_k) = \min_{\{u_j\}_{j=k}^{k+N-1}} V_N(\{u_j\}_{j=k}^{k+N-1})$$

subject to the constraints

$$\{u_j\}_{j=k}^{k+N-1} \in \mathbb{U}^N, \qquad \{x_j\}_{j=k+1}^{k+N} \in \mathbb{Y}^N$$

where the closed sets $\mathbb{U} \subseteq \mathbb{R}^m$ and $\mathbb{Y} \subseteq \mathbb{R}^N$, and the state trajectory, and

$$V_N(\{u_j\}_{j=k}^{k+N-1}) \triangleq \sum_{j=0}^{N-1} l(x_{k+j}, u_{j+k}) + F(x_{N+k}).$$

Let $\{u_{j|k}^*\}_{j=k}^{k+N-1}$ denote the solution to $(P4)$. The feedback law is given by $u_k = u_{k|k}^*$.

Following a discussion similar to [13], we have the relationship for the optimal cost function

$$V_N^*(x_{k+1}) - V_N^*(x_k) = -l(x_k, u_k) - \left[V_{N-1}^*(x_{k+1}) - V_N^*(x_{k+1})\right].$$

A sufficient condition for asymptotic stability is $V_{N-1}^*(x) \geq V_N^*(x)$ for all x. This condition is equivalent to requiring

$$F(x_k) \geq \min_{u_k} \{l(x_k, u_k) + F(x_{k+1})\}. \tag{11}$$

In other words, the terminal penalty F generates an upper bound to the cost to go. Recall assumptions $(A4)$ and $(A5)$ require the initial penalty Γ_N is a lower bound for the arrival cost. Note just as $\Gamma_N = 0$ is the trivial solution, the choice $F(x) = \infty$ for $x \neq 0$, which we interpret as the constraint $x_{N+k} = 0$, is the trivial solution for RHC.

While the arguments are slightly different (in particular, we use an M-step Lyapunov argument in MHE), RHC requires an upper bound to the backwards dynamic programming solution, whereas MHE requires a lower bound to the forward dynamic programming solution. Because the state is converging to a fixed point in RHC, we are able linearize about the fixed point and use standard local analysis to generate the upper bound. This strategy is used in [20] and [3]. However, there is no dual strategy to use local analysis in MHE.

One might expect we can enforce (11) pointwise by scaling the terminal penalty F using a strategy similar to the one proposed in Section 3.1. However, to enforce the pointwise bound in RHC, we need to scale the terminal penalty F

at time k only after seeing the optimal trajectory at time $k+1$. But, the optimal trajectory at time $k+1$ depends upon the scaling at time k, due to the transition x_k to x_{k+1}. This interdependency precludes the possibility of applying the scaling strategy to RHC. We do not have this interdependency problem in MHE, because we look backwards. Whereas in RHC, we look forward and anticipate the closed-loop response. However, by looking forward and anticipating the closed-loop response, we are able to employ local analysis in the construction of a stabilizing terminal penalty for a local invariant region in RHC.

5. Conclusion

We have demonstrated that a sufficient condition for the stability of MHE is that the initial penalty Γ_N is a global lower bound for the arrival cost. For linear models with quadratic objectives, the Kalman filter covariance yields a global lower bound for the arrival cost. This result holds regardless of whether constraints are present. For nonlinear estimation, where an explicit representation of the global lower bound is generally unavailable, we proposed an alternative strategy that yields stability. The key idea of the proposed strategy is to generate a monotone decreasing sequence that bounds the optimal cost function, thereby bypassing the need for the lower bound. We can equivalently view the decreasing sequence as the generation of a feasible estimation trajectory. Whereas in RHC there is an explicit stability constraint, we can view the stability constraint in MHE as a contraction on the bound for the optimal cost. The proposed strategy also performs in actual, noisy operation.

Acknowledgments

The authors gratefully acknowledge the financial support of the industrial members of the Texas-Wisconsin Modeling and Control Consortium and NSF through grant #CTS-9708497. The authors express their thanks to Professor D. Q. Mayne for helpful discussions and feedback about this work. The authors also acknowledge P. K. Findeisen, who studied many of the issues presented in this paper as part of a Master's research project [5] at the University of Wisconsin-Madison.

References

[1] C. Berge, *Topological spaces, including a treatment of multi-valued functions, vector spaces and convexity*, Oliver & Boyd, Edinburgh and London, 1963.

[2] A. E. Bryson and Y. Ho, *Applied optimal control*, Hemisphere Publishing, New York, 1975.

[3] H. Chen and F. Allgöwer, *A quasi-infinite horizon nonlinear model predictive control scheme with guaranteed stability*, Automatica **34** (1998), no. 10, 1205–1217.

[4] H. Cox, *On the estimation of state variables and parameters for noisy dynamic systems*, IEEE Trans. Auto. Cont. **9** (1964), no. 1, 5–12.

[5] Findeisen, P. *Moving horizon state estimation of discrete time systems*, Master's thesis, University of Wisconsin–Madison, 1997.

[6] H. Halkin, *A maximum principle of the Pontryagin type for systems described by nonlinear difference equations*, SIAM J. Cont. **4** (1966), no. 1, 90–111.

[7] S.-S. Jang, B. Joseph, and H. Mukai, *Comparison of two approaches to on-line parameter and state estimation of nonlinear systems*, Ind. Eng. Chem. Proc. Des. Dev. **25** (1986), 809–814.

[8] A. H. Jazwinski, *Stochastic processes and filtering theory*, Academic Press, New York, 1970.

[9] T. Kailath, *A view of three decades of linear filtering theory*, IEEE Trans. Inform. Theory **IT-20** (1974), no. 2, 146–181.

[10] S. S. Keerthi and E. G. Gilbert, *Optimal infinite-horizon feedback laws for a general class of constrained discrete-time systems: Stability and moving-horizon approximations*, J. Optim. Theory Appl. **57** (1988), no. 2, 265–293.

[11] D. G. Luenberger, *Observers for multivariable systems*, IEEE Trans. Auto. Cont. **11** (1966), no. 2, 190–197.

[12] ———, *An introduction to observers*, IEEE Trans. Auto. Cont. **16** (1971), no. 2, 596–602.

[13] E. S. Meadows, *Dynamic programming and model predictive control*, Proceedings of the 1997 American Control Conference, 1997, pp. 1635–1639.

[14] H. Michalska and D. Q. Mayne, *Robust receding horizon control of constrained nonlinear systems*, IEEE Trans. Auto. Cont. **38** (1993), no. 11, 1623–1633.

[15] ———, *Moving horizon observers and observer-based control*, IEEE Trans. Auto. Cont. **40** (1995), no. 6, 995–1006.

[16] P. E. Moraal and J. W. Grizzle, *Observer design for nonlinear systems with discrete-time measurements*, IEEE Trans. Auto. Cont. **40** (1995), no. 3, 395–404.

[17] K. R. Muske and T. F. Edgar, *Nonlinear state estimation*, Nonlinear Process Control (Michael A. Henson and Dale E. Seborg, eds.), Prentice Hall, 1996, pp. 311–370.

[18] K. R. Muske and J. B. Rawlings, *Nonlinear moving horizon state estimation*, Methods of Model Based Process Control (Dordrecht, The Netherlands) (Ridvan Berber, ed.), Nato Advanced Study Institute series: E Applied Sciences 293, Kluwer, 1995, pp. 349–365.

[19] K. R. Muske, J. B. Rawlings, and J. H. Lee, *Receding horizon recursive state estimation*, Proceedings of the 1993 American Control Conference, June 1993, pp. 900–904.

[20] T. Parisini and R. Zoppoli, *A receding-horizon regulator for nonlinear systems and a neural approximation*, Automatica **31** (1995), no. 10, 1443–1451.

[21] D. G. Robertson, J. H. Lee, and J. B. Rawlings, *A moving horizon-based approach for least-squares state estimation*, AIChE J. **42** (1996), no. 8, 2209–2224.

[22] H. L. Royden, *Real analysis*, 3rd ed., Prentice Hall, Englewood Cliffs, New Jersey, 1988.

[23] Y. Song and J. W. Grizzle, *The extended Kalman filter as a local asymptotic observer for discrete-time nonlinear systems*, J. Math. Sys., Estim, and Control. **5** (1995), no. 1, 59–78.

[24] E. D. Sontag and Y. Wang, *Output-to-state stability and detectability of nonlinear systems*, Sys. Cont. Let. **29** (1997), 279–290.

[25] M. Soroush, *Nonlinear state-observer design with application to reactors*, Chem. Eng. Sci. **52** (1997), no. 3, 387–404.

[26] C. Striebel, *Sufficient statistics in the optimum control of stochastic systems*, J. Math. Anal. Appl. **12** (1965), 576–592.

[27] M. L. Tyler and M. Morari, *Stability of constrained moving horizon estimation schemes, Preprint AUT96-18*, Automatic Control Laboratory, Swiss Federal Institute of Technology, 1996.

[28] B.L. Walcott, M.J. Corless, and S.H. Żak, *Comparative study of non–linear state–observation techniques*, Int. J. Control **45** (1987), no. 6, 2109–2132.

[29] G. Zimmer, *State observation by on-line minimization*, Int. J. Control **60** (1994), no. 4, 595–606.

Department of Chemical Engineering,
University of Wisconsin-Madison,
1415 Engineering Drive
Madison, WI 53706-1691
E-mail address: rao@bevo.che.wisc.edu
E-mail address: jbraw@bevo.che.wisc.edu

Progress in Systems and Control Theory, Vol. 26

Predictive Control of Constrained Hybrid Systems

Alberto Bemporad, Manfred Morari

Abstract. This paper proposes a framework for modeling and controlling systems described by interdependent physical laws, logic rules, and operating constraints, denoted as *Mixed Logical Dynamical* (MLD) systems. These are described by linear dynamic equations subject to linear inequalities involving real and integer variables. MLD systems include linear hybrid systems, finite state machines, some classes of discrete event systems, constrained linear systems, and nonlinear systems which can be approximated by piecewise linear functions. A predictive control scheme is proposed which is able to stabilize MLD systems on desired reference trajectories while fulfilling operating constraints, and possibly take into account previous qualitative knowledge in the form of heuristic rules. Due to the presence of integer variables, the resulting on-line optimization procedures are solved through *Mixed Integer Quadratic Programming* (MIQP), for which efficient solvers have been recently developed. Some examples and a simulation case study on a complex gas supply system are reported.

1. Introduction

The concept of *model* of a system is traditionally associated with differential or difference equations, typically derived by physical laws governing the *dynamics* of the system under consideration. Consequently, most of the control theory and tools have been developed for such systems, in particular for systems whose evolution is described by smooth linear or nonlinear state transition functions. On the other hand, in many applications the system to be controlled is also constituted by parts described by *logic*, such as for instance on/off switches or valves, gears or speed selectors, evolutions dependent on if-then-else rules. Often, the control of these systems is left to schemes based on heuristic rules inferred by practical plant operation.

Recently, in the literature researchers started dealing with *hybrid systems*, namely hierarchical systems constituted by dynamical components at the lower level, governed by upper level logical/discrete components [23, 3]. For this class of systems, design procedures have been proposed which naturally lead to hierarchical, hybrid control schemes, with continuous controllers at the lower level calibrated for each dynamical subsystem in order to provide regulation and tracking properties, and discrete controllers supervising, resolving conflicts, and planning strategies at a higher level [16].

However, in some applications a precise distinction between different hierarchic levels is not possible, especially when dynamical and logical facts are dramatically interdependent. For such a class of systems, not only it is not clear how to design feedback controllers, but even how to systematically obtain models.

This paper proposes a framework for modeling and controlling models of systems described by interacting physical laws, logical rules, and operating constraints. According to techniques described e.g. in [30, 5, 22], propositional logic is transformed into linear inequalities involving integer and continuous variables. This allows to arrive at *Mixed Logical Dynamical* (MLD) systems described by linear dynamic equations subject to linear mixed-integer inequalities, i.e. inequalities involving both *continuous* and *binary* (or *logical*, or *0-1*) variables. These include physical/discrete states, continuous/integer inputs, and continuous/binary auxiliary variables. MLD systems generalize a wide set of models, among which there are linear hybrid systems, finite state machines, some classes of discrete event systems, constrained linear systems, and nonlinear systems whose nonlinearities can be expressed (or, at least, suitably approximated) by piecewise linear functions.

Mixed-integer optimization techniques have been investigated in [21, 22] for chemical process synthesis. For feedback control purposes, we propose a predictive control scheme which is able to stabilize MLD systems on desired reference trajectories while fulfilling operating constraints, and possibly take into account previous qualitative knowledge in the form of heuristic rules. Moving horizon optimal control and model predictive control have been widely adopted for tracking problems of systems subject to constraints [14, 17, 20]. These methods are based on the so called *receding horizon* philosophy: a sequence of future control actions is chosen according to a prediction of the future evolution of the system and applied to the plant until new measurements are available. Then, a new sequence is determined which replaces the previous one. Each sequence is evaluated by means of an optimization procedure which take into account two objectives: optimize the tracking performance, and protect the system from possible constraint violations. In the present context, due to the presence of integer variables, the optimization procedure is a *Mixed Integer Quadratic Programming* (MIQP) problem [8, 13, 24], for which efficient solvers exist [7]. A first attempt to use on-line mixed-integer programming to control dynamic systems subject to logical conditions has appeared in [27] and reconsidered in [6]. Other attempts of combining MPC to hybrid control have also appeared in [26, 25].

This paper is organized as follows. In Section 2 some basic facts from propositional calculus, Boolean algebra, and mixed-integer linear inequalities are reviewed. These tools are used in Section 3 to motivate the definition of MLD systems and provide examples of systems which can be modeled within this framework. Stability definitions and related issues are discussed in Section 4. Section 5 deals with the optimal control of MLD systems and shows how heuristics can eventually be taken into account. These results are then used in Section 6 to develop a *Mixed-Integer Predictive Controller* (MIPC), which essentially solves on-line at each time step an optimal control problem through MIQP, and apply the optimal solution according to the aforementioned receding horizon philosophy. Finally, a brief description of available MIQP solvers is given in Section 7.

2. Propositional calculus and linear integer programming

By following standard notation [28, 5, 30], we adopt capital letters X_i to represent statements, e.g. "$x \geq 0$" or "Temperature is hot". X_i is commonly referred to as *literal*, and has a *truth value* of either "T" (true) or "F" (false). Boolean algebra enables statements to be combined in compound statements by means of *connectives*: "$\wedge$" (and), "$\vee$" (or), "$\sim$" (not), "$\rightarrow$" (implies), "$\leftrightarrow$" (if and only if), "$\oplus$" (exclusive or) (a more comprehensive treatment of Boolean calculus can be found in digital circuit design texts, e.g. [11, 10]. For a rigorous exposition see e.g. [18]). Connectives are defined by means of the *truth table* reported in Table 1. Other

TABLE 1. Truth table

X_1	X_2	$\sim X_1$	$X_1 \vee X_2$	$X_1 \wedge X_2$	$X_1 \rightarrow X_2$	$X_1 \leftrightarrow X_2$	$X_1 \oplus X_2$
F	F	T	F	F	T	T	F
F	T	T	T	F	T	F	T
T	F	F	T	F	F	F	T
T	T	F	T	T	T	T	F

connectives may be similarly defined. Connectives satisfy several properties (see e.g. [11]), which can be used to transform compound statements into equivalent statements involving different connectives, and simplify complex statements. It is known that all connectives can be defined in terms of a subset of them, for instance $\{\vee, \sim\}$, which is said to be a complete set of connectives. Below we report some properties which will be used in the sequel

$$X_1 \rightarrow X_2 \text{ is the same as } \sim X_1 \vee X_2 \tag{1a}$$

$$X_1 \rightarrow X_2 \text{ is the same as } \sim X_2 \rightarrow \sim X_1 \tag{1b}$$

$$X_1 \leftrightarrow X_2 \text{ is the same as } (X_1 \rightarrow X_2) \wedge (X_2 \rightarrow X_1) \tag{1c}$$

Correspondingly one can associate with a literal X_i a *logical variable* $\delta_i \in \{0, 1\}$, which has a value of either 1 if X_i =T, or 0 otherwise. Integer programming has been advocated as an efficient inference engine to perform automated deduction [5]. A propositional logic problem, where a statement X_1 must be proved to be true given a set of (compound) statements involving literals $X_1, \ldots, X_n$, can be in fact solved by means of a linear integer program, by suitably translating the original compound statements into linear inequalities involving logical variables

δ_i. In fact, the following propositions and linear constraints can easily be seen to be equivalent [30, p. 176]

$$X_1 \vee X_2 \text{ is equivalent to } \delta_1 + \delta_2 \geq 1 \tag{2a}$$

$$X_1 \wedge X_2 \text{ is equivalent to } \delta_1 = 1,\ \delta_2 = 1 \tag{2b}$$

$$\sim X_1 \text{ is equivalent to } \delta_1 = 0 \tag{2c}$$

$$X_1 \rightarrow X_2 \text{ is equivalent to } \delta_1 - \delta_2 \leq 0 \tag{2d}$$

$$X_1 \leftrightarrow X_2 \text{ is equivalent to } \delta_1 - \delta_2 = 0 \tag{2e}$$

$$X_1 \oplus X_2 \text{ is equivalent to } \delta_1 + \delta_2 = 1 \tag{2f}$$

We borrow this computational inference technique to model logical parts of processes (on/off switches, discrete mechanisms, combinational and sequential networks) and heuristics knowledge about plant operation as integer linear inequalities. As we are interested in systems which have both logic and dynamics, we wish to establish a link between the two worlds. In particular, we need to establish how to build statements from operating events concerning physical dynamics. As will be shown in a moment, we end up with *mixed-integer linear inequalities*, i.e. linear inequalities involving both *continuous variables* $x \in \mathbb{R}^n$ and logical (*indicator*) variables $\delta \in \{0,1\}$. Consider the statement $X \triangleq [f(x) \leq 0]$, where $f : \mathbb{R}^n \mapsto \mathbb{R}$ is linear, assume that $x \in \mathcal{X}$, where $\mathcal{X}$ is a given bounded set, and define

$$M \triangleq \max_{x \in \mathcal{X}} f(x) \tag{3a}$$

$$m \triangleq \min_{x \in \mathcal{X}} f(x) \tag{3b}$$

Theoretically, an over[under]-estimate of M [m] suffices for our purpose. However, more realistic estimates provide computational benefits [30, p. 171]. It is easy to verify that

$$[f(x) \leq 0] \ \wedge \ [\delta = 1] \text{ is true iff } f(x) - \delta \leq -1 + m(1-\delta) \tag{4a}$$

$$[f(x) \leq 0] \ \vee \ [\delta = 1] \text{ is true iff } f(x) \leq M\delta \tag{4b}$$

$$\sim [f(x) \leq 0] \text{ is true iff } f(x) \geq \epsilon \tag{4c}$$

where ϵ is a small tolerance (typically the machine precision), beyond which the constraint is regarded as violated. By (1a) and (4b), it also follows

$$[f(x) \leq 0] \ \rightarrow \ [\delta = 1] \text{ is true iff } f(x) \geq \epsilon + (m - \epsilon)\delta \tag{4d}$$

$$[f(x) \leq 0] \ \leftrightarrow \ [\delta = 1] \text{ is true iff } \left\{ \begin{array}{lcl} f(x) & \leq & M(1-\delta) \\ f(x) & \geq & \epsilon + (m-\epsilon)\delta \end{array} \right. \tag{4e}$$

Finally, we report procedures to transform products of logical variables, and of continuous and logical variables, in terms of linear inequalities, which however require the introduction of *auxiliary variables* [30, p. 178]. The product term $\delta_1\delta_2$ can be replaced by an auxiliary logical variable $\delta_3 \triangleq \delta_1\delta_2$. Then, $[\delta_3 = 1] \leftrightarrow [\delta_1 = 1]$

$\wedge$ $[\delta_2 = 1]$, and therefore

$$\delta_3 = \delta_1\delta_2 \text{ is equivalent to } \left\{ \begin{array}{rcl} -\delta_1 + \delta_3 & \leq & 0 \\ -\delta_2 + \delta_3 & \leq & 0 \\ \delta_1 + \delta_2 - \delta_3 & \leq & 1 \end{array} \right. \tag{5a}$$

Moreover, the term $\delta f(x)$, where $f : \mathbb{R}^n \mapsto \mathbb{R}$ and $\delta \in \{0,1\}$, can be replaced by an auxiliary real variable $y \triangleq \delta f(x)$, which satisfies $[\delta = 0] \rightarrow [y = 0]$, $[\delta = 1] \rightarrow [y = f(x)]$. Therefore, by defining M, m as in (3), $y = \delta f(x)$ is equivalent to

$$\left\{ \begin{array}{rcl} y & \leq & M\delta \\ y & \geq & m\delta \\ y & \leq & f(x) - m(1-\delta) \\ y & \geq & f(x) - M(1-\delta) \end{array} \right. \tag{5b}$$

Alternative methods and formulations for transforming propositional logic problems into equivalent integer programs exist. For instance [5] compare the approach above with the approach which utilizes *conjunctive normal forms* (CNF), and conclude that efficiency of a modeling approach depends on the form of logical statements. The problem of finding *minimal forms*, is also well known in the digital network design realm, where the need arises to minimize the number of gates and connections. A variety of methods exists to perform such a task. The reader is referred to [10, Chap. 5] for a detailed exposition.

3. Mixed Logical Dynamical (MLD) systems

In the previous section we have provided some tools to transform logical facts involving continuous variables into linear inequalities. These tools will be used now to express relations describing the evolution of systems where physical laws, logic rules, and operating constraints are interdependent. Before giving a general definition of such a class of systems, consider the following system

$$x(t+1) = \left\{ \begin{array}{rl} 0.8x(t) + u(t) & \text{if} \quad x(t) \geq 0 \\ -0.8x(t) + u(t) & \text{if} \quad x(t) < 0 \end{array} \right. \tag{6}$$

where $x(t) \in [-10, 10]$, and $u(t) \in [-1, 1]$. The condition $x(t) \geq 0$ can be associated to a binary variable $\delta(t)$ such that

$$[\delta(t) = 1] \leftrightarrow [x(t) \geq 0] \tag{7}$$

By using the transformation (4e), equation (7) can be expressed by the inequalities

$$-m\delta(t) \leq x(t) - m$$
$$-(M+\epsilon)\delta \leq -x - \epsilon$$

where $M = -m = 10$, and ϵ is a small positive scalar. Then (6) can be rewritten as

$$x(t+1) = 1.6\delta(t)x(t) - 0.8x(t) + u(t) \tag{8}$$

By defining a new variable $z(t) = \delta(t)x(t)$ which, by (5b), can be expressed as

$$\begin{aligned} z(t) &\leq M\delta(t) \\ z(t) &\geq m\delta(t) \\ z(t) &\leq x(t) - m(1-\delta(t)) \\ z(t) &\geq x(t) - M(1-\delta(t)), \end{aligned}$$

the evolution of system (6) is ruled by the linear equation

$$x(t+1) = 1.6z(t) - 0.8x(t) + u(t)$$

subject to the linear constraints above. This example can be generalized by describing Mixed Logical Dynamical (MLD) Systems through the following linear relations

$$x(t+1) = A_t x(t) + B_{1t}u(t) + B_{2t}\delta(t) + B_{3t}z(t) \tag{9a}$$

$$y(t) = C_t x(t) + D_{1t}u(t) + D_{2t}\delta(t) + D_{3t}z(t) \tag{9b}$$

$$E_{2t}\delta(t) + E_{3t}z(t) \leq E_{1t}u(t) + E_{4t}x(t) + E_{5t} \tag{9c}$$

where $t \in \mathbb{Z}$,

$$x = \begin{bmatrix} x_c \\ x_\ell \end{bmatrix}, \; x_c \in \mathbb{R}^{n_c}, \; x_\ell \in \{0,1\}^{n_\ell}, \; n \triangleq n_c + n_\ell$$

is the state of the system, whose components are distinguished between continuous x_c and 0-1 x_ℓ;

$$y = \begin{bmatrix} y_c \\ y_\ell \end{bmatrix}, \; y_c \in \mathbb{R}^{p_c}, \; y_\ell \in \{0,1\}^{p_\ell}, \; p \triangleq p_c + p_\ell$$

is the output vector,

$$u = \begin{bmatrix} u_c \\ u_\ell \end{bmatrix}, \; u_c \in \mathbb{R}^{m_c}, \; u_\ell \in \{0,1\}^{m_\ell}, \; m \triangleq m_c + m_\ell$$

is the command input, collecting both continuous commands u_c, and binary (on/off) commands u_ℓ (discrete commands, i.e. assuming values within a finite set of reals, can be modeled as 0-1 commands, as described later); $\delta \in \{0,1\}^{r_\ell}$ and $z \in \mathbb{R}^{r_c}$ represent respectively auxiliary logical and continuous variables.

The form (9) involves linear discrete-time dynamics. One might formulate a continuous time version by replacing $x(t+1)$ by $\dot{x}(t)$ in (9a), or a nonlinear version by changing the linear equations and inequalities in (9) to more general nonlinear functions. We restrict the dynamics to be linear and discrete-time in order to obtain computationally tractable control schemes, as will be described in the next sections. Nevertheless, we believe that this framework permits the description of a very broad class of systems.

In principle, the inequality in equation (9) might be satisfied for many values of $\delta(t)$ and/or $z(t)$. On the other hand, we wish that $x(t+1)$ and $y(t)$ were uniquely determined by $x(t)$ and $u(t)$. To this aim, we introduce the following definition

Definition 3.1. *Let $\mathcal{I}_{Bt}$ denote the set of all indices $i \in \{1,\dots,r_\ell\}$, such that $[B_{2t}]^i \neq 0$, where $[B_{2t}]^i$ denotes the i-th column of B_{2t}. Let $\mathcal{I}_{Dt}$, $\mathcal{J}_{Bt}$, $\mathcal{J}_{Dt}$ be defined analogously by collecting the positions of nonzero columns of D_{2t}, B_{3t}, and D_{3t} respectively. Let $\mathcal{I}_t \triangleq \mathcal{I}_{Bt} \bigcup \mathcal{I}_{Dt}$, $\mathcal{J}_t \triangleq \mathcal{J}_{Bt} \bigcup \mathcal{J}_{Dt}$. A MLD system (9) is said to be* well posed *if, $\forall t \in \mathbb{Z}$,*

i. $x(t)$ and $u(t)$ satisfy (9c) for some $\delta(t) \in \{0,1\}^{r_\ell}$, $z(t) \in \mathbb{R}^{r_c}$, and $x_\ell(t+1) \in \{0,1\}^{n_\ell}$;

ii. $\forall i \in \mathcal{I}_t$ there exists a mapping $\mathcal{D}_{it} : \mathbb{R}^{n+m} \mapsto \{0,1\}$ such that the i-th component $\delta_i(t) = \mathcal{D}_{it}(x(t),u(t))$, and $\forall j \in \mathcal{J}_t$ there exists a mapping $\mathcal{Z}_{jt} : \mathbb{R}^{n+m} \mapsto \mathbb{R}$ such that $z_j(t) = \mathcal{Z}_{jt}(x(t),u(t))$.

A MLD system (9) is said to be completely well posed *if in addition $\mathcal{I}_t = \{1,\dots,r_\ell\}$ and $\mathcal{J}_t = \{1,\dots,r_c\}$, $\forall t \in \mathbb{Z}$.*

Note that the functions $\mathcal{D}_{it}$, $\mathcal{Z}_{jt}$ are implicitly defined by the inequalities (9c). Note also that these functions are nonlinear, the nonlinearity being caused by the integer constraint $\delta_i \in \{0,1\}$. In the sequel, we shall say that an auxiliary variable $\delta_i(t)$ ($z_j(t)$) is *well posed* if $i \in \mathcal{I}_t$ ($j \in \mathcal{J}_t$), or *indefinite* otherwise.

Hereafter, we shall assume that system (9) is well posed. This property entails that, once $x(t)$ and $u(t)$ are assigned, $x(t+1)$ and $y(t)$ are uniquely defined, and therefore *trajectories* in the x-space and y-space for system (9) can be defined. In particular, we will denote by $x(t, t_0, x_0, u_{t_0}^{t-1})$ the trajectory generated in accordance to (9) by applying the command inputs $u(t_0)$, $u(t_0+1)$, ... , $u(t-1)$ from initial state $x(t_0) = x_0$. Although typically a model derived from a real system is well posed, a simple numerical test for checking this property is reported in Appendix 1.

In order to transform propositional logic into linear inequalities, and because of the physical constraints present during plant operation (e.g. saturating actuators, safety conditions, ...), we include in the control problem the following constraint

$$\left[\begin{smallmatrix} x \\ u \end{smallmatrix}\right] \in \mathcal{C} \triangleq \left\{ \left[\begin{smallmatrix} x \\ u \end{smallmatrix}\right] \in \mathbb{R}^{n+m} : \; Fx + Gu \leq H \right\} \tag{10}$$

Since typically physical constraints are specified on continuous components, often (10) can be expressed as the Cartesian product $\mathcal{C} = \mathcal{C}_c \times [0,1]^{n_\ell+m_\ell}$ where $\mathcal{C}_c \triangleq \left\{ \left[\begin{smallmatrix} x_c \\ u_c \end{smallmatrix}\right] \in \mathbb{R}^{n_c+m_c} : \; F_c x_c + G_c u_c \leq H_c \right\}$. Note that the constraint $Fx + Gu \leq H$ has the form (9c).

To express logical facts involving continuous state variables by using the tools presented in Section 2 we will often have to define upper- and lower-bounds as in (3), therefore from now on we assume that

Assumption 3.2. *$\mathcal{C}$ is a polytope.*

Note that assuming that $\mathcal{C}$ is bounded is not restrictive in practice. In fact, continuous inputs and states are often bounded by physical reasons, and logical input/state components are intrinsically bounded. The following developments will be meaningful if, in addition, $\mathcal{C}$ has a nonempty interior.

In the sequel, we shall denote by $\|\cdot\|$ the standard Euclidean norm. Note that for pure logical vectors v, $\|v\|^2$ is a nonnegative integer corresponding to the number of nonzero components of v. The symbol $B(x_0, \delta)$ will denote the ball $\{x : \|x - x_0\| \leq \delta\}$.

Observe that the class of MLD systems includes the following important classes of systems:

- Linear hybrid systems;
- Sequential logical systems (Finite State Machines, Automata) $(n_c = m_c = p_c = 0)$;
- Nonlinear dynamic systems, where the nonlinearity can be expressed through combinational logic $(n_\ell = 0)$;
- Some classes of discrete event systems $(n_c = p_c = 0)$;
- Constrained linear systems $(n_\ell = m_\ell = p_\ell = r_\ell = r_c = 0)$;
- Linear systems $(n_\ell = m_\ell = p_\ell = r_\ell = r_c = 0,\ E_{it} = 0,\ i = 1, 4, 5)$;

where the terms "combinational" and "sequential" are borrowed from digital circuit design jargon. The remaining part of this section is devoted to show in detail examples of systems that can be expressed as MLD systems.

3.1. Piece-Wise Linear Dynamic Systems

Consider the following *Piece-Wise Linear Time-Invariant* (PWLTI) dynamic system

$$x(t+1) = \begin{cases} A_1 x(t) + B_1 u(t) & \text{if} \quad \delta_1(t) = 1 \\ \vdots & \\ A_s x(t) + B_s u(t) & \text{if} \quad \delta_s(t) = 1 \end{cases} \tag{11}$$

where $\delta_i(t) \in \{0, 1\}$, $\forall i = 1, \ldots, s$, are 0-1 variables satisfying the exclusive-or condition

$$\bigoplus_{i=1}^{s} [\delta_i(t) = 1]\,. \tag{12}$$

System (11) is completely well posed iff $\mathcal{C}$ can be partitioned in s parts $\mathcal{C}_i$ such that

$$\mathcal{C}_i \bigcap \mathcal{C}_j = \emptyset,\ \forall i \neq j \tag{13a}$$

$$\bigcup_{i=1}^{s} \mathcal{C}_i = \mathcal{C} \tag{13b}$$

and δ_i's are defined as

$$[\delta_i = 1] \leftrightarrow \left[\left[\begin{smallmatrix} x \\ u \end{smallmatrix}\right] \in \mathcal{C}_i\right] \tag{14}$$

A frequent representation of (11) arises in gain-scheduling, where the linear model (and, consequently, the controller) is switched among a finite set of models, according to changes of the operating conditions. Several nonlinear models can be approximated by a model of the form (11), although this approximation capability is limited for computational reasons by the number s of logical variables.

When the sets $\mathcal{C}_i$ are polytopes of the form $\mathcal{C}_i = \{\left[\begin{smallmatrix} x \\ u \end{smallmatrix}\right] : S_i x + R_i u \leq T_i\}$, the $\leftarrow$ implication in (14) corresponds to

$$[\delta_i = 0] \;\rightarrow\; \bigvee_{j=1}^{n_i} \left[\, S_i^j x + R_i^j u > T_i^j \right], \tag{15}$$

where S_i^j denotes the j-th row of S_i^j. Eq. (15) cannot be easily tackled. However, it is easy to see that (15) is implied by (12) and (13), and therefore can be omitted. In fact, let $\left[\begin{smallmatrix} x \\ u \end{smallmatrix}\right] \in \mathcal{C}_i$ and $\delta_i = 0$. Then, by (12) there exists some $\delta_j = 1$, which implies $\left[\begin{smallmatrix} x \\ u \end{smallmatrix}\right] \in \mathcal{C}_j$, a contradiction by (13a). Eqs. (12)–(14) are therefore equivalent to

$$S_i x(t) + R_i u(t) - T_i \leq M_i^* [1 - \delta_i(t)] \tag{16a}$$

$$\sum_{i=1}^{s} \delta_i(t) = 1 \tag{16b}$$

where $M_i^* \triangleq \max_{x \in \mathcal{C}} S_i x(t) + R_i u(t) - T_i$. Eq. (11) can be rewritten as

$$x(t+1) = \sum_{i=1}^{s} [A_i x(t) + B_i u(t)] \delta_i(t) \tag{17}$$

Unfortunately, (17) is nonlinear, since it involves products between logical variables, states, and inputs. We adopt the procedure (5b) to translate (17) into equivalent mixed-integer linear inequalities. To this aim, set

$$x(t+1) \;=\; \sum_{i=1}^{s} z_i(t) \tag{18}$$

$$z_i(t) \;\triangleq\; [A_i x(t) + B_i u(t)] \delta_i(t) \tag{19}$$

and define the vectors $M = [M_1 \;\ldots\; M_n]'$, $m = [m_1 \;\ldots\; m_n]'$ as

$$M_j \;\triangleq\; \max_{i=1,\ldots,s} \left\{ \max_{\left[\begin{smallmatrix} x \\ u \end{smallmatrix}\right] \in \mathcal{C}} A_i^j x + B_i^j u \right\} \tag{20}$$

$$m_j \;\triangleq\; \min_{i=1,\ldots,s} \left\{ \min_{\left[\begin{smallmatrix} x \\ u \end{smallmatrix}\right] \in \mathcal{C}} A_i^j x + B_i^j .u \right\} \tag{21}$$

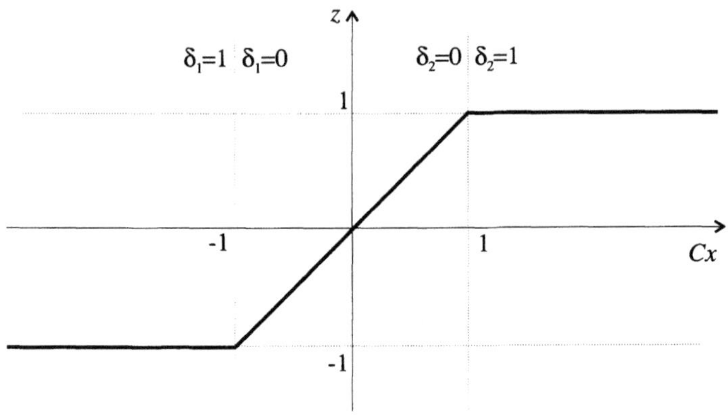

FIGURE 1. Saturation function $z = \text{sat}(Cx)$ and role of δ_1, δ_2

Note that by Assumption 3.2, M and m are finite, and can be either estimated or exactly computed by solving $2ns$ linear programs. Then, (19) is equivalent to

$$\left.\begin{array}{rcl} z_i(t) & \leq & M\delta_i(t) \\ z_i(t) & \geq & m\delta_i(t) \\ z_i(t) & \leq & A_ix(t) + B_iu(t) - m(1 - \delta_i(t)) \\ z_i(t) & \geq & A_ix(t) + B_iu(t) - M(1 - \delta_i(t)) \end{array}\right\} \tag{22}$$

Therefore, (16), (18), and (22) represent (11) in the form (9).

For $s > 2$, the number of 0-1 variables can be reduced by setting $h \triangleq \lceil \log_2 s \rceil$ ($\lceil x \rceil$ denoting the smallest integer greater than or equal to x), and

$$i \triangleq \sum_{j=0}^{h-1} 2^j \delta_j(t) \in \{0, \ldots, s-1\} \tag{23}$$

3.2. Piece-Wise Linear Output Functions

In practical applications, it frequently happens that a process can be modeled as a linear dynamic system cascaded by a nonlinear output function $y = h(x)$. When this can be approximated by a piece-wise linear function, by introducing some auxiliary logical variables δ, we obtain the MLD form (9). As an example, consider the following system

$$\left\{\begin{array}{rcl} x(t+1) & = & Ax(t) + Bu(t) \\ y(t) & = & \text{sat}(Cx(t)) \end{array}\right. \tag{24}$$

along with $x \in \mathcal{X}$, $\mathcal{X}$ bounded, where $\text{sat}(\cdot)$ is the standard saturation function (see Fig. 1)

$$\text{sat}(y) = \left\{\begin{array}{rl} -1 & \text{if} \;\; y \leq -1 \\ y & \text{if} \;\; -1 \leq y \leq 1 \\ 1 & \text{if} \;\; y \geq 1 \end{array}\right. \tag{25}$$

Introduce the following auxiliary logical variables $\delta_1(t)$, $\delta_2(t)$, defined as

$$[Cx > 1] \to [\delta_2 = 1] \tag{26a}$$
$$[Cx < -1] \to [\delta_1 = 1] \tag{26b}$$
$$[Cx < 1] \to [\delta_2 = 0] \tag{26c}$$
$$[Cx > -1] \to [\delta_1 = 0] \tag{26d}$$

By setting $M \triangleq \max_{x\in\mathcal{X}}\{Cx\}$, $m \triangleq \min_{x\in\mathcal{X}}\{Cx\}$, the logical conditions (26) can be rewritten respectively as

$$-Cx + (M-1)\delta_2 \geq -1 \tag{27a}$$
$$Cx - (m+1)\delta_1 \geq -1 \tag{27b}$$
$$Cx + (1-m)(1-\delta_2) \geq 1 \tag{27c}$$
$$Cx - (1+M)(1-\delta_1) \leq -1 \tag{27d}$$

Also, δ_1, δ_2 are related by the logical equations

$$[\delta_1 = 1] \to [\delta_2 = 0] \tag{28a}$$
$$[\delta_2 = 1] \to [\delta_1 = 0] \tag{28b}$$

which can be rewritten as

$$\delta_1 - (1-\delta_2) \leq 0 \tag{29a}$$
$$\delta_2 - (1-\delta_1) \leq 0 \tag{29b}$$

Introduce the auxiliary variable $z \triangleq \text{sat}(Cx)$. It is clear that

$$[\delta_1 = 0] \to [z \leq Cx] \tag{30a}$$
$$[\delta_2 = 0] \to [z \geq Cx] \tag{30b}$$

or, equivalently,

$$z - (M-m)\delta_1 \leq Cx \tag{31a}$$
$$z + (M-m)\delta_2 \geq Cx \tag{31b}$$

and that

$$z \geq -1 \tag{32a}$$
$$z - (M+1)(1-\delta_1) \leq -1 \tag{32b}$$
$$z \leq 1 \tag{32c}$$
$$z + (1-m)(1-\delta_2) \geq 1 \tag{32d}$$

It is easy to verify that the above relations correctly define z also in the case $Cx = \pm 1$, which is not explicitly taken into account in (26). In conclusion, the output relation in (24) can be represented by the linear inequalities (27), (29), (31), (32), and consequently (24) belongs to the class of MLD systems (9).

The modeling of non-differentiable functions by using an integer variable for each discontinuity or point of non-differentiability is also discussed in [21].

3.3. Discrete inputs and qualitative outputs

Control laws typically provide command inputs ranging on a continuum. However, in applications frequently one has to cope with command inputs which are inherently discrete. Sometimes, the quantization process can be neglected, for instance when the control law is implemented on a digital microprocessor with a sufficiently high number of bits. On the other hand, some applications present intrinsically discrete command variables, such as "on/off" switches, gears or speed selectors, number of individuals or wares, etc. In this case, the quantization error cannot be neglected, since it may lead to very poor performance or even instability. This type of commands can be easily modeled by logical variables. Consider for instance the following system

$$\left\{\begin{array}{rcl} x(t+1) & = & Ax(t)+Bu(t) \\ u(t) & \in & \{u_1,u_2,u_3,u_4\} \end{array}\right. \tag{33}$$

By defining two logical inputs $u_{\ell 1}(t), u_{\ell 2}(t) \in \{0,1\}$, and an auxiliary variable $z(t)$ such that

$$\begin{array}{ll} [u_{\ell 1}(t)=0, u_{\ell 2}(t)=0] & \rightarrow [z(t)=u_1] \\ [u_{\ell 1}(t)=0, u_{\ell 2}(t)=1] & \rightarrow [z(t)=u_2] \\ [u_{\ell 1}(t)=1, u_{\ell 2}(t)=0] & \rightarrow [z(t)=u_3] \\ [u_{\ell 1}(t)=1, u_{\ell 2}(t)=1] & \rightarrow [z(t)=u_4] \end{array}$$

it follows that (33) admits the equivalent representation (9)

$$\left\{\begin{array}{rcl} x(t+1) & = & Ax(t)+Bz(t) \\ \begin{bmatrix} 1 \\ -1 \\ 1 \\ -1 \\ 1 \\ -1 \\ 1 \\ -1 \end{bmatrix} z(t) & \leq & \begin{bmatrix} u_4-u_1 & u_4-u_1 \\ 0 & 0 \\ u_4-u_2 & u_2-u_4 \\ u_2-u_1 & u_1-u_2 \\ u_3-u_4 & u_4-u_3 \\ u_1-u_3 & u_3-u_1 \\ 0 & 0 \\ u_1-u_4 & u_1-u_4 \end{bmatrix} u_\ell(t) + \begin{bmatrix} u_1 \\ -u_1 \\ u_4 \\ -u_1 \\ u_4 \\ -u_1 \\ u_4 \\ u_4-2u_1 \end{bmatrix} \end{array}\right.$$

where $u_\ell(t) \triangleq [u_{\ell 1}(t)\ u_{\ell 2}(t)]'$. In alternative, by defining a four-dimensional logical input $u_\ell(t) \triangleq [u_{\ell 1}(t)\ u_{\ell 2}(t)\ u_{\ell 3}(t)\ u_{\ell 4}(t)]'$, (33) can be transformed as

$$\left\{\begin{array}{rcl} x(t+1) & = & Ax(t)+B\begin{bmatrix} u_1 & u_2 & u_3 & u_4 \end{bmatrix} u_\ell(t) \\ \begin{bmatrix} 0 \\ 0 \end{bmatrix} & \leq & \begin{bmatrix} 1 & 1 & 1 & 1 \\ -1 & -1 & -1 & -1 \end{bmatrix} u_\ell(t) + \begin{bmatrix} -1 \\ 1 \end{bmatrix}. \end{array}\right.$$

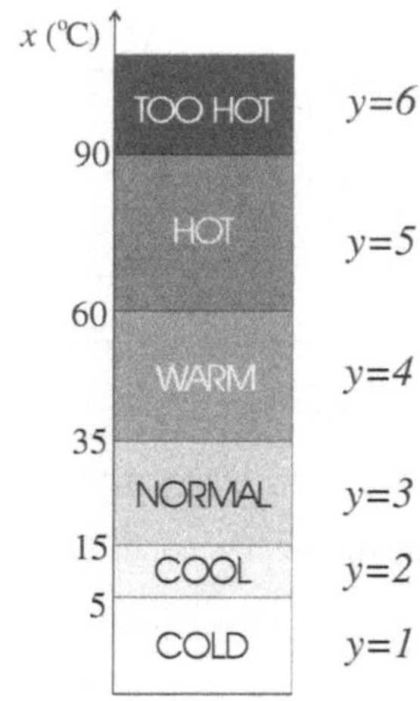

FIGURE 2. Qualitative output Y, and corresponding discrete output y, as a function of the temperature x.

In a similar manner we can treat qualitative outputs. Consider for instance the following example of a thermal system

$$\begin{cases} x(t+1) = ax(t) + bu(t) \\ Y(t) = \begin{cases} \text{COLD} & \text{if } x(t) \leq 5\ °C \\ \text{COOL} & \text{if } 5\ °C < x(t) \leq 15\ °C \\ \text{NORMAL} & \text{if } 15\ °C < x(t) \leq 35\ °C \\ \text{WARM} & \text{if } 35\ °C < x(t) \leq 60\ °C \\ \text{HOT} & \text{if } 60\ °C < x(t) \leq 90\ °C \\ \text{TOO HOT} & \text{if } x(t) > 90\ °C \end{cases} \end{cases} \tag{34}$$

Qualitative properties can be conventionally enumerated and associated with an integer y. Here we associate $y = 1$ with Y="COLD", $y = 2$ with Y="COOL", ..., $y = 6$ with Y="TOO HOT". Similarly to the procedure adopted to define the saturation function (25), define the following logical variables

$$\begin{aligned} [\delta_1(t) = 1] &\leftrightarrow [x(t) \leq 5] \\ [\delta_2(t) = 1] &\leftrightarrow [x(t) \leq 15] \\ [\delta_3(t) = 1] &\leftrightarrow [x(t) \leq 35] \\ [\delta_4(t) = 1] &\leftrightarrow [x(t) \leq 60] \\ [\delta_5(t) = 1] &\leftrightarrow [x(t) \leq 90] \end{aligned}$$

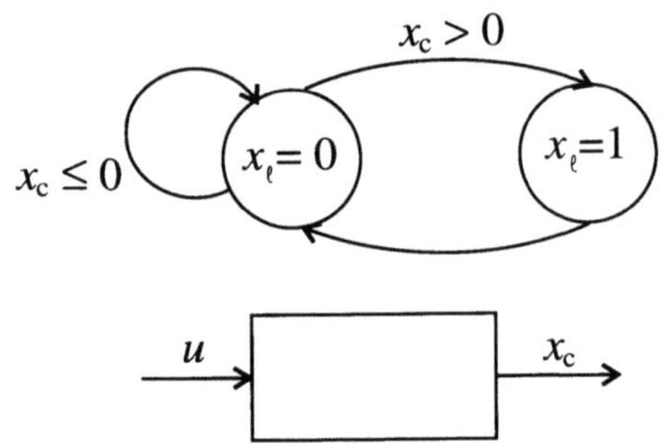

FIGURE 3. Automaton driven by conditions on an underlying dynamic system

which must satisfy the logical conditions

$$\begin{array}{rcl} [\delta_1(t)=1] & \rightarrow & [\delta_2(t)=\delta_3(t)=\delta_4(t)=\delta_5(t)=1] \\ [\delta_2(t)=1] & \rightarrow & [\delta_3(t)=\delta_4(t)=\delta_5(t)=1] \\ [\delta_3(t)=1] & \rightarrow & [\delta_4(t)=\delta_5(t)=1] \\ [\delta_4(t)=1] & \rightarrow & [\delta_5(t)=1] \end{array}$$

By using (4e), these logical conditions can be rewritten in the form (9c). Then, $y(t) \triangleq 1\delta_1(t)+2(\delta_2(t)-\delta_1(t))+3(\delta_3(t)-\delta_2(t))+4(\delta_4(t)-\delta_3(t))+5(\delta_5(t)-\delta_4(t))+6(1-\delta_5(t))$, which represents an equivalent output of the system which can take only six different values, and has the form (9b). This type of modeling is useful to include heuristics and rules of thumb in optimal control problems, as detailed later in Section 5.

3.4. Bilinear systems

Consider the class of nonlinear systems of the form

$$x(t+1) = Ax(t) + Bu(t) + \sum_{i=1}^{m} u_i(t)C_i x(t), \; x \in \mathbb{R}^n, \; u \in \mathbb{R}^m \tag{35}$$

If we assume that the input $u(t)$ is quantized, these can be transformed into MLD system (9). For the sake of simplicity, consider $m = 1$, and let

$$u(t) = D\delta(t), \; D \triangleq u_0 \left[\; 2^0 \quad \dots \quad 2^{r-1} \;\right], \; \delta(t) \in \{0,1\}^r \tag{36}$$

similarly to Eq. (23). Then, $x(t+1) = Ax(t) + BD\delta(t) + \delta'(t)D'C_1x(t)$. By introducing the auxiliary continuous vector $z(t) = \delta'(t)D'C_1x(t)$ and recalling (5b) the bilinear system (35)–(36) can be rewritten in the form (9).

3.5. Finite state machines (Automata)

We consider here finite state machines whose events are generated by an underlying LTI dynamic system. A typical and important example of systems which can be modeled within this framework are real-time systems, where physical processes are controlled by embedded digital controllers. Consider for instance the simple automaton and linear system depicted in Fig. 3, and described by the relations

$$\begin{cases} [x_\ell(t) = 0] \;\wedge\; [x_c \le 0] \;\rightarrow [x_\ell(t+1) = 0] \\ [x_\ell(t) = 0] \;\wedge\; [x_c > 0] \;\rightarrow [x_\ell(t+1) = 1] \\ [x_\ell(t) = 1] \;\rightarrow\; [x_\ell(t+1) = 0] \\ x_c(t+1) = ax_c(t) + bu(t) \end{cases} \tag{37}$$

The (0-1) finite state $x_\ell(t)$ remains in 0 as long as the continuous state $x_c(t)$ is non-positive. If $x_c(t) > 0$ at some t, then x_c generates a digital impulse, i.e. $x_\ell(t+1) = 1$, $x_\ell(t+2) = 0$. The automaton's dynamics is hence driven by events generated by the underlying linear system. Let $x \triangleq [x_c'\; x_\ell']'$, and introduce the auxiliary logical variables $\delta_1(t)$, $\delta_2(t)$ defined as

$$[\delta_1(t) = 1] \;\leftrightarrow\; [x_c(t) \le 0] \tag{38a}$$

$$[\delta_2(t) = 1] \;\leftrightarrow\; [x_\ell(t) = 0] \;\wedge\; [\delta_1(t) = 0] \tag{38b}$$

By (4e), Eq. (38a) can be rewritten as

$$x_c(t) \le M(1 - \delta_1(t)) \tag{39a}$$

$$x_c(t) \ge \epsilon + (m - \epsilon)\delta_1(t) \tag{39b}$$

where $\epsilon > 0$ is a small tolerance (machine precision), and by (5a)

$$\delta_2(t) \le (1 - \delta_1(t)) \tag{40a}$$

$$\delta_2(t) \le (1 - x_\ell(t)) \tag{40b}$$

$$\delta_2(t) \ge (1 - \delta_1(t)) + (1 - x_\ell(t)) - 1 \tag{40c}$$

The mixed-integer linear inequalities (39)-(40) along with the equality $x_\ell(t+1) = \delta_2(t)$ define the automaton part in system (37), which hence is a MLD system. As a further example, in [3] the authors describe how to associate a finite automaton similar to the one depicted in Fig. 3 with hysteresis phenomena which frequently occur in different contexts (e.g. magnetic, electrical, etc.). Finally, time dependence can be emulated in the time-invariant MLD framework by modeling time as the output of a digital clock, which is a finite state machine in free evolution.

4. Stability of MLD systems

Since we treat systems having both real and logical states evolving within a bounded set $\mathcal{C}$, we adapt here standard definitions of stability (see e.g. [12]) to MLD systems.

Definition 4.1. *A vector* $x_e \in \mathbb{R}^{n_c} \times \{0,1\}^{n_\ell}$ *is said to be an* equilibrium state *for (9) and input* $u_e \in \mathbb{R}^{m_c} \times \{0,1\}^{m_\ell}$ *if* $[x_e'\; u_e']' \in \mathcal{C}$ *and* $x(t, t_0, x_e, u_e) = x_e$, $\forall t \ge t_0$, $\forall t_0 \in \mathbb{Z}$. *The pair* (x_e, u_e) *is said to be an* equilibrium pair.

Definition 4.2. *Given an equilibrium pair* (x_e, u_e), $x_e \in \mathbb{R}^{n_c} \times \{0,1\}^{n_\ell}$ *is said to be* stable *if, given* $t_0 \in \mathbb{Z}$, $\forall \epsilon > 0\ \exists \delta(\epsilon, t_0)$ *such that* $\|x_0 - x_e\| \leq \delta \Rightarrow \|x(t, t_0, x_0, u_e) - x_e\| \leq \epsilon$, $\forall t \geq t_0$.

Definition 4.3. *Given an equilibrium pair* (x_e, u_e), $x_e \in \mathbb{R}^{n_c} \times \{0,1\}^{n_\ell}$ *is said to be* asymptotically stable *if* x_e *is stable and* $\exists r > 0$ *such that* $\forall x_0 \in B(x_e, r)$ *and* $\forall \epsilon > 0\ \exists T(\epsilon, t_0)$ *such that* $\|x(t, t_0, x_0, u_e) - x_e\| \leq \epsilon$, $\forall t \geq T$.

Definition 4.4. *Given an equilibrium pair* (x_e, u_e), $x_e \in \mathbb{R}^{n_c} \times \{0,1\}^{n_\ell}$ *is said to be* exponentially stable *if* x_e *is asymptotically stable and in addition* $\exists \delta > 0$, $\alpha > 0$, $0 \leq \beta < 1$ *such that* $\forall x_0 \in B(x_e, \delta)$ *and* $\|x(t, t_0, x_0, u_e) - x_e\| \leq \alpha \beta^{t - t_0} \|x_0 - x_e\|$.

Note that asymptotic convergence of the logical component $x_\ell(t)$ to $x_{\ell e}$ is equivalent to the existence of a finite time t_e such that $x_\ell(t) \equiv x_{\ell e}$, $\forall t \geq t_e$ [19]. Consequently, local stability properties could be restated for the continuous part x_c only, by setting $x_\ell = x_{\ell e}$. Note also that there exists a set around the continuous part x_{ce} of the equilibrium state x_e such that, by perturbing $x_c(t)$ within that set, the equations of motion are again satisfied for $x_\ell(t) = x_{\ell e}$.

For an equilibrium pair (x_e, u_e), in the time-invariant case a corresponding equilibrium value can be established for well-posed components of auxiliary variables via the functions $\mathcal{D}_i$, $\mathcal{Z}_j$ introduced earlier. In addition, for indefinite components we relax the concept of "equilibrium" through the following definition

Definition 4.5. *Let* (x_e, u_e) *be an equilibrium pair for a MLD system, and let the system be well posed. Assume that* $\mathcal{I} \triangleq \lim_{t\to\infty} \mathcal{I}_t$ *and* $\mathcal{J} \triangleq \lim_{t\to\infty} \mathcal{J}_t$ *exist. For* $i \in \mathcal{I}$, $j \in \mathcal{J}$, *let* $\delta_{e,i}$, $z_{e,j}$ *the corresponding equilibrium auxiliary variables. An auxiliary vector* δ *(or* z*) is said to be* definitely admissible *if* $\delta_i = \delta_{e,i}$, $\forall i \in \mathcal{I}$, $(z_j = z_{e,j}, \forall j \in \mathcal{J})$, *and* $\exists t_e$ *such that*

$$E_{2t}\delta + E_{3t}z \leq E_{1t}u_e + E_{4t}x_e + E_{5t}, \quad \forall t \geq t_e \tag{41}$$

Note that for time-invariant MLD systems, $\mathcal{I} \equiv \mathcal{I}_t$, $\mathcal{J} \equiv \mathcal{J}_t$, $\forall t \in \mathbb{Z}$, and (41) reduces to only one set of linear inequalities.

Example 4.6. *Consider the following system*

$$\left\{\begin{array}{rcl} x(t+1) & = & 0.8\begin{bmatrix} \cos\alpha(t) & -\sin\alpha(t) \\ \sin\alpha(t) & \cos\alpha(t) \end{bmatrix} x(t) + \begin{bmatrix} 0 \\ 1 \end{bmatrix} u(t) \\ y(t) & = & [1\ 0]x(t) \\ \alpha(t) & = & \left\{\begin{array}{ll} \frac{\pi}{3} & \text{if } [1\ 0]x(t) \geq 0 \\ -\frac{\pi}{3} & \text{if } [1\ 0]x(t) < 0 \end{array}\right. \\ x(t) & \in & [-10, 10] \times [-10, 10] \\ u(t) & \in & [-1, 1] \end{array}\right. \tag{42}$$

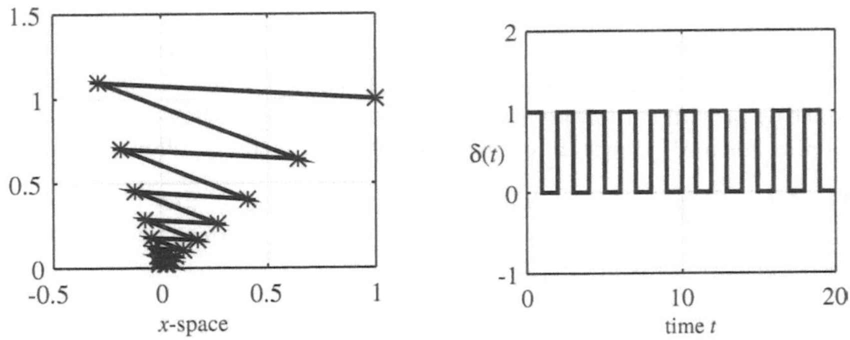

(a) State $x(t)$. (b) Logical variable $\delta(t)$.

FIGURE 4. Evolution of system (42).

According to (22), by using auxiliary variables $z(t) \in \mathbb{R}^4$ and $\delta(t) \in \{0,1\}$ such that $[\delta(t) = 1] \leftrightarrow [[1\ 0]x(t) \geq 0]$, Eq. (42) can be rewritten in the form (9) as

$$\left\{\begin{array}{l} x(t+1) = \begin{bmatrix} I & I \end{bmatrix} z(t) \\ \begin{bmatrix} 10 \\ -10-\epsilon \\ -M \\ -M \\ M \\ M \\ M \\ M \\ -M \\ -M \\ 0 \\ 0 \\ 0 \\ 0 \end{bmatrix} \delta(t) + \begin{bmatrix} 0 & 0 \\ 0 & 0 \\ I & 0 \\ -I & 0 \\ 0 & I \\ 0 & -I \\ I & 0 \\ -I & 0 \\ 0 & I \\ 0 & -I \\ 0 & 0 \\ 0 & 0 \\ 0 & 0 \\ 0 & 0 \end{bmatrix} z(t) \leq \begin{bmatrix} 0 \\ 0 \\ 0 \\ 0 \\ 0 \\ 0 \\ B \\ -B \\ B \\ -B \\ 0 \\ 0 \\ 1 \\ -1 \end{bmatrix} u(t) + \begin{bmatrix} 1\ \ 0 \\ -1\ \ 0 \\ 0 \\ 0 \\ 0 \\ 0 \\ A_1 \\ -A_1 \\ A_2 \\ -A_2 \\ I \\ -I \\ 0 \\ 0 \end{bmatrix} x(t) + \begin{bmatrix} 10 \\ -\epsilon \\ 0 \\ 0 \\ M \\ M \\ M \\ M \\ 0 \\ 0 \\ N \\ N \\ 1 \\ 1 \end{bmatrix} \end{array}\right.$$

where $B = [0\ 1]'$, A_1, A_2 are obtained by (42) by setting respectively $\alpha = \frac{\pi}{3}, -\frac{\pi}{3}$, $M = 4(1+\sqrt{3})[1\ 1]' + B$, $N \triangleq 10[1\ 1]'$, and ϵ is a properly small positive scalar. The evolution starting from $x(0) = [1\ 1]'$ for $u(t) \equiv 0$, $\forall t \geq 0$, is depicted in Fig. 4(a). It is easy to prove that the origin $[0\ 0]'$ is exponentially stable and has the whole set C as domain of attraction. However, if one defines a new system having state $\bar{x}(t) = [x'(t)\ \delta(t-1)]'$, neither $[0\ 0\ 0]'$ nor $[0\ 0\ 1]'$ are equilibria, since $\delta(t)$, as shown in Fig. 4(b), keeps oscillating as t proceeds.

5. Optimal control of MLD systems

In the previous sections we have presented a modeling framework for systems described by dynamics, logic, and constraints. In the following sections we motivate this choice by providing tools for synthesizing controllers for such a class of systems. To this aim, for a MLD system of the form (9), consider the following problem:

Problem 5.1. *Given an initial state* x_0 *and a final time* T*, find (if it exists) the control sequence* $u_0^{T-1} \triangleq \{u(0), u(1), \ldots, u(T-1)\}$ *which transfers the state from* x_0 *to* x_f *and minimizes the* performance index

$$J(u_0^{T-1}, x_0) \triangleq \sum_{t=0}^{T-1} \|u(t) - u_f\|_{Q_1}^2 + \|\delta(t, x_0, u_0^t) - \delta_f\|_{Q_2}^2 +$$
$$\|z(t, x_0, u_0^t) - z_f\|_{Q_3}^2 + \|x(t, x_0, u_0^{t-1}) - x_f\|_{Q_4}^2 + \|y(t, x_0, u_0^{t-1}) - y_f\|_{Q_5}^2 \quad (43)$$

subject to

$$x(T, x_0, u_0^{T-1}) = x_f \quad (44)$$

and the MLD system dynamics (9), where $\|x\|_Q^2 \triangleq x'Qx$, $Q_i = Q_i' \geq 0$, $i = 1, \ldots, 5$, *are given weight matrices, and* x_f, u_f, δ_f, z_f, y_f *are given offset vectors satisfying (9b)–(9c).*

Note that if $\delta_f = 0$ and Q_2 is diagonal, the second quadratic term is equivalent to the linear term $\sum_{i=0}^{r_\ell} Q_{ii}\delta_i(t, x_0, u_0^t)$.

Problem 5.1 can be solved as a *Mixed-Integer Quadratic Programming* (MIQP) problem. In fact, let $x(t)$ be a compact notation for $x(t, x_0, u_0^{t-1})$, the same convention being used for $\delta(t)$, $z(t)$. From (9a), for time-invariant systems we have the solution formula

$$x(t) = A^t x_0 + \sum_{i=0}^{t-1} A^i \left[B_1 u(t-1-i) + B_2\delta(t-1-i) + B_3 z(t-1-i)\right] \quad (45)$$

where the relation between $x(t)$ and x_0, u_0^{t-1} is only apparently linear, because $\delta(i)$, $z(i)$ hide a nonlinear dependence on x_0 and u_0^{t-1}, as observed earlier. By plugging (45) into (9c) and (43), and by defining the vectors

$$\Omega \triangleq \begin{bmatrix} u(0) \\ \vdots \\ u(T-1) \end{bmatrix}, \quad \Delta \triangleq \begin{bmatrix} \delta(0) \\ \vdots \\ \delta(T-1) \end{bmatrix}, \quad \Xi \triangleq \begin{bmatrix} z(0) \\ \vdots \\ z(T-1) \end{bmatrix}, \quad \mathcal{V} \triangleq \begin{bmatrix} \Omega \\ \Delta \\ \Xi \end{bmatrix}$$

we obtain the following equivalent formulation

$$\begin{cases} \min_{\mathcal{V}} \mathcal{V}'S_1\mathcal{V} + 2(S_2 + x_0'S_3)\mathcal{V} \\ \text{subj. to } F_1\mathcal{V} \leq F_2 + F_3 x_0 \end{cases} \quad (46)$$

where matrices S_i, F_i, $i = 1, 2, 3$ are suitably defined. Then, existence, uniqueness, and continuity with respect to x_0 of the optimal control sequence can be investigated as feasibility, uniqueness, and continuity with respect to parameters of the solution of the MIQP problem (46).

5.1. Soft constraints and constraint priority

In practical applications, it is common use to distinguish between *hard constraints*, which cannot be violated (for instance motor voltage limits), and *soft constraints*, whose violation is allowed (e.g. bounds on temperatures), even if penalized. When dealing with soft constraints, one would like to fulfill two requirements. First, if the set of feasible solutions to the hard constrained problem is nonempty, the minimizer should belong to that set. Second, if such a set is empty, one should be able to decide a trade-off between the cost and constraint violation.

Consider the following optimization problem

$$\begin{cases} \min\limits_{x \in X} x'Sx \\ \text{subj. to } Ax \leq B \end{cases} \tag{47}$$

where the set X is a bounded polyhedron. The simplest way to soften constraints is to modify (47) in the following problem

$$\begin{cases} \min\limits_{x \in X, \epsilon \geq 0} \; x'Sx + \epsilon' M_1 \epsilon \\ \text{subj. to } Ax - C\epsilon \leq B \end{cases} \tag{48}$$

where $\epsilon \in \mathbb{R}^s$ is a vector of slack variables, C is a vector whose components are 0 or 1 according if the corresponding constraint is hard or soft, and M_1 is a (large) penalty weight matrix. The problem with the formulation (48) is that the first requirement is not guaranteed. As an alternative, define a logical variable δ such that $[\delta = 0] \leftrightarrow [\epsilon_i = 0, \; \forall i = 1, \ldots, s]$, and minimize $x'Sx + \epsilon' M_1 \epsilon + M_2 \delta$ with respect to x, ϵ, and δ, where $M_2 > \max_{x \in X} x'Sx$, and M_1 decides the trade-off between cost and constraint violation, when no feasible solution exists to the hard-constrained problem.

Constraint violation can also be considered at r levels of priority, by introducing r 0-1 variables δ_i, $i = 1, \ldots, r$, by letting

$$\begin{array}{lcl} [\delta_1 = 0] & \leftrightarrow & [\epsilon_1 = \ldots = \epsilon_{i_1} = 0] \\ [\delta_2 = 0] & \leftrightarrow & [\delta_1 = 0] \wedge [\epsilon_{i_1+1} = \ldots = \epsilon_{i_2} = 0] \\ \vdots & & \vdots \\ [\delta_r = 0] & \leftrightarrow & [\delta_1 = \ldots = \delta_{r-1} = 0] \wedge \left[\epsilon_{i_{r-1}+1} = \ldots = \epsilon_{i_r} = 0\right] \end{array},$$

and by minimizing $x'Sx + \epsilon' M_1 \epsilon + M_2 \sum_{i=1}^{r} \delta_i$.

Note that soft constraints and constraint priority can be directly considered in the MLD structure (9). In fact, constraints in (9c) can be softened and/or prioritized by incorporating the slack vector $\epsilon(t)$ in the z-vector, and the auxiliary logical variables $\delta_1(t), \ldots, \delta_r(t)$ in the δ-vector.

5.2. Integrating heuristics, logic, and dynamics

As shown in [21, 22] for process synthesis, logic and heuristics can be integrated through propositional logic. This type of qualitative knowledge is useful for two purposes. First, in many cases solutions which reflect the operator's experience are simply preferred. Second, it may help to expedite the search for feasible solutions, for instance by generating a base case. On the other hand, qualitative knowledge are typically just rules of thumb which may not always hold, lead to solutions which are far away from optimality, and even be contradictory.

Heuristic rules can be expressed as "soft" logic facts, by considering instead of the clause D which expresses the rule the following

$$D \vee V \tag{49}$$

Since the clause is also a disjunction, the conversion of (49) into linear inequalities is straightforward, for instance $[\sim P_1 \vee P_2] \vee V$ yields

$$1 - \delta_1 + \delta_2 + v \geq 1 \tag{50}$$

where v can also be interpreted as a slack variable that allows the violation of the inequality. Since (50) only involves 0-1 variables, in this case the variable v can be treated as a continuous nonnegative variable, despite the fact that it will take only 0-1 values. As a further example, consider $[T \geq T_{HOT}] \rightarrow [\delta_1 = 1] \vee [v = 1]$, which is equivalent to $T - T_{HOT} - M(\delta_1 + v) \leq 0$, where M is a known upper bound on $T - T_{HOT}$ and v is a binary variable that represents the violation of the heuristics [22].

When the fulfillment of heuristic rules is impossible or destroys optimality, one should violate the weaker (more uncertain) set of rules. A discrimination between weak and strong rules can be obtained by penalizing with different weights w_i the violation variables v_i. The penalty w_i is a nonnegative number expressing the uncertainty of the corresponding logical expression. The more uncertain the rule according to the designer's experience, the lower the penalty for its violation.

For the optimal control problem at hand, one can add the linear term $w'v$ in (43) and minimize with respect to $\mathcal{V}$ and v. As an alternative, if the performance index should not be mixed with heuristics violation penalties, one can first find the vector v^* which minimizes $w'v$ subject to linear constraints involving $\mathcal{V}$, and v (a *Mixed Integer Linear Problem* (MILP)), set $v = v^*$, and then minimize (43) with respect to $\mathcal{V}$ only. This corresponds to a preprocessing of the given set of logical, dynamical, and heuristic conditions in order to obtain the feasible set which better takes into account qualitative knowledge.

6. Predictive control of MLD systems

As observed in the previous sections, a large quantity of situations can be modeled through the MLD structure. Then, it is interesting from both a theoretical and practical point of view to ask whether or not a MLD system can be stabilized to an equilibrium state or can track a desired reference trajectory, possibly via feedback control. Finding such a control law is not an easy task, being the system neither linear nor even smooth. In this section, we show how predictive control provide successful tools to perform this task. For the sake of notational simplicity, the index $_t$ will be dropped from (9), by assuming that the system is time-invariant.

Consider an equilibrium pair (x_e, u_e) and let (δ_e, z_e) be definitely admissible in the sense of Definition 4.5. Let the components $\delta_{e,i}$, $z_{e,j}$, $i \notin \mathcal{I}$, $j \notin \mathcal{J}$, correspond to desired steady-state values for the indefinite auxiliary variables. Let t be the current time, and $x(t)$ the current state. Consider the following optimal control problem

$$\min_{\{v_0^{T-1}\}} J(v_0^{T-1}, x(t)) \triangleq \sum_{k=0}^{T-1} \|v(k) - u_e\|^2_{Q_1} + \|\delta(k|t) - \delta_e\|^2_{Q_2} + \\ + \|z(k|t) - z_e\|^2_{Q_3} + \|x(k|t) - x_e\|^2_{Q_4} + \|y(k|t) - y_e\|^2_{Q_5} \tag{51}$$

$$\text{subj. to} \begin{cases} x(T|t) = x_e \\ x(k+1|t) = Ax(k|t) + B_1 v(k) + B_2\delta(k|t) + B_3 z(k|t) \\ y(k|t) = Cx(k|t) + D_1 v(k) + D_2\delta(k|t) + D_3 z(k|t) \\ E_2\delta(k|t) + E_3 z(k|t) \leq E_1 v(k) + E_4 x(k|t) + E_5 \end{cases} \tag{52}$$

where $Q_1 = Q_1' > 0$, $Q_2 = Q_2' \geq 0$, $Q_3 = Q_3' \geq 0$, $Q_4 = Q_4' > 0$, $Q_5 = Q_5' \geq 0$, $x(k|t) \triangleq x(t+k, x(t), v_0^{k-1})$, and $\delta(k|t)$, $z(k|t)$, $y(k|t)$ are similarly defined. Assume for the moment that the optimal solution $\{v_t^*(k)\}_{k=0,\dots,T-1}$ exists. According to the *receding horizon* philosophy mentioned above, set

$$u(t) = v_t^*(0), \tag{53}$$

disregard the subsequent optimal inputs $v_t^*(1), \dots, v_t^*(T-1)$, and repeat the whole optimization procedure at time $t+1$. The control law (51)–(53) will be referred to as the *Mixed Integer Predictive Control* (MIPC) law. Note that once x_e, u_e have been fixed, consistent steady-state vectors δ_e, z_e can be obtained by choosing feasible points in the domain described by (9c), for instance by solving a MILP.

Several formulations of predictive controllers for MLD systems might be proposed. For instance, the number of control degrees of freedom can be reduced to $N_u < T$, by setting $u(k) \equiv u(N_u - 1)$, $\forall k = N_u, \dots, T$. However, while in other contexts this amounts to hugely down-sizing the optimization problem at the price of a reduced performance, here the computational gain is only partial, since all the T $\delta(k|t)$ and $z(k|t)$ variables remain in the optimization. Infinite horizon formulations are inappropriate for both practical and theoretical reasons. In fact, approximating the infinite horizon with a large T is computationally prohibitive, as the number of 0-1 variables involved in the MIQP depends linearly on T. Moreover, the quadratic term in δ might oscillate, as exemplified in Example 1, Fig. 4(b), and hence "good" (i.e. asymptotically stabilizing) input sequences might be ruled out by a corresponding infinite value of the performance index; it could even happen that no input sequence has finite cost.

Theorem 6.1. *Let (x_e, u_e) be an equilibrium pair and (δ_e, z_e) definitely admissible. Assume that the initial state $x(0)$ is such that a feasible solution of problem (51) exists at time $t = 0$. Then $\forall Q_1 = Q_1' > 0$, $Q_2 = Q_2' \geq 0$, $Q_3 = Q_3' \geq 0$, $Q_4 = Q_4' > 0$, and $Q_5 = Q_5' \geq 0$ the MIPC law (51)–(53) stabilizes the system in that*

$$\begin{aligned}
\lim_{t\to\infty} x(t) &= x_e \\
\lim_{t\to\infty} u(t) &= u_e \\
\lim_{t\to\infty} \|\delta(t) - \delta_e\|_{Q_2} &= 0 \\
\lim_{t\to\infty} \|z(t) - z_e\|_{Q_3} &= 0 \\
\lim_{t\to\infty} \|y(t) - y_e\|_{Q_5} &= 0
\end{aligned}$$

while fulfilling the dynamic/relational constraints (9c).

Note that if $Q_2 > 0$ (or $Q_3 > 0$, $Q_5 > 0$), convergence of $\delta(t)$ (or $z(t)$, $y(t)$) follows as well.

Proof. The proof easily follows from standard Lyapunov arguments. Let $\mathcal{U}_t^*$ denote the optimal control sequence $\{v_t^*(0), \ldots, v_t^*(T-1)\}$, let

$$V(t) \triangleq J(\mathcal{U}_t^*, x(t))$$

denote the corresponding value attained by the performance index, and let $\mathcal{U}_1$ be the sequence $\{v_t^*(1), \ldots, v_t^*(T-2), u_e\}$. Then, $\mathcal{U}_1$ is feasible at time $t+1$, along with the vectors $\delta(k|t+1) = \delta(k+1|t)$, $z(k|t+1) = z(k+1|t)$, $k = 0, \ldots, T-2$, $\delta(T-1|t+1) = \delta_e$, $z(T-1|t+1) = z_e$, being $x(T-1|t+1) = x(T|t) = x_e$ and (δ_e, z_e) definitely admissible. Hence,

$$\begin{aligned}
V(t+1) \leq J(\mathcal{U}_1, x(t+1)) = V(t) - \|x(t) - x_e\|_{Q_4} - \|u(t) - u_e\|_{Q_1} - \\
\|\delta(t) - \delta_e\|_{Q_2} - \|z(t) - z_e\|_{Q_3} - \|y(t) - y_e\|_{Q_5} \quad (54)
\end{aligned}$$

and $V(t)$ is decreasing. Since $V(t)$ is lower-bounded by 0, there exists $V_\infty = \lim_{t\to\infty} V(t)$, which implies $V(t+1) - V(t) \to 0$. Therefore, each term of the sum

$$\begin{aligned}
\|x(t) - x_e\|_{Q_4} + \|u(t) - u_e\|_{Q_1} + \|\delta(t) - \delta_e\|_{Q_2} + \\
\|z(t) - z_e\|_{Q_3} + \|y(t) - y_e\|_{Q_5} \leq V(t) - V(t+1)
\end{aligned}$$

converges to zero as well, which proves the theorem. □

Remark 1. Despite the fact that very effective methods exist to compute the (global) optimal solution of the MIQP problem (51)–(53) (see Section 7 below), in the worst-case the solution time depends exponentially on the number of integer variables. In principle, this might limit the scope of application of the proposed method to very slow systems, since for real-time implementation the sampling time should be large enough to allow the worst-case computation. However, the proof of Theorem 6.1 does not require that the evaluated control sequences $\{\mathcal{U}_t^*\}_{t=0}^{\infty}$ are global optima. In fact, Eq. (54) just requires that

$$J(\mathcal{U}_{t+1}^*, x(t+1)) \leq J(\mathcal{U}_1, x(t+1)). \tag{55}$$

The sequence $\mathcal{U}_1$ is available from the previous computation (performed at time t), and can be used to initialize the MIQP solver at time $t+1$. The solver can then be interrupted at any intermediate step to obtain a suboptimal solution $\mathcal{U}_{t+1}^*$ which satisfies (55). For instance, when Branch & Bound methods are used to solve the MIQP problem, the new control sequence $\mathcal{U}_t^*$ can be selected as the solution to a QP subproblem which is integer-feasible and has the lowest value. Obviously in this case tracking performance deteriorates.

Remark 2. Since in general the implicitly defined functions $\mathcal{D}_i$, $\mathcal{Z}_j$ are not continuous, convergence of the well posed components of δ, z cannot be inferred by convergence of x and u. For instance, a variable δ defined as $[\delta = 1] \leftrightarrow [x > 0]$ has a corresponding $\mathcal{D}$ function which is discontinuous in $x = 0$.

Remark 3. Nothing can be inferred about the asymptotic behavior of the indefinite components of δ, z, unless Q_2, $Q_3 > 0$. However, the behavior of unweighted indefinite variables are clearly of little interest.

Remark 4. The stability result proved in Theorem 6.1 is not affected by the presence of positive linear terms in (51). For instance, if $z \in \mathbb{R}$, $z_e = 0$ and the constraint $z(t) \geq 0$ is present, a term of the form $q_3 z$, $q_3 \geq 0$ can be included in (51). Hence, soft constraints or heuristic rules can be taken into account by modifying the performance index (51) as detailed in Section 5, without corrupting the warranty of stability.

Remark 5. Note that because of its receding horizon mechanism, MIPC is a closed-loop approach, and is clearly more robust than pure open-loop optimal control. On the other hand, MIPC control can be also adopted for off-line computation of open-loop input trajectories. Let N be the duration in time steps of the batch operation to be designed. Since short horizons T can be implemented within MIPC, this would require the solution of N MIQP problems of size T. On the other hand, pure optimal control would require the solution of one MIQP problem of size N. Assuming a worst case exponential dependence on the size of the problem, the first would have a complexity of $N2^T$, while the second of 2^N. For $N = 100$, $T = 5$ this is equivalent to 3200 versus about 10^{30}. This gain in computational efficience, however, may be paid at the price of a deteriorated performance, due to the gap between the open-loop performance objective minimized at each step and the actual performance. Note that this gap increases as the prediction horizon T gets shorter.

Example 6.2. *Consider again the MDL system of Example 4.6. In order to stabilize the system to the origin, the feedback control law (51)–(53) is adopted, along with the parameters* $T = 3$, $u_e = 0$, $\delta_e = 0$, $z_e = [0\ 0\ 0\ 0]'$, $x_e = [0\ 0]'$, $y_e = 0$, $Q_1 = 1$, $Q_2 = 0.01$, $Q_3 = 0.01 I_4$, $Q_4 = I_2$, $Q_5 = 0$, $z_e = [0\ 0\ 0\ 0]'$, $\delta_e = 1$, *and* $u_e = 0$. *Fig. 5 shows the resulting trajectories. The trajectories obtained at*

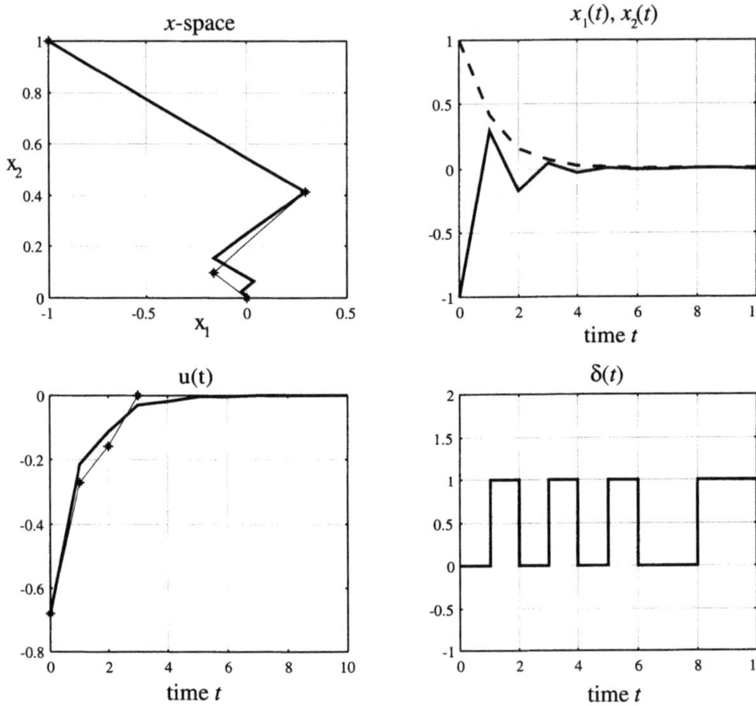

FIGURE 5. Closed-loop regulation problem for system (42). Closed-loop trajectories (thick lines) and optimal solution at $t = 0$ (thin lines, right plots).

time $t = 0$ *by solving the optimal control problem (51)–(52) are also reported in the right plots (thin lines). Consider now a desired reference* $r(t) = \sin(t/8)$ *for the output* $y(t)$*. We apply the same MIPC controller, with the exception of* $Q_4 = 10^{-8} I_2$, $Q_5 = 1$*. The steady-state parameters are selected as* $y_e = r(t)$*, and* u_e, x_e, δ_e, z_e *consistently. Fig. 6 shows the resulting closed-loop trajectories. Notice that the constraint* $-1 \leq u(t) \leq 1$ *prevents the system from tracking the peaks of the sinusoid, and therefore the output trajectory is chopped.*

7. MIQP solvers

With the exception of particular structures, mixed-integer programming problems involving 0-1 variables are classified as $\mathcal{NP}$-complete, which means that in the worst case, the solution time grows exponentially with the problem size [21]. Despite this combinatorial nature, several algorithmic approaches have been proposed and applied successfully to medium and large size application problems [9]. For MIQP problems, [8] indicates Generalized Benders' Decomposition [13], Outer

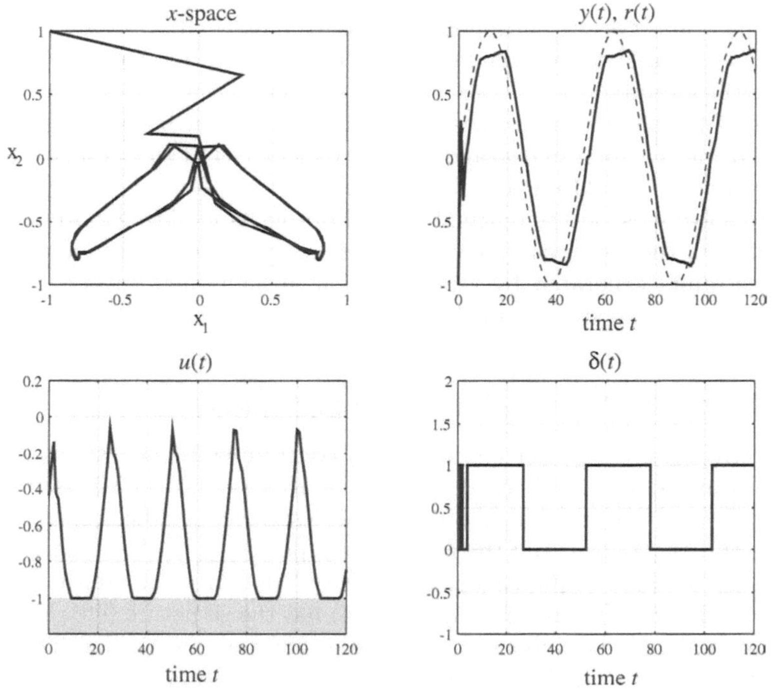

FIGURE 6. Closed-loop tracking problem for system (42), with $y(t) = x_1(t)$.

Approximation, LP/QP based branch and bound, and Branch and Bound as the major solvers. See [24] for a review of these methods.

Several authors agree on the fact that branch and bound methods are the most successful for mixed integer nonlinear programs. In [8], a numerical experience is reported which compares different approaches, and Branch and Bound results superior by an order of magnitude.

As described in [8], the Branch and Bound algorithm for MIQP consists of solving and generating new QP problems in accordance with a tree search, where the nodes of the tree correspond to QP subproblems. Branching is obtained by generating child-nodes from parent-nodes according to branching rules, which can be based for instance on a-priori specified priorities on integer variables, or on the amount by which the integer constraints are violated. Nodes are labeled as either pending, if the corresponding QP problem has not yet been solved, or fathomed, if the node has already been fully explored. The algorithm stops when all nodes have been fathomed.

The success of the branch and bound algorithm relies on the fact that whole subtrees can be excluded from further exploration by fathoming the corresponding root nodes. This happens if the corresponding QP subproblem is either infeasible or an integer solution is obtained. In the second case, the corresponding value of the cost function serves as an upper bound on the optimal solution of the MIQP problem, and is used to further fathoming other nodes having greater optimal value or lower bound.

Some of the simulation results reported in this paper have been obtained in Matlab by using the Fortran package [7] as a MIQP solver. This package can handle both dense and sparse MIQP problems. The latter has proven to be particularly effective to solve most of the optimal control problems for MLD systems. In fact, because of (45), the constraints have a triangular structure, and in addition most of the constraints generated by representation of logic facts involve only a few variables lead often to sparse matrices.

8. A case study: Control of a gas supply system

The theoretical framework for modeling and controlling MLD systems developed in the previous sections has been applied to the Kawasaki Steel Mizushima Works gas supply system described in [1]. A detailed description of the system, its translation into the MLD form, and the results are reported in [2]. We report here only some remarks about computational complexity. At each time step t, the MIQP problem which derives from (9), (43), and (44) has the structure (46), involves 121 linear constraints, 25 continuous variables, and 28 integer variables. The problem has a sparseness of around 93%. On a Sun SPARCStation 4 at time $t = 0$, for instance, the MIQP problem is solved by the sparse version of the package [7] in 1.20 s (8 QP subproblems). Great simulation computational time is also saved by exploiting the information about precedent solution, namely by shifting the previous optimal solution, which is a feasible initial condition, as observed in the proof of Theorem 6.1. Considering that for the gas supply system $\Delta T = 2$ h, real time implementation of the proposed scheme is reasonable.

9. Conclusions

Motivated by the key idea of transforming propositional logic into linear mixed-integer inequalities, and by the existence of techniques for solving mixed-integer quadratic programming, this paper has presented a framework for modeling and controlling systems described by both dynamics and logic, and subject to operating constraints, denoted as *Mixed Logical Dynamical* (MLD) systems. For these systems, a systematic control design method based on model predictive control ideas has been presented, which provides stability, tracking, and constraint fulfillment properties. The proposed strategy seems to be particularly appealing for higher-level control and optimization of complex systems.

Acknowledgments

The authors thank Sven Leyffer and Roger Fletcher for the MIQP solver.

References

[1] K. Akimoto, N. Sannomiya, Y. Nishikawa, and T. Tsuda. An optimal gas supply for a power plant using a mixed integer programming model. *Automatica*, 27(3):513–518, 1991.

[2] A. Bemporad, M. Morari. Control of systems integrating logic, dynamics, and constraints. *IFA Tech. Rep. AUT98-04*, submitted to Automatica, also appeared in Proc. 17th Benelux Meeting on Systems and Control, Mierlo, The Netherlands, March, 1998.

[3] M.S. Branicky, V.S. Borkar, and S.K. Mitter. A unified framework for hybrid control: model and optimal control theory. *IEEE Trans. Autom. Contr.*, 43(1):31–45, 1998.

[4] P.J. Campo and M. Morari. Model predictive optimal averaging level control. *AIChE Journal*, 35(4):579–591, 1989.

[5] T. M. Cavalier, P. M. Pardalos, and A. L. Soyster. Modeling and integer programming techniques applied to propositional calculus. *Computers Opns Res.*, 17(6):561–570, 1990.

[6] Y. Chikkula and J. H. Lee, Dynamically scheduled model predictive control of nonlinear processes using hinging-hyperplane models. *Proc. AIChE Annual Meeting*, 1995. Submitted AIChe Journal.

[7] R. Fletcher and S. Leyffer. A mixed integer quadratic programming package. Technical report, University of Dundee, Dept. of Mathematics, Scotland, U.K., 1994.

[8] R. Fletcher and S. Leyffer. Numerical experience with lower bounds for miqp branch-and-bound. Technical report, University of Dundee, Dept. of Mathematics, Scotland, U.K., 1995. submitted to SIAM Journal on Optimization, `http://www.mcs.dundee.ac.uk:8080/ sleyffer/miqp_art.ps.Z`.

[9] C. A. Floudas. *Nonlinear and Mixed-Integer Optimization.* Oxford University Press, 1995.

[10] J. P. Hayes. *Introduction to Digital Logic Design.* Addison-Wesley Publishing Company, Inc., 1993.

[11] D. Christiansen (Editor in chief). *Electronics Engineers' Handbook, 4th edition.* IEEE Press/ McGraw Hill, Inc., 1997.

[12] S.S. Keerthi and E. Gilbert. Optimal infinite-horizon feedback laws for a general class of constrained discrete time systems: stability and moving-horizon approximations. *Journal of Optimization Theory and Applications*, 2:265–293, 1988.

[13] R. Lazimy. Improved algorithm for mixed-integer quadratic programs and a computational study. *Mathematical Programming*, 32:100–113, 1985.

[14] J.H. Lee and B. Cooley. Recent advances in model predictive control. In *Chemical Process Control – V*, volume 93, no. 316, pages 201–216b. AIChe Symposium Series – American Institute of Chemical Engineers, 1997.

[15] S. Leyffer. *Deterministic methods for mixed integer nonlinear programming.* Ph.D. dissertation, University of Dundee, Scotland, U.K., December 1993.

[16] J. Lygeros, D.N. Godbole, and S.S. Sastry. A design framework for hierarchical, hybrid control. Technical report, Intelligent Machines and Robotic Laboratory, University of California, Berkeley, 1997. submitted to the IEEE Trans. Automat. Contr., special issue on Hybrid Systems.

[17] D.Q. Mayne. Nonlinear model predictive control: an assessment. In *Chemical Process Control – V*, volume 93, no. 316, pages 217–231. AIChe Symposium Series – American Institute of Chemical Engineers, 1997.

[18] E. Mendelson. *Introduction to mathematical logic.* Van Nostrand, 1964.

[19] K. M. Passino, A. N. Michel, and P. J. Antsaklis. Lyapunov stability of a class of discrete event systems. *IEEE Trans. Automat. Contr.*, 2(39):269–279, 1994.

[20] S.J. Qin and T.A. Badgewell. An overview of industrial model predictive control technology. In *Chemical Process Control – V*, volume 93, no. 316, pages 232–256. AIChe Symposium Series – American Institute of Chemical Engineers, 1997.

[21] R. Raman and I. E. Grossmann. Relation between milp modeling and logical inference for chemical process synthesis. *Computers Chem. Engng.*, 15(2):73–84, 1991.

[22] R. Raman and I. E. Grossmann. Integration of logic and heuristic knowledge in minlp optimization for process synthesis. *Computers chem. Engng.*, 16(3):155–171, 1992.

[23] R.L.Grossmann, A. Nerode, A.P.Ravn, and H. Rischel (Eds.). *Hybrid Systems.* Springer Verlag, New York, 1993. no. 736 in LCNS.

[24] V. A. Roschchin, O.V. Volkovich, and I.V. Sergienko. Models and methods of solution of quadratic integer programming problems. *Cybernetics*, 23:289–305, 1987.

[25] O. Slupphaug, J. Vada, and B. A. Foss. MPC in Systems with Continuous and Discrete Control Inputs. *Proc. American Control Conf.*, Albuquerque, NM, USA, 1997.

[26] O. Slupphaug and B. A. Foss. Model Predictive Control for a class of Hybrid Systems. *Proc. European Control Conf.*, Brussels, Belgium, 1997.

[27] M.L. Tyler and M. Morari. Propositional logic in control and monitoring problems. Technical Report AUT96-15, Institut für Automatik, ETH - Swiss Federal Institute of Technology, Zürich, Switzerland, 1996.

[28] H.P. Williams. Logical problems and integer programming. *Bulletin of the Institute of Mathematics and Its Applications*, 13:18–20, 1977.

[29] H.P. Williams. Linear and integer programming applied to the propositional calculus. *Int. J. Systems Research and Info. Science*, 2:81–100, 1987.

[30] H.P. Williams. *Model Building in Mathematical Programming.* John Wiley & Sons, Third Edition, 1993.

Institut für Automatik,
ETH – Swiss Federal Institute of Technology,
ETHZ - ETL, CH-8092 Zürich, Switzerland
tel.+41-1-632 7626, fax +41-1-632 1211
E-mail address: `bemporad,morari@aut.ee.ethz.ch`

Progress in Systems and Control Theory, Vol. 26

Stability, Feasibility, Optimality and the Degrees of Freedom in Constrained Predictive Control

Basil Kouvaritakis, Mark Cannon and J. Anthony Rossiter

Abstract. The characterization of the class of stabilizing predictions for linear systems allows for predictive control algorithms which interpolate linearly between some key predicted input trajectories. Such interpolation endows the closed-loop system with desirable attributes even when a small number of degrees of freedom is used. This paper explores the ideas behind interpolation and extends them to the case of predictive control of nonlinear systems by proposing an effective yet computationally undemanding strategy.

1. Introduction

Model Based Predictive Control (MBPC) algorithms developed for linear systems (eg [1, 2, 3]) cannot be extended easily to the case of nonlinear systems. The difficulty is that the degrees of freedom are taken to be future control increments, and the receding horizon cost J depends quadratically on the future errors and control increment. For linear systems J turns out to be quadratic in the future control increments, and thus the minimization of J can be performed explicitly in the unconstrained case, or via Quadratic Programming (QP) otherwise.

A recent MBPC approach [4] for linear systems has given a complete characterization of the class of stabilizing input trajectories and the parametrization of this class is in terms of a stable transfer function $F(z)$, or equivalently the vector $\boldsymbol{f}$ of its impulse response. Clearly $\boldsymbol{f}$ is infinite dimensional and so would be the QP problem of minimizing J over $\boldsymbol{f}$ in the presence of constraints. To get practicable algorithms one is then forced to consider subsets of the stabilizing class. An attractive solution is proposed in [3] which considers future control increments up to a finite horizon N to be the degrees of freedom and constraints the control law beyond N to be optimal state feedback. The theoretical appeal of this strategy lies in the fact that there exist finite (albeit large) N for which the resulting finite dimensional QP solution provides an exact solution to the infinite dimensional problem. Unfortunately feasibility considerations often imply that N has to be taken to be large, and thus the algorithm becomes impracticable. A further alternative [6] considers a different subset of the stabilizing class by selecting

$F(z) = F_1(z) + F_2(z)c(z)$ where $F_1(z)$, $F_2(z)$ are given stable transfer functions and $c(z)$ is a polynomial of a given degree, $\nu - 1$. Thus, by choosing $F_1(z)$, $F_2(z)$ and ν in a suitable manner it is possible to ensure that for different choices of $\boldsymbol{c}$, the vector of coefficients of $c(z)$, the predicted input trajectory can be made to be some desirable trajectory. Depending on this, the resulting receding horizon algorithm, which at each time minimizes J over $\boldsymbol{c}$ and implements only the first element of the computed optimal trajectory, can be endowed with different properties. Thus in [5] ν is chosen to be only 1, yet the resulting algorithm has guaranteed stability and converges to the LQ optimal solution just as soon as this becomes feasible. Alternatively in [6] consideration is given to LQ optimality in conjunction with feasibility. The key idea in these variants is that, even for a small number of degrees of freedom (ie for small ν) one is able to derive attributes that are not possible with MBPC algorithms which use future control moves as the degrees of freedom. For example it is shown in [6] that for a small number of degrees of freedom the algorithm of [3] can run into infeasibility and instability, whereas the algorithm based on $\boldsymbol{c}$ yields good closed loop responses.

Here we cast the problem of choosing $\boldsymbol{c}$ as a linear interpolation problem and determine the smallest number of degrees of freedom needed to ensure desirable properties such as stability, and feasibility. The motivation for this is that the work is to be extended to the case of nonlinear systems for which the computational load becomes prohibitive. Our particular preference is for $\nu = 1$ because then in the case of J linear systems one does not need QP, but can perform the minimization of J in closed form, whereas in the nonlinear case minimization can be performed by simple univariate search algorithms. The device used in the nonlinear case is input-state linearization according to which it is possible to: (i) derive a suitable discrete-time linear MBPC control problem; (ii) compute an unconstrained optimum solution; (iii) interpolate between this and the extension to current time of the input trajectory computed at the previous instant. This way the resulting receding horizon algorithm can be shown to be stable, and to have guaranteed feasibility providing that a feasible stabilizing input trajectory exists at start-time. In addition it is also shown that the algorithm goes to a small neighbourhood of the unconstrained optimum just as soon as this becomes feasible. The results of the paper are illustrated by means of a simple nonlinear model of an experimental rig.

The paper is in essence a tutorial paper in that it reviews earlier linear MBPC results and routinely applies input-output/state linearization to a special class of nonlinear systems. However the main point of this work is to emphasise the benefits one derives from linear interpolation and to show how, for a class of nonlinear systems, such interpolation can be used to reduce significantly the computational burden so as to make nonlinear MBPC tractable.

2. Review of earlier work

The starting point of the work is the characterization of the class of stabilising control predictions for linear time-invariant system with discrete-time model:

$$y_{t+1} = \tfrac{b(z)}{\alpha(z)}\Delta u_t,\ \Delta u_t = u_t - u_{t-1},\ \alpha(z) = \Delta(z)a(z),\ \Delta(z) = 1 - z^{-1}, \quad (1)$$
$$a(z) = a_0 + a_1 z^{-1} + \cdots + a_n z^{-n},\ b(z) = b_0 + b_1 z^{-1} + \cdots + b_{n-1} z^{-n+1}$$

where $a(z), b(z)$ are coprime, u_t, y_t denote the instantaneous input and output values, z is the z-transform variable, and z^{-1} when acting on u_t, y_t denotes the delay operator. For a constant future setpoint r, the above yields the equation:

$$E(z) = R(z) - Y(z) = \frac{Q(z) - b(z)\Delta U(z)}{\alpha(z)},\quad Q(z) = a(z)r - P(z) \quad (2)$$

where $E(z), R(z), Y(z), \Delta U(z)$ denote the z-transforms of the predictions for the error, setpoint, output, and control increment trajectories; $P(z)$ accounts for the effect of past inputs and outputs. Strictly speaking, different notation should be used for "predicted" and "actual" values, but for simplicity we shall use the same notation, it being understood that the context will remove any ambiguity.

Theorem 2.1. [4] *The class of stabilizing prediction equations for the system of eqn. (1) is given, in z-transform form, as:*

$$\left.\begin{array}{l}\Delta U(z) = G_2(z) + G_1(z)F(z) \\ E(z) = H_2(z) + H_1(z)F(z)\end{array}\right\},\ \left\{\begin{array}{l}G_2(z) = \frac{\psi(z)}{b^-(z)},\ G_1(z) = -\frac{\alpha^+(z)}{b^-(z)} \\ H_2(z) = \frac{\phi(z)}{a^-(z)},\ H_1(z) = \frac{b^+(z)}{a^-(z)}\end{array}\right. \quad (3)$$

where

$$\alpha^+(z)\phi(z) + b^+(z)\psi(z) = Q(z) \quad (4)$$

Proof. From eqn. (2) we have that the predicted error will converge to zero iff:

$$Q(z) - b(z)\Delta U(z) = \alpha^+(z)\phi(z) \quad (5)$$

where $\phi(z)$ is a polynomial or stable transfer function; $(.)^+/(.)^-$ denotes the factors of polynomials with roots outside/inside the unit circle with roots on the unit circle being included in $(.)^+$. The above equation in turn implies that the predicted control increment trajectory converges to zero if and only if eqn. (4) holds true. Eqn. (4) is a Bezout identity and has many solution pairs $\{\phi(z), \psi(z)\}$ which are given as:

$$\{\phi(z) + b^+(z)F(z),\ \psi(z) - \alpha^+(z)F(z)\}$$

where $\{\phi(z), \psi(z)\}$ denotes a particular solution pair and $F(z)$ is a stable transfer function or, equivalently and more conveniently, the infinite order polynomial corresponding to the impulse response of a stable transfer function. Substitution of the general expression for the solution pair of (4) into the relevant prediction equation defines the whole class of stable predictions, P_∞, as given in (3). □

Thus $\boldsymbol{f}_\infty$, the vector of coefficients of $F(z)$, represents the degrees of freedom available for the minimization of the usual MBPC cost, which for an appropriate matrix $S_\infty > 0$, vector $\boldsymbol{\beta}_\infty$ and constant γ_∞ can be written as:

$$J = \sum_{i=1}^{\infty}[r - y_{t+i}]^2 + \lambda \sum_{i=1}^{\infty} \Delta u_{t+i-1}^2 = \boldsymbol{f}_\infty^T S_\infty \boldsymbol{f}_\infty + \boldsymbol{\beta}_\infty^T \boldsymbol{f}_\infty + \gamma_\infty \tag{6}$$

subject to input constraints of the form:

$$|u_{t+i} - u_{centre}| \leq u_{radius}, \; |\Delta u_{t+i} - \Delta u_{centre}| \leq \Delta u_{radius} \tag{7}$$

Remark 2.2. *Earlier MBPC algorithms which did not take explicit account of constraints deployed λ as a tuning parameter through which to adjust the emphasis placed on control activity thereby attempting to maintain the control trajectories within reasonable bounds. In the framework of (6) and (7) however this is no longer necessary and, notwithstanding consideration of expended energy, tear and wear of actuators etc, the use of a non-zero λ is no longer essential. In this paper we assume that λ is some arbitrary small value chosen simply so as to avoid numerical problems in the computation of the unconstrained optimum of J.*

The minimization of the cost J of (6) subject to constraints (7) defines an *infinite* dimensional Quadratic Program (as is emphasised by the subscript $(.)_\infty$), the solution to which is not practicable. Cast in this light, MBPC can be said to be all about reducing this infinite dimensional problem to a finite dimensional one, by considering "suitable" subsets of P_∞; what is meant by "suitable" depends on design objectives. In general, use of subsets of P_∞ results in suboptimal trajectories, unless one defines the subset, P_N, that allows the first N control increments to become degrees of freedom and generates the subsequent control moves by an optimal state feedback control law [3]. N is assumed to be large enough so that the resulting optimization is feasible and the optimal state feedback control moves stay within the limits of (7). Then it can be shown that the finite dimensional optimization provides an exact solution to the infinite dimensional QP problem posed above. This is an attractive feature, but N may need to be very large [3, 6] thereby implying an inordinate computational load.

To reduce the dimensionality of the optimization, an alternative [6] considers the subset, P_ν, of the stabilizing class P_∞ that corresponds to the transfer function $F(z)$:

$$F(z) = F_{feas}(z) + G_3(z)c(z) \tag{8}$$

where $c(z)$ is a polynomial of order $\nu - 1$, $G_3(z)$ is defined in such a way that for a particular choice of $c(z)$, $F(z) = F_{opt}(z)$, where $F_{opt}(z)$ denotes the transfer function which yields the unconstrained optimal predicted control trajectory, and where $F_{feas}(z)$ yields the unconstrained optimum solution to the cost of (6) with λ arbitrarily large. For convenience we shall refer to this unconstrained optimal solution as "mean level". The computation of $F(z) = F_{opt}(z)$, $F_{feas}(z)$ is straightforward and can be performed applying optimal control theory to a model state space description, or equivalently applying optimal l_2 theory to the input-output formulation of the prediction equations given in (3). In this approach, the degrees of freedom are contained in the ν-dimensional vector $\boldsymbol{c}$ of the coefficients of $c(z)$ and the resulting optimization becomes that of minimizing over $\boldsymbol{c}$ the cost:

$$J = \sum_{i=1}^{\infty}[r - y_{t+i}]^2 + \lambda \sum_{i=1}^{\infty} \Delta u_{t+i-1}^2 = \boldsymbol{c}^T S \boldsymbol{c} + \boldsymbol{\beta}^T \boldsymbol{c} + \gamma \tag{9}$$

subject to constraints (7). The positive definite matrix S, the vector $\boldsymbol{\beta}$ and constant γ can be computed through the solution of a Lyapunov equation. Under the assumption that the mean level solution remains with the constraints of (7), the subset of stabilizing predictions above contain as members: (i) a solution which is feasible; (ii) the unconstrained optimum. As a consequence we have [6]:

Theorem 2.3. *Under the assumption that the mean level solution is feasible, the receding horizon application of the strategy which, at each time instant t minimizes over* $\boldsymbol{c}$ *(as defined in eqn. (8)) the cost J of (9), results in an algorithm with guaranteed feasibility and which converges to the unconstraint optimum, as soon as this becomes feasible.*

The proof of stability in MBPC is often based on Lyapunov arguments which establish that the cost J is a monotonically decreasing function of time. The idea is simple: if the current set of input predictions includes as a member the extension to current time of the input trajectory that was optimized at the previous time, then the current cost, J_t will be

$$J_t = J_{t-1}^* - e_{t-1}^2 - \lambda \Delta u_{t-2}^2 \leq J_{t-1}^* \tag{10}$$

where J_{t-1}^* denotes the optimal value of the cost at the previous time instant. The equality sign in the above can only hold for *all* future t if the error and control increments have reached their steady state values of zero. If steady state is not already achieved, optimization at time t will produce a optimal cost J_t^* which will be even smaller than J_{t-1}^*. For convenience we shall refer to the extension to current time of the input trajectory that was calculated at the previous time instant as the "tail". An alternative strategy therefore to those described above is to define the finite dimensional subset of P_∞ so as to include the unconstrained optimum and the "tail" [5]. This approach gives surprisingly good results [5], despite the fact that the number of degrees of freedom is actually reduced to one. The remarkable feature of this observation is that with only one degree of freedom, the problem of minimizing a quadratic cost subject to linear constraints is no longer a QP problem and can in fact be solved in closed form, thereby producing a very significant reduction in computation time. The difficulty with this approach is that in the presence of uncertainty (either in the form of model mismatch, or disturbances) infeasibility may become a problem.

The conclusion of the discussion presented in this section is that in defining a finite dimensional subset of L_∞, one should aim to include: (i) the unconstrained optimum; (ii) the "mean level" solution; (iii) the "tail" of the control trajectory computed at the previous time instant. In the sequel, we use simple linear interpolation to combine these three objectives and show that the minimum number of degrees of freedom required is 2.

3. Linear interpolation in MBPC

3.1. Capturing the "tail"

In this section we need to distinguish between "previous" and "current" values, and this we achieve by using the notation for current $(.)'$. Accordingly "previous" and "current" input prediction equations can be written as:

$$\Delta U(z) = \frac{\psi(z)}{b^-(z)} - \frac{\alpha^+(z)}{b^-(z)}F(z),\ \Delta U'(z) = \frac{\psi'(z)}{b^-(z)} - \frac{\alpha^+(z)}{b^-(z)}F'(z) \tag{11}$$

with $F(z) = n(z)/d(z), F'(z) = n'(z)/d'(z)$, where $n(z), n'(z), d(z), d'(z)$ are four polynomials the last two of which are strictly Hurwitz, and where:

$$a^+(z)\phi(z) + b^+(z)\psi(z) = Q(z),\ a^+(z)\phi'(z) + b^+(z)\psi'(z) = Q'(z) \tag{12a,b}$$

A necessary condition for the class current input prediction $\Delta U'(z)$ to include the "tail" of the input prediction computed at the previous instant is that, $\Delta U(z)$, must share common poles. This in turn implies that $d'(z) = d(z)$ which therefore yields the prediction equations:

$$\Delta U(z) = \frac{\psi(z)}{b^-(z)} - \frac{\alpha^+(z)}{b^-(z)d(z)}n(z),\ \Delta U'(z) = \frac{\psi'(z)}{b^-(z)} - \frac{\alpha^+(z)}{b^-(z)d(z)}n'(z) \tag{13a,b}$$

From the definition of $Q(z)$ it is easy to show that:

$$Q'(z) = z[Q(z) - \alpha(z)e - b(z)\Delta u] \tag{14}$$

where $e, \Delta u$ denote the actual (as opposed to predicted) values of the error and control increment at the previous time instant. Furthermore, from the error prediction equation at the previous time instant we have:

$$\alpha(z)E(z) = Q(z) - b(z)\Delta U(z) \implies \alpha_0 e + b_0 \Delta u = Q_0 \tag{15}$$

where α_0, b_0, Q_0 denote the coefficient of z^0 in $\alpha(z), b(z), Q(z)$, respectively.

Theorem 3.1. *For a given $d(z)$ let the degrees of freedom of the prediction class (13b) defined by the coefficients of $n'(z)$ form the elements of a vector $\mathbf{n}'$. Then the class of (13b) captures the "tail" of (13a) if and only if $\mathbf{n}'$ is allowed to assume the value defined by the vector $\mathbf{n}_{tail}$ of coefficients of the polynomial*

$$n'_{tail}(z) = z\left[n(z) - \frac{n_0}{d_0}d(z)\right] \tag{16}$$

Proof. From the definition of the "tail" and eqn. (11), it follows that the z-transform of the "tail" is given as:

$$\Delta U'_{tail}(z) = z[\Delta U(z) - \Delta u] = z\frac{\psi(z)d(z) - \alpha^+(z)n(z) - b^-(z)d(z)\Delta u}{b^-(z)d(z)} \tag{17}$$

Therefore a necessary and sufficient condition for capturing the "tail" is that a polynomial $n'_{tail}(z)$ exists such that:

$$z[\psi(z)d(z) - \alpha^+(z)n(z) - b^-(z)d(z)\Delta u] = \psi'(z)d(z) - \alpha^+(z)n'_{tail}(z) \tag{18}$$

or

$$n'_{tail}(z) = \frac{\psi'(z)d(z) - z[\psi(z)d(z) - \alpha^+(z)n(z) - b^-(z)d(z)\Delta u]}{\alpha^+(z)} \tag{19}$$

Multiplying (12a) by z, subtracting it from (12b) and using (14) gives:

$$\alpha^+(z)[\phi'(z) - z\phi(z)] + b^+(z)[\psi'(z) - \psi(z)z] = -z[\alpha(z)e + b(z)\Delta u] \tag{20}$$

or

$$\alpha^+(z)[\phi'(z) - z\phi(z) + z\alpha^-(z)e] + b^+(z)[\psi'(z) - z\psi(z) + zb^-(z)\Delta u] = 0 \tag{21}$$

The minimal order solution for the square-bracketed quantities above yields:

$$\phi'(z) = z\left[\phi(z) - \alpha^-(z)e - b^+(z)\frac{\phi_0 - \alpha_0^- e}{b_0^+}\right], \tag{22a}$$

$$\psi'(z) = z\left[\psi(z) - b^-(z)\Delta u - \alpha^+(z)\frac{\psi_0 - b_0^- \Delta u}{\alpha_0^+}\right]. \tag{22b}$$

Now by eqn. (13a) we have:

$$\frac{\psi_0 - b_0^- \Delta u}{\alpha_0^+} = \frac{n_0}{d_0} \tag{23}$$

which in conjunction with (22a), after multiplication by $d(z)$ and substitution into (19) yields the polynomial of (16). □

3.2. Introducing optimality and feasibility

Capturing the tail ensures the monotonic decrease of cost, but gives no guarantee of optimality. To get this it is necessary to include the unconstrained optimum as a member of the prediction class, so that it is selected just as soon as it becomes feasible. The condition for inclusion of the unconstrained optimum is given below.

Theorem 3.2. *Let $\Lambda(z,\lambda)$ the closed loop polynomial of the feedback control law that minimizes the cost of (6) in the absence of constraints. Then the prediction class of (11b) comprises the unconstrained optimal input trajectory as a member if and only if $d'(z) = \Lambda(z,\lambda)$ and $n'(z)$ is allowed to assume the polynomial solution $n'(z,\lambda)$ of the equation:*

$$\alpha^+(z)n'(z,\lambda) = \Lambda(z,\lambda)\psi'(z) - b^-(z)m'(z,\lambda) \tag{24}$$

where $m'(z,\lambda)$ is defined from the minimal order solution pair $[m'(z,\lambda), k'(z,\lambda)]$ of the Bezout identity:

$$\alpha(z)k'(z,\lambda) + b(z)m'(z,\lambda) = \Lambda(z,\lambda)Q(z) \tag{25}$$

Proof. The optimal predicted error and input trajectories must have the form:

$$E'(z,\lambda) = \frac{k'(z,\lambda)}{\Lambda(z,\lambda)}, \quad \Delta U'(z,\lambda) = \frac{m'(z,\lambda)}{\Lambda(z,\lambda)} \tag{26a,b}$$

where $m'(z,\lambda), k'(z,\lambda)$ satisfy (25). Then $n'(z,\lambda), \Lambda(z,\lambda)$ define the numerator and denominator of the transfer function $F'(z)$ that causes the prediction of (11b) to become the optimal prediction of (26b). Finally a polynomial solution to (24) always exists because the unstable factor $\alpha^+(z)$ is also a factor of the right-hand side; this follows from (4,25). □

The results of Theorems 3.1 and 3.2 combine to give an algorithm which optimizes over a class that contains both the unconstrained optimum and the "tail" of the previously computed control trajectory. Such an algorithm is endowed with three desirable attributes: (i) guarantee of a monotonically decreasing cost; (ii) convergence to the optimal solution just as soon as this becomes feasible; (iii) guarantee of feasibility, given feasibility at start-time. This last property follows from the simple observation that if the previously computed control trajectory is within the input limits then its "tail" will also satisfy the constraints. However during transients, the previously computed trajectory will lie on the boundary of the feasible region and thus the "tail" itself is likely to be close to the boundary; thus the guarantee of feasibility may not be robust with respect to uncertainty (be it in the form of model mismatch and/or disturbances).

3.3. Including the "mean level" solution together with the "tail" and unconstrained optimum

To maximise robustness of the feasibility condition we need to include in the prediction class a feasible element which is as far away from the boundaries of the feasible region as possible. The computation of such an element in the general case is not practicable: it involves an infinite dimensional l_∞ minimization. Instead it is possible to use the solution which minimizes the l_2 norm as an approximation. The computation and inclusion of this solution, the "mean level" solution, is easy to effect by invoking the procedure of Theorem 3.2 but with λ replaced by an arbitrarily large value λ^*. In the light of the approximation involved it is now necessary to assume that the "mean level" solution is feasible. This is rather weak assumption since in the case of no uncertainty it will always hold true providing that the prediction class includes the "tail" and providing that at start-time there exists (as indeed there must exist) at least one stabilizing feasible solution. The theorem below provides an obvious way for the inclusion of the unconstrained optimum, the "tail" and the "mean level" solution.

Theorem 3.3. *The minimum number of degrees of freedom required to define a prediction class that includes the unconstrained optimum, the "tail" and the "mean level" solution is 2 and the resulting class can be written as:*

$$E'(z,\lambda) = \frac{k'(z,\lambda)}{\Lambda(z,\lambda)} + \frac{b^+(z)}{a^-(z)}\frac{f(z)}{d'(z)}, \quad \Delta U'(z,\lambda) = \frac{m'(z,\lambda)}{\Lambda(z,\lambda)} - \frac{\alpha^+(z)}{b^-(z)}\frac{f(z)}{d'(z)} \tag{27}$$

with

$$\left.\begin{array}{l} d'(z) = \Lambda(z,\lambda)\Lambda(z,\lambda^*) \\ f(z) = [n'_{tail}(z) - n'(z,\lambda) \;\; n'(z,\lambda^*) - n'(z,\lambda)]\mathbf{c} \end{array}\right\}, \; \mathbf{c} = \left[\begin{array}{c} c_1 \\ c_2 \end{array}\right] \qquad \text{(28a–c)}$$

where $n'_{tail}(z)$ *is as defined in Theorem 3.1 but for* $d(z) = d'(z) = \Lambda(z,\lambda)\Lambda(z,\lambda^*)$ *and where* $n'(z,\lambda), n'(z,\lambda^*)$ *are the polynomials solutions of:*

$$\begin{aligned} \alpha^+(z)n'(z,\lambda) &= [\Lambda(z,\lambda)\psi'(z) - b^-(z)m'(z,\lambda)]\Lambda(z,\lambda^*) \\ \alpha^+(z)n'(z,\lambda^*) &= [\Lambda(z,\lambda^*)\psi'(z) - b^-(z)m'(z,\lambda^*)]\Lambda(z,\lambda) \end{aligned} \qquad (29)$$

$\Lambda(z,\lambda^*), m'(z,\lambda^*)$ *are defined as* $\Lambda(z,\lambda), m'(z,\lambda)$ *but with* λ *replaced by* λ^*.

Proof. To include the "mean level" solution, $d(z)$ must have $\Lambda(z,\lambda^*)$ as a factor, whereas to include the unconstrained optimum it must have $\Lambda(z,\lambda)$ as a factor. Therefore to include both, $d(z)$ must be as given in (28a). For this choice of $d'(z)$ the condition on $n'(z)$ that ensures the inclusion of the "tail", unconstrained optimum and "mean level" is that $n'(z)$ interpolates through the polynomials $n'_{tail}(z), n'(z,\lambda), n'(z,\lambda^*)$ which are given as the solutions of (16), (24) and again (24) when λ is replaced by λ^* with the difference that the right-hand side of the last two equations must be multiplied by $\Lambda(z,\lambda^*)$, $\Lambda(z,\lambda)$ respectively. Therefore $n'(z)$ must have the form:

$$n'(z) = f(z) + n'(z,\lambda) \qquad (30)$$

where $f(z)$ is as given in (28b). By the definition of $k'(z,\lambda), m'(z,\lambda)$ we have

$$\begin{aligned} \frac{\psi'(z)}{b^-(z)} - \frac{\alpha^+(z)n'(z,\lambda)}{b^-(z)\Lambda(z,\lambda)\Lambda(z,\lambda^*)} &= \frac{m'(z,\lambda)}{\Lambda(z,\lambda)} \\ \frac{\phi'(z)}{a^-(z)} - \frac{b^+(z)n'(z,\lambda)}{a^-(z)\Lambda(z,\lambda)\Lambda(z,\lambda^*)} &= \frac{k'(z,\lambda)}{\Lambda(z,\lambda)} \end{aligned} \qquad (31)$$

and thus it is possible to regroup the prediction equation as per (27). □

Corollary 3.4. *The minimization of the cost of (6) over the prediction class of Theorem 3.3, namely over the* $\mathbf{c}$, *subject to constraints (7) is a 2-dimensional QP problem. Furthermore the predictive control law that arises out of the receding horizon application of this optimization, ie the computation and implementation at each sampling instant of the first element of the optimal predicted control trajectory, has guaranteed stability and feasibility and will converge to the unconstrained optimum just as soon as this becomes feasible.*

Proof. The predictions of Theorem 3.3 are linear in $\mathbf{c}$ and so the cost of (6) is quadratic in $\mathbf{c}$, while the constraints of (7) are linear in $\mathbf{c}$; thus the minimization problem is a 2-dimensional QP. The inclusion in the prediction class of the "tail", "mean level" and unconstrained optimum respectively guarantee stability, feasibility and convergence to the unconstrained optimum. □

The optimization of the corollary above constitutes a semi-infinite program. However given the stable nature of the prediction class, it is easy to show (following arguments similar to those used in [3])that constraints (7) need only be invoked over a finite horizon, say n_{con}.

4. Extension to the control of nonlinear systems

4.1. One dimensional interpolation

In this paper we shall consider the class of nonlinear systems whose state vector $\boldsymbol{x}$, input u and output y are related by dynamic model equations:

$$\dot{\boldsymbol{x}} = f(\boldsymbol{x}) + g(\boldsymbol{x})u, \; y = h(\boldsymbol{x}) \tag{32a}$$

$$|u(t) - u_{centre}| \leq u_{radius}, \; |\dot{u}(t) - \dot{u}_{centre}| \leq \dot{u}_{radius} \tag{32b}$$

and shall extend the ideas of the earlier sections to derive a predictive control algorithm which: (i) is easy to implement; (ii) has the guarantee of stability and (iii) converges to the optimal solution as soon as this becomes feasible. Nonlinear dynamics are notorious for the difficulties they create with respect to predictions and cost minimisation in the presence of constraints; the use of a large number of degrees of freedom often results in computation times which are excessive. Our aim here is to invoke the interpolation techniques discussed earlier in order to retain the attributes (ii) and (iii) above, while keeping computation down to a bear minimum. It is precisely for this reason that we first state a special case of Theorem 3.3 which (for the linear discrete time case) interpolates between the unconstrained optimum and the "tail". Such an interpolation requires only one degree of freedom on account of which the QP optimization problem collapses to a problem which can be solved in closed form.

Theorem 4.1. *The prediction class*

$$\Delta U(z) = \frac{m'(z,\lambda)}{\Lambda(z,\lambda)} + \frac{\alpha^+(z)(n'_{tail}(z) - n'(z,\lambda))}{b^-(z)\Lambda(z,\lambda)}c \tag{33}$$

includes both the "tail" of the control trajectory computed at the previous time instant and the unconstrained optimum. Thus use of this class in a receding horizon application will result in a closed-loop control law which has guaranteed stability and which converges to the optimum as soon as this becomes feasible.

Proof. This is the convenient form that the combination of Theorems 3.1 and 3.2 assumes for the special case when the transfer function $F(z)$ of the stabilizing class of (3) is replaced by $a^-(z)b^-(z)F(z)$. The values $c = 0$, 1 produce the two interpolation points, the unconstrained optimum and the "tail" respectively. □

Dropping the "mean level" solution from the interpolation implies that the guarantee of feasibility is no longer robust with respect to uncertainty, but offers the very significant advantage that it trivialises the QP problem into one that can be solved in closed form. Infeasibility due to uncertainty can be avoided if

$u_{radius}, \Delta U_{radius}$ are scaled down; the choice of scaling factors can be based on the information available on the size of uncertainty. The use of input-output/state linearization for the control of constrained nonlinear processes is by no means new (e.g. see [9]), nor is its use in the context of constrained MBPC. Furthermore the steps involved here constitute a straightforward application of well known results (e.g. [8]). What is new is however is the introduction of linear interpolation as a means of avoiding the very significant computational difficulties that are posed by the nonlinear map between the linearising input (and states) and the actual input which in general implies a non-convex constrained optimization problem.

5. The nonlinear algorithm

The nonlinear dynamics of eqn. (32a) are in continuous time and direct application of the results discussed thus far is not possible. One avenue is to extend Theorem 4.1 to the continuous-time case; this can be done using the ideas developed in [7] but will not be undertaken here. Instead discretization and subsequent application of Theorem of 4.1 will be achieved through the process of input-state linearization. For brevity we only consider the case when the nonlinear systems of (32a) have relative degree n, equal to the system order; in this case one need not be concerned about whether the zero dynamics are nonminimum phase or not, nor does one have to establish the stability of internal dynamics.

Theorem 5.1. *In a region Ω of $\mathbb{R}^n$ let $\boldsymbol{f}, \boldsymbol{g}, h$ have continuous derivatives with respect to $\boldsymbol{x}$, and let the relative degree of the system (32a) be n. Then the law*

$$u(t) = \frac{1}{L_g L_f^{n-1} h}(-L_f^n h + v(t)) \tag{34}$$

where $L_{(.)}$ denotes Lie derivatives, provides a linear dynamic relation between v and y which, upon discretization and incorporation of integral action yields the input output relation:

$$y_{k+1} = \frac{b(z)}{\alpha(z)} \Delta v_k \tag{35}$$

Furthermore let c_k denote the smallest value of c for which the discrete-time sequence v derived from the inverse z-transform of the prediction class

$$\Delta V(z) = \frac{m'(z,\lambda)}{\Lambda(z,\lambda)} + \frac{\alpha^+(z)(n'_{tail}(z) - n'(z,\lambda))}{b^-(z)\Lambda(z,\lambda)} c \tag{36}$$

makes the $u(t)$ of equation (34) stay within the limits of (32b) for all future times t. Then the receding horizon application of the implied control law has guaranteed stability and feasibility, providing a stabilizing sequence v exists at start-time. For arbitrarily small λ this receding horizon algorithm converges to a solution that is *arbitrarily close to the solution that minimizes the cost*

$$J = \sum_{i=1}^{\infty} [r - y_{k+i}]^2 + \lambda \sum_{i=1}^{\infty} \Delta u_{k+i-1}^2 \tag{37}$$

as soon as this solution becomes feasible.

Proof. This is a consequence of Theorem 4.1 used in conjunction with input-state linearization (e.g. see [8]). Thus under the assumptions of the theorem the control law of (34) yields the linear state space equations

$$\dot{\boldsymbol{z}} = A\boldsymbol{z} + Bv, \ y = C\boldsymbol{z} \tag{38}$$

where

$$z_1 = y, \ z_i = L_f^{i-1} h(\boldsymbol{x}) \ i > 1 \tag{39}$$

and where the state space model of (38) is in controllable canonical form. Therefore, the linearized dynamics of (38) can be discretized and upon the introduction of an integrator the resulting model will have the form of (35).

For this system, the prediction class of (36) is stabilizing and includes the "tail" of the sequence v computed at the previous discrete time instant. Thus

$$J = \sum_{i=1}^{\infty} [r - y_{k+i}]^2 + \lambda \sum_{i=1}^{\infty} \Delta v_{k+i-1}^2 \tag{40}$$

will be monotonically decreasing and this establishes stability. Finally for $c = 0$ the prediction class of (36) yields the unconstrained optimum with respect to the cost of (40), which for arbitrarily small λ is arbitrarily close to that of (37). Thus the receding horizon law of the algorithm will converge to a small neighbourhood of the minimizing solution to (37) just as soon as this becomes feasible. □

Remark 5.2. *The controllability form of (38) implies that all the poles of the linearized model are at the origin and this leads to numerical problems in the computation of the prediction class of (36). Such problems can be avoided by simply replacing v by $v - K\boldsymbol{z}$, where K is a stabilizing state feedback controller.*

Remark 5.3. *The determination of the smallest permissible value for c involves a test that (32b) is not violated for all positive t. Following arguments similar to those for the linear discrete-time case, it is possible to show that one need only invoke the constraints up to a finite time, say t_{con}. Furthermore interpolation techniques can be used to ensure that invoking constraints at a finite number of points, say N, implies satisfaction of constraints at all points in the interval $[0, t_{con}]$. To accommodate the error in interpolating approximations, it is possible to artificially tighten the constraints, but this tightening can be made as insignificant as desired by increasing N. This can be achieved at minimal computation cost because the optimization problem solved is not demanding (and can be performed by any simple numerical univariate search), especially when compared with QP.*

Remark 5.4. *As per remark 2.2, taking explicit account of constraints obviates the need for nonzero λ; arbitrarily small λ are still useful in avoiding numerical problems as well as ensuring that $\boldsymbol{x}$ remains in Ω.*

6. Design study

To illustrate the advantages of the predictive control strategy presented in this paper we use the model of a coupled tank system which is given as:

$$\begin{aligned} \frac{d}{dt}\begin{bmatrix} x_1 \\ x_2 \end{bmatrix} &= \begin{bmatrix} -a_1 & a_2 \\ a_1 & -a_3 \end{bmatrix}\begin{bmatrix} |x_1|^{1/2}\text{sign}(x_1) \\ |x_2|^{1/2}\text{sign}(x_2) \end{bmatrix} + \begin{bmatrix} 0 \\ 5 \end{bmatrix} u, \quad \left\{ \begin{array}{l} a_1 = 0.366187 \\ a_2 = 0.394968 \\ a_3 = 0.789936 \end{array} \right. \\ y &= x_1 \end{aligned} \tag{41}$$

with the constraint

$$|u - 0.5| \leq 0.5 \tag{42}$$

The input u drives (through a D-A converter) a water pump whose output is fed into the first tank, and affects (through a connecting orifice) the level in a second tank; a second orifice lets water out of the second tank. The state variables x_1, x_2 denote the excess in level of the first tank over the second tank, and the height of the level in the second tank over the outlet valve. The numerical values of the model correspond to the physical parameters of an actual coupled-tanks rig.

Physical considerations imply that $x_1, x_2 \geq 0$ and it is obvious that everywhere in the quadrant, say Ω, defined by $x_1, x_2 > 0$, $\boldsymbol{f}, \boldsymbol{g}, h$ are smooth and the relative degree of the model is 2. Therefore it is easy to show that the control law:

$$u = 0.2\sqrt{x_2}\left[(a_2 a_3 - a_1^2) + a_1 a_2\left(\sqrt{\frac{x_2}{x_1}} - \sqrt{\frac{x_1}{x_2}}\right) + 2v\right] \tag{43}$$

brings the nonlinear dynamic model of (41) into the form

$$\begin{aligned} \frac{d}{dt}\begin{bmatrix} z_1 \\ z_2 \end{bmatrix} &= \begin{bmatrix} 0 & 1 \\ 0 & 0 \end{bmatrix}\begin{bmatrix} z_1 \\ z_2 \end{bmatrix} + \begin{bmatrix} 0 \\ 5 \end{bmatrix} v, \quad \left\{ \begin{array}{l} z_1 = x_1 \\ z_2 = -a_1\sqrt{x_1} + a_2\sqrt{x_2} \end{array} \right. \\ y &= [1 \;\; 0] \end{aligned} \tag{44}$$

Replacing v by $v - 5 \times 10^{-5} z_1 - 1.5 \times 10^{-2} z_2$ has the effect of shifting the poles of the above from the origin to -0.005 and -0.01 and subsequent discretization for a sampling interval of 1 sec results in the input-output model:

$$y_{k+1} = \frac{0.4975 + 0.495z^{-1}}{(1 - z^{-1})(1 - 1.9851z^{-1} + 0.985z^{-2})}\left[\frac{\Delta v_k}{325}\right] \tag{45}$$

The scaling factor of 325 is introduced to balance the contribution of error and control activity in the cost. Invoking the algorithm of Theorem 5.1 to this model for a setpoint of 75 with $\lambda = 0.05$ yields the closed loop level responses shown in Fig. 1. The corresponding closed loop responses of z_1, z_2 are shown in Fig. 2 whereas those of c, v, u are shown in Fig. 3; the plot of v is scaled down by a factor of

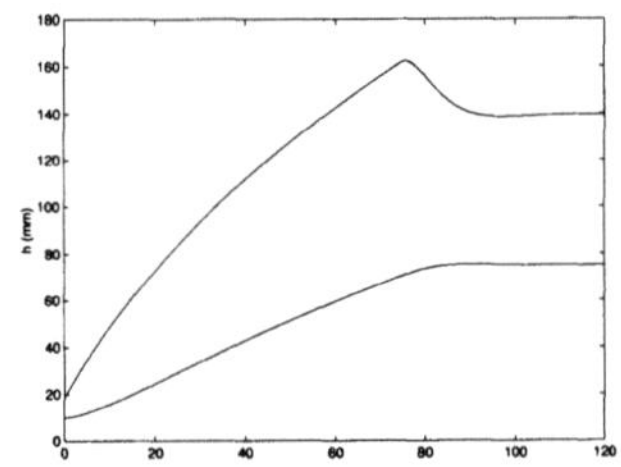

FIGURE 1. Level response of tanks 1 and 2

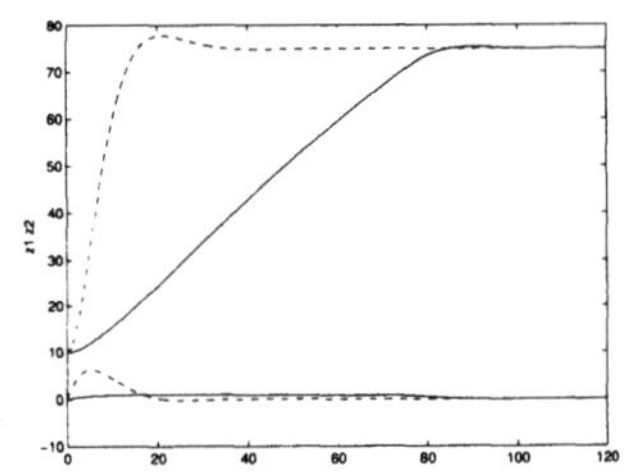

FIGURE 2. Response of z_1 and z_2 (dashed lines for unconstrained response)

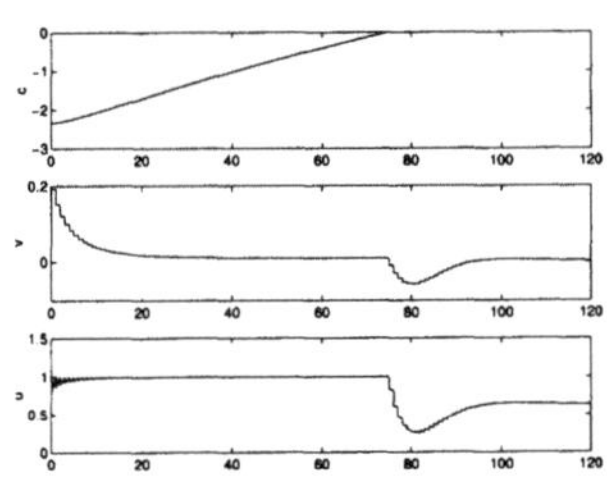

FIGURE 3. Response of c, v and u

325. The response of the second level is excellent: it is as fast as possible (this being limited by the pump saturation, ie the constraint on u) and is non-overshooting. The small seesaw-like effect on the input is the effect of discretization of v in combination with the continuous time variations of u (as per eqn. 43) and can be removed, if desired, by faster sampling or by a simple linear interpolation between two consecutive optimal u predictions. Alternatively, one can use a continuous-time implementation of the predictive algorithms, but this falls beyond the scope of the current work.

7. Acknowledgements

The work was funded by the Engineering and Physical Sciences Research Council.

References

[1] D. W. Clarke, C. Mohtadi and P. S. Tuffs, *Generalized predictive control. Parts 1 and 2*, Automatica, **23** (1987), 137–160.

[2] J. B. Rawlings and K. R. Muske, *The stability of constrained receding horizon control*, IEEE Trans. Automat. Contr., **38 (19)** (1993), 1512–1516.

[3] P. O. M. Scokaert and J. B. Rawlings, *Infinite horizon linear quadratic control with constraints*, Proceedings of the 13th IFAC World Congress (1996), 109–114.

[4] J. A. Rossiter, J. R. Gossner and B. Kouvaritakis, *Infinite horizon stable predictive control*, IEEE Trans. Automat. Contr., **41 (10)** (1996), 1522–1527.

[5] M. Cannon and B. Kouvaritakis, *Fast suboptimal predictive control with guaranteed stability*, Control and Systems Letters, to appear.

[6] B. Kouvaritakis, J. A. Rossiter and M. Cannon, *Linear quadratic feasible predictive control*, Automatica, to appear.

[7] M. Cannon and B. Kouvaritakis, *Infinite horizon predictive control of constrained continuous-time linear systems*, submitted to Automatica.

[8] J. E. Slotine and W. Li, *Applied nonlinear control*, Prentice Hall, New Jersey, **1991**.

[9] M. J. Kurtz and M. A. Henson, *Input-output linearizing control of constrained non-linear processes*, J. Proc. Cont., **7 (1)** (1997), 3–17.

Department of Eng. Science,
University of Oxford,
Parks Road
Oxford OX1 3PJ, UK
E-mail address: `basil.kouvaritakis@eng.ox.ac.uk`

Progress in Systems and Control Theory, Vol. 26

A Predictive Command Governor for Nonlinear Systems under Constraints

David Angeli, Alessandro Casavola and Edoardo Mosca

Abstract. A command governor (CG) is described for solving set-point tracking problems for nonlinear systems when pointwise-in-time input and/or state related inequality constraints have to be fulfilled. A CG is a nonlinear device which is added to a primal compensated system. The CG action, based on the current state, set-point and prescribed constraints, is finalized to select, at any time, a command sequence under which the constraints are possibly fulfilled with acceptable tracking performance. It is shown how to use off-line an iterative algorithm so as to restrict to a fixed finite integer the infinite number of time-instants over which the prescribed constraints must be checked in order to test admissibility of command sequences. An analysis based on a Lyapunov function argument shows that, if the reference becomes constant, the CG output converges to the closest admissible approximation to the set-point. An example is presented to illustrate the method.

1. Introduction

This paper deals with the design of command governors by mainly exploiting the conceptual tools of the predictive control methodology. A CG is a nonlinear device which is added to a primal compensated control system. The latter, in the absence of the CG, is designed so as to perform satisfactorily in the absence of constraints. Whenever necessary, the CG modifies the input to the primal control system so as to possibly avoid violation of the constraints. However, in contrast with predictive control, CG action is not directed to modify the primal compensated system dynamics. In fact, the CG design procedure consists of enforcing separation between compensation requirements and constraint prescriptions. In this way, a system equipped with a CG takes a special simplified structure at the cost typically of performance degradation with respect to less restrictive approaches, e.g. general constrained predictive control. CG adoption can be however justified in typical industrial applications wherein a massive amount of flops per sampling time is not allowed, and/or one is only commissioned to add to existing standard PID-like compensators peripheral units which, as CG's, do not change the primal compensated control system.

Studies along these lines, limited to primal linear systems, have already appeared in [10]. More mature assessments of the related state of art for linear systems can be found in [5, 6]. For previous results on CG applied to nonlinear systems see also [2, 4]. For CGs approached from different perspectives see [8] and [7]. In most cases, predictive control computations amount to solving on-line high-dimensional mathematical programming problems. For nonlinear systems, non-convex, math programming problems result. This implies quite a formidable computational burden if on-line solutions are required, e.g. see [9] and references therein. The main goal of the present paper is to better specify and extend the design of CG's to nonlinear systems presented in [2].

2. CG problem formulation and design for nonlinear systems

Consider the following discrete-time nonlinear dynamic system

$$\left\{ \begin{array}{rcl} x(t+1) & = & f(x(t), g(t)) \\ y(t) & = & \eta(x(t)) \\ c(t) & = & [x'(t)\ g'(t)]' \end{array} \right. \tag{1}$$

In (1): the prime denotes transpose; $t \in \mathbb{Z}_+ := \{0, 1, \dots\}$; $x(t) \in \mathbb{R}^n$ is the state vector; $g(t) \in \mathbb{R}^m$ the manipulable command input which, if no CG was present, would coincide with the output reference $r(t) \in \mathbb{R}^m$; $y(t) \in \mathbb{R}^m$ the output which is required to track $r(t)$; and $c(t) \in \mathbb{R}^{n_c}$, $n_c := n + m$, a vector which has to fulfill the pointwise-in-time set-membership constraint

$$c(t) \in \mathcal{C}, \ \forall t \in \mathbb{Z}_+ \tag{2}$$

with $\mathcal{C} \subset \mathbb{R}^{n_c}$, a prescribed constraint set. It is assumed

$$\mathcal{C} \text{ is compact, convex and } 0_{n_c} \in \text{int}\{\mathcal{C}\} \tag{3}$$

Further, we assume that $f(x, g)$ in (1) satisfies

$$\left\{ \begin{array}{l} 0_n = f(0_n, 0_m) \\ x_w = f(x_w, w) \text{ has a unique solution } x_w, \end{array} \right. \tag{4}$$

for all $w \in \mathbb{R}^m$ such that $c_w := [x'_w\ w']' \in \mathcal{C}$. Condition (4) appears to be very restrictive. In fact, it excludes nonlinear systems which have limit cycles or multiple equilibrium points for each value of the command w. However, because we aim at ensuring global tracking to the controlled system, such an assumption is quite natural.

Hereafter, we shall denote by

$$x(\cdot, x, g(\cdot)) := \{x(t, x, g(\cdot))\}_{t=0}^{\infty} \tag{5}$$

the state response of (1) to the input sequence $g(\cdot) := \{g(t)\}_{t=0}^{\infty}$ from the state x at time 0. Similar meaning will be associated to $c(\cdot, x, g(\cdot))$.

If the next inclusion holds

$$c(\cdot, x, g(\cdot)) \subset \mathcal{C} \tag{6}$$

for a given $g(\cdot)$, the state x is said to be *admissible*, the command sequence $g(\cdot)$ *feasible* and the pair $(x, g(\cdot))$ is said to be *executable*. For the sake of simplicity, throughout the paper, (x, w) will denote the pair made up of the state x and the constant sequence $g(t) = w$, $\forall t \in \mathbb{Z}_+$. Similar meaning will be associated to $x(\cdot, x, w)$ and $c(\cdot, x, w)$.

The problem we wish to consider next is how to characterize the set of all sequences which are feasible for $x(0)$ and, among them, how to select the one which is optimal w.r.t. some performance criterion. In particular, we are interested in an on-line CG strategy based on the present state and reference

$$g(t) := \underline{g}(x(t), r(t)) \tag{7}$$

such that, under suitable conditions, the constraints (2) are fulfilled and, possibly, $g(t) \approx r(t)$.

It is also assumed that system (1) enjoys the following properties:

Property 1: - Let $(x, g(\cdot))$ be executable and $[x'_{g(t)}\ g'(t)]' \in \mathcal{C}, \forall t \in \mathbb{Z}_+$, where $x_{g(t)}$ is the solution of the algebraic equation $x_{g(t)} = f(x_{g(t)}, g(t))$. Then, we assume that

$$\lim_{t \to \infty} [g(t) - g(t-1)] = 0_p \Longrightarrow \lim_{t \to \infty} [x(t, x, g(\cdot)) - x_{g(t)}] = 0_n. \tag{8}$$

Eq. (8) implies that for asymptotically slowly-varying input signals $g(t)$, the system tends to behave as a static one. Indeed this is always the case for stable linear systems. On the contrary, (8) is not implied by global asymptotic stability in a non-linear context. Nevertheless the stronger notion of input-to-state stability (ISS) [11] happens to suffice.

Under (8), the following extra implications follow at once

$$\lim_{t \to \infty} g(t) = w \Longrightarrow \lim_{t \to \infty} x(t, x, g(\cdot)) = x_w = f(x_w, w) \tag{9}$$

$$(x, w) \text{ executable} \Longrightarrow c_w \in \mathcal{C} \tag{10}$$

Property 2: - Let $[x'_w\ w']' \in \mathcal{C}$ and $(x_w + \tilde{x}, w + \tilde{w})$ be executable. Then, there exists a function of class $\mathcal{K}$ [1], call it $M(\cdot)$, independent of w, such that

$$\|x(k, x_w + \tilde{x}, w + \tilde{w}) - x_w\| \leq M\left(\|\tilde{x}\| + \|\tilde{w}\|\right) \tag{11}$$

where $\|x\| := \sqrt{x'x}$. Condition (11) is a special instance of the more general notion of Global Input Stability (GIS), e.g. see [1]. It consists of assuming robust marginal Lyapunov stability for x_w under state and input perturbations $\tilde{x}$ and $\tilde{w}$, with a boundedness condition on the extents of the perturbed state evolutions.

---

[1] A function $\gamma(x)$, $x \in \mathbb{R}_+ \mapsto \gamma(x) \in \mathbb{R}_+$, is said of class $\mathcal{K}$ if it is continuous, strictly increasing and $\gamma(0) = 0$. A function $\beta(x, t)$ is said of class $\mathcal{KL}$ if $x \mapsto \beta(x, t)$ is of class $\mathcal{K}$ and $t \in \mathbb{R}_+ \mapsto \beta(x, t) \in \mathbb{R}_+$, is a continuous, strictly decreasing map such that $\lim_{t \to \infty} \beta(x, t) = 0$.

In general, Properties 1 and 2 are quite difficult to check because they require the solution of (1). However, a sufficient condition based on Lyapunov-like arguments can be stated by exploiting the existing connections between Properties 1 and 2 and the notion of ISS. A nonlinear system is said to be ISS iff, for each equilibrium point x_w corresponding to any $w \in \mathbb{R}^m$, there exist a function $\beta_w(\cdot,\cdot)$ of class $\mathcal{KL}$ and two functions $\gamma_w(\cdot)$ and $\lambda_w(\cdot)$ both of class $\mathcal{K}$, such that for all $(x, g(\cdot))$ it results

$$\|x(t,x,g(\cdot)) - x_w\| \le \beta_w(\|x - x_w\|, t) + \gamma_w\left(\sum_{i=0}^{t-1} e^{-(t-i)}\lambda_w(\|g(i) - w\|)\right) \tag{12}$$

By a standard "comparison principle" argument [11], it can be shown that a nonlinear system (1) is ISS if there exist a positive definite radially unbounded function $V(x)$ and a function $\lambda(\cdot)$ of class $\mathcal{K}$ such that, for all pairs $(x, g(\cdot))$, it results

$$V(x(t+1)) \le \alpha V(x(t)) + \lambda(\|g(t)\|), \; \forall t \in \mathbb{Z}_+ \tag{13}$$

with $0 \le \alpha < 1$. The following Proposition 2.1, whose proof is reported in the Appendix, clarifies the connections between ISS and Properties 1 and 2.

Proposition 2.1. *Let (1) be ISS along all executable pairs $(x, g(\cdot))$ with bounded commands $g(t)$. Then, Properties 1 and 2 hold.*

Remark 2.2. *It is worth pointing out that, though ISS is a global notion of stability, it only needs to be verified for executable pairs for the purpose of establishing Properties 1 and 2. However, it is usually more easily verifiable globally via (13). Notice that global ISS is also sufficient for global asymptotic stability of (1) and fulfilment of (4).*

For reasons that will became clear, it is convenient to give the following definition for a given $\delta > 0$

$$\mathcal{C}_\delta := \{c \in \mathbb{R}^{n_c} : c + \tilde{c} \in \mathcal{C}, \; \forall \tilde{c} \in \mathcal{B}_\delta^{n_c}(0)\} \tag{14}$$

where $\mathcal{B}_\delta^{n_c}(0) := \{c \in \mathbb{R}^{n_c} \; : \; \|c\| \le \delta\}$. In addition, let

$$\mathcal{V}_\delta := \{w \in \mathbb{R}^m : c_w \in \mathcal{C}_\delta\} \tag{15}$$

We shall assume that a possibly vanishingly small $\delta > 0$ can be chosen so as to make $\mathcal{V}_\delta$ non-empty. Choose next

$$\mathcal{W}_\delta \subset \mathcal{V}_\delta \text{ non-empty, closed and convex.} \tag{16}$$

Hereafter, we shall restrict our attention to constant command sequences

$$g(k) = w, \forall \; k \in \mathbb{Z}_+, \; w \in \mathcal{W}_\delta \tag{17}$$

The idea here, inspired by receding horizon control, is to determine at each time t a command vector w which, if applied for all subsequent time instants, would produce no future constraint violation for the related predicted evolution of the system. This command is then applied and the procedure is repeated at the next time instant. Of course, if more commands w have the stated property, a selection rule has to be adopted.

To this end, consider the following quadratic selection index

$$J(r(t), g(t-1), w) := \|w - r(t)\|_\Psi^2 + \|w - g(t-1)\|_\Gamma^2 \tag{18}$$

where $\|w\|_\Psi^2 := w'\Psi w$, $\|w\|_\Gamma^2 := w'\Gamma w$, $\Psi = \Psi' > 0$ and $\Gamma = \Gamma' > 0$. The weights Ψ and Γ are free design knobs. For instance, in Example 2 of [5] Ψ was chosen in order to penalize the component of w orthogonal to $r(t)$. Notice that strictly positive definite Γ's are required for ensuring that $g(t) - g(t-1) \to 0_p$ as $t \to \infty$ for any constant r. This, together to Proposition 1, ensures the required asymptotic properties. Further, let $\mathcal{V}(x)$ be, for a given x, the set of all $w \in \mathcal{W}_\delta$ such that (x, w) is executable

$$\mathcal{V}(x) := \{w \in \mathcal{W}_\delta : \; c(k, x, w) \subset \mathcal{C}, \; \forall k \in \mathbb{Z}_+\} \tag{19}$$

By continuity of $c(k, x, \cdot)$, this set is closed. Assume for a while that $\mathcal{V}(x(t))$ is non-empty for every $t \in \mathbb{Z}_+$. This means that, though not necessarily unique, there exists a minimizer

$$\begin{aligned} w^*(t) &= \arg \min_{w \in \mathcal{W}_\delta} \{J(r(t), g(t-1), w) : \; c(\cdot, x(t), w) \subset \mathcal{C}\} \\ &= \arg \min_{w \in \mathcal{V}(x(t))} J(r(t), g(t-1), w) \end{aligned} \tag{20}$$

Then, we select the actual CG action as follows

$$g(t) = w^*(t) \tag{21}$$

Important issues to be addressed next are: 1) Viability, viz. conditions under which $\mathcal{V}(x(t))$ is non-empty for every $t \in \mathbb{Z}_+$; 2) Stability and offset-free response, viz. state boundedness and zero steady-state tracking error; 3) On-line computability of (20).

2.1. Viability and asymptotic properties

Consider $w \in \mathcal{W}_\delta$ and $w + \tilde{w} \in \mathcal{W}_\delta$. Then, from Property 2 it follows that there are two positive reals δ_x, δ_w such that $\|\tilde{x}\| \leq \delta_x$ and $\|\tilde{w}\| \leq \delta_w$ imply that $[x'(k), w' + \tilde{w}']' \in \mathcal{C}, \forall\, k \in \mathbb{Z}_+$, where $x(k) := x(k, x_w + \tilde{x}, w + \tilde{w})$. Therefore, being $w + \tilde{w} \in \mathcal{W}_\delta$, the trajectory $x(k)$ can be made admissible for every $k \in \mathbb{Z}_+$ by taking $\|\tilde{x}\|$ and $\|\tilde{w}\|$ both nonzero, though possibly small. In fact, it suffices that

$$M(\|\tilde{x} + \tilde{w}\|) \leq \delta$$

where $M(\cdot)$ is the function of class $\mathcal{K}$ in (11). The conclusion is that starting sufficiently close to an equilibrium state x_w, $w \in \mathcal{W}_\delta$, by using a feasible constant command, one can arrive in a finite time as close as desired to any state $x_{w+\tilde{w}}$, $w + \tilde{w} \in \mathcal{W}_\delta$, at a nonzero, though possibly small, distance from x_w. Then, we can move out from any state $x(0)$, such that $(x(0), \bar{w})$ is executable for some $\bar{w} \in \mathcal{W}_\delta$, to reach asymptotically x_w, any $w \in \mathcal{W}_\delta$, by concatenating a finite number of feasible constant commands by switching from one to another, the last switching taking place at a finite, though possibly large, time.

Proposition 2.3. (Viability property) – *Consider the nonlinear dynamic system (1). Let (3) and (4) hold true and Properties 1 and 2 be satisfied. Let the initial state $x(0)$ be such that $(x(0), \bar{w})$ is executable for some $\bar{w} \in \mathcal{W}_\delta$. Then, there exists a concatenation of a finite number of feasible constant command sequences $g(\cdot) \equiv w_i$, $w_i \in \mathcal{W}_\delta$, with finite switching times, capable of asymptotically driving the system state from $x(0)$ to x_w, any $w \in \mathcal{W}_\delta$.*

Proposition 2.3 ensures that $\mathcal{V}(x)$ non-empty implies that $\mathcal{V}(\bar{x})$ is non-empty provided that (x, w) is executable and $\bar{x} = f(x, w)$. As a consequence,

$$\mathcal{V}(x(0)) \text{ non-empty} \tag{22}$$

is a sufficient condition for ensuring that $\mathcal{V}(x(t))$, $\forall t \in \mathbb{Z}_+$, is non-empty along the executable trajectories of the system. Consequently, no danger occurs of being trapped into a blind alley if (1) is driven by a feasible constant command or its input switched from one to another feasible constant command.

Consider now the question whether the CG yields an offset-free control system. Assume that the reference becomes constant, $r(t) \equiv r$ for all $t \geq t^*$, $\mathcal{V}(x(t))$ is non-empty at each $t \in \mathbb{Z}_+$, and let $V(t) := \|g(t) - r\|^2_\Psi$, $\forall t \geq t^* \in \mathbb{Z}_+$. Then, it follows that

$$\begin{aligned} V(\infty) \quad &:= \quad \lim_{\tau \to \infty} V(\tau) \leq V(t+1) \leq \\ &\leq \quad \|g(t+1) - r\|^2_\Psi + \|g(t+1) - g(t)\|^2_\Gamma \leq V(t) \qquad \forall t \geq t^* \end{aligned} \tag{23}$$

In (23), the last equality holds true because the pair $(x(t+1), g(\cdot) \equiv w^*(t))$ is executable at time $t+1$ but $w^*(t)$ is not necessarily optimal for (20) and it results $J(r, w^*(t), w^*(t) = V(t)$. As a consequence, the CG output exhibits asymptotically vanishing variations in that $\lim_{t\to\infty}[g(t) - g(t-1)] = 0_p$. Further, because of Property 1, it follows that

$$\lim_{t \to \infty} [x(t) - x_{g(t)}] = 0_n \tag{24}$$

where $x_{g(t)}$ is the solution of the algebraic equation $x_{g(t)} = f(x_{g(t)}, g(t))$. We are now ready to prove that the output of the system controlled by the CG converges to the best possible approximation to the reference.

Proposition 2.4. *Consider the nonlinear dynamic system (1) controlled by the CG (20) and (21). Let $r(t) \equiv r$, $\forall t \geq t^* \in \mathbb{Z}_+$. Then, under Property* 1 *and the same conditions as in Proposition 2.3, the prescribed constraints are satisfied at every $t \in \mathbb{Z}_+$, and the CG output asymptotically converges to the vector w_r in $\mathcal{W}_\delta$ at the minimum distance from r*

$$\lim_{t \to \infty} g(t) = w_r := \arg \min_{w \in \mathcal{W}_\delta} \|w - r\|_\Psi \tag{25}$$

Further,

$$\lim_{t \to \infty} y(t) = w_r \tag{26}$$

provided that the following offset-free condition holds for the primal system (1)

$$y_w := \eta(x_w) = w, \qquad \forall\, w \in \mathcal{W}_\delta \tag{27}$$

Proof - Let $V(\infty)$ be as defined before (23), and $V_r := \|w_r - r\|^2_\Psi$. By contradiction, assume that there is an $\delta > 0$, such that $V(\infty) = V_r + \delta$. Taking into account (24) and the viability property, there exists $\bar{t} \in \mathbb{Z}_+$ and $\beta \in (0,1)$ such that,

$$\|w(t) - w(t-1)\| < \frac{\underline{\lambda}(\Psi)\underline{\lambda}(\Psi+\Gamma)}{\bar{\lambda}(\Gamma)\bar{\lambda}(\Psi+\Gamma)}\beta\delta,\ \forall t \geq \bar{t} \tag{28}$$

where $\underline{\lambda}(\Psi)$ $\left(\bar{\lambda}(\Psi)\right)$ denotes the minimum (maximum) eigenvalue of $\Psi = \Psi' > 0$. Further, the pair $\left(x(t), w_\alpha(t)\right)$ is executable $\forall \alpha \in (0, \bar{\alpha}]$, for some $\bar{\alpha} > 0$, where

$$w_\alpha(t) := (1-\alpha)w(t) + \alpha w_r \tag{29}$$

It is easy to check that, subject to (28) and (29)

$$\|w_\alpha(t) - r\|^2_\Psi + \|w_\alpha(t) - w(t-1)\|^2_\Gamma < \|w(t) - r\|^2_\Psi + \|w(t) - w(t-1)\|^2_\Gamma \tag{30}$$

provided that

$$\alpha < 2\left[(1-\beta)\frac{\underline{\lambda}(\Psi)}{\bar{\lambda}(\Psi+\Gamma)} + \frac{(w_r - w)'\Psi(r - w_r)}{\|w - w_r\|^2_{\Psi+\Gamma}}\right] \tag{31}$$

Note that the right-most additive term in (31) is nonnegative by convexity of $\mathcal{W}_\delta$ and definition of w_r. The conclusion is that (30) contradicts optimality of $w(t)$.

2.2. Computability

Eq. (20) embodies an infinite number of constraints. For practical implementation, we must find out if and how these constraints can be reduced to a finite number of constraints at prefixed time locations. To this end, it is convenient to introduce some extra notation. We express the response of (1) from the event $(0, x)$ to the constant command sequence $g(\cdot) \equiv w$ as follows

$$\begin{cases} c(k+1) &= F\left(c(k)\right), \text{ with } c(0) = \begin{bmatrix} x \\ w \end{bmatrix} \in \mathcal{C} \bigcap \{\mathbb{R}^n \times \mathcal{W}_\delta,\} \\ c(k) &:= c(k, x, w) \end{cases} \tag{32}$$

with obvious definition of $F(\cdot)$.
For $i \in \mathbb{Z}_1 := \{1, 2, 3, \dots\}$, consider the following sets

$$\mathcal{Z}_i := \{c \in \mathbb{R}^n \times \mathcal{W}_\delta : F^{k-1}(c) \in \mathcal{C},\ k \in \underline{i} := \{1, 2, \dots, i\}\}. \tag{33}$$

$$\mathcal{Z} := \bigcap_{i=1}^{\infty} \mathcal{Z}_i \tag{34}$$

where $F^k := \underbrace{F \circ F \circ \cdots F}_{k \text{ times}}$ denotes the k-fold composition of F, and $\mathcal{Z}_i$ is the set of initial vectors c which give rise for $w \in \mathcal{W}_\delta$ to evolutions fulfilling the constraints over the first i-th time steps $k = 0, 1, \dots, i-1$, while $\mathcal{Z}$ is the set of all executable pairs (x, w). Notice that $\mathcal{Z}_{i+1} \subset \mathcal{Z}_i$, $\forall i \in \mathbb{Z}_1$. Further, by continuity of F^{k-1}, $\forall k \in \mathbb{Z}_1$, all $\mathcal{Z}_i$'s, and hence $\mathcal{Z}$, are closed. Moreover, by the viability property of Proposition 2, $\mathcal{Z}$ is non-empty, and we have the following result which can be easily proved by induction [5]

$$\mathcal{Z}_i = \mathcal{Z}_{i+1} \Longrightarrow \mathcal{Z}_i = \mathcal{Z} \tag{35}$$

Proposition 2.5. *Let (16) be fulfilled, and Property 1 hold uniformly in (x, w). Then, there exists an integer $i_o \in \mathbb{Z}_1$ such that $\mathcal{Z}_{i_o} = \mathcal{Z}$.*

Proof - To show that $\mathcal{Z}$ is finitely determinable, note that $\lim_{i\to\infty} x(i, x, w) = x_w$. Then, by uniform convergence for every executable pair (x, w) $\forall \epsilon > 0$, $\exists i_o > 0$ such that $\forall i > i_o$, $||x(i, x, w) - x_w|| < \epsilon$. Then, if δ is as in (14), there is an integer $i_o \in \mathbb{Z}_1$, such that

$$i \geq i_o \Longrightarrow ||x(i, x, w) - x_w|| \leq \epsilon, \quad \forall c \in \mathcal{Z}$$

It follows that $\mathcal{Z}$ is finitely determinable, viz. it suffices to check the constraints over the initial i_o time steps in order to ensure constraint fulfillment over $\mathbb{Z}_+$. Consequently, problem (20) is equivalent to the following finite dimensional constrained optimization problem at each $t \in \mathbb{Z}_+$:

$$\begin{array}{rcl} w(t) & := & \arg\min_{w\in\mathcal{W}_\delta} J(r(t), g(t-1), w) \\ & & \text{subject to } c(i-1, x(t), w)) \in \mathcal{C},\ i \in \underline{i}_o \end{array} \tag{36}$$

Assume now

$$\begin{array}{rcl} \mathcal{C} & = & \{c \in \mathbb{R}^{n_c} : q_j(c) \leq 0, j \in \underline{n}_q\} \\ & & q_j : \mathbb{R}^{n_c} \to \mathbb{R} \text{ continuous} \end{array} \tag{37}$$

Then, the Gilbert and Tan (1991) algorithm can be adapted to the present case to find $i_o = \min_{i\geq 1}\{i \mid \mathcal{Z}_i = \mathcal{Z}\}$. To this end, let $\mathcal{C}^\delta := \{(x, w) \in \mathcal{C},\ w \in \mathcal{W}_\delta\}$. $\mathcal{C}^\delta$ is compact and the following optimization problem is well posed.

$$\begin{array}{rcl} G_i(j) & := & \max_{(x,w)\in\mathcal{C}_\delta}\{q_j(c(i, x, w))\},\ j \in \underline{n}_q, i = 1, 2, \ldots \\ & & \text{subject to } q_j(c(k-1, x, w)) \leq 0,\ j \in \underline{n}_q,\ k \in \underline{i} \end{array} \tag{38}$$

Then, the constraint horizon i_o can be computed off-line via the following algorithm:

$$\left.\begin{array}{ll} 1. & i \leftarrow 1; \\ 2. & \text{Solve } G_i(j),\ \forall j \in \underline{n}_q; \\ 3. & \text{If } G_i(j) \leq 0,\ \forall j \in \underline{n}_q, \text{ let } i_o = i \text{ and stop;} \\ 4. & \text{Otherwise } i \leftarrow i+1, \text{ and go to 2.} \end{array}\right\} \tag{39}$$

In conclusion, we have found that our initial optimization problem having an infinite number of constraints is equivalent to a constrained optimization problem with a finite number of constraints. Notice, however, that the assumption that $\mathcal{C}$ is convex does not ensure in general that every $G_i(j)$ is a convex optimization problem. Therefore, the determination of the control horizon i_o can be a difficult task also in simple cases.

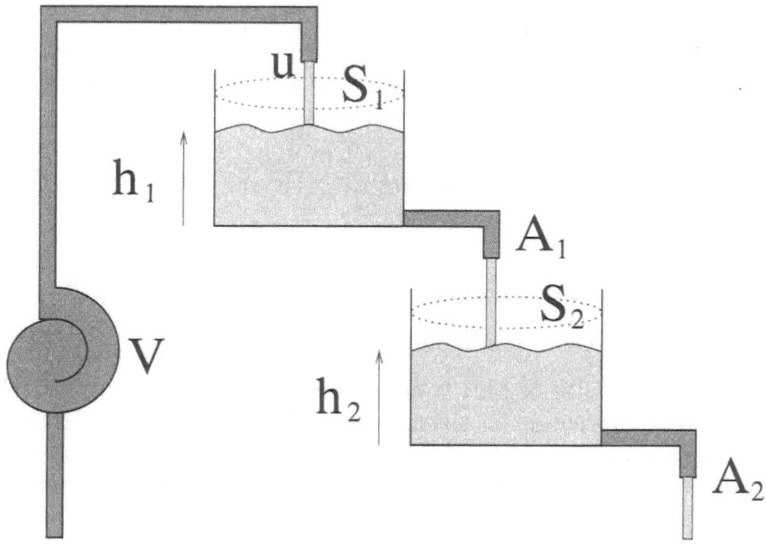

FIGURE 1. Plant arrangement

Example 1 - Consider the plant shown in Fig. 1, which is described by the following nonlinear differential equations

$$\begin{aligned} \rho S_1 \dot{h}_1 &= -\rho A_1 \sqrt{2gh_1} + u \\ \rho S_2 \dot{h}_2 &= \rho A_1 \sqrt{2gh_1} - \rho A_2 \sqrt{2gh_2} \end{aligned} \qquad (40)$$

where u is the water flow supplied by the pump whose command is the voltage V, S_i, $i = 1, 2$, are the tank sections, h_i the water levels in the tanks, A_i the sections of the pipes connecting the tanks, and g and ρ the gravity constant and, respectively, the water density. The problem is to control the water level $h_2(t)$ by acting on the incoming water flow by $u(t)$.

A simple static equation is used to model the relationship between the input voltage $V(t)$ and the incoming mass of water

$$u(t) = \begin{cases} V(t) & \text{if } V(t) \geq 0 \\ 0 & \text{if } V(t) < 0 \end{cases} \qquad (41)$$

In order to design a globally asymptotically stable and offset-free closed-loop system, consider that in steady state one has:

$$h_{1\infty} = \left(\frac{A_2}{A_1}\right)^2 h_{2\infty}, \qquad u_\infty = \rho A_2 \sqrt{2gh_{2\infty}}, \quad M_\infty = \rho S_1 h_{1\infty} + \rho S_2 h_{2\infty} \quad (42)$$

where $M(t) = \rho S_1 h_1(t) + \rho S_2 h_2(t)$ is the total mass of water in the two tanks. Then, Lyapunov analysis shows that the nonlinear control

$$u(t) = \sigma\left(-\lambda[M(t) - M_\infty] + u_\infty\right), \ h_{2\infty} = r(t), \qquad (43)$$

Tank 1	Value	Tank 2	Value	Parameters	Value
S_1	0.25 m^2	S_2	0.16 m^2	g	9.80 m/(s^2)
A_1	$9 \cdot 10^{-4}$ m^2	A_2	$4 \cdot 10^{-4}$ m^2	ρ	1000 Kg/(m^3)
$\bar{h}_1$	0.5 m	$\bar{h}_2$	1.2 m	λ	0.3
$\underline{h}_1$	0.01 m	$\underline{h}_2$	0.1 m	T_c	0.5 s

TABLE 1. Parameter values of the plant

globally asymptotically stabilizes the system and ensures offset free responses. In (43), $\lambda > 0$ is a free scalar parameter, and $\sigma(r) := \begin{cases} r \text{ if } r \geq 0 \\ 0 \text{ if } r < 0 \end{cases}$. Notice that global asymptotic stability of (40) is not sufficient for ensuring global ISS to the closed-loop system. Moreover, for the actual problem it seems hard to find an ISS-Lyapunov function which satisfies (13). However, such an ISS property actually holds for the closed-loop system (40)–(43) as it can be indirectly proved by using the arguments reported in the Remark 2 of Sect. 3 of [3].

Here, the following constraints must be enforced.

$$\begin{array}{rcl} \underline{h}_1 \leq & h_1 & \leq \bar{h}_1 \\ \underline{h}_2 \leq & h_2 & \leq \bar{h}_2 \\ 0 \leq & V & \leq V_{\max} \end{array} \tag{44}$$

where $V_{\max} = 10$ Volt and all other quantities are shown in Tab. 1.

Fig. 2 reports the responses of the variables to be constrained reference step change $\begin{cases} 100 & 0 \leq t < 750 \\ 20 & 750 \leq t \leq 1500 \end{cases}$ (dash line) from the initial state $(h_1(0), h_2(0)) = (5, 20)$ without the use of CG, where the dot lines delimit the admissible ranges. Notice that the prescribed constraints on h_1 and V are violated.

We can keep unaltered the controller structure provided that we include a command governor between the feedforward gain and the closed-loop system as shown in Fig. 3 where $\gamma(\cdot)$ is the nonlinear transformation

$$\gamma(r) = \lambda \left[\rho S_1 \left(\frac{A_2}{A_1} \right)^2 + \rho S_2 \right] r + \rho A_2 \sqrt{2gr} \tag{45}$$

The plant model is converted in discrete-time by substituting $\dot{h}$ with the increment $[h(t+1) - h(t)]/T_c$ for a short sampling interval T_c. We repeat the previous experiment with the use of the nonlinear CG. It is found, that all constant references in the range $10 \leq r \leq 120$ are admissible in steady-state. We set $\delta = 5$, that is $\mathcal{W}_\delta = \{w \in \mathbb{R} \,:\, 15 \leq w \leq 115\}$, and we determined that $i_o = 120$ was a sufficient control horizon for the prescribed reference signals. In particular, such a value was achieved by solving problems $G_i(j)$ in (39) only along the state trajectory corresponding to the prescribed reference signal. Notice that this approach is effective only when the reference is pre-calculated. On the contrary, the global maximum of all $G_i(j)$ in (39) needs to be determined when the reference is generated on-line.

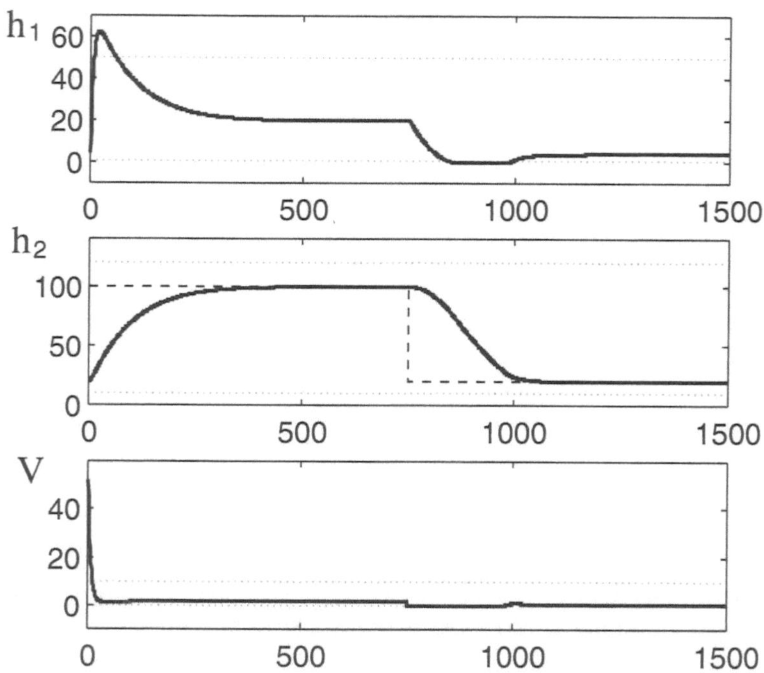

FIGURE 2. Unconstrained state and input evolution to a set-point change without CG: (thick) - c-signal; (dash) - reference.

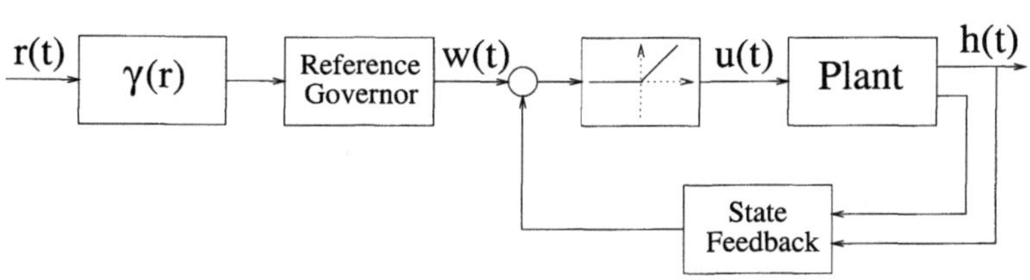

FIGURE 3. Controlled system with CG

Further, $\Psi = \Gamma = 1$ were selected and, because the minimization problem is scalar, a bisection algorithm was used to get the minimum at an average computational burden of 10.373 flops for step.

In Fig. 4, the use of the CG on the same reference step change is depicted. It can be seen that the filtering action of the governor is quite evident both in the upward and downward fronts of the reference signal due to the presence of active constraints.

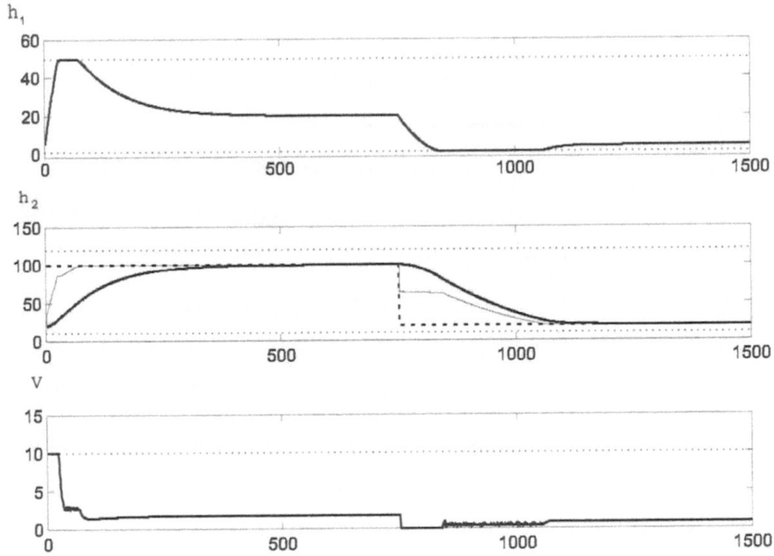

FIGURE 4. Evolutions of the system with the nonlinear CG: (thick) - c-signal (thin) - CG action (dash) - reference

3. Conclusions

The CG problem, viz. the one of on-line selecting a command input in such a way that a primal compensated nonlinear control system can operate with a stable set-point tracking behaviour and no constraint violation, has been addressed by exploiting some ideas originating from predictive control. In this connection, the concept of a "virtual" command sequence is instrumental to synthesize CG's having the stated properties.

It has been shown how to use off-line an iterative algorithm so as to restrict to a fixed finite integer the infinite number of time-instants over which the prescribed constraints must be checked in order to test admissibility of command sequences. An analysis based on a Lyapunov function argument shows that, if the reference becomes constant, the CG output converges to the closest admissible approximation to the reference.

References

[1] V. Andriano, A. Bacciotti and G. Beccari, "Global Stability and External Stability of Dynamical Systems", *Nonlinear Analysis., Theory, Methods and Appl.*, Vol. 28, pp. 1167–1185, 1997.

[2] D. Angeli and E. Mosca, "Command Governor for Nonlinear Systems under Constraints", *IEEE Trans. Automat. Control*, to be published.

[3] D. Angeli, A. Casavola and E. Mosca, "Command Governors for Constrained Nonlinear Systems: Direct Nonlinear vs. Linearization-based Strategies", submitted for publication.

[4] A. Bemporad, "Reference Governor for Constrained Nonlinear Systems", *IEEE Trans. Automat. Control*, Vol. 43, n. 3, pp. 415–419, March 1998.

[5] A. Bemporad, A. Casavola and E. Mosca, "Nonlinear Control of Constrained Linear Systems via Predictive Reference Management," *IEEE Trans. Automat. Control*, Vol. 42, pp. 340–349, 1997.

[6] A. Casavola and E. Mosca, "Reference Governor for Constrained Uncertain Linear Systems Subject to Bounded Input Disturbances", *Proceedings of the 35th IEEE CDC*, pp. 3531–3536, Kobe, Japan, 1996.

[7] E.G. Gilbert, I. Kolmanovsky and K. Tin Tan, "Discrete-time Reference Governors and the Nonlinear Control of Systems with State and Control Constraints", *Proceedings of the 33rd IEEE CDC*, pp. 144–194, Lake Buena Vista, FL., U.S.A., 1994.

[8] P. Kapasouris, M. Athans and G. Stein, "Design of feedback control systems for unstable plants with saturating actuators," *Proc. IFAC Symp. on Nonlinear Control System Design*, pp. 302–307, Pergamon Press, 1990.

[9] M.V. Kothare, V.Nevistić and M.Morari, "Robust Constrained Model Predictive control for Nonlinear Systems: A Comparative Study", in *Proceedings of the 34th IEEE CDC*, New Orleans, LO, Dec. 1995.

[10] E. Mosca, "Nonlinear Predictive Command Governor for Constrained Tracking", in C. Bonivento, G. Marro and R. Zanasi (Eds.), *Colloquium on Automatica Control*, Springer Verlang, pp. 55–76, 1996

[11] E. Sontag and Y. Wang, "On Characterizations of the Input-to-State Stability Property" , *Systems and Control Letters*, Vol. 24 , pp. 351–359, 1995.

4. Appendix: proof of Proposition 2.1

Observe first that $x_{g(t)}$ is bounded because $g(t)$ is such and the map $w \mapsto x_w$ can be easily shown to be continuous under the ISS assumption. Assume, forall $t \in \mathbb{Z}_+$, that $\|g(t)\| \leq G_1$ and $\|x_{g(t)}\| \leq G_2$. As a consequence, $\|g(t_1) - g(t_2)\| \leq 2G_1$, $\forall t_1, t_2 \in \mathbb{Z}_+$. Then, the assumption that (1) is ISS along all admissible state evolutions corresponding to bounded command sequences implies that

$$\begin{aligned} &\|x(t, x(0), g(\cdot)) - x_{g(t)}\| \\ &\leq \beta_{g(t)}(\|x(0) - x_{g(t)}\|, t) + \gamma_{g(t)}\left(\sum_{i=0}^{t-1} e^{-(t-i)}\lambda_{g(t)}(\|g(i) - g(t)\|)\right) \end{aligned} \tag{46}$$

In particular, define

$$\tilde{\gamma}(r) = \max_{\|w\| \leq G_1} \gamma_w(r), \; \tilde{\lambda}(r) = \max_{\|w\| \leq G_1} \lambda_w(r), \; \tilde{\beta}(r,t) = \max_{\|w\| \leq G_1} \beta_w(r,t) \tag{47}$$

Then, it is easy to see that both $\tilde{\gamma}(\cdot)$, $\tilde{\lambda}(\cdot) \in \mathcal{K}$ and $\tilde{\beta}(\cdot,\cdot) \in \mathcal{KL}$. Moreover, w.l.o.g. we can assume that $\tilde{\gamma}(\cdot)$ and $\tilde{\lambda}(\cdot)$ are concave and, respectively, convex functions (Praly and Wang, 1995). Then, by direct manipulations, it results that

$$(46) \;\leq\; \tilde{\beta}(\|x(0)\| + G_2, t) + \tilde{\gamma}\left(\sum_{i=0}^{t-1} e^{-(t-i)} \sum_{j=i}^{t-1} \|g(j+1) - g(j)\| \; \tilde{\lambda}(2G_1)/G_1\right)$$

Further, because $\beta(\|x(0)\|+G_2, t)$ tends to 0 as $t \to \infty$, in order to show that Prop. 1 holds true we only need to proof that $\sum_{i=0}^{t-1} e^{-(t-i)} \sum_{j=i}^{t-1} \|g(j+1) - g(j)\| \to 0$ too. To this end, observe that

$$\begin{aligned}\sum_{i=0}^{t-1} e^{-(t-i)} \sum_{j=i}^{t-1} \|g(j+1) - g(j)\| &= \sum_{j=0}^{t-1} e^{-t} \|g(j+1) - g(j)\| \frac{e^{j+1} - 1}{e - 1} \\ &\leq \sum_{j=0}^{t-1} e^{-(t-j)} \|g(j+1) - g(j)\| \frac{e}{e-1}\end{aligned}$$

Then, because the last inequality can be interpreted as the convolution between the impulsive response of an asymptotically stable linear system and the signal $\|g(t+1) - g(t)\|$, standard linear system theory arguments are enough to show that $\|g(t+1) - g(t)\| \to 0$ as $t \to \infty$ implies that $\sum_{j=0}^{t-1} e^{-(t-j)} \|g(j+1) - g(j)\| \to 0$ as well. Hence, Property 1 follows. Property 2 follows from ISS by selecting $\tilde{\beta}(\cdot, 0) + \tilde{\gamma}(\cdot) \circ \tilde{\lambda}(\cdot)$ as function the $M(\cdot)$ function in (11).

Dip. Sistemi e Informatica,
University of Florence,
3, Via di S. Marta,
50139 Firenze, Italia
E-mail address: angeli,casavola,mosca@dsi.dsi.unifi.it

Progress in Systems and Control Theory, Vol. 26

Some Practical Issues and Possible Solutions for Nonlinear Model Predictive Control

Alex Zheng

Abstract. Several practical issues (e.g., excessive on-line computational time) associated with a conventional Nonlinear Model Predictive Control algorithm are illustrated through simulation studies on two chemical processes – a binary distillation column and the Tennessee Eastman Challenge Process. We show why it is generally not a good idea to use a small control horizon to reduce the on-line computational time. The effectiveness of the Nonlinear Model Predictive Control algorithm recently proposed by Zheng in resolving these issues is illustrated.

1. Introduction

Many Nonlinear Model Predictive Control (NMPC) algorithms have been proposed for control of nonlinear systems with constraints [6, 7, 16, 5, 10, 21, 13, etc.]. Several excellent reviews have been published (see, for example, [1, 11, 17]). Strong and non-conservative nominal stability results have been proven [8, 13, 2, etc.]. In particular, the quasi-infinite horizon NMPC algorithm proposed by Chen and Allgöwer [2] allows one to approximate the infinite horizon, which is difficult to implement for NMPC, by a finite horizon and appropriate terminal weight.

Despite of the flurry of research activity on NMPC, practical implementations of NMPC algorithms on large nonlinear systems have not been as forth coming. Most of the implementations are based on successive linearization with a small control horizon (typically one) on systems with a small number of input and output (typically one). The purposes of this paper are to compare a conventional NMPC algorithm which solves the on-line optimization problem exactly and an NMPC algorithm that has been proposed by Zheng [22] through simulations on several large chemical processes. The major difference between the two algorithms is that Zheng's algorithm computes the first control move, which is implemented, exactly and approximates the rest of control moves, which are not implemented, while a conventional NMPC algorithm computes all of the control moves exactly. Thus, the resulting on-line optimization for Zheng's algorithm has only n_u decision variables, where n_u is the number of inputs, regardless of the control horizon (H_c), versus n_uH_c decision variables for the conventional NMPC algorithm. This reduction in the number of decision variables is significant since the on-line optimization problem is generally non-convex and H_c should be chosen reasonably large for reasons of stability and performance.

Several practical chemical process examples, including the Tennessee Eastman Challenge process, are used for this comparison. We show why it is generally not a good idea to use a small control horizon for a conventional NMPC algorithm and why a conventional NMPC algorithm may result in a steady-state offset. The difference between finding a local solution and a "global" solution is also discussed.

The paper is organized as follows: Section 2 describes the basic ideas behind a conventional NMPC algorithm and the NMPC algorithm proposed by Zheng. Sections 3 and 4 compare the two algorithms on a binary distillation column and the Tennessee Eastman Challenge process, respectively. We show through an example that it is possible for a conventional NMPC algorithm to result in a steady-state offset in Section 5. While model uncertainty has not been considered in this paper, Section 6 briefly discusses the difference in robustness properties of the two algorithms. Section 7 concludes the paper.

2. NMPC algorithms

We consider a continuous-time nonlinear system described below.

$$\left\{\begin{array}{ll} \dot{x}(t) & = f(x(t), u(t), d(t)) \\ y(t) & = g(x(t), d(t)) \end{array}\right. \tag{1}$$

where x of dimension n_x is the state, u of dimension n_u is the input, y of dimension n_y is the output, and d of dimension n_d is the disturbance.

The objective function at sampling time k is defined as follows:

$$\Phi_k = \int_0^{H_p} ||r(k+t) - y(k+t|k)||^2_{\Gamma_y} dt + \sum_{i=0}^{H_c-1} ||\Delta u(k+i|k)||^2_{\Gamma_u} \tag{2}$$

where H_p is the output horizon, H_c is the input (or control) horizon, k is the kth sampling time (the sampling period is assumed to be one for simplicity), r is the desired output trajectory, $\Delta u(k) = u(k) - u(k-1)$, and $\Gamma_y > 0$ and $\Gamma_u > 0$ are diagonal weights. $||x||_P = \sqrt{x^T P x}$. Notice that the inputs are discrete (i.e., $u(t)$ is constant for $t \in [k, k+1)$) while the outputs are continuous. Δu is penalized instead of u to generate integral control, at least when the system is *linear* (see discussion in Section 5). A conventional NMPC algorithm, which directly extends the DMC algorithm [3] for linear systems, is defined as follows.

Definition 2.1 (Controller NMPC #1). *At each sampling time k, $\Delta u(k)$ equals the first control move $\Delta u^{\text{opt}}(k|k)$ of the sequence $\{\Delta u^{\text{opt}}(k|k), \ldots, \Delta u^{\text{opt}}(k+H_c-1|k)\}$*

which is the solution to the following optimization problem.

$$\min_{\Delta u(k|k),\dots,\Delta u(k+H_c-1|k)} \quad \Phi_k \qquad \text{subject to}$$

$$\begin{cases} \dot{\hat{x}}(k+t|k) = f(\hat{x}(k+t|k), u(k+t|k), \hat{d}(k+t|k)) \\ y(k+t|k) = g(\hat{x}(k+t|k), \hat{d}(k+t|k)), t \geq 0 \\ u(k+i|k) \in \mathcal{U}, i = 0, \dots, H_c - 1 \\ \Delta u(k+i|k) = 0, i \geq H_c \end{cases} \tag{3}$$

where $\bullet(k+i|k)$ *denotes the variable at time* $k+i$ *predicted at time* k, $\hat{\bullet}(k)$ *denotes the estimated variable at time* k, *and the objective function* Φ_k *is defined by (2).* $\mathcal{U} = \{u : u^{\min} \leq u \leq u^{\max}\}$.

Under certain assumptions, it is straightforward to prove that asymptotic stability of the closed-loop system can be guaranteed if the optimization problem is feasible. The optimization problem (3) has $n_u H_c$ decision variables. In general, for reasons of stability and performance, H_c should be chosen reasonably large [18, 25]. Since the optimization problem (3) is generally non-convex, the on-line computational demand grows exponentially with the number of decision variables. Thus, practical implementation of `Controller NMPC #1` becomes difficult, if not impossible, for any reasonably large nonlinear system.

Motivated by this difficulty, Zheng [23] recently proposed a NMPC algorithm which combines the best features of NMPC algorithms which attempt to solve the problem exactly and NMPC algorithms which reduce the optimization problem to a quadratic program via linearization. The basic idea is to compute the first control move $\Delta u(k|k)$, which is implemented, exactly, while approximating the rest of control moves, which are implemented, via linearization. Furthermore, in contrast to most of the NMPC algorithms which use an open-loop control strategy, a closed-loop control strategy (i.e., future control moves depend on future predicted output) is incorporated into the algorithm. The algorithm is stated below.

Definition 2.2 (Controller NMPC #2). *At each sampling time* k, $\Delta u(k)$ *is the solution to the following optimization problem.*

$$\min_{\Delta u(k|k)} \quad \Phi_k \qquad \text{subject to}$$

$$\begin{cases} \dot{\hat{x}}(k+t|k) = f(\hat{x}(k+t|k), u(k+t|k), \hat{d}(k+t|k)) \\ y(k+t|k) = g(\hat{x}(k+t|k), \hat{d}(k+t|k)) \\ u(k|k) \in \mathcal{U} \\ \mathbf{\Delta u(k+i|k) = \text{sat}(C(x(k+i|k), e(k+i|k))), i = 1, \dots, H_p - 1} \end{cases} \tag{4}$$

where sat *is the saturation function and* $e(k+i|k) = r(k+i) - y(k+i|k)$. *For simplicity,* C *is a linear controller corresponding to the tuning parameters and a linearized model.*

Major differences between the two algorithms are summarized in Table 1 and discussed below. The on-line optimization problem for `Controller NMPC #2` has

	Controller NMPC #1	Controller NMPC #2
Optimization Variables	$n_u H_c$	n_u
Control Horizon	H_c	H_p
Optimization Strategy	Open-loop	Closed-loop

TABLE 1. Major differences between Controllers NMPC #1 and #2.

only n_u decision variables (versus $n_u H_c$ for Controller NMPC #1), *regardless* of the control horizon H_c. This results in significant savings in on-line computational time as the on-line computational time grows exponentially with the number of decision variables in the worst case. Another major difference is that $H_c = H_p$ for Controller NMPC #2. This is because the control horizon has no effect on on-line computational demand for Controller NMPC #2. Also the on-line optimization strategy differs: Controller NMPC #1 uses an open-loop control strategy while Controller NMPC #2 uses a closed-loop control strategy. In the nominal case (i.e., perfect model), there is no difference between the open-loop and closed-loop strategies. In the presence of modeling error, however, there is a significant difference between the two and a closed-loop control strategy is preferred [9, 24, 12]. This difference is further discussed in Section 6.

The rest of this paper focuses on comparing the two algorithms on two chemical processes—a binary distillation column and the Tennessee Eastman (TE) challenge process [4]. We *emphasize* that the sole purpose of these examples is to compare Controllers NMPC #1 and #2; it is *not* to show that nonlinear controllers are needed to control these systems.

3. Distillation column dual composition control

3.1. Descriptions

Skogestad [20] developed a tray-by-tray model for a binary distillation column, which has 41 trays, and implemented the model in MATLAB. The model has 82 states and the pressure is assumed to be constant. The model does not account for energy balances. We use the model both as the nominal model and as the actual plant (i.e., no model uncertainty) for Controllers NMPC #1 and #2. Furthermore, the sump and reflux drum levels are controlled by the bottoms and distillation flows, respectively, with P-only controllers. NMPC algorithms are used to control the distillate and bottoms compositions by manipulating reflux flow and vapor boilup (i.e., the control system has 2 inputs and 2 outputs).

3.2. On-line computations

A sampling period of 5 minutes is chosen. The on-line optimization problem seems to be convex for this example, at least for $H_c = 1$ (Figure 1). The on-line optimization problem is solved using the Powell's method in MATLAB on a Pentium 266

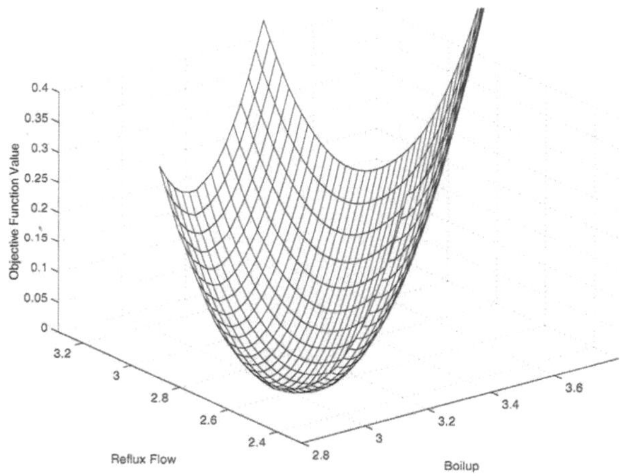

FIGURE 1. The objective function versus decision variables for $H_c = 1$.

mHz PC. The integration routine **Gear** is used. Figure 2 shows the on-line computational time per sampling time for determining one *local* solution versus H_c for the two algorithms based on *actual* simulations for $H_p = 400$ sampling periods (i.e., 2000 minutes). While the absolute values in the figure depend on the initial guess, software, hardware, etc., the qualitative trend does not. Regardless of H_c, the computational time for **Controller NMPC #2** is about 3.5 minutes, which is larger than the computational time for **Controller NMPC #1** with $H_c = 1$ (about 3 minutes) because the inclusion of the controller C increases the time for evaluating the objective function. In general, since the nonlinear system is much more complex than the controller C, simulation time for the nonlinear system with or without C should be very similar. Therefore, including the controller C should have minimal impact on the on-line computational time.

The on-line computational time increases *linearly* with the output horizon H_p, if an integration algorithm with a fixed step size (e.g., Euler) is used; and much more slowly than the linear growth if an integration algorithm with variable-step size (e.g., Gear) is used. In general, one would like to choose H_p to be at least the process settling time plus the control horizon and to be as large as possible so that nominal stability of the closed-loop system may be ensured. In the limiting case of $H_p = \infty$, which is not feasible to implement, nominal stability is guaranteed.

3.3. On the choice of H_c

With the existing software and hardware, it is clearly impossible to implement **Controller NMPC #1** for any $H_c > 1$ (i.e., on-line computational time is larger than the sampling time). One may be tempting to use a smaller control horizon

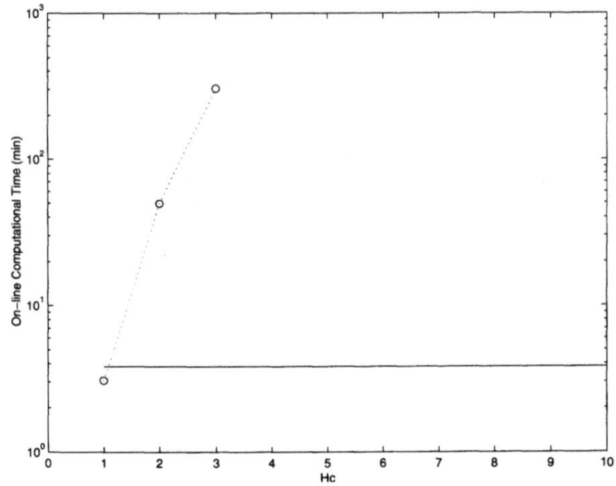

FIGURE 2. Comparison of on-line computational time per sampling time for `Controllers NMPC #1` (dotted) and `#2` (solid).

that is feasible (e.g., $H_c = 1$). While this approach has been popular in the literature, it may *not* be desirable. Figure 3 compares performance for `Controller NMPC #1` with $H_c = 1$ and a linear controller that corresponds to `Controller NMPC #1` with $H_c = H_p = 400$, no constraints, and a linearized model, for a setpoint change in the distillate composition from 0.99 to 0.999. The weights for both controllers are $\Gamma_y = I$ and $\Gamma_u = 0.1I$. `Controller NMPC #1` with $H_c = 1$ gives much *worse* performance than the linear controller, which requires almost no on-line computation and is much simpler to implement. Therefore, simply reducing H_c to make the implementation of `Controller NMPC #1` feasible may not be a good idea.

3.4. Performance comparison

Figure 4 compares performance for `Controller NMPC #2` and `Controller NMPC #1` with $H_c = 1$ for a setpoint change in the distillate composition from 0.99 to 0.999. $H_p = 400$ sampling periods, $\Gamma_y = I$, and $\Gamma_u = 0.1I$. While the on-line computational times for `Controller NMPC #2` and `Controller NMPC #1` with $H_c = 1$ are about the same, the performance for `Controller NMPC #2` is significantly better.

The controller C for `Controller NMPC #2` is linear and is determined analytically off-line based on the following linear model obtained at the nominal operating conditions (i.e., distillation composition: 0.99; bottom composition: 0.01):

$$G(s) = \frac{1}{190s+1}\begin{bmatrix} 0.878 & -0.864 \\ 1.082 & -1.096 \end{bmatrix}$$

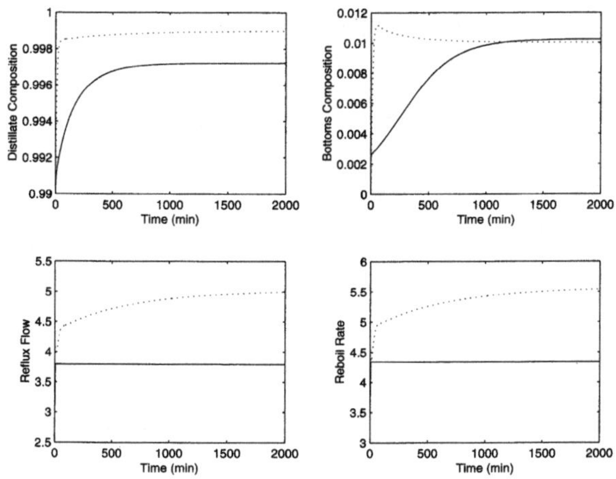

FIGURE 3. Closed-loop performance for **Controller NMPC #1** with $H_c = 1$ (solid) and a linear controller corresponding to linear unconstrained MPC with $H_c = H_p = 400$ (dotted). The setpoints are 0.999 (distillate composition) and 0.01 (bottoms composition).

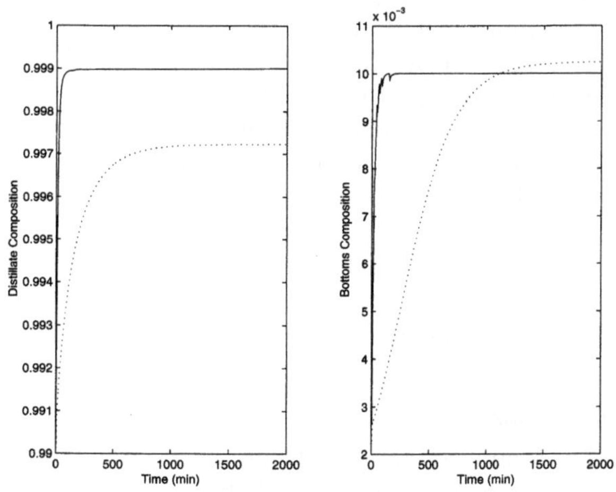

FIGURE 4. Closed-loop performance for **Controller NMPC #2** (solid) and **Controller NMPC #1** with $H_c = 1$ (dotted).

which is discretized with a sampling period of 5 minutes and the following tuning parameters:

$$\Gamma_y = I, \Gamma_u = 0.1I, H_c = 400, H_p = 400$$

To speed up the simulations, a continuous controller is used for C.

4. Tennessee-Eastman challenge process

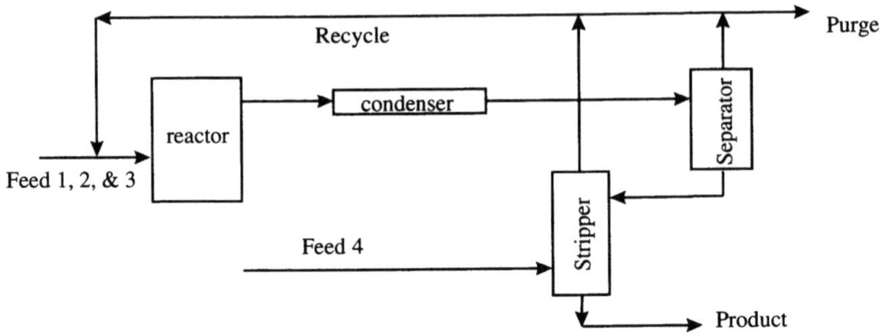

FIGURE 5. Simplified block diagram of the TE problem.

4.1. Descriptions

The TE challenge process was proposed by Downs and Vogel [4] for testing alternative control and optimization strategies for continuous chemical processes (Figure 5). A simplified model implemented in MATLAB, which has 26 states and includes two PI controllers,[1] developed by Ricker [19] is used as both the nominal model and the actual plant (i.e., no model uncertainty).

Ten inputs are selected as manipulated variables and six outputs as controlled variables (Table 2). Three level controllers, with parameter values from [19], are implemented to reduce the size of the on-line optimization problem for `Controller NMPC #1` with large values of H_c. Thus, NMPC algorithms handle a subsystem with seven inputs and three outputs, which are italicized in Table 2. Space limitation prevents us from any further discussion on the TE process; but the interested readers are referred to the papers by Downs and Vogel [4] and Ricker [19] for details.

4.2. On-line computations

A sampling period of 5 minutes is used. Again all the simulations are carried out in MATLAB on a Pentium 266 mHz PC and the on-line optimization problem is solved using the Powell's method. Unlike the distillation column control problem, the solution of the on-line optimization problem depends on the initial guess, implying a *non-convex* optimization problem. Solving for the global solution by searching over the allowable space was not "feasible" for any value of H_c larger than one.[2] We decide to approximate the global solution by the best solution

[1] The two PI controllers control reactor and separator temperatures by manipulating reactor and condenser coolant valves, respectively.

[2] "Feasible" means that a simulation can be finished within a few days on a Pentium 266 mHz PC.

Manipulated Variables	Controlled Variables
Feed 1 (pure A)	*Reactor pressure*
Feed 2 (pure D)	Reactor liquid level
Feed 3 (pure E)	Separator liquid level
Feed 4 (A & C)	Stripper reboiler level
Recycle flow	*H in product*
Purge	*Production rate*
Separator underflow	
Stripper underflow	
Reactor temperature	
Separator temperature	

TABLE 2. Summary of manipulated and controlled variables.

among the local solutions obtained from different initial guesses. The number of initial guesses needed to ensure that *on average* the best local solution is within 5% of the global solution is determined by trial-and-error. It is found to depend on H_c (Figure 6); for example, for $H_c = 1$, three different initial guesses are used while fifteen different initial guesses are needed for $H_c = 5$. Each point in the figure is obtained by at least 5 different simulations with $H_p = 200$ hours. The on-line computational time per sampling period increases significantly with H_c as both the number of decision variables and the number of initial guesses needed increase with H_c. Figure 7 depicts this trend based on actual simulations for $H_p = 200$ hours.

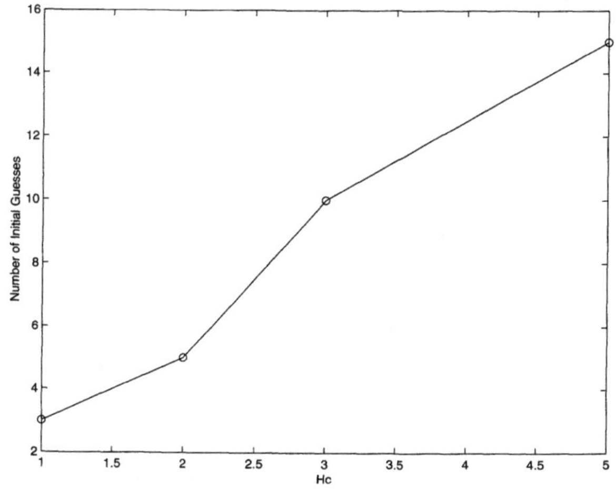

FIGURE 6. Number of initial guesses necessary versus H_c.

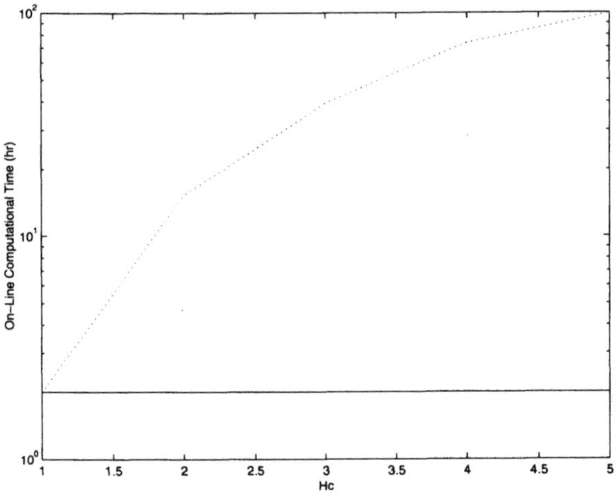

FIGURE 7. On-line computational time per sampling time versus H_c for `Controllers NMPC #1` (dotted) and `#2` (solid).

4.3. Local versus "global" solution

Finding only one local solution can potentially cause many problems. For example, it can result in a steady-state offset. This is shown in Figure 8, where the setpoint is 53.8% H in the product stream. Trying many different initial guesses greatly minimizes the occurrence of this problem: Suppose that the number of different guesses is such that *on average* the solution is within 5% of the global solution, at *each* sampling time. The chance that the solution would not converge to within 5% of the global solution for N sampling time is about 0.05^N. Thus, a steady-state offset is extremely unlikely, if the global solution results in zero steady-state offset.

4.4. Performance comparison

The following tuning parameters are used.

$$\begin{aligned} H_c &= 2 \text{ sampling periods} \\ H_p &= 200 \text{ hours} \\ \Gamma_u &= \lambda \, \text{diag}([90 \;\; 15 \;\; 12 \;\; 4 \;\; 1 \;\; 66 \;\; 100]) \\ \Gamma_y &= \text{diag}([1 \;\; 400 \;\; 20]) \end{aligned}$$

The input and output weights are chosen by scaling all the variables so that their acceptable values are within ± 1 and λ is an adjustable parameter trading off nominal performance and robust stability. The controller C for `Controller NMPC #2` is again linear and obtained for $H_c = 2$. We have chosen this value of H_c for two reasons:

1. Ideally we want to use $H_c = H_p$; however, this value was too big for the MPC Toolbox [15] to handle.

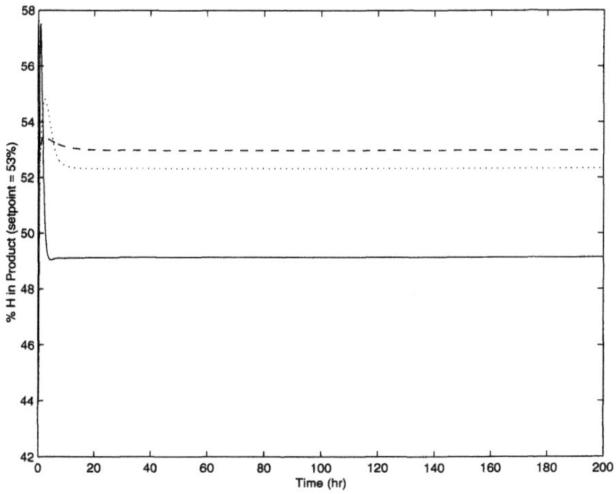

FIGURE 8. Steady-state offset due to local optima.

2. To illustrate the performance loss of `Controller NMPC #2` as compared to `Controller NMPC #1` for the same value of H_c.

The controller is further approximated by a continuous controller to speed up simulations.

Figure 9 shows performance for the two controllers with $\lambda = 40$ for a setpoint change in % H in the product stream from 43.83 to 53.83 while maintaining the production rate at 211.3 kmoles/hr and reactor pressure at 2700 kPa. As one can see, it is difficult to judge which controller performs better, although the on-line computational time for `Controller NMPC #2` is about one tenth of that for `Controller NMPC #1`.

Figure 10 compares performance for the two controllers for 10% decrease in the composition of component B while maintaining feed ratio of component A to C constant for Feed 4. `Controller NMPC #2` performs better than `Controller NMPC #1`. In fact, the closed-loop system with `Controller NMPC #1` has a steady-state offset, albeit a small one. It is entirely possible that the steady-state offset may be due to local minima. However, as we will show in the next section, this undesirable behavior can occur for `Controller NMPC #1` even if a global solution is found.

5. Integral control

The importance of integral control in process control is well recognized [14]. In this section, we show, through a simple example, that it is possible for the closed-loop system with `Controller NMPC #1` to have a steady-state offset. Thus, the steady-state offset observed in the previous section may not be due to local minima (i.e.,

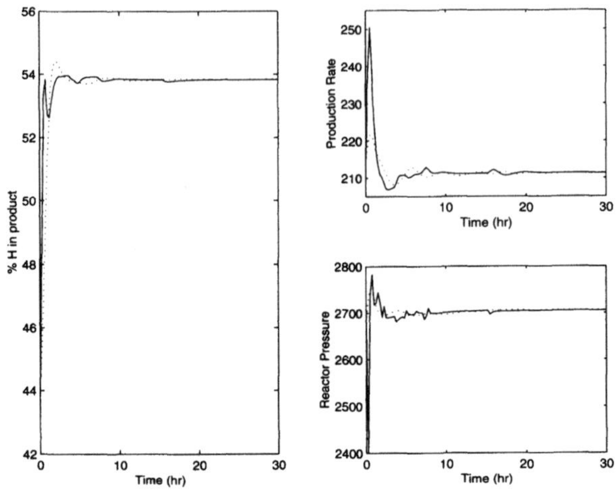

FIGURE 9. Servo responses for `Controller NMPC #1` (solid) and `Controller NMPC #2` (dotted).

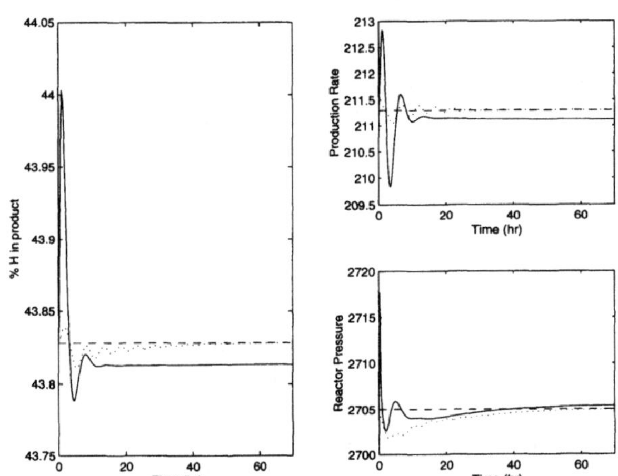

FIGURE 10. Regulatory responses for `Controller NMPC #1` (solid) and `Controller NMPC #2` (dotted).

the offset would have existed even if a global solution had been found). This is not possible for linear systems as `Controller NMPC #1` reduces to linear MPC which guarantees integral control for *almost* all values of tuning parameters. However, this is *not* the case for nonlinear systems, as illustrated by the following example.

Example 5.1. *Consider a simple system,*

$$y(k+1) = u(k)^3, k \geq 0, y(0) = u(0) = 0$$

Without loss of generality, let $H_c = 1$. The objective function becomes

$$\Phi_k = \sum_{i=1}^{H_p} (r - y(k+i|k))^2 + \lambda \Delta u(k)^2$$

where r is the setpoint. At $k = 1$, we have

$$\Phi_1 = H_p \left[(r - \Delta u(1)^3)^2 + \frac{\lambda}{H_p} \Delta u(1)^2 \right]$$

The optimal solution is obtained by solving the following optimization problem.

$$\min_{\Delta u(1)} \left[(r - \Delta u(1)^3)^2 + \bar{\lambda} \Delta u(1)^2 \right]$$

where $\bar{\lambda} = \frac{\lambda}{H_p}$. For any $\bar{\lambda} > 0$, then there exists $\bar{r} > 0$ such that $\Delta u(1) = 0$ is the global solution for $|r| < \bar{r}$, implying a steady-state offset. In fact, $\bar{r} = (\frac{16\bar{\lambda}^3}{27})^{1/4}$. Thus, for this system, offset-free tracking is not guaranteed by `Controller NMPC #1` *if $\lambda > 0$ and H_p is finite.*

6. Robustness

The on-line optimization strategy for the two control algorithms differs: `Controller NMPC #1` uses an open-loop control strategy while `Controller NMPC #2` uses a closed-loop control strategy. In the nominal case (i.e., perfect model), there is no difference between the open-loop and closed-loop strategies. In the presence of modeling error, however, there is a significant difference between the two and a closed-loop control strategy is preferred [9, 24, 12]. To see this difference, let us consider a simple example.

Example 6.1. *Consider the following system*

$$y(k) = au(k-1), 0.8 \leq a \leq 1, y(0) = 0, u(t) = 0 \;\forall\; t < 0$$

The setpoint is 1 and the objective to be minimized at each sampling time is

$$\Phi_k = \max_a \sum_{i=1}^{\infty} (1 - y(k+i))^2$$

At sampling time k, an open-loop optimal control strategy would determine a fixed sequence of controls such that the objective function is minimized. Regardless *of what the sequence is, Φ_k is not bounded. However, a simple closed-loop control strategy of $u(k+i) = u(k+i-1) + b(1 - y(k+i)), i \geq 0$, would result in a bounded Φ_k with the constant b properly chosen. Notice that $u(k+i), i \geq 1$, depends on future output $y(k+j), 0 \leq j \leq i$, but is unknown since $y(k+j)$ is unknown.*

7. Conclusions

We have compared performance for a conventional NMPC algorithm which attempts to solve the on-line optimization problem exactly and the NMPC algorithm proposed by Zheng on a binary distillation column and the TE process. We showed that Zheng's algorithm provides a significant amount of reduction in on-line computational time and that Zheng's algorithm may perform better than the conventional algorithm because of large control horizon (i.e., $H_c = H_p$). We also discussed why the conventional NMPC algorithm may not guarantee integral control for certain systems. Finally, the difference in robustness for the two algorithm was briefly discussed.

References

[1] L. T. Biegler and J. B. Rawlings. Optimization approaches to nonlinear model predictive control. In *Conf. Chemical Process Control (CPC-IV)*, pages 543–571, South Padre Island, Texas, 1991. CACHE-AIChE.

[2] H. Chen and F. Allgöwer. A quasi-infinite horizon nonlinear model predictive control scheme with guaranteed stability. *Automatica*, 1997. In press.

[3] C. R. Cutler and B. L. Ramaker. Dynamic matrix control — A computer control algorithm. In *Joint Automatic Control Conf.*, San Francisco, California, 1980.

[4] J. J. Downs and E. F. Vogel. A plant-wide industrial-process control problem. *Computers and Chemical Engineering*, 17(3):245–255, 1993.

[5] J. W. Eaton and J. B. Rawlings. Feedback control of nonlinear processes using on-line optimization techniques. *Comp. and Chem. Eng.*, 14:469–479, 1990.

[6] C. E. García. Quadratic dynamic matrix control of nonlinear processes. An application to a batch reactor process. In *AIChE Annual Meeting*, San Francisco, CA, 1984.

[7] S. Jang, B. Joseph, and H. Mukai. On-line optimization of constrained multivariable chemical processes. *AIChE Journal*, 33:26–42, 1987.

[8] S. S. Keerthi and E. G. Gilbert. Optimal infinite-horizon feedback laws for a general class of constrained discrete-time systems: Stability and moving-horizon approximations. *Journal of Optimization Theory and Applications*, 57(2):265–293, 1988.

[9] J. H. Lee and B. Cooley. Robust model predictive control of multi-variable systems using input-output models with stochastic parameters. In *Proceedings of American Control Conf.*, Seattle, WA, 1995.

[10] J. H. Lee and N. L. Ricker. Extended Kalman filter based on nonlinear model predictive control. *Ind. Eng. Chem. Res.*, 33:1530–1541, 1994.

[11] D. Q. Mayne. Nonlinear model predictive control: An assessment. In *CPC V*, Lake Tahoe, CA, 1996.

[12] D. Q. Mayne. Nonlinear model predictive control: Challenges and opporunities. In F. Allgower and A. Zheng, editors, *International Symposium on the Assessment and Future Research Directions of Nonlinear Model Predictive Control*, Ascona, Switzerland, 1998.

[13] H. Michalska and D. Mayne. Robust receding horizon control of constrained nonlinear systems. *IEEE Trans. Aut. Control*, 38(11):1623–1633, November 1993.

[14] M. Morari. Robust stability of systems with integral control. *IEEE Trans. Autom. Control*, 30:574–577, Jun 1985.

[15] M. Morari and N. L. Ricker. *Predictive Methods Toolbox for Matlab.* Test Version, 1993.

[16] A. M. Morshedi. Universal dynamic matrix control. In *Proceedings of CPC III*, New York, 1986.

[17] G. D. Nicolao, L. Magni, and R. Scattolini. Stability and robustness of nonlinear receding horizon control. In F. Allgower and A. Zheng, editors, *International Symposium on Nonlinear Model Predictive Control: Assessment and Future Directions*, Ascona, Switzerland, 1998.

[18] J. B. Rawlings and K. R. Muske. The stability of constrained receding horizon control. *IEEE Trans. Aut. Control*, 38(10):1512–1516, October 1993.

[19] N. L. Ricker. Decentralized control of the tennessee eastman challenge process. *Journal of Process Control*, 6:205–221, 1996.

[20] S. Skogestad and I. Postlethwaite. *Multivariable feedback control.* John Wiley & Sons, 1996.

[21] T. H. Yang and E. Polak. Moving horizon control of nonlinear systems with input saturation, disturbances and plant uncertainty. *Int. J. Control*, 58(4):875–903, September 1993.

[22] A. Zheng. A computationally efficient nonlinear model predictive control algorithm. In *Proceedings of American Control Conf.*, Albuquerque, NM, 1997.

[23] A. Zheng. A computationally efficient nonlinear model predictive control algorithm with guaranteed stability. In *NATO-ASI*, Antalya, Turkey, 1997.

[24] A. Zheng. Nonlinear model predictive control of the tennessee eastman process. In *Proceedings of American Control Conf.*, Philadelphia, PA, 1998.

[25] A. Zheng, V. Balakrishnan, and M. Morari. Constrained stabilization of discrete-time systems. *International Journal of Robust and Nonlinear Control*, 5(5):461–485, Aug. 1995.

Department of Chemical Engineering
University of Massachusetts
Amherst, MA 01003-3110, USA
E-mail address: zzheng@ecs.umass.edu

Progress in Systems and Control Theory, Vol. 26

Nonlinear Model Predictive Control for Index–one DAE Systems

Rolf Findeisen and Frank Allgöwer

Abstract. In this paper the nonlinear model predictive control (NMPC) problem for systems described by *index–one* differential algebraic equations (DAEs) is considered. It is shown, that with no or only minor modifications most of the NMPC schemes for ordinary differential equations can be used in the index–one DAE case and that they also guarantee stability. For the quasi-infinite horizon NMPC concept and the dual–mode control, however, the procedures to compute appropriate terminal penalty terms and terminal regions must be modified for the index–one DAE case and are presented here. The resulting quasi-infinite horizon NMPC scheme for index–one DAEs is demonstrated using a nontrivial process control example, namely the control of a high-purity binary distillation column with 40 trays. To make real–time application feasible, a nonlinear reduced order DAE model for the column based on nonlinear wave propagation phenomena is introduced. Simulations show promising closed–loop behavior while the on–line computational load permits real-time implementation.

Over the last two decades linear model predictive control has emerged as a powerful and widely used control technique, especially in the process industry [25, 28]. Recently there is growing interest in model predictive control for nonlinear systems (NMPC) in academia and in the industrial process control community. By now several NMPC schemes have been proposed, that guarantee stability [16, 21, 24, 22, 7, 11] or even robust stability [14, 10, 8, 31, 32] for systems described by ordinary differential or difference equations (ODEs). See for instance [6] for a review. Quite frequently real processes are, however, described by means of nonlinear differential *algebraic* equations (DAEs) instead of ODEs. The question arises, if and how the NMPC schemes for ODE systems must be modified so that stability can be also guaranteed in the DAE case. In this paper systems described by nonlinear index–one DAEs are considered and it is shown, that most of the NMPC schemes for ODEs can be directly applied in the index–one DAE case and that they do lead to guaranteed stability of the closed–loop system. In the case of the so called dual–mode [21] and the quasi–infinite [5, 4, 7] horizon NMPC scheme, however, modifications concerning the calculation of the terminal region and the terminal penalty term, which are utilized to guarantee stability, must be made.

The paper is organized as follows:

First the considered index–one DAE system class is introduced. In Section 2 NMPC schemes for systems described by ODEs are briefly reviewed and the main stability results are summarized. The main attention is placed on the so called quasi–infinite horizon NMPC concept. In the subsequent section it is shown, that many NMPC schemes for ODE systems do without any modifications lead to guaranteed stability for index–one DAE systems. For the quasi–infinite horizon NMPC scheme [7] and the dual–mode control [24], however, the procedure to calculate appropriate terminal regions and terminal penalty terms must be modified. Section 4 describes an application of the modified quasi–infinite horizon NMPC strategy to the control of a high purity distillation column. To make the solution of the open–loop optimal control problem computationally feasible, a reduced order index–one DAE model based on the nonlinear wave propagation phenomena is developed and employed.

1. Considered DAE system class

In what follows it is assumed, that the system can be described by the following semi–explicit set of ordinary differential and algebraic equations:

$$\dot{\mathbf{x}}(t) = \mathbf{f}(\mathbf{x}(t), \mathbf{z}(t), \mathbf{u}(t)), \qquad \mathbf{x}(0) = \mathbf{x}_0 \tag{1a}$$

$$\mathbf{0} = \mathbf{g}(\mathbf{x}(t), \mathbf{z}(t), \mathbf{u}(t)), \tag{1b}$$

with the differential variables $\mathbf{x}(t) \in \mathbb{R}^n$, the algebraic variables $\mathbf{z}(t) \in \mathbb{R}^p$ and the inputs $\mathbf{u}(t) \in \mathbb{R}^m$. Due to safety consideration and physical limitations it is additionally often required, that constraints on the inputs and differential variables of the form

$$\mathbf{u}(t) \in \mathbf{U}, \quad \mathbf{x}(t) \in \mathbf{X}, \quad \forall t \geq 0 \tag{2}$$

must be satisfied. Additionally we need that the following assumptions hold:

A1 $\mathbf{f} : \mathbb{R}^n \times \mathbb{R}^p \times \mathbb{R}^m \to \mathbb{R}^n$ and $\mathbf{g} : \mathbb{R}^n \times \mathbb{R}^p \times \mathbb{R}^m \to \mathbb{R}^p$ are twice continuously differentiable on $\mathbf{X} \times \mathbf{U}$ and, w.l.o.g., $\mathbf{f}(\mathbf{0},\mathbf{0},\mathbf{0}) = \mathbf{0}$ and $\mathbf{g}(\mathbf{0},\mathbf{0},\mathbf{0}) = \mathbf{0}$.

A2 The set $\mathbf{U} \subset \mathbb{R}^m$ is compact and convex, $\mathbf{X} \subseteq \mathbb{R}^n$ is open and simple connected with $(\mathbf{0},\mathbf{0})$ in the interior of $\mathbf{U} \times \mathbf{X}$.

Thus, $(\mathbf{x} = \mathbf{0}, \mathbf{z} = \mathbf{0})$ with the input $\mathbf{u} = \mathbf{0}$ is an equilibrium of the system. Since we consider DAE systems, we recall the notion of the differentiation index [13, 2]:

Definition 1.1. *The differentiation index of a DAE system of the form* (1) *is the minimum number of times the algebraic equations* (1b) *must be differentiated to obtain explicit (differential) equations for* $\mathbf{z}$, *i.e. to obtain* $\dot{\mathbf{z}} = \mathfrak{z}(\mathbf{x}, \mathbf{z}, \mathbf{u}, t)$.

Moreover, for the rest of the presentation we assume, that

A3 the DAE system given by (1) has uniform index–one in $\mathbf{X} \times \mathbf{U}$ and for each $(\mathbf{x}, \mathbf{u}) \in \mathbf{X} \times \mathbf{U}$ $\exists!$ $\mathbf{z}$ s.t. $\mathbf{0} = \mathbf{g}(\mathbf{x}, \mathbf{z}, \mathbf{u})$.

Notice, that A3 implies that $\mathbf{0} = \mathbf{g}(\mathbf{x}, \mathbf{z}, \mathbf{u})$ can be solved in principle for $\mathbf{z}$, i.e. there exists a function $\mathbf{g}^{-1} : \mathbb{R}^n \times \mathbb{R}^m \to \mathbb{R}^p$ with $\mathbf{z} = \mathbf{g}^{-1}(\mathbf{x}, \mathbf{u})$, $\forall (\mathbf{x}, \mathbf{u}) \in \mathbf{X} \times \mathbf{U}$. Additionally to Assumptions A1–A3 we require, that

A4 $\mathbf{g}^{-1}(\mathbf{x}, \mathbf{u})$ is twice continuously differentiable in $\mathbf{X} \times \mathbf{U}$.

A5 the system (1) has an unique solution for any initial condition $\mathbf{x}_0 \in \mathbf{X}$ and any piecewise continuous and right–continuous $\mathbf{u}(\cdot) : [0, \infty) \rightarrow \mathbf{U}$.

Given Assumptions A1–A5 the DAE system can in principle be rewritten as an ODE system substituting $\mathbf{z} = \mathbf{g}^{-1}(\mathbf{x}, \mathbf{u})$ into (1a). In what follows we will make use of the following Lemma:

Lemma 1.2. *Suppose that Assumptions A1–A5 hold. Then the DAE system* (1) *is equivalent to*

$$\dot{\mathbf{x}}(t) = \mathbf{f}(\mathbf{x}(t), \mathbf{g}^{-1}(\mathbf{x}(t), \mathbf{u}(t)), \mathbf{u}(t)), \quad \mathbf{x}(0) = \mathbf{x}_0 \tag{3a}$$

$$\mathbf{z}(t) = \mathbf{g}^{-1}(\mathbf{x}(t), \mathbf{u}(t)), \tag{3b}$$

and the explicit ODE system (3a) *does satisfy Assumptions A1 and A5.*

Proof. The ODE (3a) describes the dynamics of the differential variables, since Assumptions A3 and A4 are satisfied. (3a) does further satisfy Assumption A1, as A4 assures that $\mathbf{f}(\mathbf{x}, \mathbf{g}^{-1}(\mathbf{x}, \mathbf{u}), \mathbf{u})$ is twice continuously differentiable with respect to $\mathbf{x}$ and $\mathbf{u}$. Assumption A5 is true, since (3) is equivalent to (1). □

It is the goal of this paper to derive NMPC concepts which do not require explicit knowledge of $\mathbf{z} = \mathbf{g}^{-1}(\mathbf{x}, \mathbf{u})$, as for many practically important control problems it is difficult to obtain an analytic expression for $\mathbf{g}^{-1}$.

2. NMPC for ODE systems

In this section we shortly review some NMPC schemes for ODE systems described by

$$\dot{\mathbf{x}}(t) = \mathbf{f}(\mathbf{x}(t), \mathbf{u}(t)), \quad \mathbf{x}(0) = \mathbf{x}_0. \tag{4}$$

In a model predictive control strategy at every samplig instance an open-loop optimal control problem is solved on–line and the optimal solution is applied to the system until new measurements are available. We denote the time between two sampling instances as the sampling time δ. The mathematical formulation of the open–loop optimal control problem, which has to be solved at every sampling instance $t = i\delta$, $i = 0, 1 \ldots$ is:

Problem 1: *Find*

$$\min_{\bar{\mathbf{u}}(\cdot)} J(\mathbf{x}(t), \bar{\mathbf{u}}(\cdot)) \tag{5}$$

subject to:

$$\dot{\bar{\mathbf{x}}}(\tau) = \mathbf{f}(\bar{\mathbf{x}}(\tau), \bar{\mathbf{u}}(\tau)), \quad \bar{\mathbf{x}}(t; \mathbf{x}(t), t, \bar{\mathbf{u}}) = \mathbf{x}(t) \tag{6a}$$

$$\bar{\mathbf{u}}(\tau) \in \mathbf{U}, \quad \tau \in [t, t + T_P] \tag{6b}$$

$$\bar{\mathbf{x}}(\tau; \mathbf{x}(t), t, \bar{\mathbf{u}}) \in \mathbf{X}, \quad \tau \in [t, t + T_P]. \tag{6c}$$

with:

$$J(\mathbf{x}(t), \bar{\mathbf{u}}(\cdot)) := \int_t^{t+T_P} \left(\|\bar{\mathbf{x}}(\tau; \mathbf{x}(t), t, \bar{\mathbf{u}})\|_Q^2 + \|\bar{\mathbf{u}}(\tau)\|_R^2 \right) d\tau. \tag{7}$$

T_P is the control/prediction horizon, $\|\mathbf{x}\|$ denotes the 2–norm of a vector $\mathbf{x} \in \mathbb{R}^n$; $\|\mathbf{x}\|_P$ is defined as $\|\mathbf{x}\|_P^2 = \mathbf{x}^T P \mathbf{x}$ for any positive–semidefinite matrix P.

$Q \in \mathbb{R}^{n \times n}$ and $R \in \mathbb{R}^{m \times m}$ denote positive-definite, symmetric weighting matrices; $\bar{\mathbf{x}}(\cdot; \mathbf{x}(t), t, \bar{\mathbf{u}})$ is the solution of (6) *driven by the input $\bar{\mathbf{u}}(\cdot) : [t, t+T_P] \to \mathbf{U}$ with initial conditions $\mathbf{x}(t)$ at time t. Internal controller variables are denoted by a bar.* The system input during the sampling intervals is given by the optimal[1] input $\bar{\mathbf{u}}^\star(\cdot; \mathbf{x}(t), t, T_P) : [t, t+T_P] \to \mathbf{U}$ resulting from Problem 1 (or variations thereof):

$$\mathbf{u}^\star(\tau) = \bar{\mathbf{u}}^\star(\tau; \mathbf{x}(t), t, T_P), \quad \tau \in [t, t+\delta). \tag{8}$$

In what follows we will call the nominal system driven by the input (8) the *nominal closed–loop system*, e.g. in the ODE case the nominal closed–loop system is:

$$\dot{\mathbf{x}}(t) = \mathbf{f}(\mathbf{x}(t), \mathbf{u}^\star(t)), \quad \mathbf{x}(0) = \mathbf{x}_0. \tag{9}$$

In the following section some NMPC schemes that guarantee stability in the ODE case are shortly reviewed.

2.1. Infinite horizon NMPC and zero terminal constraint NMPC for ODE systems

The most intuitive way to achieve stability is the use of an infinite horizon cost, i.e. T_P in Problem 1 is set to ∞ [1, 17]. In the nominal case feasibility at one sampling instant also implies feasibility at the next sampling interval and the following holds:

Theorem 2.1. (Infinite horizon NMPC for ODEs) *Suppose a) A1, A2, A5 are satisfied* [2]*; b) the open–loop optimal control Problem 1 with T_P set to ∞ is feasible at $t=0$. Then the nominal closed–loop system* (9) *is asymptotically stable for small enough smapling times δ and the region of attraction contains the set of states for which the optimization problem described in b) has a solution.*

Proof: The basic idea of essentially all stability proofs in the state space NMPC literature is, to use the optimal cost function, i.e. the so called value function, as a Lyapunov function. First it is shown, that feasibility at one sampling instance does imply feasibility at the next sampling instance for the nominal case and short enough sampling intervals δ. In a second step it is established, that the value function is strictly decreasing and by this, that the state and input do converge to the origin. Utilizing the continuity of the value function at the origin and the monotonicity property, stability is established in the third step. As feasibility thus implies asymptotic stability the region of attraction consists of all states, for which the open loop optimal control problem has a solution. The particular proof of Theorem 2.1 can be derived from [21, 4] as a limiting case. □

The major drawback of an infinite horizon is, that at every sampling instance an infinite dimensional optimization problem has to be solved. For nonlinear systems the solution of such an optimization problem is very difficult, if not impossible to obtain. To avoid this problem, it has been suggested to use a finite prediction

---

[1] Optimal values are denoted with a superscript star

[2] To shorten the presentation the same assumptions for ODE and DAE systems are used. For the ODE case the algebraic variables and the algebraic equations have to be disregarded.

horizon T_P, but to add a so called *zero terminal equality constraint* at the end of the horizon, i.e. to add the equality constraint

$$\bar{\mathbf{x}}(t+T_P;\mathbf{x}(t),t,\bar{\mathbf{u}}) = \mathbf{0}. \tag{10}$$

to Problem 1. Now, only an optimization problem over a finite horizon must be solved. The resulting control law does lead to asymptotic stability of the closed–loop:

Theorem 2.2. (Zero terminal constraint NMPC for ODEs) *Suppose a) A1, A2, A5 are satisfied; b) the open–loop optimal control Problem 1 together with* (10) *is feasible at* $t=0$. *Then the nominal closed–loop system* (9) *is asymptotically stable for small enough* δ *and the region of attraction contains the set of states for which the optimization problem described in b) has a solution.*

Proof: See for example [21, 17, 22]. □

One disadvantage of a zero terminal constraint is, that the system must be brought to the origin in finite time. This does, in general, lead to feasibility problems for short horizon lengths T_P, i.e. a small region of attraction. Additionally, from a computational point of view, an exact satisfaction of a zero terminal equality constraint does require an infinite number of iterations in the nonlinear case. On the other hand, the main advantages are the straightforward application and the conceptual simplicity.

2.2. Dual–mode NMPC for ODE systems

One way to avoid the problems resulting from a zero terminal constraint is the so called dual–mode variable horizon NMPC scheme as suggested in [24]. This scheme does utilize a terminal inequality constraint (a terminal region Ω) and a local controller together with a variable horizon. Instead of requiring that the state has to be zero at the end of the horizon, it is required, that the state at the end of the variable horizon has to lay on the border of a precalculated terminal region. Within the terminal region a local linear state feedback law $\mathbf{u} = K\mathbf{x}$ is employed (i.e. the controller is switched – thus the name dual–mode), such that this region is positive invariant and asymptotically stable for the nonlinear system controlled by this linear controller. Additionally it must be assured that the state and input constraints are satisfied with this controller in Ω. Based on these properties it is proposed in [24] to use a receding horizon strategy to bring the system to the terminal region and to switch to the locally stabilizing controller once the state has reached Ω. The open–loop optimal control Problem 1 is modified as follows: The length of the prediction horizon is an additional minimizer, i.e. (5) is replaced by

$$\min_{\bar{\mathbf{u}}(\cdot),T_P} J(\mathbf{x}(t),\bar{\mathbf{u}}(\cdot),T_P) \tag{11}$$

and the state at the end of the horizon has to lay on the border Ω_B of Ω, i.e.

$$\bar{\mathbf{x}}(t+T_P;\mathbf{x},t,\bar{\mathbf{u}}) \in \Omega_B. \tag{12}$$

is added to Problem 1. A simple procedure to calculate a suitable region Ω and a linear feedback matrix K under the assumption that

A6 the Jacobian linearization of (4) at the origin, $\dot{\mathbf{x}} = A\mathbf{x} + B\mathbf{u}$, with $A := \frac{\partial \mathbf{f}}{\partial \mathbf{x}}(\mathbf{0},\mathbf{0})$, $B := \frac{\partial \mathbf{f}}{\partial \mathbf{u}}(\mathbf{0},\mathbf{0})$, is stabilizable

was proposed in [24]. A similar procedure is also used in the quasi–infinite horizon NMPC scheme [5, 4, 7], see Section 2.3. For the dual–mode control strategy with a suitable terminal region Ω and a local linear controller K the following theorem is valid:

Theorem 2.3. (Dual–mode variable horizon NMPC for ODEs) *Suppose that a) A1, A2, A5, A6 hold; b) the open–loop optimal control Problem 1 with* (5) *replaced by* (11) *and the additional constraint* (12) *is feasible at* $t=0$. *Then the nominal closed–loop system* (9) *with the input given by*

$$\mathbf{u}^\star(\tau) = \begin{cases} \bar{\mathbf{u}}^\star(\tau; \mathbf{x}(t), t, T_P), & \tau \in [t, \min\{t+\delta, t+T_P\}] \quad \textit{if} \;\; \mathbf{x}(t) \notin \Omega \\ K\mathbf{x}(\tau), & \tau \in [t, \infty] \quad \textit{if} \;\; \mathbf{x}(t) \in \Omega \end{cases} \tag{13}$$

is asymptotically stable for sufficiently small δ *and the region of attraction contains the set of states for which the optimization problem in b) has a solution.*

Proof: See [21]. □

The major advantage of this scheme is, that no equality constraint at the end of the horizon must be satisfied. One disadvantage is, that the terminal region can be small, depending on the nonlinearity of the system; this is also valid for the quasi–infinite horizon NMPC scheme. Additionally the switching to another control law once the terminal region is reached can complicate the implementation.

2.3. Quasi–infinite horizon NMPC for ODE systems

The basic idea of the quasi–infinite horizon NMPC scheme for ODE systems [7, 4] is motivated by the dual–mode control. However, instead of switching to the local linear control law inside of Ω the local linear controller is only used to provide an upper bound for the infinite horizon cost in form of a terminal penalty term inside of the terminal region. Similar to the dual–mode control the terminal penalty term and the terminal region are determined off–line using the local state feedback law. Instead of (7) the following, modified cost function is used

$$J(\mathbf{x}(t), \bar{\mathbf{u}}(\cdot)) := \int_t^{t+T_P} \left(\|\bar{\mathbf{x}}(\tau; \mathbf{x}(t), t, \bar{\mathbf{u}})\|_Q^2 + \|\bar{\mathbf{u}}(\tau)\|_R^2\right) d\tau + \|\bar{\mathbf{x}}(t+T_P; \mathbf{x}(t), t, \bar{\mathbf{u}})\|_P^2 \tag{14}$$

and a terminal (inequality) constrained is added to Problem 1.

$$\bar{\mathbf{x}}(t+T_P; \mathbf{x}(t), t, \bar{\mathbf{u}}) \in \Omega \tag{15}$$

The matrix P in (14) is the precomputed, positive definite terminal penalty matrix. Inside the terminal region Ω the virtual, i.e. never really applied, linear controller $\bar{\mathbf{u}} = K\bar{\mathbf{x}}$ has to guarantee that the following four conditions are fulfilled: a) the constraints (2) are satisfied; b) the nonlinear system with $\bar{\mathbf{u}} = K\bar{\mathbf{x}}$ is asymptotically stable; c) the set Ω is positive invariant for the nonlinear system with $\bar{\mathbf{u}}=K\bar{\mathbf{x}}$; d) for any $\mathbf{x}_1 \in \Omega$ the optimal infinite horizon cost

$$V^{\infty}(\mathbf{x}_1(t_1)) = \min_{\bar{\mathbf{u}}(\cdot)} J^{\infty}(\mathbf{x}_1(t_1), \bar{\mathbf{u}}(\cdot)) \tag{16}$$

with

$$J^{\infty}(\mathbf{x}_1(t_1), \bar{\mathbf{u}}(\cdot)) := \int_{t_1}^{\infty} \left(\|\bar{\mathbf{x}}(\tau; \mathbf{x}_1(t), t, \bar{\mathbf{u}})\|_Q^2 + \|\bar{\mathbf{u}}(\tau)\|_R^2\right) d\tau \tag{17}$$

can be bounded from above by

$$V^{\infty}(\mathbf{x}_1(t_1)) \leq \mathbf{x}_1^T P \mathbf{x}_1. \tag{18}$$

The key steps in this scheme is the off–line calculation of the terminal region Ω and the terminal penalty matrix P such that the conditions a)–d) are satisfied. Given that Assumption A6 holds, Ω and P can be obtained by the following procedure [4]:

Step 1 Find a locally stabilizing linear state feedback $\mathbf{u} = K\mathbf{x}$, based on the Jacobian linearization of (4).

Step 2 Choose a constant $\kappa \in [0, \infty)$ with $\kappa < -\lambda_{\max}(A_K)$, where $A_K = A + BK$ and solve the Lyapunov equation $(A_K + \kappa I)^T P + P(A_K + \kappa I) = -(Q + K^T RK)$ for the positive definite and symmetric terminal weight P.

Step 3 Find the largest possible α_1 such that $\Omega_1 \subseteq \mathbf{X}$ and $K\mathbf{x} \in \mathbf{U}, \forall \mathbf{x} \in \Omega_1$ with $\Omega_1 := \{\mathbf{x} \in \mathbb{R}^n | \mathbf{x}^T P \mathbf{x} \leq \alpha_1\}$.

Step 4 Determine $\Omega = \{\mathbf{x} \in \mathbb{R}^n | \mathbf{x}^T P \mathbf{x} \leq \alpha\}$ via iterations of

$$\max_{\mathbf{x}} \{\mathbf{x}^T P \phi(\mathbf{x}) - \kappa \mathbf{x}^T P \mathbf{x} | \mathbf{x}^T P \mathbf{x} \leq \alpha\} \tag{19}$$

by reducing α from α_1, until the optimal value of (19) is nonpositive. $\phi(\mathbf{x}) := \mathbf{f}(\mathbf{x}, K\mathbf{x}) - A_K \mathbf{x}$ is the difference between the linearized and the nonlinear system with the input $\mathbf{u} = K\mathbf{x}$.

With the terminal region Ω and the terminal penalty P obtained by this procedure the following theorem can be stated [4]:

Theorem 2.4. (Quasi–infinite horizon NMPC for ODEs) *Suppose that a) A1, A2, A5, A6 are satisfied; b) the open–loop optimal control Problem 1 with* (14) *and* (15) *is feasible at* $t=0$. *Then the nominal closed–loop system* (9) *is asymptotically stable for small enough* δ *and the region of attraction contains the set of states for which the optimization problem described in b) has a solution.*

Proof: See [4]. □

Schemes for discrete time nonlinear systems that also utilize an approximation of the infinite cost function inside a region can be found in [12, 11]. The major advantage of this NMPC concept is, that the region of attraction is enlarged compared to other schemes, e.g. the zero terminal constraint NMPC scheme. Also the needed computational time is decreased since a) the horizon length T_P necessary for feasibility can be decrease b) the equality constraint in the optimization problem is replaced by an inequality constraint.

3. NMPC for index–one DAE systems

So far the NMPC problem for ODE systems has been reviewed. In this section we show, that most of the NMPC concepts for ODE systems can be also used in the index–one DAE case and that they lead to guaranteed stability of the close loop system. It is the goal of this section to devise NMPC schemes that do not require explicit knowledge of $\mathbf{g}^{-1}$. The basic NMPC setup follows the ODE case and is not repeated here. The only difference is that (6a) is replaced by the nonlinear DAE model:

$$\dot{\bar{\mathbf{x}}}(\tau) = \mathbf{f}(\bar{\mathbf{x}}(\tau), \bar{\mathbf{z}}(\tau), \bar{\mathbf{u}}(\tau)), \quad \bar{\mathbf{x}}(t; \mathbf{x}(t), t, \bar{\mathbf{u}}) = \mathbf{x}(t) \tag{20a}$$

$$\mathbf{0} = \mathbf{g}(\bar{\mathbf{x}}(\tau), \bar{\mathbf{z}}(\tau), \bar{\mathbf{u}}(\tau)). \tag{20b}$$

The nominal closed–loop system with the input $\mathbf{u}$ given as the solution of the optimal control problem $\mathbf{u}^\star$ is:

$$\dot{\bar{\mathbf{x}}}(t)(= \mathbf{f}(\bar{\mathbf{x}}(t), \bar{\mathbf{z}}(t), \mathbf{u}^\star(t)), \quad \mathbf{x}(0) = \mathbf{x}_0 \tag{21a}$$

$$\mathbf{0} = \mathbf{g}(\bar{\mathbf{x}}(t), \bar{\mathbf{z}}(t), \mathbf{u}^\star(t)), \tag{21b}$$

compare also equation (8). For simplicity reasons we consider no weighting of the algebraic variables in the cost function and no constraints on the algebraic variables.

Remark 3.1. *Direct weighting of the algebraic variables in the cost function does lead to a non–quadratic stage cost function when expressed in the differential variables alone. For the infinite horizon NMPC and the zero state constraint NMPC scheme with a non–quadratic stage cost function stability can still be guaranteed, providing that certain additional assumptions are satisfied (see for example* [17, 22, 18]*). In the case of the dual–mode control and the quasi–infinite horizon NMPC scheme it is, however, difficult to obtain an upper bound of the "remaining" infinite horizon cost for the general nonlinear case* [4] *for ODEs and thus an extension to the DAE case with direct weighting of the algebraic variables in the stage cost is not easily possible.*

3.1. Infinite horizon and zero terminal constraint NMPC for index–one DAE systems

Due to Lemma 1.2 it is immediately clear that the infinite horizon NMPC scheme for the considered class of index–one DAE systems guarantees stability provided that the optimal control problem is feasible:

Theorem 3.2. (Infinite horizon NMPC for index–one DAEs) *Suppose a) A1–A5 are satisfied; b) the open–loop optimal control Problem 1 with T_P set to ∞ and* (6a) *replaced by* (20) *is feasible at $t=0$. Then the nominal closed–loop system* (21) *is asymptotically stable for sufficiently small δ and the region of attraction contains the set of states for which the optimization problem in b) has a solution.*

Proof: The proof follows directly from Theorem 2.1 and Lemma 1.2. □

The same is valid for the zero terminal equality constraint NMPC scheme. However, only the differential states have to be forced to the origin at the end of the horizon, since A4 assures, that $\mathbf{z} \to \mathbf{0}$ as $(\mathbf{u}, \mathbf{x}) \to \mathbf{0}$.

Theorem 3.3. (Zero terminal constraint NMPC for index–one DAEs) *Suppose a) A1–A5 are satisfied; b) the open–loop optimal control Problem 1 with* (6a) *replaced by* (20) *and with* (10) *is feasible at* $t=0$. *Then the nominal closed–loop system* (21) *is asymptotically stable for sufficiently small* δ *and the region of attraction contains the set of states for which the optimization problem in b) has a solution.*

Proof: The proof follows directly from Theorem 2.1 and Lemma 1.2. □

Both, the infinite horizon NMPC scheme and the finite horizon NMPC scheme for index–one DAE systems do suffer the same feasibility and computational problems as in the ODE case.

3.2. Dual–mode NMPC for index–one DAE systems

The principle idea of the the dual–mode NMPC concept does also carry over to the index–one DAE case. Only the procedure for the calculation of the linear controller that is applied in the terminal region Ω and the calculation of the terminal region have to be adapted. Similar to the ODE case the following assumptions about the linearization of the DAE–system are required:

A7 the Jacobian linearization of (1) at the origin

$$\dot{\mathbf{x}} = A_{11}\mathbf{x} + A_{12}\mathbf{z} + B_1\mathbf{u} \tag{22a}$$

$$\mathbf{0} = A_{21}\mathbf{x} + A_{22}\mathbf{z} + B_2\mathbf{u}, \tag{22b}$$

is stabilizable.

Here A_{11}, A_{12}, A_{21}, A_{22}, B_1, B_2 define the partial derivatives of $\mathbf{f}$, $\mathbf{g}$ with respect to $\mathbf{x}$, $\mathbf{z}$ and $\mathbf{u}$ at the origin: $A_{11} = \frac{\partial \mathbf{f}}{\partial \mathbf{x}}(\mathbf{0},\mathbf{0},\mathbf{0})$, $A_{12} = \frac{\partial \mathbf{f}}{\partial \mathbf{z}}(\mathbf{0},\mathbf{0},\mathbf{0})$, $A_{21} = \frac{\partial \mathbf{g}}{\partial \mathbf{x}}(\mathbf{0},\mathbf{0},\mathbf{0})$, $A_{22} = \frac{\partial \mathbf{g}}{\partial \mathbf{z}}(\mathbf{0},\mathbf{0},\mathbf{0})$, $B_1 = \frac{\partial \mathbf{f}}{\partial \mathbf{u}}(\mathbf{0},\mathbf{0},\mathbf{0})$, $B_2 = \frac{\partial \mathbf{g}}{\partial \mathbf{u}}(\mathbf{0},\mathbf{0},\mathbf{0})$. Since the DAE system considered satisfies Assumption A4, this is equivalent to

$$\dot{\mathbf{x}} = (A_{11} + A_{12}A_{22}^{-1}A_{21})\mathbf{x} + (B_1 + A_{12}A_{22}^{-1}B_2)\mathbf{u} \tag{23}$$

being stabilizable. The main problem lays in the calculation of the terminal region and the local linear controller without requiring explicite knowledge about $\mathbf{g}^{-1}$. Since the procedure for the calculation of these quantities is similar for the quasi–infinite horizon NMPC strategy, it will be given in Section 3.3. Assuming that the terminal region Ω and the linear controller K satisfy the assumptions given in Section 2.2, the following theorem is valid:

Theorem 3.4. (Dual–mode variable horizon NMPC for index–one DAEs) *Suppose that a) A1–A5,A7 hold; b) the open–loop optimal control Problem 1 with* (5) *replaced by* (11), (6a) *replaced by* (20) *and with the additional constraint* (12) *is* *feasible at* $t=0$. *Then the nominal closed-loop system* (21) *with*

$$\mathbf{u}^\star(\tau) = \begin{cases} \bar{\mathbf{u}}^\star(\tau;\mathbf{x}(t),t,T_P), & \tau \in [t, \min\{t+\delta, t+T_P\}] \quad \textit{if} \;\; \mathbf{x}(t) \notin \Omega \\ K\mathbf{x}(\tau), & \tau \in [t,\infty] \quad \textit{if} \;\; \mathbf{x}(t) \in \Omega \end{cases} \tag{24}$$

is asymptotically stable for small enough δ *and the region of attraction contains the set of states for which the optimization problem in b) has a solution.*

Proof: Utilizing Lemma 1.2 the proof follows the exact same steps as in [24]. □

3.3. Quasi–infinite horizon NMPC for index–one DAE systems

Similar to the dual–mode case the quasi–infinite horizon NMPC scheme for ODE systems can be directly used in the index–one case. Only the ODE system description has to be replaced by the corresponding DAE description. The key step lays, as mentioned, in the calculation of the terminal region Ω and the terminal penalty P, for which the following procedure is given:

Step 1 Find a locally stabilizing linear feedback gain $\mathbf{u} = [\,K_1\; K_2\,][\begin{smallmatrix}\mathbf{x}\\ \mathbf{z}\end{smallmatrix}]$ for DAE system (1) based on the Jacobian linearization (22) such that $(A_{22}+B_2K_2)$ is invertible. The input $\mathbf{u}$ can be written as a function of $\mathbf{x}$ alone, i.e. $\mathbf{u} = K\mathbf{x}$, with $K = K_1 + K_2(A_{22}+B_2K_2)^{-1}(A_{21}+B_2K_1)$.

Step 2 Choose a constant $\kappa \in [0,\infty)$ with $\kappa < -\lambda_{\max}(A_K)$, where $A_K = (A_{11} + A_{12}A_{22}^{-1}A_{21}) + (B_1 + A_{12}A_{22}^{-1}B_2)K$ and solve the Lyapunov equation $(A_K + \kappa I)^T P + P(A_K+\kappa I) = -(Q + K^TRK)$ for the positive definite and symmetric terminal weight P.

Step 3 Find the largest possible α_1 such that $\Omega_1 \subseteq \mathbf{X}$ and $K\mathbf{x} \in \mathbf{U}, \forall \mathbf{x} \in \Omega_1$ with $\Omega_1 := \{\mathbf{x} \in \mathbb{R}^n | \mathbf{x}^T P \mathbf{x} \leq \alpha_1\}$.

Step 4 Determine $\Omega = \{\mathbf{x} \in \mathbb{R}^n | \mathbf{x}^T P \mathbf{x} \leq \alpha\}$ via iterations of

$$\max_{\mathbf{x}} \{\mathbf{x}^T P \phi(\mathbf{x},\mathbf{z}) - \kappa \mathbf{x}^T P \mathbf{x} | \mathbf{x}^T P \mathbf{x} \leq \alpha, \mathbf{0} = \mathbf{g}(\mathbf{x},\mathbf{z},K\mathbf{x})\} \tag{25}$$

by reducing α from α_1, until the optimal value of (19) is nonpositive. $\phi(\mathbf{x},\mathbf{z}) := \mathbf{f}(\mathbf{x},\mathbf{z},K\mathbf{x}) - A_K\mathbf{x}$ is the difference between the linearized and the nonlinear system with $\mathbf{u} = K\mathbf{x}$.

Remark 3.5. *Explicit knowledge about* $\mathbf{g}^{-1}(\mathbf{x},\mathbf{u})$ *is neither required during the off–line calculation of the terminal region and terminal penalty nor during the solution of the open–loop optimal control problem. The maximization* (25) *is in general carried out using a nonlinear optimizer, the algebraic equations* $\mathbf{0} = \mathbf{g}(\mathbf{x},\mathbf{z},\mathbf{u})$ *can be considered as additional equality constraints, that are solved implicitly.*

Remark 3.6. *The design of the linear stabilizing controller in Step 1 can either be done using* (23) *and standard ODE techniques or using descriptor methods, see for example* [23].

With the terminal region Ω and the terminal penalty P calculated as given, the following theorem can be stated.

Theorem 3.7. (Quasi–infinite horizon NMPC for index–one DAEs) *Suppose that a) A1–A5, A7 are satisfied; b) the open–loop optimal control Problem 1 with* (5) *replaced by* (11) *and the additional constraint* (14), (15) *is feasible at* $t=0$. *Then the nominal closed–loop system* (21) *is asymptotically stable for small enough* δ *and the region of attraction contains the set of states for which the optimization problem in b) has a solution.*

Proof: As for the dual–mode case this can be shown in analogy to to the ODE case [4] utilizing Lemma 1.2. □

4. Application example: Control of a distillation column

In this section the proposed quasi–infinite horizon NMPC algorithm for index–one DAE systems is examplary used for the control of a high purity binary distillation column with 40 trays for the separation of Methanol and n-Propanol. This example does also provide some insight into the robustness of the proposed algorithm, since different models for control and plant simulation are used (model plant mismatch) and the model states are estimated via a simple observer from plant measurements. A schematic representation of the process is given in Figure 1. The binary mixture

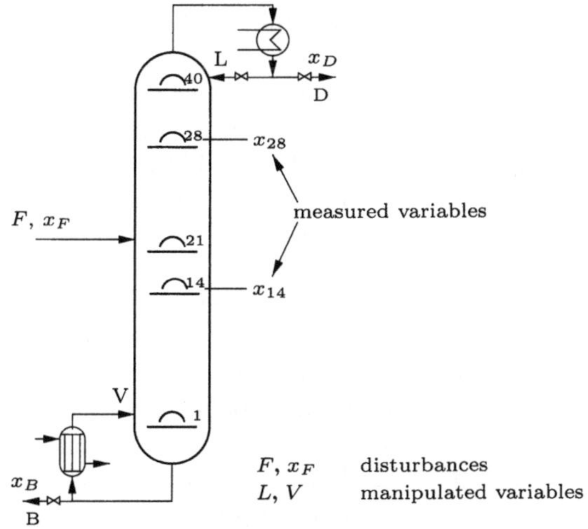

FIGURE 1. Distillation column scheme.

is fed in the column with flow rate F and molar feed composition x_F given in molar fractions. The separation products are removed at the top and bottom of the column with the concentrations x_B and x_D and the flowrates B and D respectively. The column is considered in LV configuration, i.e. the liquid flow rate L and the vapor flow rate V are considered to be the control inputs. It is assumed that only the concentrations x_{14} and x_{28} on trays 14 and 18 are available

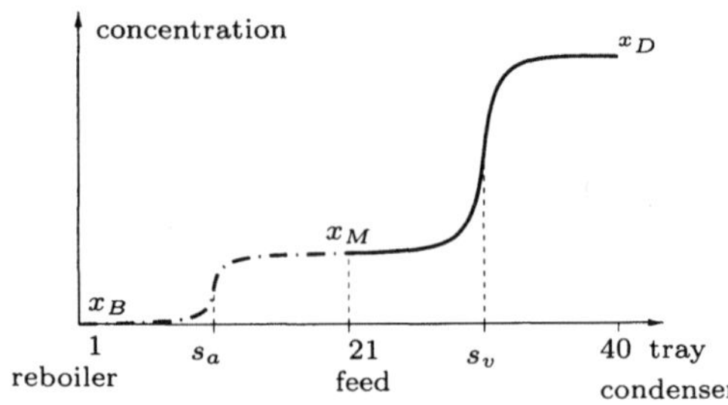

FIGURE 2. Illustration of the wave profiles in the stripping and rectifying section of the column.

as measurements. The control problem is to maintain the specifications on the product concentrations x_B and x_D despite disturbances in the feed flow F and the feed concentration $\mathbf{x}_F$.

4.1. Reduced order model

Modeling of the distillation column under the assumptions of constant relative volatility, constant molar overflow, no pressure losses, neglect of energy balances and hydrodynamics leads to a 42nd order ODE model. This rather simple model describes the behavior of the distillation column sufficiently well for control purposes. However, since the model is of 42nd order, it does not seem advisable to use it for NMPC, as at every control step a nonlinear optimal control problem involving possibly many simulations of the ODE model must be solved. To allow reasonable solution time a lower order model is needed. This model is devised on the basis of the so called wave propagation phenomena [19, 15]. The concentration profiles as a function of the trays can be approximately described by two wave fronts, one in the rectifying section and one in the stripping section (see Figure 2). Based on this observation and mass balances for the reboiler, condenser, feed tray, rectifying and stripping section a set of 5 differential equations for the feed tray concentration x_M, the distillate concentration x_D, the boiler concentration x_B and the positions of the inclination points s_a, s_v of the waves was derived in [30]. A drawback of this model is, that it contains very fast dynamics and very slow dynamics, due to the fact that the total mass in the rectifying and stripping sections is much larger than the one in the boiler, condenser and feed tray. Three fast eigenmodes, approximately corresponding to the concentrations x_B, x_M and x_D and two by a factor of at least 20 slower eigenmodes, approximately corresponding to the wave positions s_a and s_v can be observed. In order to circumvent numerical problems, e.g.possible slow solution of the optimal control problem caused by the stiffness of the ODEs, a quasi steady state assumption for x_B, x_M and x_D is used. The resulting model is a DAE system with differential variables $\mathbf{z} = (x_B, x_M, x_D)^T$, algebraic variables $\mathbf{x}=(s_a, s_v)^T$, and manipulated inputs $\mathbf{u}=(L, V)^T$:

$$\begin{aligned} 0 &= g_B(x_B, s_a, x_M, L, V, F) && \text{(26a)}\\ \dot{s}_a &= f_a(x_B, s_a, x_M, s_v, x_D, L, V, F, x_F) && \text{(26b)}\\ 0 &= g_M(x_B, s_a, x_M, s_v, x_D, L, V, F, x_F) && \text{(26c)}\\ \dot{s}_v &= f_v(x_B, s_a, x_M, s_v, x_D, L, V, F, x_F) && \text{(26d)}\\ 0 &= g_D(x_M, s_v, x_D, L, V, F). && \text{(26e)} \end{aligned}$$

This DAE system is generic index–one, since $det\left(\frac{\partial(g_B,g_M,g_D)^T}{\partial(x_B,x_M,x_D)}\right) \neq 0$ for all practically important values of $(s_a, s_v, L, V, F,\ x_F)$. Also the right–hand side of the differential equations and the algebraic equations are twice continuously differentiable, such that Assumptions A1 and A3 are satisfied. Furthermore Assumption A7 is valid at the nominal operating point, i.e. the linearization of (26) is controllable via L and V. It is also easy to verify, that Assumptions A4 and A5 hold.

4.2. Controller synthesis

Due to physical limitations the manipulated variables are limited to $4.37\times10^{-5}\frac{\text{kmol}}{\text{s}} < V < 6.5\times10^{-5}\frac{\text{kmol}}{\text{s}}$ and $1.75\times10^{-5}\frac{\text{kmol}}{\text{s}} < L < 4.3\times10^{-5}\frac{\text{kmol}}{\text{s}}$. The nominal product specifications are $x_B^{\text{set}} = 0.00016$ and $x_B^{\text{set}} = 0.999967$. As usual in the wave propagation framework x_B and x_D are not controlled directly. Instead an inferential control scheme which does control the deviation of the wave positions from the set point ($s_a^{\text{set}}=0$, and $s_v^{\text{set}}=0$) is used[3]. Based on these considerations the following cost function

$$J = \int_t^{t+T_P} \left(\left\| \begin{matrix} s_a \\ s_v \end{matrix} \right\|_Q^2 + \left\| \begin{matrix} L - L^{\text{set}} \\ V - V^{\text{set}} \end{matrix} \right\|_R^2 \right) d\tau + \left\| \begin{matrix} s_a(t+T_P) \\ s_v(t+T_P) \end{matrix} \right\|_P^2 \tag{27}$$

with the weighting matrices chosen as

$$Q = \begin{bmatrix} 1 & 0 \\ 0 & 10 \end{bmatrix}, \quad R = \begin{bmatrix} 5000 & 0 \\ 0 & 5000 \end{bmatrix} \tag{28}$$

is used. The terminal region Ω and the terminal penalty matrix P are calculated according to the scheme given in Section 3.3. In a first step a LQR controller K on the basis of equation (23) utilizing the weights given in (28) is designed. The tuning factor κ was chosen to $\kappa=-0.85\lambda_{\max}(A_K)$. In Steps 3 and 4 the size of the terminal region must be determined as a function of α. Since the DAE system has only two differential variables, this can be achieved by simple graphical means.

[3]As is customary in distillation control, we do not require that the closed–loop has zero steady state offset.

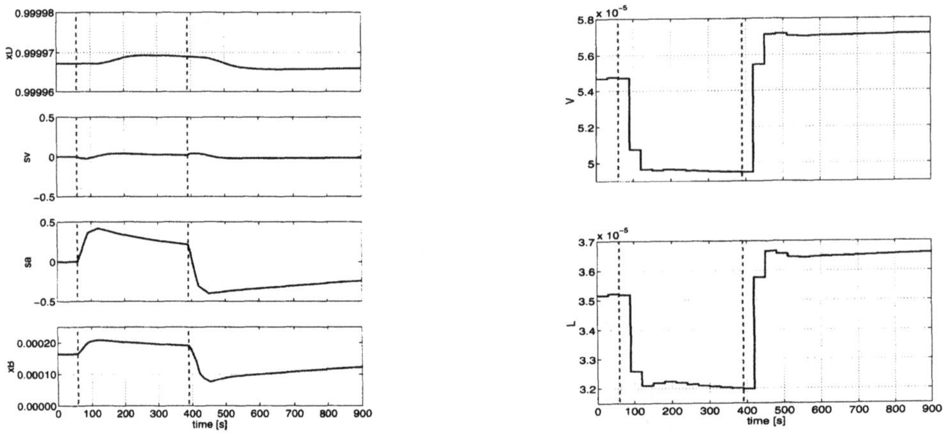

FIGURE 3. Simulation of the controlled column for feed flow disturbances at $t_1 = 90$s by $\Delta F_1 = -0.1F^{\text{set}}$ and $t_2 = 390$s by $\Delta F_2 = 0.15F^{\text{set}}$ (time instances are marked by dashed vertical lines). The wave inflection points s_a and s_v are interpolated from the tray concentrations of the 42nd order model.

4.3. Simulation results

The plant is simulated with the full 42nd order ODE model, whereas for the NMPC controller the reduced order index–one DAE model is used, thus testing for robustness of the controller. The horizon T_P is set to 30 minutes. For the solution process the input signal is parameterized by six control intervals of length 5 minutes with constant input. The relative rough parameterization is necessary to allow sufficiently fast solution of the open–loop optimization problem while looking far enough in the future. To partially compensate for the rough paramterization and to allow for fast reaction to disturbances a sampling time of $\delta = 0.5$ minutes is used, i.e. all 30 seconds a new input is computed utilizing new measurements. For the controller the wave positions s_a, s_v and the feed disturbances F and x_F are needed. Estimates for these variables are obtained using a simple observer as suggested in [30]. This observer is driven by the concentration measurements x_{14}, x_{28}. The simulations are carried out under Matlab [20] using the NAG [27] E04UCF optimizer. The plant and system model is integrated using DDASEC [3], which is connected to Matlab via mex–functions. The solution of one open–loop control problem takes approximately 1.5 minutes. The computational time could be reduced significantly by avoiding Matlab as "communication" layer, which would make real–time implementation feasible. Exemplary results for the following disturbance scenario are shown in Figure 3: At time $t_1 = 90$s a feed flow disturbance of $\Delta F_1 = -0.1F^{\text{set}}$ occurs. This disturbance is over compensated at $t_2 = 390$s by a second feed flow disturbance with $\Delta F_2 = +0.15F^{\text{set}}$. The two figures on the right show the manipulated variables L and V. The figures on the left show the wave positions s_a and s_v and the product concentrations x_B and x_D as a function of time. As can be seen the controller keeps the product concentration in reasonable bounds considering the rather large feed disturbances. The maximal change in the wave positions is less than half a tray. Also the input signals stay within reasonable bounds. The delay between the change of the manipulated variables L and V and the disturbances is a result from the fact, that the feed flow is estimated by an observer.

Remark 4.1. *In the disturbance case the controller does not bring the product concentrations nor the wave positions back to the set point. This results from the fact, that the controller is designed for stabilization and not disturbance rejection. A change in the system parameters, e.g. the feed flow rate, does lead to a new steady state, if the set points of the manipulated variables are not changed. Possible remedies of this problem could be the augmentation of the plant by additional integrators, the adaption of the set points of the manipulated variables or the usage of a disturbance rejecting NMPC scheme as for example* [8].

This example shows, that the proposed quasi–infinite horizon NMPC scheme for index–one DAE system together with the reduced order model does lead to good control performance, even in the presence of model plant mismatch and output feedback via an observer. The required computation time even allows real–time implementation. Also the computation of the terminal region and the terminal penalty term is straightforward, using the procedure outlined in Section 3.3. A direct comparison with the zero terminal constraint NMPC scheme and an approximation of the infinite horizon NMPC using a long enough control horizon was not possible due to numerical problems.

5. Conclusions

In this paper it was shown, that most of the NMPC schemes for ODE systems can also be used for index–one DAE systems and it was shown that they establish guaranteed stability of the closed–loop. For the quasi–infinite horizon NMPC scheme it was outlined how the terminal region and the terminal penalty term can be calculated in the index–one case. The same procedure can also be used for the dual–mode control problem. The main advantage of this control scheme is, that at no point explicit knowledge of the inverse of the algebraic equations is needed and that the computational advantage of the corresponding NMPC scheme for ODEs can be preserved. As an application example the control of a high purity distillation column was considered. For this column, a previously published wave propagation distillation model was extended to allow real–time solutions. The simulation results show promising behavior of the closed–loop system while the online computational load is restricted enough to allow real–time implementation. Future research will focus on the control of higher index DAE systems and the use of algebraic variables in the cost function.

References

[1] R. R. Bitmead, M. Gevers, and V. Wertz. *Adaptive Optimal Control – The Thinking Man's GPC.* Prentice Hall, New York, 1990.

[2] K.E. Brenan, S.L. Campbell, and L.R. Petzold. *Numerical Solution of Initial–Value Problems in Differential–Algebraic Equations.* Classics in Applied Mathematics. SIAM, Philadelphia, 1996.

[3] M. Caracotsios and W.E. Stewart. Sensitivity analysis of initial value problems with mixed ODEs and algebraic equations. *Comp. & Chem. Eng.*, 9(4):359–365, 1985.

[4] H. Chen. Stability and robustness considerations in nonlinear model predictive control. PhD thesis, Universität Stuttgart, 1997.

[5] H. Chen and F. Allgöwer. A quasi-infinite horizon predictive control scheme for constrained nonlinear systems. In *Proc. 16th Chinese Control Conference*, pages 309–316, Qindao, 1996.

[6] H. Chen and F. Allgöwer. Nonlinear model predictive control schemes with guaranteed stability. In C. Berber, R. und Kravaris, editor, *Nonlinear Model Based Process Control*, pages 465–494. Kluwer Academic Publishers, Dordrecht, 1998.

[7] H. Chen and F. Allgöwer. A quasi-infinite horizon nonlinear model predictive control scheme with guaranteed stability. *Automatica*, 34(10):1205–1218, 1998.

[8] H. Chen, C.W. Scherer, and F. Allgöwer. A game theoretic approach to nonlinear robust receding horizon control of constrained systems. In *Proc. Amer. Contr. Conf.*, pages 3073–3077, Albuquerque, 1997.

[9] L. Dai. *Singular Control Systems.* Lecture Notes in Control and Information Sciences. Springer, Berlin, 1989.

[10] G. De Nicolao, L. Magni, and R. Scattolini. On the robustness of receding horizon control with terminal constraints. *IEEE Trans. Automat. Contr.*, 41(3):451–453, 1996.

[11] G. De Nicolao, L. Magni, and R. Scattolini. Stabilizing nonlinear receding horizon control via a nonquadratic terminal state penalty. In *Symposium on Control, Optimization and Supervision, CESA'96 IMACS Multiconference*, pages 185–187, Lille, 1996.

[12] R.H. Findeisen and J.B. Rawlings. Suboptimal infinite horizon nonlinear model predictive control for discrete time systems, 1997. IfA Technical Report AUT 97-13, ETH Zürich.

[13] C.W. Gear. Differential–algebraic equation index transformation. *SIAM J. Sci. Stat. Comput.*, 9:39–47, 1988.

[14] H. Genceli and M. Nikolaou. Design of robust constrained model-predictive controllers with Volterra series. *AIChE J.*, 41(9):2098–2107, 1995.

[15] F.H. Helfferich and G. Klein. *Multicomponent chromatography. Theory of interference.* M. Dekker, New York, 1970.

[16] S.S. Keerthi and E.G. Gilbert. An existence theorem for discrete-time infinite-horizon optimal control problems. *IEEE Trans. Automat. Contr.*, 30(9):907–909, 1985.

[17] S.S. Keerthi and E.G. Gilbert. Optimal infinite-horizon feedback laws for a general class of constrained discrete-time systems: Stability and moving-horizon approximations. *J. Opt. Theory and Appl.*, 57(2):265–293, 1988.

[18] L. Magni and R. Sepulchre. Satbility margins of nonlinear receeding–horizon control via inverse optimality. *Syst. Contr. Lett.*, (32):241–245, 1997.

[19] W. Marquardt. *Nichtlineare Wellenausbreitung – Ein Weg zu reduzierten dynamischen Modellen von Stofftrennprozessen.* VDI Verlag, Düsseldorf, 1988.

[20] MathWorks, Inc. *MATLAB Reference Guide.* Natick, Mass., 1998.

[21] D. Q. Mayne and H. Michalska. Receding horizon control of nonlinear systems. *IEEE Trans. Automat. Contr.*, AC-35(7):814–824, 1990.

[22] E.S. Meadows, M.A. Henson, J.W. Eaton, and J.B. Rawlings. Receding horizon control and discontinuous state feedback stabilization. *Int. J. Contr.*, 62(5):1217–1229, 1995.

[23] V.L. Mehrmann. *The Autonomous Linear Quadratic Control Problem.* Lecture Notes in Control and Information Sciences. Springer, Berlin, 1991.

[24] H. Michalska and D.Q. Mayne. Robust receding horizon control of constrained nonlinear systems. *IEEE Trans. Automat. Contr.*, AC-38(11):1623–1633, 1993.

[25] M. Morari and J.H. Lee. Model predictive control: Past, present and future. In *Proc. PSE'97-ESCAPE-7 Symposium*, Trondheim, 1997.

[26] P.C. Müller. Optimal control of mechanical descriptor systems. In D.H. van Campen, editor, *IUTAM Symposium on Interaction between Dynamics and Control in Advanced Mechanical Systems*, pages 247–254, Dordrecht, 1995. Kluwer Academic Publishers.

[27] Numerical Algorithms Group, Oxford. *NAG Fortran Library*, 1993. Mark 16.

[28] S.J. Qin and T.A. Badgwell. An overview of industrial model predictive control technology. In J.C. Kantor, C.E. Garcia, and B. Carnahan, editors, *Fifth International Conference on Chemical Process Control – CPC V*, pages 232–256. American Institute of Chemical Engineers, 1996.

[29] P.J. Rabier and W.C. Rheinboldt. A geometric treatment of implicite differential–algebraic equations. *Journal of Differential Equations*, 109:110–146, 1994.

[30] A. Rehm and F. Allgöwer. Nonlinear H_∞-control of a high purity distillation column. In *UKACC International Conference on CONTROL'96*, pages 1178–1183, Exeter, 1996.

[31] O. Slupphaug and B.A. Foss. Bilinear matrix inequalities and robust stability of nonlinear multi-model MPC. In *Proc. Amer. Contr. Conf.*, pages 1689–94, Philadelphia, 1998.

[32] A. Zheng and W.H. Zhang. Robust model predictive control of large-scale nonlinear constrained systems. To appear in *Proc. 14th IFAC World Congress*, paper N-7a-02-2, 1999.

Institut für Automatik,
ETH – Swiss Federal Institute of Technology,
CH–8092 Zürich, Switzerland
tel.: +41-1-632 3557, fax: +41-1-632 1211
E-mail address: `{allgower,findeise}@aut.ee.ethz.ch`

Progress in Systems and Control Theory, Vol. 26

Analytical Model Predictive Control

Masoud Soroush and Kenneth R. Muske

Abstract. This presentation deals with the model predictive controllers that admit an analytical solution. In particular, short horizon and long (infinite) horizon controllers are considered. A general model predictive control (MPC) law is formulated, and it is then shown that in special cases the MPC law leads to the following:

- input-output linearizing control laws that inherently include optimal windup and directionality compensators,
- model state feedback control and modified internal model control laws that inherently include an optimal directionality compensator,
- proportional–integral (PI) and proportional–integral–derivative (PID) controllers that inherently include optimal windup and directionality compensators.

Thus, model predictive control allows for solving the important problems of *optimal* windup and directionality compensations in the existing analytical controllers. Other implications of this work are:

- A modified internal model controller is a model predictive controller only when the characteristic (decoupling) matrix of the process under consideration can be made diagonal by row or column rearrangements, and/or the process does not have a constraint.
- An original internal model controller is a model predictive controller only when the process under consideration does not have a constraint.

Special cases of long horizon MPC that lend themselves an exact or an approximate analytical solution are also discussed.

1. Introduction

Model predictive control is a very broad controller-synthesis methodology. This is due to the optimization-based nature of MPC and the great number of tunable parameters that a model predictive controller can possess. Its optimization-based nature allows one to use any desirable performance index irrespective of the type of process model available. Its considerable number of tunable parameters provide many degrees of freedom to obtain a desirable closed-loop response. The tunable parameters include model horizon, prediction horizons, control horizons, reference trajectories (set-point filters), and penalty matrices in the performance index. Each of the parameters has a distinct effect on the closed-loop performance and this effect in general is not "monotonic". The performance of a model predictive controller can be as non-robust and aggressive as that of a deadbeat controller, or as robust and slow as that of a steady state (open-loop) controller.

TABLE 1. Unconstrained single-input single-output model predictive controllers with no penalty on manipulated input magnitude or rate of change (r=relative order).

Prediction Horizon=P	Control Horizon=M	Reference Trajectory	Resulting Controller
r	1	No	Deadbeat
r	1	Yes	I-O Linearization
r	1	Yes	Modified IMC
r	1	Yes	MSFC
$r = 1$	1	Yes	PI, PID
$P >> r$	$M > 1$	Yes/No	Long Horizon MPC
∞	1	Yes/No	Steady State

Much of the literature on model predictive control has been on practical issues in the implementation of MPC, controller tuning, and stability analysis. On the other hand, there has been little work on establishing the connections between MPC and existing analytical control methods. There are many who still believe that MPC and analytical control methods are not related, while as we will see, analytical control methods are special cases of MPC. More precisely, in special cases MPC leads to analytical control methods that are identical to the existing analytical control methods when there are no constraints, but that do not exhibit the poor performance of the existing analytical control methods when a constraint is present. A few special cases of unconstrained single-input single-output (SISO) MPC are listed in Table 1. As we increase the prediction horizon and decrease the control horizon, we obtain a more robust but less aggressive controller.

In this work, the connections between MPC and several analytical control methods – such as input-output (I-O) linearization, model state feedback control (MSFC) (Coulibaly et al., 1995), modified internal model control (IMC) (Zheng et al., 1994), PI control, and PID control – are established. By establishing these connections, the important problems of integral windup, process directionality, and controller tuning in the analytical control methods are solved directly. The class of nonlinear processes described by a discrete-time state-space nonlinear model are considered. A general MPC law is formulated, and it is then shown that in special cases the MPC law leads to analytical control methods with very appealing theoretical features. Special cases of long horizon MPC that can take an analytical form are also discussed.

Section 2 describes the scope of the study and presents some mathematical preliminaries. A general formulation of model predictive control is given in Section 3. Several special cases of the general model predictive control law are considered in Section 4. In particular, the control law is applied to time-invariant linear systems to solve the problems of optimal windup and/or directionality compensation(s) in the modified internal model control, the model state feedback control, PI control, and PID control. Section 5 presents a discussion of the special cases of long horizon model predictive control where the control law has an analytical solution. Finally concluding remarks are given in Section 6.

2. Scope and preliminaries

We consider nonlinear multivariable processes described by a mathematical model of the form

$$\begin{aligned} x(k+1) &= \Phi[x(k), u(k)], \qquad x(0) = 0 \\ y(k) &= h[x(k)] \end{aligned} \tag{1}$$

with the magnitude and rate-of-change input constraints

$$\begin{aligned} u_{l_j} \le u_j(k) \le u_{h_j}, \ j = 1, \ldots, q \\ \Delta u_{l_j} \le u_j(k) - u_j(k-1) \le \Delta u_{h_j}, \ j = 1, \ldots, q \end{aligned}$$

where $x = [x_1 \cdots x_n]^T$, $u = [u_1 \cdots u_q]^T$, and $y = [y_1 \cdots y_m]^T$ denote the vectors of state variables, manipulated inputs, and controlled outputs respectively, all in the form of deviation variables [i.e. $(x_{ss}, u_{ss}) = (0,0)$ is the nominal equilibrium pair], $\Phi(x,u)$ and $h(x)$ are smooth vector functions, and u_{l_j}, u_{h_j}, Δu_{l_j}, and Δu_{h_j}, $j = 1, \ldots, q$, are constant scalar quantities that satisfy $u_{l_j} < 0 < u_{h_j}$ and $\Delta u_{l_j} < 0 < \Delta u_{h_j}$, $j = 1, \ldots, q$. The vector of measured controlled outputs, denoted by $\bar{y}$, is related to the vector of measured state variables, represented by $\bar{x}$, according to

$$\bar{y}(k) = h[\bar{x}(k)] + d \tag{2}$$

where d is the vector of unmeasurable constant disturbances.

For a process of the form of (1),

- relative order (degree) of an output y_i with respect to the vector of manipulated inputs is denoted by r_i. A relative order r_i is the smallest integer for which $y_i(k + r_i)$ depends explicitly on a manipulated input u_j (Nijmeijer and van der Schaft, 1990).
- relative order (degree) of an output y_i with respect to a manipulated input u_j is represented by r_{ij}. A relative order r_{ij} is the smallest integer for which $y_i(k + r_{ij})$ depends explicitly on the manipulated input u_j (Soroush and Kravaris, 1996).
- the characteristic (decoupling) matrix

$$\Theta(x,u) \triangleq \begin{bmatrix} \dfrac{\partial}{\partial u} h_1^{r_1}(x,u) \\ \vdots \\ \dfrac{\partial}{\partial u} h_m^{r_m}(x,u) \end{bmatrix}$$

where

$$\begin{aligned} h_i^0(x) &\triangleq h_i(x), \quad i = 1, \ldots, m \\ h_i^\ell(x) &\triangleq h_i^{\ell-1}[\Phi(x,u)], \quad \ell = 1, \ldots, r_i - 1, \qquad i = 1, \ldots, m \\ h_i^{r_i}(x,u) &\triangleq h_i^{r_i-1}[\Phi(x,u)], \quad i = 1, \ldots, m \end{aligned}$$

A relative order r_{ij} is indeed the minimum number of sampling periods that takes to observe the effect of a change in the manipulated input u_j on the controlled output y_i. A relative order r_i is related to the relative orders $r_{i1}, \ldots, r_{im}$ according to $r_i = \min(r_{i1}, \ldots, r_{im})$. The characteristic matrix of a process includes important information on systems properties of the process. For example, generic singularity of the characteristic matrix of a process represents a structural process deficiency which complicates the synthesis of input-output linearizing controllers, internal model controllers, and shortest prediction horizon model predictive controllers. If the characteristic matrix of a process is generically singular, then to the process no controller can induce a closed-loop response that is time-optimal in every controlled output and that is completely input-output decoupled (Soroush, 1996).

3. Optimization problem: Model predictive control law

We consider a constrained minimization problem of the form:

$$\min_{U(k)} \left\{ \sum_{j=1}^{m} \sum_{\ell=r_j}^{P_j} a_{j\ell} \left[y_{d_j}(k+\ell) - \hat{y}_j(k+\ell) \right]^2 \right. \\ \left. + \sum_{j=1}^{q} \sum_{\ell=0}^{M_j-1} b_{j\ell} \left[u_j(k+\ell) - u_j(k+\ell-1) \right]^2 + \sum_{j=1}^{q} \sum_{\ell=0}^{Q_j} c_{j\ell} \left[u_j(k+\ell) \right]^2 \right\} \tag{3}$$

subject to

$$\begin{aligned} x(k+1) &= \Phi[x(k), u(k)] \\ \bar{y}(k) &= h[\bar{x}(k)] + d \\ u_{l_j} \le u_j(k) &\le u_{h_j}, \; j = 1, \ldots, q \\ \Delta u_{l_j} \le u_j(k) - u_j(k-1) &\le \Delta u_{h_j}, \; j = 1, \ldots, q \\ u_j(k+\ell) = u_j(k+\ell-1), &\;\; \ell = M_j, \ldots, Q_j, \; j = 1, \ldots, q \end{aligned}$$

where P_j is the prediction horizon of a controlled output y_j, M_j is the control horizon of a manipulated input u_j, $a_{j\ell}$, $j = 1, \ldots, m$, $\ell = r_j, \ldots, P_j$, are weights on the deviations of the controlled outputs from their corresponding reference trajectories, $b_{j\ell}$, $j = 1, \ldots, q$, $\ell = 0, \ldots, M_j - 1$, are weights on the size of manipulated inputs, $c_{j\ell}$, $j = 1, \ldots, q$, $\ell = 0, \ldots, Q_j$, are weights on the rates of change of manipulated inputs, and

$$Q_j = \max[(P_1 - r_{1j}), \ldots, (P_m - r_{mj})], \quad j = 1, \ldots, q$$

$$U(k) = [u_1(k) \cdots u_1(k+Q_1) \quad \cdots \quad u_q(k) \cdots u_q(k+Q_q)]^T$$

Each reference trajectory y_{d_i} is given by

$$y_{d_i}(k+P_i) = (1 + \beta_{i1} + \cdots + \beta_{iP_i}) y_{sp_i} - \beta_{i1}\hat{y}_i(k+P_i-1) - \cdots - \beta_{iP_i}\hat{y}_i(k), \quad i = 1, \ldots, m$$

$$y_{d_i}(k+\ell) = \hat{y}_i(k+\ell), \qquad \ell = 0, \ldots, P_i - 1, \; i = 1, \ldots, m$$

where $\beta_{i1}, \ldots, \beta_{iP_i}$, $i = 1, \ldots, m$, are tunable scalar parameters chosen such that the diagonal matrix

$$Q = \operatorname{diag}\left\{ 1 + \sum_{\ell=1}^{r_i} \beta_{i\ell} \right\}$$

is not singular, and $y_{sp} = [y_{sp_1} \cdots y_{sp_m}]^T$ is the vector of output set-points. Each predicted output $\hat{y}_i$ is given by

$$\begin{aligned} \hat{y}_i(k+\ell) &\triangleq \bar{y}_i(k) + h_i^\ell[x(k)] - h_i[x(k)], \qquad \ell = 1, \ldots, r_i - 1 \\ \hat{y}_i(k+r_i) &\triangleq \bar{y}_i(k) + h_i^{r_i}[x(k), u(k)] - h_i[x(k)] \\ \hat{y}_i(k+\ell) &\triangleq \bar{y}_i(k) + h_i^\ell[x(k), U(k)] - h_i[x(k)], \qquad \ell = r_i + 1, \ldots, P_i \end{aligned} \tag{4}$$

where

$$\begin{aligned} h_i^{r_i+1}(x, U) &\triangleq h_i^{r_i}[\Phi(x, u), u] \\ h_i^\ell(x, U) &\triangleq h_i^{\ell-1}[\Phi(x, u), U], \quad \ell = r_i + 2, \ldots, P_i \end{aligned}$$

When measurements of the state variables are available, in the preceding prediction equations we set $x = \bar{x}$. Otherwise, the values of the state variables have to be estimated, for example, via on-line simulation of the process model (use of an open-loop state estimator) or by using an extended Kalman filter.

4. Shortest prediction horizon MPC

4.1. I-O linearizing state feedback

For the class of processes with equal number of manipulated inputs and controlled outputs ($q = m$), no manipulated-input rate-of-change constraints, finite relative orders $r_1, \ldots, r_m$, and a locally nonsingular characteristic matrix $\Theta(x, u)$, let us use the following settings for the general model predictive control law of (3): $P_j = r_j$, $M_j = 1$, $b_{j0} = c_{j0} = 0$, $j = 1, \ldots, m$.

In other words, we seek nonlinear model predictive controllers that are the solution to the constrained m-dimensional minimization problem:

$$\min_{u(k)} \left\{ \sum_{j=1}^{m} a_{jr_j} \left[y_{d_j}(k + r_j) - \hat{y}_j(k + r_j) \right]^2 \right\} \tag{5}$$

subject to the input constraints $u_{l_\ell} \leq u_\ell(k) \leq u_{h_\ell}$, $\ell = 1, \ldots, m$

Solving the constrained minimization problem of (5) leads to the static state feedback:

$$u(k) = \Psi[x(k), y_{sp}(k) - d(k)] \tag{6}$$

where $\Psi[x(k), v(k)]$ denotes the solution to the constrained m-dimensional minimization problem:

$$\min_{u(k)} ||\text{diag}\{a_{ir_i}\}\{\phi[x(k), u(k)] - Qv(k)\}||^2 \tag{7}$$

subject to the input constraints $u_{l_\ell} \leq u_\ell(k) \leq u_{h_\ell}$, $\ell = 1, \ldots, m$

Here $|| \cdot ||$ is the Euclidean norm and

$$\phi[x,u] \triangleq \begin{bmatrix} h_1^{r_1}[x,u] + \sum_{\ell=1}^{r_1} \beta_{1\ell} h_1^{r_1-\ell}[x] \\ \vdots \\ h_m^{r_m}[x,u] + \sum_{\ell=1}^{r_m} \beta_{m\ell} h_m^{r_m-\ell}[x] \end{bmatrix} \tag{8}$$

In the absence of the constraints, the optimizing state feedback of (6) can force all the controlled outputs to match their corresponding linear reference trajectories exactly provided that the vector of constant disturbances, d, is known. Thus, in this case the optimizing state feedback of (6) induces the following offset-free, linear, input-output, closed-loop response:

$$\begin{bmatrix} \bar{y}_1(k+r_1) + \sum_{\ell=1}^{r_1} \beta_{1\ell} \bar{y}_1(k+r_1-\ell) \\ \vdots \\ \bar{y}_m(k+r_m) + \sum_{\ell=1}^{r_m} \beta_{m\ell} \bar{y}_m(k+r_m-\ell) \end{bmatrix} = Q y_{sp}(k) \tag{9}$$

that is, the state feedback of (6) is an input-output linearizing state feedback.

4.2. I-O linearizing feedback control

In order for the state feedback of (6) to induce the offset-free, linear, input-output, closed-loop response of (9), the unmeasurable disturbances should be estimated from the available measurements. The vector of the disturbances $d = \bar{y} - y$, where y is the vector of "disturbance-free" controlled outputs given by $y = h(x)$. Thus, if we can estimate the disturbance-free controlled outputs, then we can estimate the unmeasurable disturbances. There are many approaches to estimating the disturbance-free controlled outputs. Here we use only two of these to derive feedback controllers that can induce the offset-free, linear, input-output, closed-loop response of (9).

The first approach to obtaining estimates of the disturbance-free controlled outputs, $y_1, \ldots, y_m$, involves using measurements of state variables (when measurements of all of the state variables are available), values of manipulated inputs, and the identity:

$$\begin{bmatrix} y_1(k+r_1) + \sum_{\ell=1}^{r_1} \beta_{1\ell} y_1(k+r_1-\ell) \\ \vdots \\ y_m(k+r_m) + \sum_{\ell=1}^{r_m} \beta_{m\ell} y_m(k+r_m-\ell) \end{bmatrix} = \phi[x(k), u(k)] \tag{10}$$

We use the linear system

$$\begin{aligned} \eta(k+1) &= A_c \eta(k) + B_c \phi[x(k), u(k)], \quad \eta(0) = 0 \\ y(k) &= C_c \eta(k) \end{aligned} \tag{11}$$

to represent a minimal-order state-space realization of the system of (10), where $\eta \in \Re^{r_1+\cdots+r_m}$, and A_c, B_c and C_c are constant matrices. Thus, in this case the disturbance estimate $d = \bar{y} - C_c\eta$.

The second approach to obtaining estimates of the disturbance-free controlled outputs is to simulate the process model:

$$\begin{aligned} x(k+1) &= \Phi[x(k), u(k)] \\ y(k) &= h[x(k)] \end{aligned} \tag{12}$$

Thus, in this case the disturbance estimate $d = \bar{y} - h(x)$.

The two approaches to estimating the disturbance-free controlled outputs are different in the following aspects:

- The system of (11) is of order $(r_1 + \cdots + r_m)$, while that of (12) is of order n; that is, the system of (11) is of a lower order.
- The first approach requires measurements of all the state variables, while the second approach does not need any of the measurements.
- The first approach is applicable to processes operating at an unstable steady state, while the second approach is not (see the controllers of Theorems 1 and 2 and the conditions for closed-loop stability given in Subsubsection 4.2.3).
- In the case that all the state variables of the process under consideration can be measured on-line and that the process is to be operated at a stable steady state, the dynamic controller resulting from using the first approach (that of Theorem 1) will be a reduced-order realization of the dynamic controller resulting from using the second approach (that of Theorem 2).

The estimation of d by the two approaches and use of the estimates along with the state feedback of (6) lead to feedback controllers that have integral action (in the presence of constant disturbances and model errors, induce an offset-free closed-loop response) but that do not exhibit integral windup (because they are model predictive controllers). Several definitions of windup can be found in (Valluri et al., 1998). The following two theorems describe two nonlinear controllers that have the aforementioned properties.

4.2.1. I-O LINEARIZING MIXED ERROR- AND STATE-FEEDBACK CONTROL

Theorem 1 (Grantz et al., 1998): For a process of the form of (1) with complete state measurements ($\bar{x}$), finite relative orders, $r_1, \ldots, r_m$, and a locally nonsingular characteristic matrix $\Theta(x, u)$, the dynamic mixed error- and state-feedback control law:

$$\begin{aligned} \eta(k+1) &= A_c\eta(k) + B_c\phi[\bar{x}(k), u(k)], \quad \eta(0) = 0 \\ u(k) &= \Psi\{\bar{x}(k), C_c\eta(k) + e(k)\} \end{aligned} \tag{13}$$

where $e = y_{sp} - \bar{y}$, is the solution to the minimization problem of (5), and in the absence of constraints induces the linear input-output closed-loop response of (9).

The block diagram of the controller of (13) is depicted in Figure 1. The term $C_c\eta$ simply represents the vector of the estimated values of disturbance-free

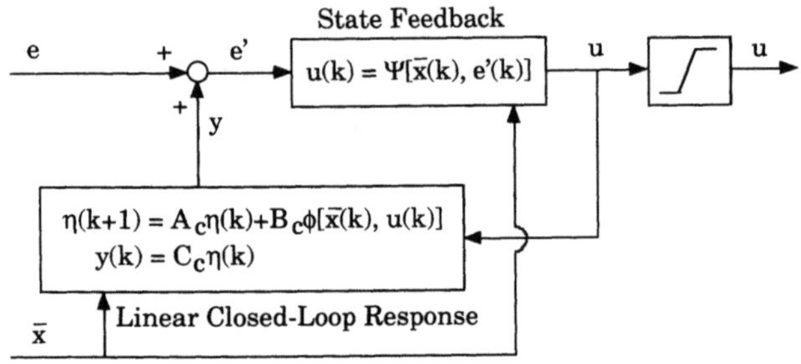

FIGURE 1. Mixed error- and state-feedback control structure.

controlled outputs, obtained via on-line simulation of the nominal linear input-output closed-loop response. Since $C_c\eta + e = y_{sp} - d$, the preceding controller can be interpreted as a nonlinear model state feedback (modified internal model) controller for processes with complete state measurements. In the continuous-time setting, these connections have been established in (Valluri et al., 1998).

4.2.2. I-O LINEARIZING ERROR-FEEDBACK CONTROL

Theorem 2 (Grantz et al., 1998): For a process of the form of (1) with incomplete state measurements, finite relative orders, $r_1, \ldots, r_m$, and a locally nonsingular characteristic matrix $\Theta(x, u)$, the dynamic error-feedback control law:

$$\begin{aligned} x(k+1) &= \Phi\left[x(k), u(k)\right], \qquad\qquad x(0) = 0 \\ u(k) &= \Psi\{x(k), h[x(k)] + e(k)\} \end{aligned} \tag{14}$$

is the solution to the minimization problem of (5), and in the absence of constraints induces the linear input-output closed-loop response of (9).

The block diagram of the controller of (14) is depicted in Figure 2. In this case, the estimated values of the disturbance-free controlled outputs are obtained via on-line simulation of the process model. Since $h(x) + e = y_{sp} - d$, the preceding controller can be interpreted as a nonlinear model state feedback (modified internal model) controller for processes with incomplete state measurements.

Note that the integral action of the controllers of Theorems 1 and 2 ensures offset-free closed-loop responses provided that (a) the closed-loop system is asymptotically stable and (b) the process is subjected to "rejectable" constant disturbances and "feasible" set-point step changes (rejectable and feasible in the sense that at steady state the manipulated inputs always take values that are within the manipulated input bounds).

4.2.3. CONDITIONS FOR CLOSED-LOOP STABILITY

In the absence of constraints, model-errors, and disturbances, the closed-loop systems under the controllers of Theorems 1 and 2 will be asymptotically stable, if the following conditions are satisfied. The closed-loop systems will be input-output stable, if the parameters $\beta_{i\ell}$, $i =, \ldots, m$, $\ell = 1, \ldots, r_i$, are chosen such that all the roots of the characteristic equation

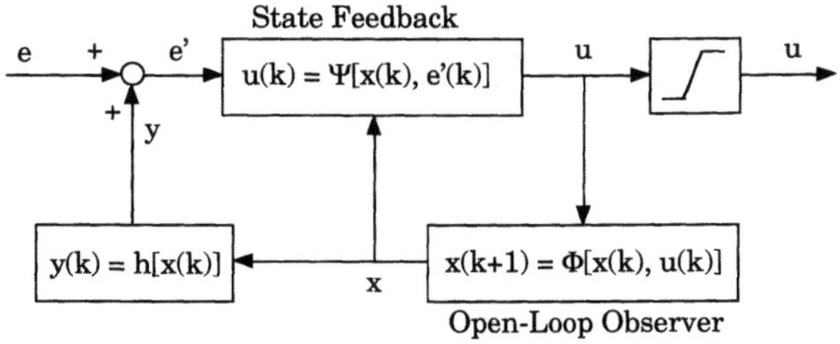

FIGURE 2. Error-feedback control structure.

$$\left(z^{r_1} + \sum_{\ell=1}^{r_1} \beta_{1\ell} z^{r_1-\ell}\right) \cdots \left(z^{r_m} + \sum_{\ell=1}^{r_m} \beta_{m\ell} z^{r_m-\ell}\right) = 0 \tag{15}$$

lie inside the unit circle. For an input-output stable closed-loop system, the internal stability will be ensured [a] under the controller of Theorem 1, if the process is hyperbolically minimum-phase; and [b] under the controller of Theorem 2, if the process is hyperbolically minimum-phase, and the process is asymptotically open-loop stable. A process is said to be hyperbolically minimum-phase, if all the zeros of the linear approximation of the process (obtained by a first-order truncated Taylor series expansion of the process model around the nominal steady state) lie strictly inside the unit circle (Soroush and Kravaris, 1996).

4.2.4. NONLINEAR SPECIAL CASES

Case 1. In the case that the vector function $\phi(x,u)$ is affine-in-u [i.e. $\phi(x,u) = f(x)+\Theta(x)u$, where $f(x) \in \Re^{n\times 1}$], and that the characteristic matrix, $\Theta(x)$, can be made diagonal by row or column rearrangements, the minimization problem of (5) will have an analytical solution, and thus there is no need for on-line optimization. The corresponding explicit analytical forms of the control laws of Theorems 1 and 2 respectively are:

$$\begin{aligned} \eta(k+1) &= A_c\eta(k) + B_c\left\{f[\bar{x}(k)] + \Theta[\bar{x}(k)]u(k)\right\}, \quad \eta(0)=0 \\ u(k) &= \text{sat}\left\{\text{diag}\left\{\frac{1}{\Theta_{ii}}\right\}(Q\{C_c\eta(k)+e(k)\} - f[\bar{x}(k)])\right\} \end{aligned} \tag{16}$$

and

$$\begin{aligned} x(k+1) &= \Phi\left[x(k), u(k)\right], \qquad x(0)=0 \\ u(k) &= \text{sat}\left\{\text{diag}\left\{\frac{1}{\Theta_{ii}}\right\}(Q\{h[x(k)]+e(k)\} - f[x(k)])\right\} \end{aligned} \tag{17}$$

where Θ_{ii} is the iith entry of the characteristic matrix, and

$$\text{sat}_i(\theta) \triangleq \begin{cases} u_{l_i}, & \theta \le u_{l_i} \\ \theta, & u_{l_i} \le \theta \le u_{h_i} \\ u_{h_i}, & u_{h_i} \le \theta \end{cases} \qquad i=1,\ldots,m \tag{18}$$

Thus, in this case the optimal controller action is obtained exactly by "clipping" the unconstrained controller outputs.

Case 2: Nonlinear PI controllers. In the case of first-order single-input single-output (SISO) processes, the controllers of Theorems 1 and 2 respectively take the following simple forms:

$$\begin{aligned} \eta(k+1) &= \beta_1[h[\bar{x}(k)] - \eta(k)] + h^1[\bar{x}(k), u(k)] \\ u(k) &= \Psi\{\bar{x}(k), \eta(k) + e(k)\} \end{aligned} \tag{19}$$

$$\begin{aligned} x(k+1) &= \Phi[x(k), u(k)] \\ u(k) &= \Psi\{x(k), h[x(k)] + e(k)\} \end{aligned} \tag{20}$$

which are simply two nonlinear PI controllers. Unlike the PI controller of (19), that of (20) cannot operate the SISO processes at an unstable steady state (see the conditions for closed-loop stability under the control law of (14), given in Subsubsection 4.2.3). This inability of the model-based controller of (20) is a consequence of the fact that the controller includes an open-loop state estimator; the number of the unstable modes of the closed-loop system under the controller of (20) are the same as that of the open-loop process.

4.2.5. Linear special cases

Consider the class of time-invariant, linear processes described by a state-space model of the form

$$\begin{aligned} x(k+1) &= Ax(k) + Bu(k), \qquad x(0) = 0 \\ y(k) &= Cx(k) \end{aligned} \tag{21}$$

with the same rate-of-change and magnitude input constraints of the system of (1), where A, B and C are $n\times n$, $n\times m$ and $m\times n$ constant matrices respectively. This class of systems is a special case of (1). It is assumed that (a) the delay-free part of the process is minimum-phase, (b) the process has finite relative orders $r_1, \ldots, r_m$, and (c) the process has a nonsingular characteristic matrix, given by $\Theta = \left[c_1 A^{r_1-1}B \cdots c_m A^{r_m-1}B\right]^T$, where c_i is the ith row of the matrix C.

Application of the control law of (13) to the linear processes of the form of (21) leads to the linear control law:

$$\begin{aligned} \eta(k+1) &= A_c\eta(k) + B_c\left[D\bar{x}(k) + \Theta u(k)\right] \\ w(k) &= \Theta^{-1}\left[Qe'(k) - D\bar{x}(k)\right] \\ u(k) &= F\left\{\Theta, w(k)\right\} \end{aligned} \tag{22}$$

where $e' = e + C_c\eta$,

$$D = \left[\left(c_1A^{r_1} + \sum_{\ell=1}^{r_1}\beta_{1\ell}c_1A^{r_1-\ell}\right) \cdots \left(c_mA^{r_m} + \sum_{\ell=1}^{r_m}\beta_{m\ell}c_mA^{r_m-\ell}\right)\right]^T$$

and $F[\Theta, w]$ represents the solution to the m-dimensional quadratic program:

$$\min_u ||diag\{a_{ir_i}\}\Theta(u-w)||^2 \tag{23}$$

subject to $u_{l_\ell} \le u_\ell \le u_{h_\ell}$, $\ell = 1, \ldots, m$.

The quadratic program of (23) is trivially solvable. For example, one can use the following computationally efficient, simple iterative method (Soroush and Valluri, 1998):

$$u^{\ell+1} = \tilde{\Phi}\left(u^\ell\right), \quad \ell = 0, 1, 2, 3, \ldots \tag{24}$$

where

$$\tilde{\Phi}(u) = sat\left[\delta^{-1}P(w-u)+u\right], \qquad P = [p_{ij}] \triangleq \Theta^T \mathrm{diag}\{a_{ir_i}^2\}\Theta,$$

$$\delta = ||P||_F \triangleq \sqrt{\sum_{i=1}^{m}\sum_{j=1}^{m} {p_{ij}}^2}, \qquad \text{(Frobenius norm of P)}$$

The function $\tilde{\Phi}(u)$ has the following very attractive properties:

(i) $u^+ = \tilde{\Phi}(u^+)$, where u^+ denotes the solution to the quadratic program of (23),

(ii) the fixed-point iteration sequence $\{u^\ell\}$ generated by

$$u^{\ell+1} = \tilde{\Phi}\left(u^\ell\right), \quad \ell = 0, 1, 2, 3, \ldots$$

converges to u^+ for every initial guess $u^0 \in \Re^{m\times 1}$.

The expression $F[\Theta, w]$ is indeed an *optimal directionality compensator* that calculates optimal feasible process inputs on the basis of unconstrained controller outputs and the characteristic matrix of the process. This optimal directionality compensator is the discrete-time analog of the optimal directionality compensator for continuous-time processes presented in (Soroush and Valluri, 1998).

Remark 1: For the processes whose characteristic matrix can be made diagonal by row or column rearrangements, the optimal directionality compensator is identical to clipping:

$$F[\Theta, w(k)] = \mathrm{sat}\{w(k)\}$$

Application of the control law of (14) to the linear processes of the form of (21) results in the linear control law:

$$\begin{aligned} x(k+1) &= Ax(k) + BF\{\Theta, w(k)\} \\ w(k) &= \Theta^{-1}\left[Qe'(k) - Dx(k)\right] \\ u(k) &= F\{\Theta, w(k)\} \end{aligned} \qquad (25)$$

where $e' = e + Cx$.

Case 1: Modified internal model control. Modified IMC parameterization of the linear control law of (25) is depicted in Figure 3, and the corresponding controller components are given in Table 2. The component Q_3 is the optimal directionality compensator. When the control law of (25) is applied to a time-invariant linear system whose characteristic (decoupling) matrix can be made diagonal by row or column rearrangements, the optimal directionality compensator will be identical to clipping and the resulting linear controller will be a minimal-order state-space realization of a modified internal model controller. These results have the following implications:

- A modified internal model controller is a model predictive controller only when the characteristic matrix of the process under consideration can be made diagonal by row or column rearrangements, and/or the process is not constrained.
- An original internal model controller is a model predictive controller only when the process under consideration is not constrained.

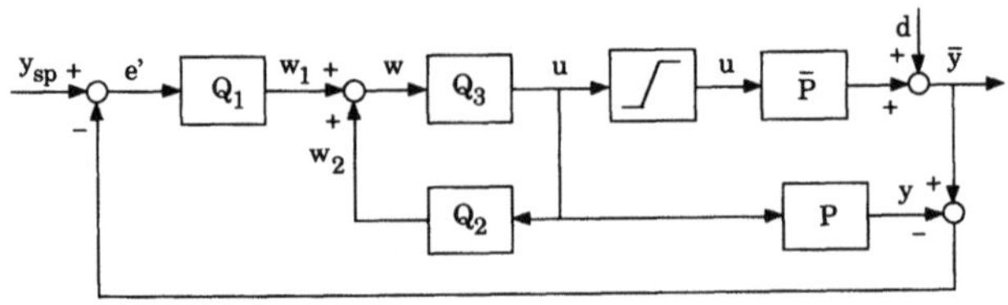

FIGURE 3. Modified IMC parameterization of the linear controller of (25).

TABLE 2. Modified IMC parameterization of the linear controller of (25).

P	$x(k+1) = Ax(k) + Bu(k)$ $y(k) = Cx(k)$
Q_1	$w_1(k) = \Theta^{-1}Qe'(k)$
Q_2	$x(k+1) = Ax(k) + Bu(k)$ $w_2(k) = \Theta^{-1}Dx(k)$
Q_3	$u(k) = F\{\Theta, w(k)\}$

Case 2: Model state feedback control. MSFC parameterizations of the linear control laws of (25) and (22) are respectively depicted in Figures 4 and 5, and the corresponding controller components are given in Tables 3 and 4. In these parameterizations the components Q_3' and Q_3'' represent the optimal directionality compensator that is not included in the MSFC structure developed by Brosilow and coworkers (Coulibaly et al., 1995; Mhatre and Brosilow, 1996). Mhatre and Brosilow (1996) have addressed the problem of process directionality in MSFC by detuning their proposed model state feedback controller: when a manipulated input saturates the controller is detuned so that every controller output falls within the feasible range of the corresponding plant input.

Case 3: 1st-order SISO linear processes. Consider the class of first-order linear processes described by

$$\begin{aligned} x(k+1) &= ax(k) + bu(k) \\ y(k) &= x(k) \end{aligned} \tag{26}$$

where $x \in \Re$ and $b \neq 0$. For such a process, the model predictive controllers of (22) and (25) respectively take the forms (after some manipulations)

$$\begin{aligned} \eta(k+1) &= \eta(k) + (1+\beta_1)e(k) - b\,[w(k) - \text{sat}\{w(k)\}] \\ w(k) &= \frac{1+\beta_1}{b}[e(k) + \eta(k)] - \frac{\beta_1 + a}{b}y(k) \end{aligned} \tag{27}$$

$$\begin{aligned} x(k+1) &= x(k) + (1+\beta_1)e(k) - b\,[w(k) - \text{sat}\{w(k)\}] \\ w(k) &= \frac{1+\beta_1}{b}e(k) + \frac{1-a}{b}x(k) \end{aligned} \tag{28}$$

which are discrete-time PI controllers with only one tunable parameter, β_1. Each of the PI controllers inherently includes an optimal windup compensator. While the controller of (27) is suitable for unstable and stable first-order processes of

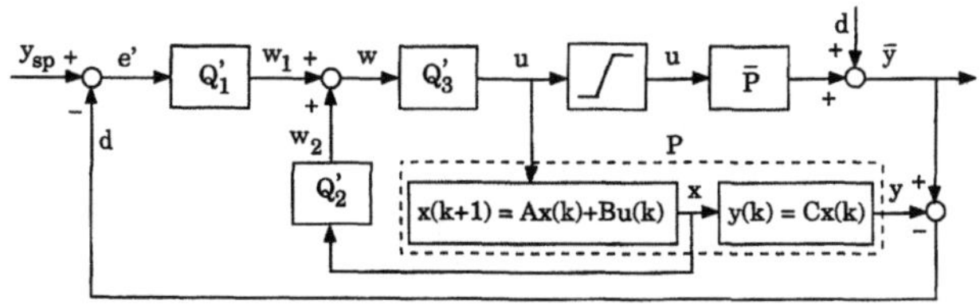

FIGURE 4. MSFC parameterization of the linear controller of (25).

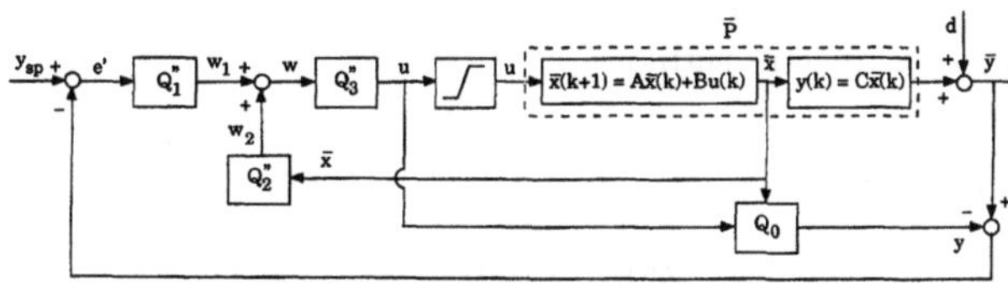

FIGURE 5. MSFC parameterization of the linear controller of (22).

TABLE 3. MSFC parameterization of the linear controllers of (25) and (22).

P	$x(k+1) = Ax(k) + Bu(k)$ $y(k) = C_c\eta(k)$	Q_0	$\eta(k+1) = A_c\eta(k) + B_c\left[D\bar{x}(k) + \Theta u(k)\right]$ $y(k) = C_c\eta(k)$
Q_1'	$w_1(k) = \Theta^{-1}Qe'(k)$	Q_1''	$w_1(k) = \Theta^{-1}Qe'(k)$
Q_2'	$w_2(k) = \Theta^{-1}Dx(k)$	Q_2''	$w_2(k) = \Theta^{-1}D\bar{x}$
Q_3'	$u(k) = F\{\Theta, w(k)\}$	Q_3''	$u(k) = F\{\Theta, w(k)\}$

the form of (26) (whether the process is stable or unstable, choosing $|\beta_1| < 1$ ensures the closed-loop stability), that of (28) cannot be used for unstable first-order processes (see the conditions for closed-loop stability under the control law of (14), given in Subsubsection 4.2.3). Parameterizations of the PI controllers of (27) and (28) in the z-domain are depicted in Figure 6. This type of parameterization is a very common for PI controllers with the classical anti-integral windup scheme. For a review of the available methods of anti-windup schemes, the reader can refer to (Kothare et al., 1994; Bernstein and Michel, 1995; Kendi and Doyle, 1997; Kapoor and Daoutidis, 1998). In the case of multivariable linear processes with $r_1 = \cdots = r_m = 1$, the controllers of (22) and (25) lead to multivariable PI controllers with optimal windup and directionality compensators.

Case 4: 2nd-order SISO linear processes. For the class of second-order SISO linear processes of the form

$$\begin{aligned} x_1(k+1) &= x_2(k) \\ x_2(k+1) &= a_1x_1(k) + a_2x_2(k) + bu(k) \\ y(k) &= cx_1(k) + x_2(k) \end{aligned} \tag{29}$$

the model predictive control laws of (22) and (25) lead to two discrete-time PID controllers with only one tunable parameter, β_1. Each of these PID controllers includes inherently an optimal windup compensator.

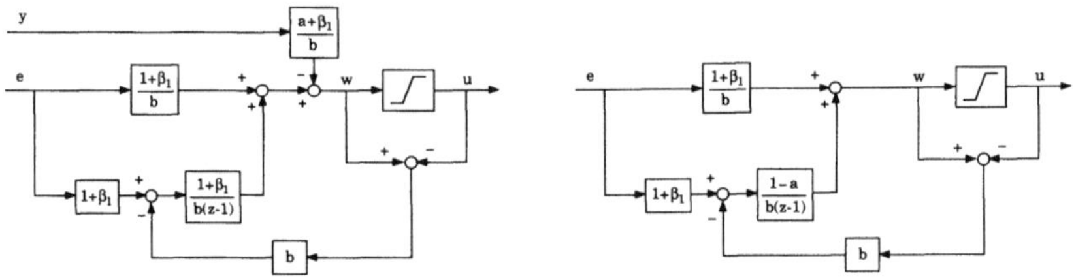

FIGURE 6. Parameterization of the PI controllers of (27), on the left, and (28), on the right.

5. Long horizon MPC

In this section, special cases of long horizon model predictive control where the optimization problem admits an analytical solution are discussed. In long horizon model predictive control, the prediction horizons are much greater than the relative orders of the process and the control horizons are greater than one. An advantage of long horizon MPC is that it is applicable to a much broader class of processes, for example, to non-minimum-phase processes. Recall that the short horizon MPC previously discussed is inadequate for these processes. However, an analytical form of solution is generally not available for long horizon model predictive controllers.

The general control law in (3) with the following settings:

$$\begin{array}{l} P_j = N-1,\ j=1,\ldots,m;\ \ Q_j = N-1,\ j=1,\ldots,q \\ a_{j\ell} = a_j,\ j=1,\ldots,m;\ \ell=1,\ldots,n-1;\ \ b_{j\ell}=0,\ \forall\, j,\ell \\ c_{j\ell} = c_j,\ j=1,\ldots,q;\ \ell=1,\ldots,N-1 \\ y_{sp_j} \equiv 0,\ \beta_{j\ell}=0,\ j=1,\ldots,m;\ \ell=1,\ldots,N-1 \end{array}$$

and with a terminal state penalty added to the objective function, takes the form:

$$\min_{\tilde{U}(k)} \left\{ \sum_{j=1}^{m}\sum_{\ell=0}^{N-1} a_j\left[\hat{y}_j(k+\ell)\right]^2 + \sum_{j=1}^{q}\sum_{\ell=0}^{N-1} c_j\left[u_j(k+\ell)\right]^2 + \sum_{j=1}^{n} g_j\left[x_j(k+N)\right]^2 \right\} \tag{30}$$

where $g_1,\ldots,g_n$ are positive scalars (weights on the terminal values of the state variables), and $\tilde{U}(k) = [u_1(k)\cdots u_q(k)\ \ \cdots\ \ u_1(k+N)\cdots u_q(k+N)]^T$.

In this case, the controlled outputs are controlled to the origin, the prediction and control horizons are N, and the rate of change penalties on the manipulated inputs are ignored.

5.1. Linear quadratic regulator

For the class of unconstrained, linear, time-invariant, discrete-time processes described by a model of the form of (21) without the assumptions (a) through (c), the optimization problem of (30) admits an analytical solution in the form of the linear state feedback (Bitmead et al., 1990)

$$u(k) = K(N)x(k) \tag{31}$$

$$K(N) = -\left(B^T P(N) B + \bar{R}\right)^{-1} B^T P(N) A \tag{32}$$

where $P(N)$ is the Nth iteration of the Riccati difference equation and

$$\bar{Q} = \mathrm{diag}\{a_j\}, \quad \bar{R} = \mathrm{diag}\{c_j\}, \quad \bar{Q}_N = \mathrm{diag}\{g_j\}$$

When (A, B) is stabilizable and $\left(A, (C^T\bar{Q}C)^{1/2}\right)$ is detectable, nominal stability of the state feedback controller can be achieved by choosing a value of N such that $P(N+1) \leq P(N)$. If, in addition, $\bar{R}$ is positive definite, the Riccati difference equation converges to a constant matrix $P(\infty)$ yielding a nominally stabilizing infinite horizon feedback gain $K(\infty)$ (Bitmead et al., 1990).

In the case of processes with incomplete state measurements, the state feedback linear quadratic regulator in (31) is implemented in conjunction with a state estimator, leading to an output feedback controller (Muske and Rawlings, 1993).

5.2. Constrained linear MPC

Minimization of a constrained, infinite horizon objective function is an infinite-dimensional optimization problem. Therefore, the infinite control horizon must be abandoned in favor of an approach that results in a finite-dimensional optimization problem. Sufficient conditions on the horizon length N and terminal penalty matrix $\bar{Q}_N$ to ensure stability of the linear quadratic regulator for a finite horizon are discussed in (Kwon and Byun, 1989; Bitmead et al., 1990). A nominally stabilizing regulator that forces the state to the origin at the end of the control horizon is presented in (Kwon and Pearson, 1978). A feedback gain for this regulator can be determined by iterating the inverse of the discrete Riccati matrix starting with $P(0)^{-1} = \bar{Q}_N^{-1} = 0$ in order to determine $P(N)$ (Bitmead et al., 1990).

One can use a finite control horizon and infinite prediction horizon to guarantee nominal stability (Rawlings and Muske, 1993). This choice guarantees stability without tuning and is a less aggressive controller than forcing the state to the origin. For stable systems, it leads to a terminal state penalty that can be computed as the solution to a discrete Lyapunov equation. For unstable systems, also a terminal state penalty is defined for the stable modes while the unstable modes are forced to the origin. The determination of the manipulated variable values for constrained, linear model predictive control is typically performed by solving a quadratic program (QP) to determine the future manipulated variable profile.

5.3. Analytic constrained linear MPC

Although an exact analytical solution to the QP does not exist for general multivariable processes, approximate solutions to the constrained optimization problem are possible. An approximate analytic solution to the unconstrained long horizon MPC problem can be obtained by "clipping" the unconstrained controller outputs. This approach usually leads to a very poor closed-loop performance for multivariable processes. A more accurate approximate solution can be obtained by minimizing the deviation between the unconstrained solution and the constraints (Zheng, 1998). This still may not provide an acceptable closed-loop performance.

6. Conclusions

The very broad controller-synthesis methodology of model predictive control was used to solve the important fundamental problems of windup and/or directionality in input-output linearization, modified internal model control, model state feedback control, PI control, and PID control. Indeed it was MPC that led us to define the fundamental notion of directionality in input-constrained processes and develop a general optimal directionality compensator in (Soroush and Valluri, 1998).

It was shown that depending on the process model available and requested desired closed-loop response, a model predictive controller can be an input-output linearizing controller with optimal windup and directionality compensators, a model state feedback controller with an optimal directionality compensator, a modified internal model controller with an optimal directionality compensator, a PI controller with optimal windup and directionality compensators, and/or a PID controller with optimal windup and directionality compensators. Furthermore, MPC allows for developing guidelines for tuning analytical controllers. For example, in this presentation "optimal" PI and PID controllers with only one tunable parameter were derived. All the results presented here carry over to continuous-time processes. In the continuous-time setting, MPC in the same special cases leads to the same but continuous-time analytical controllers (for the continuous-time analog of the results presented here, see Valluri et al., 1998).

A short-prediction-horizon model predictive controller may have an analytical form and is of low computational cost. However, it may not be applicable to a process with a nonminimum-phase delay-free part, and in the presence of a constraint its closed-loop performance may suffer from the "short-sighted-ness" of the controller. Furthermore, when process deadtimes are not balanced, a short-prediction-horizon model predictive controller may be ill-defined (the corresponding optimization problem may have infinite number of solutions). Long prediction horizon model predictive controllers do not suffer from these restrictions. However, these controllers generally do not have an analytical form, and thus an optimization problem has to be solved numerically on-line.

Acknowledgment

The first author would like to acknowledge financial support from the National Science Foundation through the grant CTS-970-3278. The second author would like to thank Los Alamos National Laboratory for financial support through DOE W-7405-ENG-36 and the DoD/DOE Memorandum of Understanding.

References

[1] Bernstein, D. S., and A. N. Michel, "A Chronological Bibliography on Saturating Actuators," *Internat. J. of Robust and Nonlinear Contr.*, **5** 375 (1995).

[2] Bitmead, R. R., M. Gevers, and V. Wertz, *Adaptive Optimal Control, The Thinking Man's GPC,* Prentice-Hall, N. J. (1990).

[3] Coulibaly, E., S. Maiti, and C. Brosilow, "Internal Model Predictive Control (IMPC)," *Automatica*, **31**, 1471 (1995).

[4] Grantz, J., S. Valluri, and M. Soroush, "Discrete-Time Nonlinear Control of Processes with Actuator Saturation Nonlinearities," *AIChE J.*, **44(7)**, 1701 (1998).

[5] Kapoor, N., and P. Daoutidis, "An Observer-based Anti-Windup Scheme for Nonlinear Systems with Input Constraints," *Inter. J. of Contr.*, in press (1998).

[6] Kendi, T. A., and F. J. Doyle, "An Anti-Windup Scheme for Multivariable Nonlinear Systems," *J. of Process Contr.*, **7(5)**, 329 (1997).

[7] Kothare, M. V., P. J. Campo, M. Morari, and C. N. Nett, "A Unified Framework for the Study of Antiwindup Designs," *Automatica*, **30**, 1869 (1994).

[8] Kwon, W. H., and D. G. Byun, "Receding Horizon Tracking Control as a Predictive Control and its Stability Properties," *Int. J. of Contr.*, **50**, 1807 (1989).

[9] Kwon, W. H., and A. E. Pearson, "On Feedback Stabilization of Time-Varying Discrete Linear Systems," *IEEE Trans. Auto. Contr.*, **23**, 479 (1978).

[10] Mhatre, S., and C. Brosilow, "Multivariable Model State Feedback," *Proc. IFAC WC,* San Francisco, **M**, 139 (1996).

[11] Muske, K. R., and J. B. Rawlings, "Model Predictive Control with Linear Models," *AIChE J.*, **39**, 262 (1993).

[12] Nijmeijer, H., and A. J. van der Schaft, *Nonlinear Dynamical Control Systems*, Springer-Verlag, N.Y. (1990).

[13] Rawlings, J. B., and K. R. Muske, "Stability of Constrained Receding Horizon Control," *IEEE Trans. Auto. Control,* **38** 1512 (1993).

[14] Soroush, M., "Evaluation of Achievable Control Quality in Nonlinear Processes," *Computers & Chem. Eng.*, **20(4)**, 357 (1996).

[15] Soroush, M., and C. Kravaris, "Discrete-Time Nonlinear Feedback Control of Multivariable Processes," *AIChE J.*, **42**, 187 (1996).

[16] Soroush, M., and S. Valluri, "Optimal Directionality Compensation in Processes with Input Saturation Constraints," *Inter. J. of Contr.*, in press (1998).

[17] Valluri, S., M. Soroush, and M. Nikravesh, "Shortest-Prediction-Horizon Nonlinear Model Predictive Control," *Chem. Eng. Sci.*, **53**, 273 (1998).

[18] Zheng, A., "Model predictive Control of Large-Scale Constrained Systems," Submitted to *J. Process Control* (1998).

[19] Zheng, A., M. V. Kothare, and M. Morari, "Anti-windup Design for Internal Model Control," *Int. J. Contr.,* **60**, 1015 (1994).

Department of Chemical Engineering
Drexel University
Philadelphia, PA 19104, U.S.A.
E-mail address: masoud.soroush@coe.drexel.edu

Department of Chemical Engineering
Villanova University
Villanova, PA 19085, U.S.A.
E-mail address: krmuske@kayak.che.vill.edu

Progress in Systems and Control Theory, Vol. 26

Integrating Predictive and Switching Control: Basic Concepts and an Experimental Case Study

J. M. Lemos, L. M. Rato and E. Mosca

Abstract. The paper is concerned with the integration of predictive and switching control in order to achieve a structure suitable for application in wide range control of a class of nonlinear processes. The switching structure considered consists in a bank of predictive controllers, each tuned to a given operating subrange of the process. The use of this structure in the control of the oil outlet temperature in a distributed collector solar field is described.

1. Introduction

A wide class of plants with interest for industrial control applications may be described by the nonlinear autoregressive model with exogenous input given by

$$y(t+1) = f(u(t),\, s(t),\, s_w(t)) + \epsilon(t) \tag{1}$$

where f is a nonlinear piecewise smooth function, u and y are, respectively, the plant input and output, ϵ is a residue modeled as a white sequence, $s(t)$ is given by a vector of past and present samples of plant input and output and $s_w(t)$ is a vector of samples of the variables $w_1, \ldots, w_{n_w}$. These variables affect the plant behavior but may not be manipulated by the controller. By defining intervals in the range of the argument variables of $f(.)$, the notion of operation regime [1] may be defined as corresponding to each of the products of these intervals. As an example, consider the control of the oil outlet temperature in a distributed collector solar field, where a heat transfer medium (oil) is passed through a pipe located at the focus of mirrors concentrating solar radiation. The objective consists of adjusting the oil flow (manipulated variable) so that the oil temperature at the outlet of the pipe is constant [4]. The dynamic relation between the plant input (oil flow) and output (oil outlet temperature) is affected by several variables. Under normal conditions, two important ones are solar radiation and inlet oil temperature. The operating regime depends on these two variables and also on the level of the temperature set-point around which the field is being operated. It also depends on whether some collector is in fault. To each operating regime, a linear model, hereafter referred as a *local model* is associated. For each of the local models a linear controller (referred as *local controller*) is designed. The local controllers are designed here using predictive methods [5]. A block named *supervisor* [2, 3] decides which local

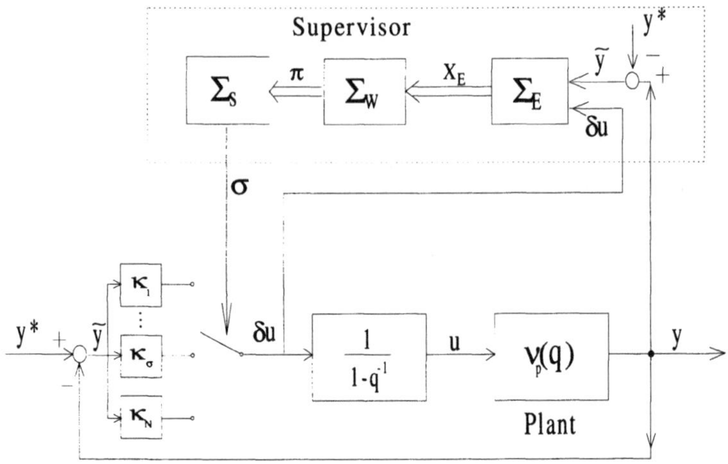

FIGURE 1. Feedback control loop and the supervisor.

controller is to be applied to the plant at each time. This decision ought to be made in the absence of *a priori* knowledge about the "shape" on the plant nonlinearity f or the way in which the various operating regimes are visited. Furthermore, the switching among local controllers is to be performed by the supervisor in such a way that the closed-loop system is stable and the reference is tracked with a reasonable error. Refs. [2, 3] concern stability and robustness properties of the class of algorithms considered.

2. The control structure

Fig. 1 shows a block diagram of the control system. Here, δu denotes increments of u; y^* denotes the reference to be tracked by y; $k_\sigma(q)$ denotes the transfer function of the σ-th controller; q denotes the forward shift operator. An integrator has been included in the loop, in series with the controller, in order to ensure zero steady-state position error. When the plant operating condition changes, the index σ of the active controller, $k_\sigma(q)$, should be adjusted in order to match the operating conditions. This adjustment is performed by the *supervisor* described below.

2.1. The supervisor

The supervisor follows the structure proposed in [2, 3] and is detailed in Fig. 2. It consists of three stages: *i)* A shared-state estimator, Σ_E; *ii)* A performance weight generator, Σ_W; *iii)* A switching logic scheme, Σ_S. As local models, consider the 1-step ahead linear predictors of the output for the situation in which the plant is working in the p-th operating regime. These are given by

$$\hat{\tilde{y}}_p(t) = [\theta_1^p \;\; \psi_1^{p'}][\delta u(t-1) \;\; s'(t-1)]' \tag{2}$$

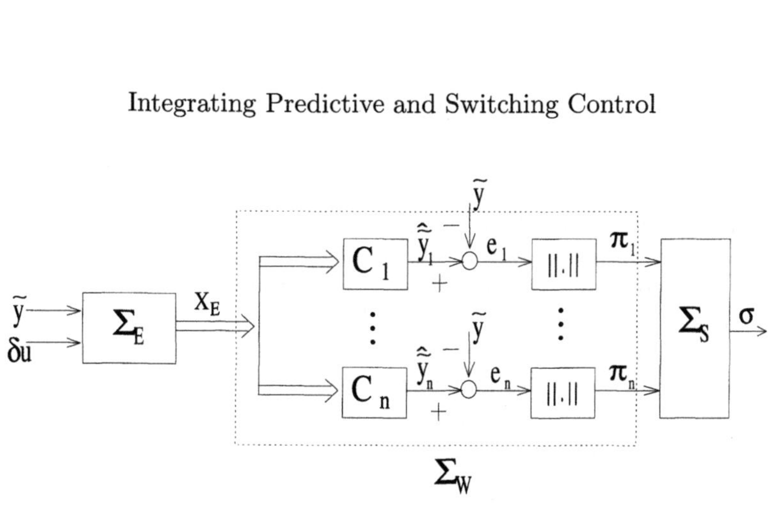

FIGURE 2. Shared-state estimator based supervisor.

the coefficients θ_1^p and ψ_1^p being obtained by least squares projection [5] and

$$s(t) = [\tilde{y}(t)\ \tilde{y}(t-1)\ \ldots\ \tilde{y}(t-n_a+1)\ \delta u(t-1)\ \delta u(t-2)\ \ldots\ \delta u(t-n_b)]' \quad (3)$$

The estimator error of the $p-th$ model is $e_p(t) = \tilde{y}_p(t) - \hat{\tilde{y}}_p(t)$ The shared-state estimator generates a vector X_E to be used by a bank of models, each corresponding to one operating point. It is defined by

$$X_E(t+1) = \begin{bmatrix} A_E & 0 \\ 0 & A_E \end{bmatrix} X_E(t) + \begin{bmatrix} b_E \\ 0 \end{bmatrix} y(t) + \begin{bmatrix} 0 \\ b_E \end{bmatrix} \delta u(t) \quad (4)$$

$$\hat{\tilde{y}}_p = C_p X_E(t) \quad (5)$$

where A_E and b_E are given by

$$A_E = \begin{bmatrix} 0 & 0 & 0 & \cdots \\ 1 & 0 & 0 & \cdots \\ 0 & 1 & 0 & \cdots \\ \vdots & \vdots & \ddots & \ddots \end{bmatrix} \qquad b_E = \begin{bmatrix} 1 \\ 0 \\ 0 \\ \vdots \end{bmatrix} \quad (6)$$

and C_p depends on the parameters of the p-th plant model, being given by $C_p = [\theta_1^p\ \psi_1^p]$. The shared state $X_E(t)$ is thus $X_E(t) = [\delta u(t-1)\ s'(t-1)]'$. The performance weight generator computes a low pass filtering of the squared estimator errors by

$$\pi_p(t+1) = \pi_p(t)\lambda + (\hat{\tilde{y}}_p(t) - \tilde{y}(t))^2 \quad (7)$$

where $0 < \lambda < 1$ and π_p is the performance signal associated to the p-th local process model. A value of $\lambda = 0.9$ is used in this paper. The lower π_p, the better the matching between the p-th model and the operating conditions. The switching logic scheme used consists in choosing the controller corresponding to the local model with the lowest performance signal, and assuring a minimum period of time between each switch. This period of time is referred as the *dwell time* and has been settled as 10 samples in the solar collector field example below.

2.2. Controller design

Several methods may be used for designing the local controllers $k_\sigma(q)$. The approach followed hereafter allows an efficient tuning of the (reduced complexity) local controllers and, furthermore, fits well the implementation using the shared-state realization. In this paper the MUSMAR adaptive control algorithm [5] is applied off-line to each of the local models $\nu_p(q)$ and run till convergence is obtained. The resulting controller gains are then recorded, defining $k_p(q)$. For each operating point p, this design aims at minimizing the quadratic cost functional defined over a T-steps horizon:

$$J_T(t) = \frac{1}{T} E \left\{ \sum_{k=t}^{t+T-1} \left[\tilde{y}^2(k+1) + \rho \delta u^2(k) \mid I(t) \right] \right\} \tag{8}$$

where $\rho \geq 0$ is a weight in the control action, $\tilde{y}(t)$ is the tracking error, $\tilde{y}(t) = y(t) - y^*(t)$ and $E\{ . | I(t) \}$ denotes the mean conditioned on the information available up to time t (projection)[5]. Let the input to the integrator when the p-th controller is active, $\delta u_p(t)$, be given by

$$\delta u(t) = F_p' s(t) \tag{9}$$

where F_p is the vector of controller gains minimizing the cost functional $J_T(t)$, assuming that the plant is in the p-th operating regime. This vector of gains is yielded by performing the following steps at each discrete time t:

MUSMAR Algorithm [5]

1. For $i = 1, \ldots, T$, compute the recursive-least-square estimate of the parameters θ_i^p, ψ_i^p, μ_i^p and ψ_i^p in the predictive regression models

$$\tilde{y}(t+i-T) \approx \theta_i^p \delta u(t-T) + \psi_i^{p\prime} s(t-T) \tag{10}$$

$$\delta u(t+i-T-1) \approx \mu_{i-1}^p \delta u(t-T) + \phi_{i-1}^{p}{}' s(t-T) \tag{11}$$

where $\approx$ denotes equality in least squares sense.

2. Update the local controller gains according to

$$F_p = - \frac{\sum_{i=1}^{T} \theta_i^p \psi_i^p + \rho \sum_{i=1}^{T-1} \mu_i^p \phi_i^p}{\sum_{i=1}^{T} \theta_i^{p2} + \rho(1 + \sum_{i=1}^{T-1} \mu_i^{p2})} \tag{12}$$

3. Simulation example

In order to illustrate the above ideas, consider the plant modeled by

$$A^*(q^{-1}) y(t) = \Gamma B^*(q^{-1}) u(t) + e(t) \tag{13}$$

where y is the output, e is a white noise sequence of unit variance and u is the manipulated variable. The scalar gain Γ changes in time, assuming values in the set $G = \{\Gamma_1 = 1, \Gamma_2 = 3.64, \Gamma_3 = -3.64, \Gamma_4 = 8\}$. Polynomials A^* and B^* are given by (q^{-1} is the backward shift operator):

$$A^*(q^{-1}) = 1 - 3.1q^{-1} + 4.0496q^{-2} - 2.9736q^{-3} + 1.356q^{-4} - 0.37^{-5} + 0.046q^{-6}$$

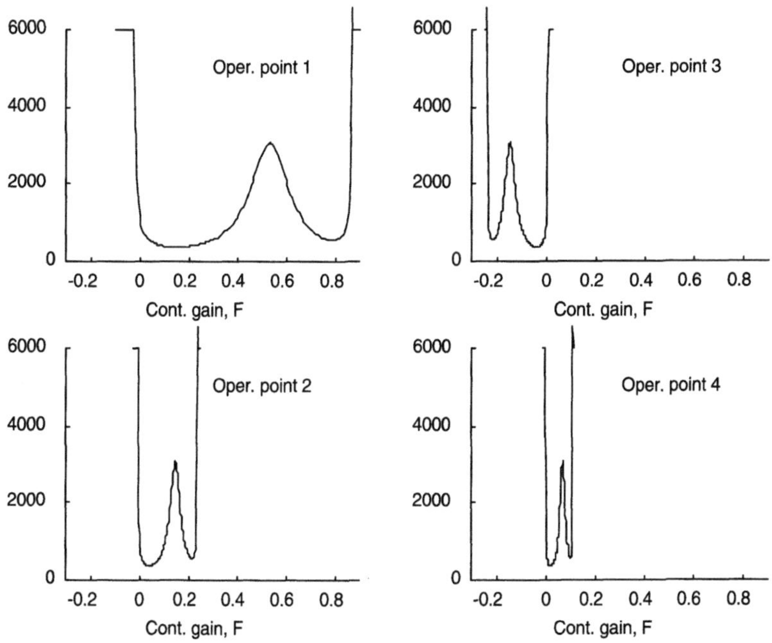

FIGURE 3. The loss around the 4 possible operating points.

$$B^*(q^{-1}) = -0.01q^{-1} - 0.983q^{-2} + 1.646q^{-3} - 1.1788q^{-4} + 0.334q^{-5} - 0.0353q^{-6}$$

This plant is such (see Fig. 3) that for $\Gamma = \Gamma_1$ the loss $E[y_t^2]$ has a local maximum when controlled with the proportional controller

$$u(t) = Fy(t)$$

The gain minimizing the loss in operating point (0. p.) 1 yields a maximum in o. p. 2. Furthermore, the gain minimizing the loss in o. p. 3 yields an unstable system if used in the other o. p.'s. Fig. 4 shows an example of plant operation and Fig. 5 shows details of operation between o. p. 1 and 2.

4. Experimental results

This section describes experimental results obtained in the example of temperature control of a distributed collector solar field. The influence from the radiation can be roughly described as a change in the plant static gain. Thus, linear models were identified from plant experimental data under two different operating conditions: 1 model with the flow around 5 l/s (model no. 0); 2 models with the flow around 7 l/s obtained with different values of radiation (models no. 4 and 7); 1 model with the flow around 3.5 l/s (model no. 8). From the above models, other models were obtained by multiplying their static gain by a constant C. In this way models

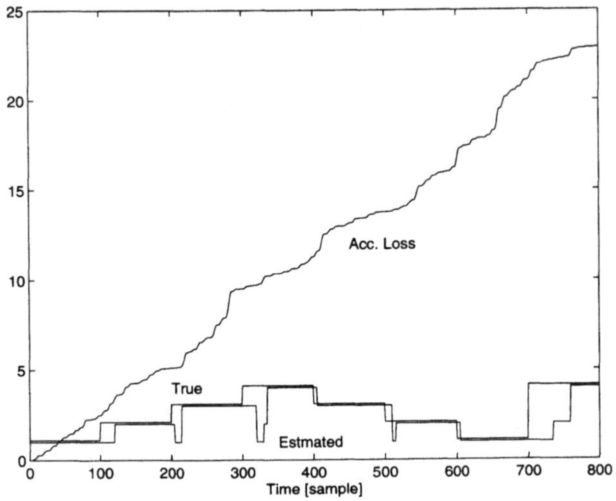

FIGURE 4. Operation with a time varying operating point.

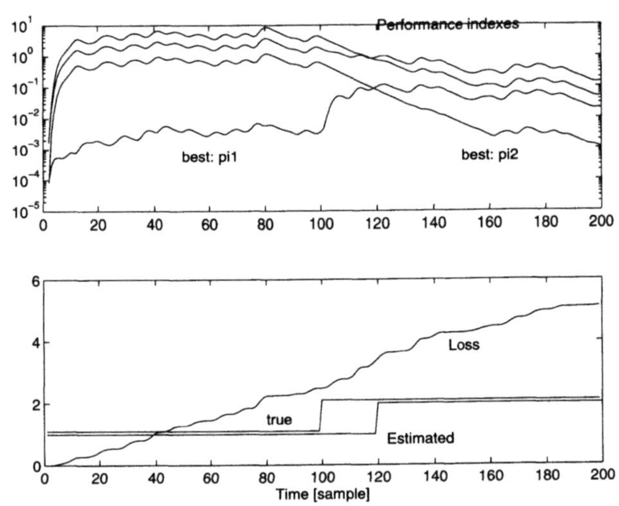

FIGURE 5. Detail of commutation between operating points 1 and 2.

1, 2, and 3, were obtained from model 0; model 5, from 4; and model 6 from 7. This is summarized in table 1. The adaptive MUSMAR algorithm has been run for each model, with the structure of $s(t)$ defined by $n_a = 3$ and $n_b = 2$, and $T = 10$. The values of ρ used with each model are listed in table 1. It is remarked that the controllers corresponding to working conditions 0, 4, 7 and 8 may not

Model no.	Flow	C	ρ
0	5 l/s	1.00	100
1	5 l/s	1.50	100
2	5 l/s	0.66	50
3	5 l/s	0.33	10
4	7 l/s	1.00	70
5	7 l/s	1.50	150
6	7 l/s	1.50	150
7	7 l/s	1.00	100
8	3.5 l/s	1.00	1000

TABLE 1. Operating conditions for the local models.

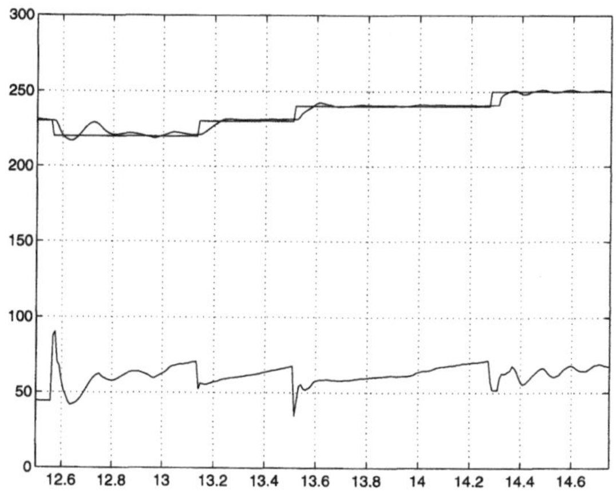

FIGURE 6. Experiment 1: Reference, output, and *flow* $\times$ 10.

be obtained from each other by just changing a static gain. The sampling time has been settled as $T_S = 30\,s$. The above settings result from previous knowledge obtained by running MUSMAR on the plant [4]. Experiment 1 concerns a situation in which there are no clouds disturbing solar radiation. Figs. 6 and 7 document this experiment. Experiment 2 is shown in Figs. 8 to 10. It illustrates the response of the controller to a fast and strong disturbance in the solar radiation caused by a cloud. The radiation drops in a few minutes from more than $800\,W/m^2$ to $300\,W/m^2$ and recovers. In order to respond to this situation the active controller is changed by the supervisor, first from 7 to 5 and then to 2. By looking at table 1, this is seen as switching from a controller matching a higher level of radiation to a controller matching a lower level radiation.

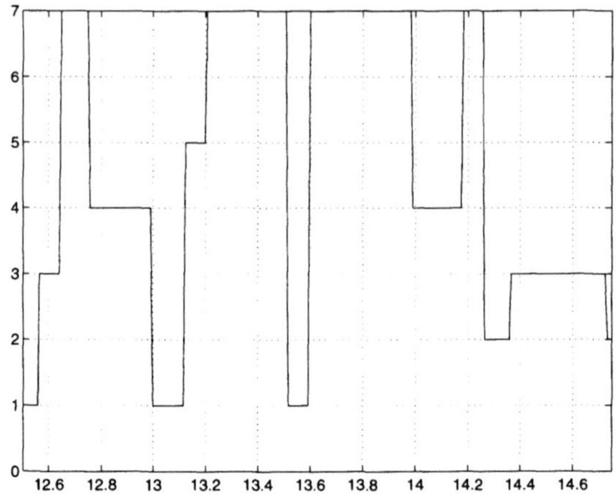

FIGURE 7. Experiment 1: Supervisor output, σ.

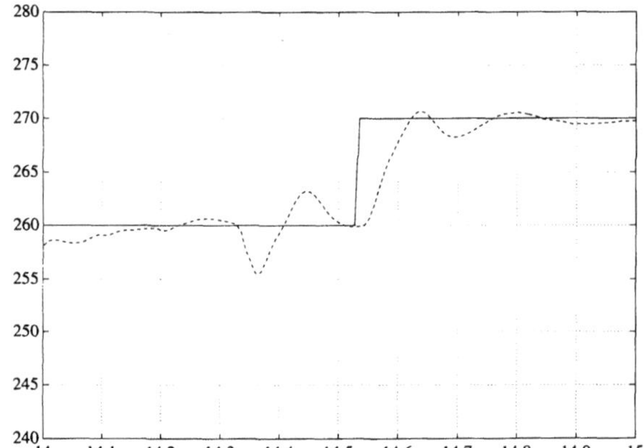

FIGURE 8. Experiment 2: Reference and output.

The results for experiment 3 can be seen in Fig. 11. In this experiment only two controllers were used. These are n.7 and n.8 in table 1. Despite that controller no. 7 is often elected by the switching mechanism in experiment 1, a feature which also happens here, the resulting performance in this case is poor, due to a persistent oscillatory behavior. This illustrates the importance of the use of a number of controllers covering the various operating conditions. In the above graphics, time is in [*hour*].

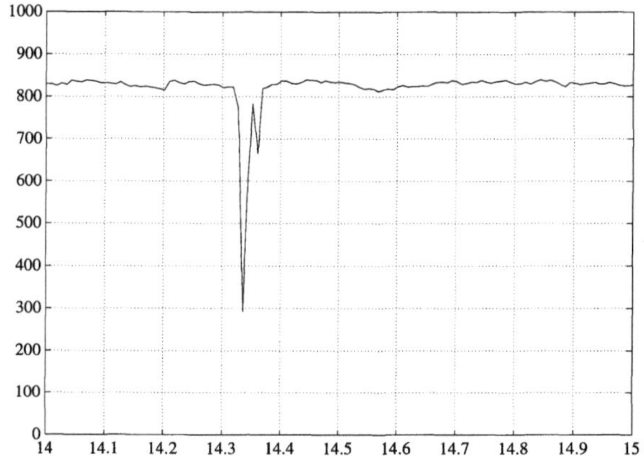

FIGURE 9. Experiment 2:Solar radiation.

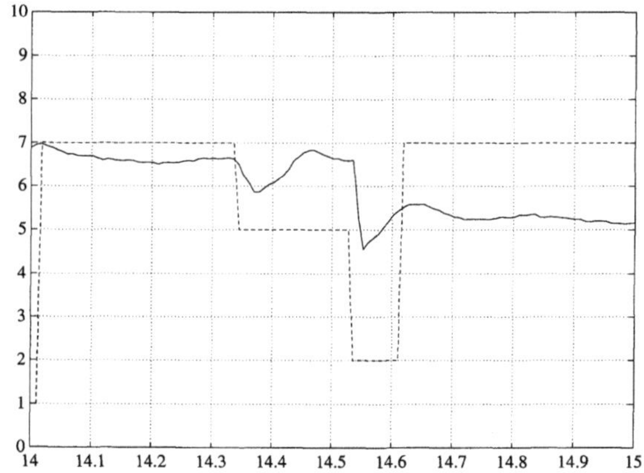

FIGURE 10. Experiment 2: Supervisor output, σ, and flow.

5. Conclusion

Integration of predictive and switching control in order to achieve a structure suitable for application in wide range control of nonlinear processes is considered. Two examples are provided, including an experimental case study in the control of the oil outlet temperature in a distributed collector solar field. These examples illustrate the way in which the control structure considered reacts to plant operating conditions changes.

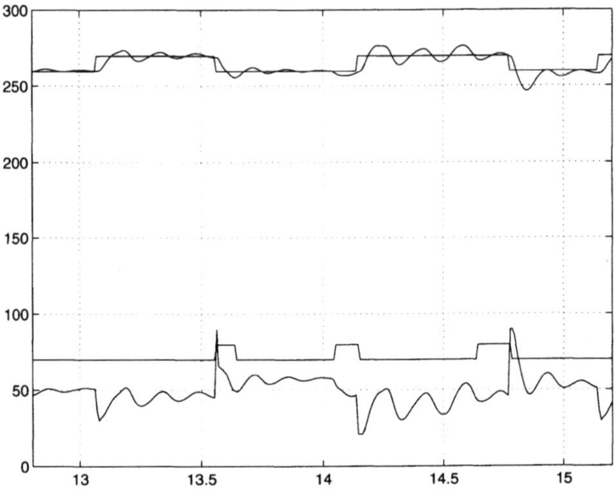

FIGURE 11. Experiment 3:Reference, output, $\sigma \times 10$, and $flow \times 10$.

References

[1] K. Hunt and T. Johansen, *Design and analysis of gain-scheduled control using local controller networks*, Int. J. Control, **66** (1997), 619–651.

[2] A. S. Morse, *Supervisory Control of Families of Linear Set-Point Controllers – Part 1: Exact matching*, IEEE Trans. Automat. Contr., **41** (1996), 1413–1431.

[3] A. S. Morse, *Supervisory control of Families of Linear Set-Point Controllers – Part 2: Robustness*, IEEE Trans. Automat. Contr., **42** (1997), 1500–1515.

[4] F. Coito, J. M. Lemos, R. N. Silva and E. Mosca, *Adaptive Control of a Solar Energy Plant: Exploiting Accessible Disturbances*, Int. J. of Adaptive Control & Signal Processing, **11** (1997), 327–342.

[5] E. Mosca, *Optimal Predictive and Adaptive Control*, Prentice-Hall, (1995).

J. M. Lemos
INESC/IST,
R. Alves Redol 9
1000 Lisboa, Portugal
E-mail address: `jlml@inesc.pt`

E. Mosca
DSI, Universitá de Firenze,
Via de S. Marta 3
50139 Firenze, Italia
E-mail address: `mosca@dsi.dsi.unifi.it`

Progress in Systems and Control Theory, Vol. 26

Exploring the Potentiality of Using Multiple Model Approach in Nonlinear Model Predictive Control

J. O. Trierweiler and A. R. Secchi

Abstract. The problem of controlling processes that operate within a wide range of operating conditions is addressed. The operating space is decomposed into a set of local model forming a network. These are combined into a global model structure using an interpolation method based on normalized basis functions. Unknown local model parameters are identified using empirical data. The control problem is solved using a model predictive controller based on the local model network (LMN). The van de Vusse reaction is used as example. The MPC based on LMN is compared with an exact nonlinear process model and others MPC algorithms based on linear models. Finally, several future works in this area are pointed out.

1. Introduction

Any model or controller has one limited interval of operation or validity, i.e., a region where it represents the system's behavior and works well. Many factors (such as: use of linear models to represent nonlinear systems, hypotheses made during the modeling, experimental conditions, etc.) contribute to reduction of the interval of validity of a model or controller. A model (or controller) that is valid in a given operating condition is called local model or submodel (or local controller). Certainly, the final objective is the attainment of a global model or controller. A way of reaching this objective is to represent all the regions of operation through a network of local models also called multiple models.

Since all chemical processes are nonlinear at least to some extent, extending the standard Model Predictive Control (MPC) concept to systems described explicitly by nonlinear models has been the subject of intense research. If one considers the basic concepts of MPC, there is nothing in the MPC structure that fundamentally forbids the use of a nonlinear model. However, some very serious practical issues arise that make the nonlinear extension of MPC a nontrivial exercise:

1. *Availability of nonlinear models.* Good nonlinear models are much more difficult to construct and identify than the linear models.

2. *Optimization problem.* The nonlinear MPC problem becomes a nonlinear programming (NLP) problem which is much more difficult to solve than the usually quadratic programming (QP) method used in the linear case.

The use of local model network (LMN) can be a quite attractive approach to solve these problems. A LMN is able to represent a nonlinear systems in an efficient and systematic way (see, e.g., [8]). An extension of this idea to a local control network has also shown to be a promising area. The present work aims to exploit the powerful advantages of this paradigm by the proposition of a Nonlinear Model Predictive Control based on a linear local model network (LLMN). Unconstrained problems and input-constrained problems are addressed on this formulation.

2. Background

2.1. Model predictive control

A MPC algorithm employs a model of the process to predict its future behavior over an extended prediction horizon. A performance objective to be minimized is defined over the prediction horizon, usually a sum of quadratic set point tracking errors and control effort terms. This cost function is minimized by evaluating a profile of manipulated input moves to be implemented at successive sampling instants over the control horizon. The feedback behavior is achieved by implementing only the first manipulated input move and repeating the complete sequence of steps at the subsequent sampling time.

The various MPC algorithms propose different cost functions for obtaining the control law. A quite general expression for the objective function is:

$$J = \sum_{j=P_0}^{P} ||\hat{y}(t+j|t) - r(t+j)||_Q^2 + \sum_{j=1}^{M} ||\Delta u(t+j-1)||_R^2 \tag{1}$$

where $\hat{y}(t+j|t)$ is the predicted output j steps into the future based upon information available at time t, $r(t+j)$ is the reference signal j steps ahead, $\Delta u(t) = (1-z^{-1})u(t) = u(t) - u(t-1)$, and $||x||_W^2$ is the weighted Euclidean norm of $x \in \mathcal{R}^n$ defined as $||x||_W^2 = x^T W x$ with $W \in \mathcal{R}^{n \times n}$ positive definite.

The tuning parameters are the minimum costing horizon (P_0), the maximum costing horizon (P), the control horizon (M), the sampling time (t_S), the controlled variable weight (Q), and the move suppression weight (R). The weighting matrices Q and R can be chosen as time-varying (i.e., functions of j). Here, for simplicity, they are assumed to be time-invariant. A tuning procedure for MIMO systems can be found in [13].

2.2. Local model networks

The basic idea behind the local models is the decomposition of the operating space in possibly overlapping subspaces. Each subspace corresponds to a local model. The construction of local model networks can be divided in:

1. Selection and range specification of the operating space variables. These variables define the operating space.
2. Division of the space in subspaces.
3. Choose of the local variables used in the definition of the local models. These variables can be other than the operating space variables.
4. Choose of the local model structure and the corresponding parameter identification.
5. A method to combine the local models into a global process model.

In industrial processes it is expected that system's properties can be represented properly by a soft interpolation of the information of specific points distributed in the space (also known as operating points). Here, as in many other works in this area (see e.g., [1], [10], [9], among others), the intermediary properties will be interpolated in the following way:

$$y = f(x_l, x_r) = \sum_{i=1}^{n} f_i(x_l)\rho_i(x_r) \tag{2}$$

where the weighting functions $\rho_1(x_r), \rho_2(x_r), \ldots, \rho_n(x_r)$ are responsible for the soft transition among the operating points, which are described by the local models $f_i(x_l)$ that, after properly weighted by the weighting functions, will compose the system's output represented by y; x_l corresponds to the group of variables used in the local models and x_r, to the group that defines the operating space, X_r. Naturally, the vectors x_r and x_l can have elements in common. In the extreme case, they can be identical.

2.2.1. Weighting functions. Typically, the weighting functions satisfy:

$$\sum_{i=1}^{n} \rho_i(x_r) = 1, \forall x_r \in X_r \quad \text{with} \quad \rho_i(x_r) \in [0,1] \tag{3}$$

which is known as the normalization condition. Weighting functions that satisfy (3) guarantee a homogeneous distribution in the definition space, being, in this way, less sensitive to the location of its centers. When the functions that quantify the influence of a given center in the space do not satisfy the normalization condition, they can easily be normalized through

$$\rho_i(x_r) = \frac{\mu_i(x_r)}{\sum_{k=1}^{n} \mu_k(x_r)} \tag{4}$$

where $\mu_i(x_r)$ represents a function that is being normalized and n the total number of centers. The weighting functions are also known as basis functions.

A usual choice for $\mu_i(x_r)$ is a gaussian function, i.e., $\mu_i(x_r) = \exp(\frac{-||x_r - c_i||_2^2}{2\sigma_i^2})$, which assumes after the normalization the following form:

$$\rho_i(x_r) = \frac{\exp \frac{-||x_r - c_i||_2^2}{2\sigma_i^2}}{\sum_{k=1}^{n} \exp \frac{-||x_r - c_k||_2^2}{2\sigma_k^2}} \tag{5}$$

In (5), $||\cdot||_2$ denotes the vector 2-norm, c_k is the location of the center k, and σ_k is the corresponding standard deviation. σ_k quantifies the degree of expansion (i.e., width) of the center k along of each dimension defined by the variables that characterize the operating space.

Depending on the location of the center and of its width, the normalization leads to a number of important side effects which can have important consequences for the resulting local model network. For example, when Gaussians are used as basis function, the main side effects are ([12]):

1. *Loss of independence and change of the basis function shape.* The shape of the basis functions is usually quite different from the unnormalized basis functions. In the extreme case occurs the reactivation of the center. Ideally, the basis functions must decrease monotonically with increasing distance to the center . The reactivation happens when the basis function increases from some point outside of the region where the local model should be active.
2. *Shift and reduction of basis function maximum.* The maximum of the basis functions may no longer be at its center and the maximum can be much smaller than the unnormalized basis function.

The above mentioned side effects happen manly when the centers are not uniformly spaced, or when the basis functions have distinct widths (see, [11]).

2.2.2. LOCAL MODELS. In principle, the local models can be of any type (e.g., state space model, transfer function). They can be both linear as nonlinear. Preferentially, they should be linear in the adjustable parameters to allow its fast and efficient determination. Here, the ARX-model structure is used as local models.

3. Model predictive control algorithms

3.1. Earlier work

Based on the objective function given by (1), the main differences among the existing strategies are the choice of the model used to predict the process outputs and the algorithm chosen to solve the optimization problem. In [4] three model predictive controllers are compared to control a terpolymerization reactor, using as process model: a linearized model around a reference steady-state, a LLMN with the weighting function based on the distance between the current point and each reference steady-state of the network, and a locally linearized model at every sampling time. In all cases the predicted outputs are obtained by a single linear model along the prediction horizon (linear prediction). Foss et al. [7] also compared three model predictive controller for a batch fermentation process, using as process model: a global linear model, a LLMN with overlapped gaussian weighting functions, and a nonlinear model. The last two strategies use nonlinear models for prediction. In [3] a LLMN is used to define a multiple model controller in the

MPC framework (similar to the gain scheduling approach) with a linear interpolation of the controller's zeros and gain as function of the outputs. The outputs are predicted by a successive one-step-ahead prediction scheme, which is a nonlinear prediction. This strategy is compared with a global linear model and a LLMN with linear output predictions.

3.2. Proposed approach

The results presented in the above references demonstrated the workability of the LLMN to derive a MPC strategy. It is also clear that the good performance of these controllers depends on a good decomposition of the operating space, which is direct related to the process knowledge, and how the local models are combined. In this work, six MPC strategies are compared regarding to the prediction model. The objective function (1) and the same constraints are used for all controllers. The constraints in the manipulated variables are given by $u_{\min} \leq u(t) \leq u_{\max}$ and $|\Delta u(t)| \leq \Delta u_{\max}$, $\forall t$.

The prediction models for the controllers are the following:

(a) a nonlinear model without model-plant mismatch;
(b) only one linear model, identified with input-output data;
(c) only one linearized model calculated for a given steady-state point;
(d) locally linearized models at every sampling time with linear predictions;
(e) a LLMN with linear predictions; and
(f) a LLMN with nonlinear predictions.

The first strategy solves the optimization problem using a SQP algorithm. All the others strategies use either the corresponding analytical solution for the unconstrained case with saturation in $u(t)$ and $\Delta u(t)$, or use a QP algorithm for the constrained case. The last strategy (i.e., (f)) can also be solved in the above manner because it was implemented through an iterative algorithm. That is, it is started by a linear prediction, i.e., the solution of the case (e). For this linear model it is calculated an initial moving trajectory for the control action u. Then several interpolated models are calculated along of this trajectory for each sampling time of the prediction horizon. With these interpolated models a new moving trajectory is optimized. After that the interpolated model can be updated and a new trajectory can again be calculated. These two steps are performed until to reach the convergence or a specified number of iterations. This idea is based on the fact that all linear prediction model can be represented by $y = Gu + d$ where the vector d contains the terms depending on known values and Gu depends on future control inputs. For example, when applying the LLMN based on discrete state-space linear models: $x_{k+j} = A_j x_{k+j-1} + B_j u_{k+j-1}$ and $y_{k+j} = C x_{k+j}$ where A_j and B_j are the interpolated matrices based on the operating condition at the future sampling time $k+j-1$, then the vector d becomes:

$$d = [CA_1, CA_1A_2, \cdots, CA_1A_2..A_P]^T x_k \tag{6}$$

and the matrix G has the form:

$$G = \begin{bmatrix} CB_1 & 0 & 0 & \cdots & 0 \\ CA_2B_1 & CB_2 & 0 & \cdots & 0 \\ CA_3A_2B_1 & CA_3B_2 & CB_3 & \cdots & 0 \\ \vdots & \vdots & \vdots & \ddots & \vdots \\ CA_P..A_2B_1 & CA_P..A_3B_2 & CA_P..A_4B_3 & \cdots & CB_M \end{bmatrix} \tag{7}$$

For the unconstrained problem, the MPC gain matrix is given by:

$$K = (G^T Q_d G + R_d)^{-1} G^T Q_d \tag{8}$$

where $Q_d = \mathrm{diag}(Q, \ldots, Q)$ and $R_d = \mathrm{diag}(R, \ldots, R)$. Therefore, the MPC gain matrix depends on the operating space variables along the prediction horizon, which are unknown but may be estimated through the linear models interpolated by the control moving trajectory of the previous iteration step.

The iterative approach is a direct substitution technique to solve the nonlinear prediction model. This strategy is a good compromise between the simplicity of a simple linear predictive model (case (e)) and the numerically intensive calculation of a nonlinear predictive model (case (a)).

4. Example: CSTR with van de Vusse reaction scheme

The van de Vusse reaction has been used as a benchmark problem for nonlinear process control algorithms ([6], [2]). In [14] the dynamics of three operating points were analyzed. Here, it will be analyzed an operating point that is almost the same as the one proposed by Chen et al. [2]. In [14] this problem was solved with the development of a new control structure and in [5], through a NMPC algorithm. The same problem can be solved using a local model network.

4.1. Description of the process

The van de Vusse reaction consists of the following reaction scheme:

$$A \xrightarrow{k_1} B \xrightarrow{k_2} C$$
$$2A \xrightarrow{k_3} D$$

Here, B is the wanted product and C and D are the undesired byproducts. The van de Vusse reaction is carried out in an ideal isothermal continuous stirred tank reactor (CSTR) which it is modeled by:

$$\frac{dc_A}{dt} = \frac{F_{in}}{V_R}(c_{Ain} - c_A) - [k_1(T)c_A + k_3(T)c_A^2] \tag{9}$$

$$\frac{dc_B}{dt} = -\frac{F_{in}}{V_R}c_B + [k_1(T)c_A - k_2(T)c_B] \tag{10}$$

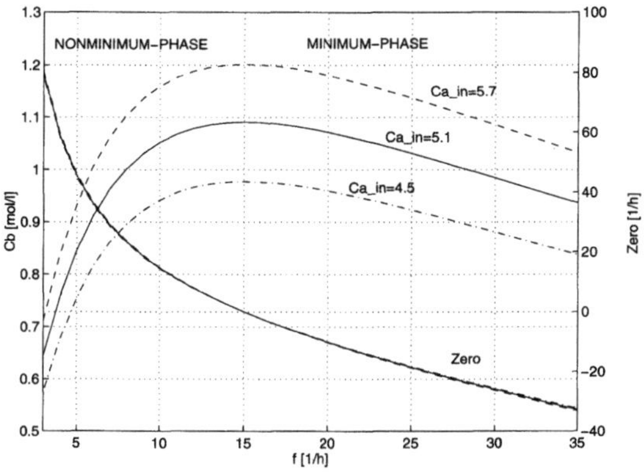

FIGURE 1. c_B vs. f (left scale) and *zero* vs. f (right scale) for $T = 114.5^oC$ and distinct concentrations of c_{Ain}

where k_i is given by the equation of Arrhenius, i.e., $k_i(T) = k_{i0}\exp(\frac{E_i}{T[^oC]+273.15})$, for $i = 1, 2$, and 3. In the present study, it is used the same kinetic parameters proposed in [6], i.e., $k_{10} = k_{20} = 1.287 \times 10^{12}\ h^{-1}, k_{30} = 9.043 \times 10^9\ \frac{liter}{molh}, E_1 = E_2 = -9758.3\ K$, and $E_3 = -8560\ K$. We assume that the CSTR works at 114.5^oC.

The manipulated and controlled variables are $f = \frac{F_{in}}{V_R}$ (inverse of the residence time) and c_B (concentration of component B), respectively. f is assumed to vary in the range $3 \leq f[h^{-1}] \leq 35$. The inlet concentration of A, c_{Ain}, is the main unmeasured disturbance and occurs in the interval $4.5 \leq c_{Ain}[mol/liter] \leq 5.7$.

4.2. System's dynamics

The plot of the steady state concentration c_B over f for a given reactor temperature, T, and inlet composition c_{Ain} (Fig. 1) reveals an interesting behavior of the system. The reactor exhibits a change in gain at the peak of the reactor yield (i.e., where the concentration c_B achieves its maximum value), and displays nonminimum-phase behavior (inverse response) for operating points on the left of this peak and minimum-phase behavior with overshoot for operating points on the right. The three curves shown in Fig. 1 were obtained by three different values of c_{Ain}. In Fig. 1 it is also shown the zero of the system, which is null for the maximum obtainable c_B for a given c_{Ain} and reactor temperature. Observe that, for the case where $k_1 = k_2$, the value of f corresponding to the maximum of c_B is independent of c_{Ain} and is given by $f_p = k_2$. For the more general case, the following expression can be used:

$$f_p = k_2 + (k_1 - k_2)\frac{-k_1 - f_p + \sqrt{(k_1 + f_p)^2 + 4k_3 c_{Ain} f_p}}{2c_{Ain}k_3} \tag{11}$$

which can easily be derived from the condition $zero = 0$ for the maximum c_B.

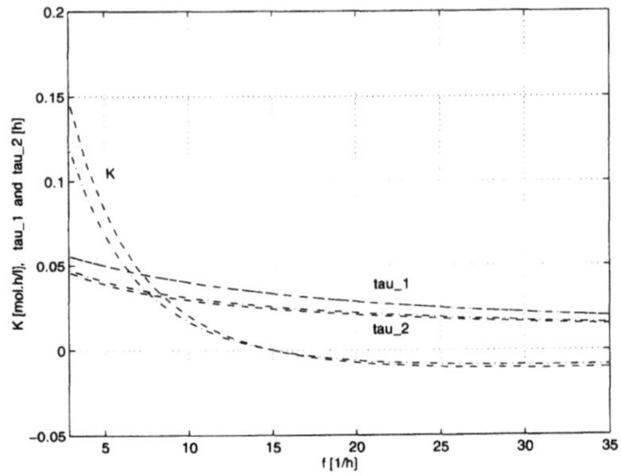

FIGURE 2. The transfer function parameters K, τ_1 and τ_2 vs. f for $c_{Ain} = 4.5\ mol/l$ (dashdot line) and $c_{Ain} = 5.7\ mol/l$ (dashed line)

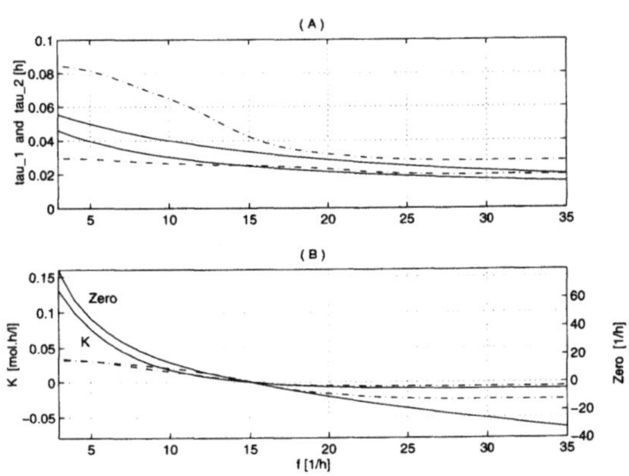

FIGURE 3. Comparison between linearized model (solid line) and the interpolated linear model based on LMN (dashdot line): (A) τ_1 and τ_2 and (B) K (on the left side) and $Zero$ (on the right side)

The equations (9) and (10) can easily be linearized at each steady-state point shown in Fig. 1. Based on this linearized model, the transfer function of f to c_B can be obtained. The transfer function can be represented in the following form: $\frac{\Delta c_B(s)}{\Delta f(s)} = G(s) = \frac{K(\beta s+1)}{(\tau_1 s+1)(\tau_2 s+1)}$. The parameters K, τ_1, and τ_2 of this transfer function are shown in Fig. 2 for the extreme c_{Ain} concentrations. Note that the dominant time constant, τ_1, is independent of c_{Ain}, whereas K, τ_2, and β ($\beta = -1/zero$, see Fig. 1) present a small variation with c_{Ain}.

TABLE 1. Comparison of the identified local models (ML) and the linearized model s (LIN) at the corresponding centers obtained for $c_{Ain} = 5.1 mol/l$

$f[h^{-1}]$	$\sigma[h^{-1}]$		$zero[h^{-1}]$	$K[\frac{mol.h}{liter}]$	$\tau_1[h]$	$\tau_2[h]$
3.84	3.6	ML	16.8	0.0369	0.087	0.03
		LIN	57.2	0.1034	0.054	0.044
11.02	3.6	ML	10.4	0.0159	0.066	0.026
		LIN	9.6	0.0115	0.039	0.029
18.20	3.6	ML	-7.5	-0.0049	0.026	0.023
		LIN	-7.7	-0.0059	0.030	0.023
25.37	3.6	ML	-12.3	-0.0063	0.022	0.020
		LIN	-19.7	-0.0096	0.025	0.019

4.3. Model identification

Based on the last discussion, one can conclude that the local model centers can be completely defined using *only* the manipulated variable f. Including c_{Ain} and c_B in x_r it will only marginally improve the LMN. Additionally, to be able to properly describe the system's dynamics, it is necessary at least one model at each side of the peak, i.e., a nonminimum-phase and a minimum-phase model.

Usually for a real application of multi-model approach in NMPC is necessary to identify LMN from process data. The algorithm LOLIMOT proposed in [10] was used to determine the structure of LMN. To shown the potentiality of this simple algorithm in the identification procedure, it was not used the knowledge about the system. The centers and the variables used in the identification of the local models have the same dimension, i.e., $x_r = [c^{\#}_{Ain}(t), c^{\#}_{Ain}(t-1), f^{\#}(t), f^{\#}(t-1), c^{\#}_B(t), c^{\#}_B(t-1)]$ and $x_l = [\Delta c_{Ain}(t), \Delta c_{Ain}(t-1), \Delta f(t), \Delta f(t-1), \Delta c_B(t), \Delta c_B(t-1)]$ where $\Delta z(t) = z(t) - z^{\#}(t)$. The training data and the training procedure (which was based on a one-step ahead identification criterion) are discussed in [15]. In [15] was shown that $f^{\#}(t)$ was the only variable used in the operating space decomposition. The algorithm was also able to produce local models very close to the equivalent linearized models. The identified four centers LMN is presented in Tab. 1. Note that only the model centered in $f^{\#} = 3.84$ has a significant discrepancy. Fig. 3 compares the LMN and the corresponding linearized model. As it could be expected, the identified LLMN can better describe the system's dynamic between $f = 10$ and 20, since the APRBS training data were concentrated between $f = 5$ and 25 (see Fig.5 in [15]).

4.4. Results

Figures 4 and 5 show the simulation results of the six MPC strategies defined in section 3 when applied to this example. The tuning parameters of the controllers are: $Q = 1$, $R = 0.01$, $P_0 = 1$, $P = 20$, $M = 3$, and $t_S = 0.01h$. The linearized

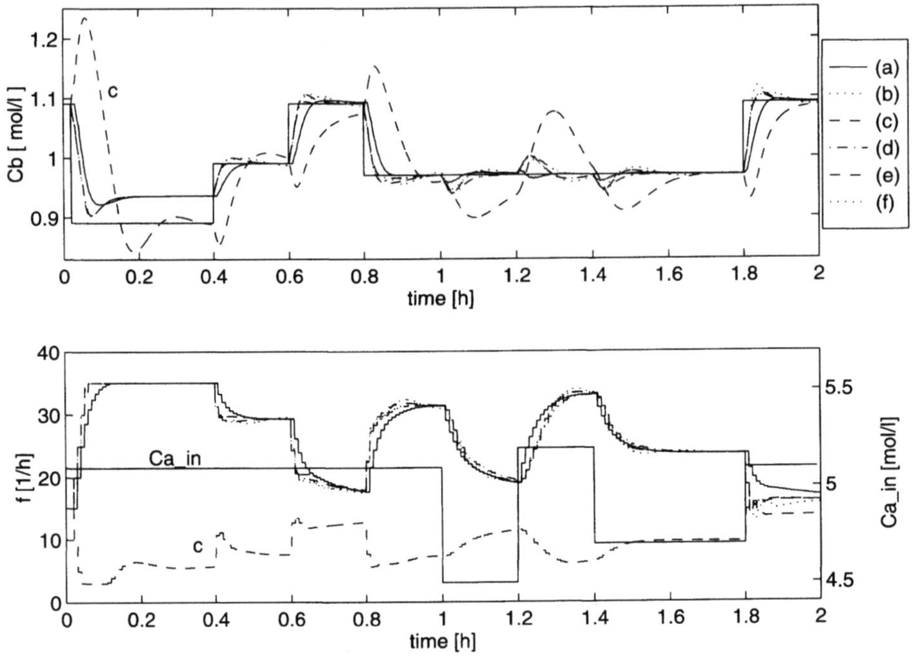

FIGURE 4. Performance of several MPC strategies *without* changing between nonminimum-phase and minimum-phase models

model used in case (c) corresponds to the operating point at $f = 11h^{-1}$ and $c_{Ain} = 5.1$. Operating points on the minimum phase side has similar behavior of case (b) where a linear model was identified by a standard output error procedure applied to input-output data. In Fig. 4, the final steady-state corresponding to the first set-point change, i.e., $c_B^{set} = 0.89\ mol/l$, falls outside of the operating range of the minimum-phase region (cf. Fig. 1). Consequently, the controllers, which were initially at the peak of the reactor yield, drove the manipulated variable, f, to its saturation. Only the case (c) was able to reach this set-point. It is easy to understand, since the model used in the case (c) belongs to the left side of the peak. As it can be seen in Fig. 1 the nonminimum-phase side includes $c_B^{set} = 0.89$ in the reachable operating region.

Due to the weak nonlinearities of the process, almost all controllers, except (c), present the same performance when there is no saturation on the controlled variable (c_B). In Fig. 5 this saturation occurs when the unmeasured disturbance c_{Ain} change to 4.5 mol/l, where according to Fig. 1 the set-point of 1.09 mol/l cannot be reached. It is evident that the controllers (b) and (c), based on linear models for the minimum- and nonminimum-phase regions, respectively, will not work properly. Perhaps, in a practical situation, (b) would be the first trial control

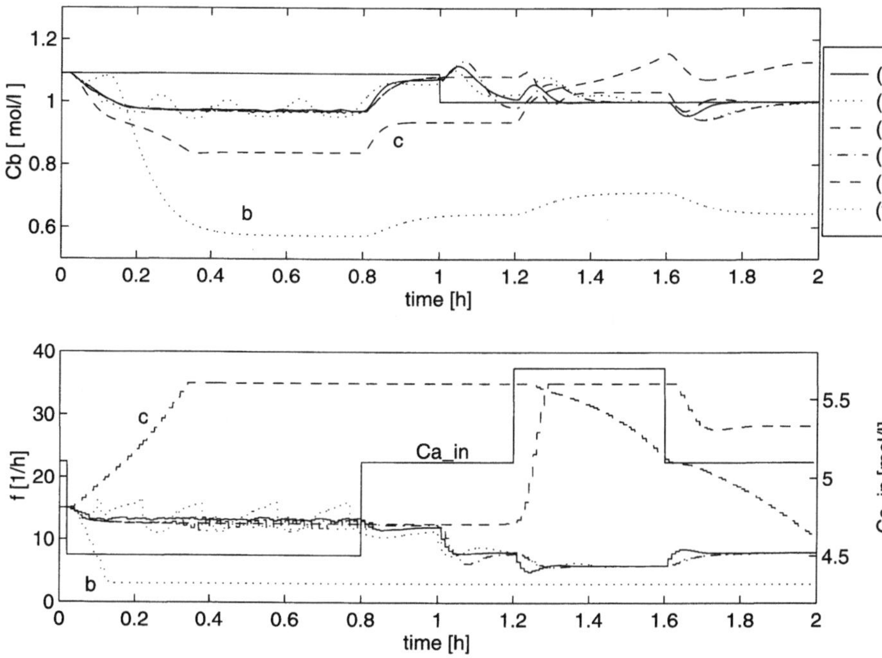

FIGURE 5. Performance of several MPC strategies *with* changing between nonminimum-phase and minimum-phase models due to the *output saturation*

solution, since the model for the case (b) was obtained via a usual system identification procedure. Fig. 5 also reveals several interesting points which are not visible in the case without output saturation. First of all, the LLMN with nonlinear prediction, (f), behaves closer to the nonlinear MPC, (a), and the locally linearized MPC, (d). The oscillations during the output saturation reflect the effects of the standard deviations used in the local models, which can be adjusted, but also show the ability of the network to switch the region. Another interesting point is related to the performance of the LLMN with a linear prediction, i.e., case (e). This solution also has a reasonable performance, while the system is in the output saturation region. Note that out of this region, i.e., after 1h, this MPC solution takes more time to 'decide' to which side it will follow. Finally, at $1.2h$ it only decides to go to the minimum-phase side, when c_{Ain} is changed to 5.7 mol/l. This decision makes the new set-point unattainable due to the input saturation. Certainly, this input saturation problem will not occur if the MPC algorithm could deal with local and global minima. In this case the algorithm would "decide" for the nonminimum-phase side where the set-point can be achieved.

5. Conclusions

The van de Vusse reaction analyzed in this work has a weak nonlinear behavior on each side of maximal concentration of the component B (cf. Fig. 1 and 2). The main nonlinearity presented by this process occurs when the gain changes its sign. To model the system's dynamics for control purposes, it is enough to use a linear model at each side of the peak, such as the local models corresponding to the centers $f = 11.02h^{-1}$ and $f = 18.20h^{-1}$ of Tab.1. Therefore, any control strategy that make possible to change the system's dynamic at the maximum reactor yield shall work.

The algorithm LOLIMOT [10] used in this paper to construct the LMN seems to be appropriate to place the centers in the operating space. The main advantage of the algorithm is to identify the relevant inputs. On the other side, the main drawbacks are related to the side effects produced by the normalization procedure. The weighting functions based on gaussian function alone do not seems to be the best basis function. Some modifications must be done to avoid the normalization side effects and to produce local models closer to the corresponding linearized one.

The performance of MPC depends critically on the predictive capabilities of the underlying process model. Here, LMN have a promising future, specially when local models are linear. The calculation simplicity of the local model can be explored given numerical efficient algorithm. The paper compared some of the simple possibilities. Due to the system's characteristic, it could not be possible to make a clear difference between all of them. For more nonlinear and complex problems the convergence of the proposed iterative approach (case (f)) must be further analyzed.

References

[1] Brown, M.D., Lightbody, G. and G.W. Irwin, *Nonlinear Internal Model Control using Local Model Networks*, ADCHEM'97 IFAC-congress, (1997), 13–18.

[2] Chen, H.; Kremling, A.; Allgöwer, F.; *Nonlinear Predictive Control of a Benchmark CSTR*, Proc. of 3rd ECC, Rome, Italy,(1995), 3247–3252.

[3] Chow, C.M., A.G. Kuznetsov and D.W. Clarke *Using Multiple Models in Predictive Control,* In Proceedings of 3rd. European Control Conference, (1995), 1732–1737.

[4] Di Marco, R., D. Semino and A. Brambilla, *From Linear to Nonlinear Model Predictive Control: Comparison of Different Algorithms,* Ind. Eng. Chem. Res., **36** (1997), 1708–1716.

[5] De Oliveira, S. L., *Model predictive control for constrained nonlinear systems,* (1996), vdf Hochschulverlag AG. an der ETH Zürich.

[6] Engell, S., Klatt, K.-U., *Nonlinear Control of a Non-Minimum-Phase CSTR*, Proc. of American Control Conference, Los Angeles, (1993), 2041–2045.

[7] Foss, B. A., T. A. Johansen, and A. V. Sørensen , *Nonlinear Predictive Control Using Local Models - Applied to a Batch Process,* Control Engineering Practice, **3** (1997),

389–396 (available at http://www.itk.ntnu.no/ansatte/Foss_Bjarne/publikasjoner/-journals/fos94.ps.gz)

[8] Johansen, T. A. and Murray-Smith, R. *The Operating Regime Approach to Nonlinear Modelling and Control,* Chapter 1 of the book Multiple Model Aproaches to Nonlinear Modelling and Control, (eds.) Murray-Smith and Johansen, (1997), Taylor & Francis.

[9] R. Murray-Smith, K. J. Hunt, *Local Model Architectures for nonlinear Modelling and Control,* in Advances in Neural Networks for Control Systems (Hunt, Irwin, and Warwick eds.), Advances in Industrial Control, (1995), Springer-Verlag (available at ftp://eivind.imm.dtu.dk/pub/rod-lcn.ps.gz)

[10] Nelles, O., *LOLIMOT- Lokale, lineare Modelle zur Identifikation nicht-linearer, dynamischer Systeme,* Automatisierungstechnik, **45** (1997), 163–174

[11] Neumann, U., *Modellierung und Regelung nichtlinearer Systeme mit interpolierten lokalen Modellen,* Diplomarbeit at Lehrstuhl für Anlagensteuerungstechnik, Chemietechnik, University of Dortmund, (1998)

[12] Shorten, R. E Murray-Smith, R., *On normalising radial basis functions networks,* in Irish Neural Networks Conf. (1994) (available at ftp://eivind.imm.dtu.dk/pub/rod-norm_dr1.ps.gz)

[13] Trierweiler, J.O., *A Tuning Strategy for Unconstrained MIMO Model Predictive Control,* Proc. of COBEQ'98, (1998), (available at http://www.enq.ufrgs.br/~jorge)

[14] Trierweiler, J.O. and S. Engell, *Controllability Analysis via the RPN for a CSTR with van de Vusse Reaction,*Proc. of European Control Conference 97, (1997), TU-A-H3.

[15] Trierweiler, J.O. and U. Neumann, *Rede de Modelos locais: Uma solução Simples para Problemas Complexos,* Proc. of COBEQ'98, (1998), (available at http://www.-enq.ufrgs.br/~jorge)

Department of Chemical Engineering,
Federal University of Rio Grande do Sul (UFRGS),
Rua Eng. Luiz Englert s/n
CEP: 90040-040 - Porto Alegre - RS
BRAZIL
E-mail address: {jorge,arge}@enq.ufrgs.br

Progress in Systems and Control Theory, Vol. 26

Continuous-time Predictive Control of Constrained Nonlinear Systems

Mark Cannon and Basil Kouvaritakis

Abstract. This paper describes a receding horizon control strategy for constrained nonlinear systems which uses adaptive techniques to minimize continuously an infinite horizon performance cost subject to constraints. Predictions of future plant behaviour evolve as the outputs of an autonomous nonlinear system. An upper bound, obtained via a passivity assumption, on the standard quadratic cost is used in conjunction with a characterization of constraints in terms of an admissible set for the prediction system state to derive an asymptotically optimal control law. Closed-loop stability and convergence are guaranteed without the need for end-point constraints.

1. Introduction

Conventional predictive control strategies for nonlinear systems (eg. [1, 2, 3]) rely on nonlinear programming methods to minimize online a predicted performance cost over future control inputs. Although recent progress has been made in specializing numerical optimization methods to the structure of the receding horizon problem [4], receding horizon control laws derived from this approach remain fundamentally limited in applicability due to their high online computational burden. Furthermore, conventional formulations of predictive control recompute predictions each time future performance is optimized, and, as a result of the intrinsic difficulties of predicting nonlinear system behaviour on an infinite horizon, predictions can only be incorporated explicitly in the receding horizon performance cost on a finite horizon. The artificial constraints that are then needed to ensure closed-loop stability can cause infeasibility with respect to system constraints and place fundamental limits on closed-loop performance.

This paper describes a new nonlinear system predictive control strategy which avoids these limitations by integrating the optimization of future performance with a continuous-time control law. By invoking conditions for future feasibility, given current feasibility, to a finite-dimensional class of input predictions, expressed as a perturbation of an L^2-stabilizing law, we obtain predictions of future plant behaviour as outputs of an autonomous system. This allows an analytic description of the boundary of the maximal admissible set for the prediction system state.

Only the local shape of this boundary is needed in order to ensure that continuously adapted predictions satisfy constraints. The adaptation of the degrees of freedom in predictions is based on the gradient of a closed-form expression for an upper bound, derived from L^2-stability theory, on the standard quadratic cost. Continuous estimates of the location and orientation of the admissible set boundary result in an asymptotically optimal continuous-time control law which requires only simple computational operations at each instant in time. Moreover this approach allows the use of infinite prediction horizons since the predictions of future plant behaviour evolve concurrently with the closed-loop system.

2. Infinite horizon predictions and performance index

We consider regulation of nonlinear systems of the form

$$\dot{\mathbf{x}}(t) = \mathbf{f}_m(\mathbf{x}(t)) + \mathbf{g}_m(\mathbf{x}(t))u(t), \; e(t) = h_m(\mathbf{x}(t)), \tag{1}$$

where $e, u \in \mathbb{R}$, $\mathbf{x} \in \mathbb{R}^{n_s}$, and $\mathbf{f}_m$, $\mathbf{g}_m$, h_m are smooth functions of $\mathbf{x}$ with $\mathbf{f}_m(0) = 0$ and $h_m(0) = 0$. The input u and output e are subject to constraints $u \in \mathcal{U}_i$, $e \in \mathcal{E}_i$, $i = 0, 1$, where

$$\mathcal{U}_0 = \{u;\ \underline{u}_0 \le u(t) \le \overline{u}_0 \ \forall t \ge 0\},\ \ \mathcal{U}_1 = \{u;\ \underline{u}_1 \le \dot{u}(t) \le \overline{u}_1 \ \forall t \ge 0\},$$

and $\mathcal{E}_i$ is defined analogously for $i = 0, 1$. Control laws are developed with the objective of minimizing the performance index

$$J_Q(t) = \int_0^\infty [\hat{e}_t^2(\tau) + \lambda \hat{u}_t^2(\tau)]\, d\tau \tag{2}$$

subject to $\hat{u}_t \in \mathcal{U}_i$, $\hat{e}_t \in \mathcal{E}_i$ $i = 0, 1$, where $\hat{u}_t$, $\hat{e}_t$ are predictions at time t of u, e on the interval $[t, \infty)$. The system (1) is assumed to be globally observable with state $\mathbf{x}$ measurable.

To make the constrained optimization problem tractable we restrict input-output predictions to lie within a finite-dimensional class. As in [5] for the case of linear discrete-time systems, and [6, 7] for the linear continuous-time case, the form of the prediction class is determined by the conditions for stable predictions, ie. the conditions under which J_Q is finite and $\hat{u}_t$ and $\hat{e}_t$ are uniformly bounded. Since $\hat{u}_t$ and $\hat{e}_t$ are necessarily uniformly continuous in order to satisfy rate constraints, these conditions are equivalent to $\hat{u}_t, \hat{e}_t \in L^2 \cap C^1$. Unlike [5, 6, 7], which consider necessary and sufficient conditions for stable predictions, we use here a sufficient condition, namely that input predictions have the form $\hat{u}_t = \hat{w}_t + \hat{v}_t$, where $\hat{w}_t$ is a continuous L^2-stabilizing feedback law and $\hat{v}_t$ is an element of $L^2 \cap C^1$.

The cost $J_Q(t)$ is positive definite and radially unbounded in $\mathbf{x}(t)$ due to the observability of (1). Hence the stability of a receding horizon control law based on a finite-dimensional prediction class can be ensured if at each instant $t \ge 0$ the class is feasible (contains $\hat{u}_t \in \mathcal{U}_i$, $\hat{e}_t \in \mathcal{E}_i$, $i = 0, 1$ with finite associated cost). Assuming the initial prediction class to be feasible with respect to constraints, future feasibility is ensured if the prediction class at each $t > 0$ contains the truncation on $[t,\infty)$ of the prediction corresponding to the input $\lim_{\delta\downarrow 0} u(t-\delta)$, since this satisfies constraints by assumption. This additional constraint can be used to determine the structure of a finite dimensional class for $\hat{v}_t$, as follows.

Let $\hat{v}_t$ lie in the span of a fixed set of basis functions $\{(\phi_i)_{i=1}^{n_c}\} \in L^2 \cap C^1$,

$$\hat{u}_t(\tau) = \hat{w}_t(\tau) + \boldsymbol{\phi}^T(\tau)\mathbf{c}(t), \ \boldsymbol{\phi} = [\phi_1 \ \cdots \ \phi_{n_c}]^T. \tag{3}$$

Here $\mathbf{c}(t) \in \mathbb{R}^{n_c}$ is a vector of prediction parameters containing the degrees of freedom of the prediction class at time t. Also define $\hat{p}^+_{t-\delta}$ by $\hat{p}^+_{t-\delta}(\tau) = \hat{p}_{t-\delta}(\tau+\delta)$ for all $\tau \geq 0$, for any signal $\hat{p}_{t-\delta}$ predicted at time $t-\delta$. Since $\hat{w}_t$ is a time-invariant feedback law of the form $\hat{w}_t(\tau) = w(\hat{\mathbf{x}}_t(\tau))$ (where $\hat{\mathbf{x}}_t$ is the prediction at time t of $\mathbf{x}$ on the interval $[t,\infty)$), in the absence of disturbances we have $\hat{w}^+_{t-\delta} = \hat{w}_t$, and $\hat{u}^+_{t-\delta}(\tau) = \hat{w}_t(\tau) + \boldsymbol{\phi}^T(\tau+\delta)\mathbf{c}(t-\delta)$. It follows that $(\hat{u}^+_{t-\delta}, \hat{e}^+_{t-\delta})$ is a member of the prediction class at time t if $K : \mathbb{R} \to \mathbb{R}^{n_c \times n_c}$ exists such that

$$\boldsymbol{\phi}(\tau+\delta) = K(\delta)\boldsymbol{\phi}(\tau) \tag{4}$$

for all $\tau \geq 0$.

From the preceding discussion, future feasibility is guaranteed if $\epsilon > 0$ exists such that $(\hat{u}^+_{t-\delta}, \hat{e}^+_{t-\delta})$ is a member of the prediction class at time t, for all $\delta \in [0,\epsilon)$. This requires that $K(\delta)$ exists for all $\delta \in [0,\epsilon)$. Since $\boldsymbol{\phi} \in C^1$, Taylor's Theorem applied to (4) yields

$$\boldsymbol{\phi}(\tau) + \dot{\boldsymbol{\phi}}(\tau)\delta + \ldots = [K(0) + \dot{K}(0)\delta + \ldots]\boldsymbol{\phi}(\tau),$$

for all $\delta \in [0,\epsilon)$. From (4) $K(0) = \mathrm{I}$, and it follows that $\dot{\boldsymbol{\phi}}(\tau) = \dot{K}(0)\boldsymbol{\phi}(\tau)$ for all $\tau \geq 0$, so that

$$\boldsymbol{\phi}^T(\tau) = \boldsymbol{\phi}^T(0)\mathrm{e}^{A_\phi \tau}, \ \ \forall \tau \geq 0,$$

where $A_\phi = \dot{K}^T(0)$. The exponential basis functions thus implied allow predictions to be expressed as the outputs of an autonomous system of order $n = n_s + n_c$: $\hat{u}_t(\tau) = g(\Phi^{\mathbf{f}}_\tau(\mathbf{z}))$, $\hat{e}_t = h(\Phi^{\mathbf{f}}_\tau(\mathbf{z}))$, where $g(\mathbf{z}) = w(\mathbf{x}) + \boldsymbol{\phi}^T(0)\mathbf{c}$, $h(\mathbf{z}) = h_m(\mathbf{x})$, and

$$\frac{\partial}{\partial \tau}\Phi^{\mathbf{f}}_\tau(\mathbf{z}) = \mathbf{f}(\Phi^{\mathbf{f}}_\tau(\mathbf{z})), \ \Phi^{\mathbf{f}}_0(\mathbf{z}) = \mathbf{z}$$
$$\mathbf{z}(t) = \begin{bmatrix} \mathbf{c}(t) \\ \mathbf{x}(t) \end{bmatrix}, \ \mathbf{f}(\mathbf{z}) = \begin{bmatrix} A_\phi \mathbf{c} \\ \mathbf{f}_m(\mathbf{x}) + \mathbf{g}_m(\mathbf{x})(w(\mathbf{x}) + \boldsymbol{\phi}^T(0)\mathbf{c}) \end{bmatrix}. \tag{5}$$

The matrix A_ϕ in (5) can be viewed as a design parameter which can be freely chosen subject to the requirement that all of its eigenvalues have strictly negative real part to ensure that $\boldsymbol{\phi} \in L^2$. In fact it can be shown [8] that for sufficiently large n_c, any choice of strictly Hurwitz A_ϕ allows arbitrarily close approximation of any square-integrable function by a function in the span of the associated basis set $\{(\phi_i)_{i=1}^{n_c}\}$. Since n_c is limited in practice by finite computational resources however, the choice of A_ϕ is critical in determining the degree of optimality achievable by a receding horizon control strategy based on the predictions generated by (5). Here we simply note that A_ϕ can be chosen on the basis of prior information such as a physical limit on the bandwidth of the input u.

The design of the L^2-stabilizing control law w is facilitated if the system (1) constitutes a strictly passive mapping from u to e. In this case, any feedback law which is itself a strictly passive mapping from $-\hat{e}_t$ to $\hat{w}_t$ results in L^2-stability of the overall dynamic map from $\hat{v}_t$ to $\hat{e}_t$ (see eg. [9]). Specifically, suppose that functions $V_1(t)$ and $V_2(t)$ can be found satisfying

$$\begin{aligned} V_1(t+\tau) &\geq \epsilon_1 \|\hat{u}_t\|_{2,\tau}^2, \quad \dot{V}_1(t+\tau) \leq \hat{u}_t(\tau)\hat{e}_t(\tau) \\ V_2(t+\tau) &\geq \epsilon_2 \|\hat{e}_t\|_{2,\tau}^2, \quad \dot{V}_2(t+\tau) \leq -\hat{e}_t(\tau)\hat{w}_t(\tau) \end{aligned} \tag{6}$$

for all $\tau \geq 0$, where $\|p\|_{2,t}$ denotes the L^2-norm of the truncation of p on $[0,t]$. Then it is easy to show that $\|\hat{e}_t\|_{\tau,2}$ is bounded by:

$$\|\hat{e}_t\|_{\tau,2} \leq \gamma \|\hat{v}_t\|_{\tau,2} + \beta, \; \gamma = \frac{(4\epsilon_1\epsilon_2 + 1)^{1/2} + 1}{2\epsilon_2}, \tag{7}$$

for some finite β. Furthermore, by integrating the bound on $\dot{V}_1(t+\tau) + \dot{V}_2(t+\tau)$ in (6) over $\tau \in [0,\infty)$, $V_1(t+\tau) + V_2(t+\tau)$ can be shown to be bounded uniformly in τ, and it follows that $\hat{u}_t, \hat{e}_t \in L^2$ for any $\hat{v}_t \in L^2$.

In the design of adaptation laws for $\mathbf{c}(t)$ we assume that the predicted performance index is available in closed-form as a function of the prediction state $\mathbf{z}(t)$. In general this will not be the case for the quadratic prediction cost $J_Q(t)$, but the assumption of strict passivity allows a convenient upper bound $J(t)$ on $J_Q(t)$ to be determined in terms of $V_1(t)$ and $V_2(t)$. Moreover, since $V_1(t) + V_2(t)$ can be interpreted as a measure of the stored energy in (1) with feedback law $u = w(\mathbf{x})$, the minimization of the bound $J(t)$ in place of $J_Q(t)$ is physically meaningful.

Theorem 2.1. *If V_1 and V_2 satisfy (6) with $\epsilon_1 \geq \lambda$ and $\epsilon_2 \geq 1$, then $J_Q(t)$ is uniformly bounded by:*

$$J_Q(t) \leq J(t) = V_1(t) + V_2(t) + \gamma \mathbf{c}^T(t) P_c \mathbf{c}(t) \tag{8}$$

where $P_c = P_c^T > 0$ is the solution of

$$P_c A_\phi + A_\phi^T P_c = -\boldsymbol{\phi}(0)\boldsymbol{\phi}^T(0). \tag{9}$$

Proof. From (6), the derivative of a function $V(t+\tau) = V_1(t+\tau) + V_2(t+\tau) - \|\hat{e}_t\|_{2,\tau}^2 - \lambda\|\hat{u}_t\|_{2,\tau}^2$ is bounded by $\dot{V}(t+\tau) \leq \hat{v}_t(\tau)\hat{e}_t(\tau) - \hat{e}_t^2(\tau) - \lambda\hat{u}_t^2(\tau)$, for all $\tau \geq 0$. Integrating over $\tau \in [0,\infty)$, we have

$$V(t) - \lim_{\tau\to\infty} V(t+\tau) \geq J_Q(t) - \int_0^\infty \hat{v}_t(\tau)\hat{e}_t(\tau)\, d\tau,$$

and since $V(t+\tau)$ is bounded and nonnegative for all $\tau \geq 0$, the bound (7) therefore implies that $J(t) = V(t) + \gamma\|\hat{v}_t\|_2^2$ satisfies $J(t) \geq J_Q(t)$. Furthermore $\|\hat{v}_t\|_2^2 = \mathbf{c}(t)P_c\mathbf{c}(t)$, where P_c is symmetric positive definite and the unique solution to (9), since $\hat{v}_t(\tau) = \boldsymbol{\phi}^T(0)\mathrm{e}^{A_\phi\tau}\mathbf{c}(t)$ for strictly Hurwitz A_ϕ, and $(A_\phi, \boldsymbol{\phi}^T(0))$ can be assumed observable without loss of generality. □

The bound (8) is restricted in application to a small class of systems due to the assumption of strict passivity in Theorem 2.1. Note however that other approaches to the derivation of closed-form expressions bounding the cost $J_Q(t)$ are available. In particular, if w is chosen as an input-output feedback linearizing control, and $\lambda = 0$ in (2), then an exact expression for J_Q in terms of the prediction system state can be obtained in closed-form. This formulation has the additional advantage that a state transformation then exists such that J_Q is quadratic (and therefore convex) in the transformed state of the prediction system (5).

3. Admissible set for the prediction system state

This section develops a characterization of the set $\mathcal{S}$ of admissible prediction system states, defined as the set of states $\mathbf{z}(t)$ such that the predictions evolving according to (5) satisfy the constraints $\hat{u}_t \in \mathcal{U}_i$, $\hat{e}_t \in \mathcal{E}_i$, $i = 0, 1$. The treatment given below is applicable to any set of time-invariant polytopic constraints on the prediction system state. For simplicity, but without loss of generality, in the following development we consider only input constraints of the form

$$u \in \mathcal{U} = \{u;\ u(t) \leq 1\ \forall t \geq 0\}. \tag{10}$$

Let $\mathcal{G}_\tau$ denote the set $\{\mathbf{z};\ g(\Phi_\tau^{\mathbf{f}}(\mathbf{z})) \leq 1\}$. Since $\mathcal{S} = \bigcap_{\tau\in[0,\infty)} \mathcal{G}_\tau$, it is clear that every point in the boundary $\partial\mathcal{S}$ of $\mathcal{S}$ also lies in the boundary $\partial\mathcal{G}_\tau$ of $\mathcal{G}_\tau$ for some $\tau \geq 0$. Lemma 3.1 below shows that every point in $\partial\mathcal{S} \cap \partial\mathcal{G}_\tau$ for some $\tau > 0$ lies on a trajectory of (5) tangent to $\partial\mathcal{G}_0$. Every point in $\partial\mathcal{S}$ therefore lies either in the set $\partial\mathcal{G}_0$, or on a trajectory of (5) that passes through the manifold

$$M_0 = \{\mathbf{z};\ g(\mathbf{z}) = 1,\ L_{\mathbf{f}}g(\mathbf{z}) = 0\}$$

(where $L_{\mathbf{f}}g$ denotes the Lie derivative of g with respect to $\mathbf{f}$). This is equivalent to the intuitive requirement that every prediction state $\mathbf{z}(t)$ in the boundary of the admissible set corresponds to a prediction $\hat{u}_t$ satisfying either $\hat{u}_t(0) = 1$, or $\hat{u}_t(\tau) = 1$ and $\dot{\hat{u}}_t(\tau) = 0$ for some $\tau > 0$.

Lemma 3.1. *If $\mathbf{z} \in \partial\mathcal{S} \cap \partial\mathcal{G}_{\tau_0}$ for some $\tau_0 > 0$, then $\Phi_\tau^{\mathbf{f}}(\mathbf{z}) \in \partial\mathcal{S} \cap \partial\mathcal{G}_{\tau_0-\tau}$ for all $\tau \in [0, \tau_0]$, and $\Phi_{\tau_0}^{\mathbf{f}}(\mathbf{z}) \in M_0$.*

Proof. From $\mathbf{z} \in \partial\mathcal{G}_{\tau_0}$ we have $g(\Phi_{\tau_0-\tau} \circ \Phi_\tau^{\mathbf{f}}(\mathbf{z})) = 1$, so that $\Phi_\tau^{\mathbf{f}}(\mathbf{z}) \in \partial\mathcal{G}_{\tau_0-\tau}$ for all $\tau \in [0, \tau_0]$. Moreover $\mathbf{z} \in \partial\mathcal{S}$ implies that $\Phi_\tau^{\mathbf{f}}(\mathbf{z}) \in \mathcal{S}$ for all $\tau \geq 0$. Therefore $\Phi_\tau^{\mathbf{f}}(\mathbf{z}) \in \partial\mathcal{S} \cap \partial\mathcal{G}_{\tau_0-\tau}$ for all $\tau \in [0, \tau_0]$, and $L_{\mathbf{f}}g(\Phi_{\tau_0}^{\mathbf{f}}(\mathbf{z})) = 0$ since $\Phi_\tau^{\mathbf{f}}(\mathbf{z})$ is necessarily tangent to $\partial\mathcal{G}_0$ at $\tau = \tau_0$. □

Define M as the set of points lying on trajectories of (5) passing through M_0. Then M can be expressed

$$M = \{\mathbf{z};\ g(\Phi_\tau^{\mathbf{f}}(\mathbf{z})) = 1,\ L_{\mathbf{f}}g(\Phi_\tau^{\mathbf{f}}(\mathbf{z})) = 0,\ \tau \in \mathbb{R}\} \tag{11}$$

and the tangent space to M at a point $\mathbf{z} \in M$ is therefore given by:

$$\triangle_M(\mathbf{z}) = \operatorname{span}\{\mathbf{f}(\mathbf{z}),\ \left(\operatorname{null}\{\mathbf{d}g(\Phi_\tau^{\mathbf{f}}(\mathbf{z}))(\Phi_\tau^{\mathbf{f}})_*(\mathbf{z}),\ \mathbf{d}L_{\mathbf{f}}g(\Phi_\tau^{\mathbf{f}}(\mathbf{z}))(\Phi_\tau^{\mathbf{f}})_*(\mathbf{z})\}\right)^T\} \tag{12}$$

where $\mathbf{d} = [\partial/\partial z_1 \ \cdots \ \partial/\partial z_n]$ and $(\Phi_\tau^{\mathbf{f}})_*(\mathbf{z})$ is the Jacobian matrix of $\Phi_\tau^{\mathbf{f}}(\mathbf{z})$. Considering a point $\mathbf{z} \in \partial\mathcal{S} \cap \partial\mathcal{G}_\tau$ with $\tau > 0$, it follows from Lemma 3.1 that the orthogonal subspace to $\triangle_M(\mathbf{z})$ in $\mathbb{R}^n$, $\triangle_M^\perp(\mathbf{z})$, is normal to $\partial\mathcal{S}$ at $\mathbf{z}$. If $L_{\mathbf{f}}^2 g(\Phi_\tau^{\mathbf{f}}(\mathbf{z})) \neq 0$ (so that $\ddot{\hat{u}}_t(\tau) \neq 0$), then $\triangle_M(\mathbf{z})$ has dimension $n-1$ and $\triangle_M^\perp(\mathbf{z})$ is given by

$$\triangle_M^\perp(\mathbf{z}) = \left(\mathbf{d}g(\Phi_\tau^{\mathbf{f}}(\mathbf{z}))(\Phi_\tau^{\mathbf{f}})_*(\mathbf{z})\right)^T.$$

In the general case of $L_{\mathbf{f}}^i g(\Phi_\tau^{\mathbf{f}}(\mathbf{z})) = 0$, $i = 1, \ldots r$, it can similarly be shown that

$$\triangle_M^\perp(\mathbf{z}) = \left(\operatorname{span}\{\left(\mathbf{d}L_{\mathbf{f}}^i g(\Phi_\tau^{\mathbf{f}}(\mathbf{z}))(\Phi_\tau^{\mathbf{f}})_*(\mathbf{z})\right)_{i=0}^{r-1}\}\right)^T. \tag{13}$$

If a prediction $\hat{u}_t \in \mathcal{U}$ reaches the limit $\hat{u}_t(\tau) = 1$ at several distinct times $(\tau_k)_{k\in\mathcal{K}}$, then the corresponding prediction state $\mathbf{z}(t)$ lies in the intersection of $\partial\mathcal{S}$ with $\bigcap_{k\in\mathcal{K}} \partial\mathcal{G}_{\tau_k}$. In this case, the sum of the subspaces $\triangle_M^\perp(\mathbf{z})$ associated with each instant τ_k, $k \in \mathcal{K}$, is normal to $\partial\mathcal{S}$ at $\mathbf{z}$. The general form of the normal to $\partial\mathcal{S}$ can therefore be stated as follows.

Theorem 3.2. *At a point $\mathbf{z} \in \partial\mathcal{S} \bigcap_{k\in\mathcal{K}} \partial\mathcal{G}_{\tau_k}$, the subspace orthogonal to $\partial\mathcal{S}$ is spanned by*

$$\triangle_{\partial\mathcal{S}}^\perp(\mathbf{z}) = \sum_{k\in\mathcal{K}} \left(\operatorname{span}\{\left(\mathbf{d}L_{\mathbf{f}}^i g(\Phi_{\tau_k}^{\mathbf{f}}(\mathbf{z}))(\Phi_{\tau_k}^{\mathbf{f}})_*(\mathbf{z})\right)_{i=0}^{r_k-1}\}\right)^T \tag{14}$$

where $L_{\mathbf{f}}^i g(\Phi_{\tau_k}^{\mathbf{f}}(\mathbf{z})) = 0$, $i = 1, \ldots r_k$ for each $k \in \mathcal{K}$.

Proof. Since $\mathbf{z} \in \partial\mathcal{S}$ implies that $\mathbf{z} \in M \cup \partial\mathcal{G}_0$, (14) follows directly from (13) and the normal, $\mathbf{d}g(\mathbf{z})$, to $\partial\mathcal{G}_0$ at $\mathbf{z}$. □

4. Prediction parameter adaptation

Theorem 3.2 shows that the boundary of the admissible set for the prediction system state is piecewise smooth. Given a closed-form expression for the predicted performance index of the form discussed in Section 2, it is therefore possible to solve the optimization problem:

$$\mathbf{c}^*(t) = \arg\inf_{\mathbf{c}}\{J(t) \text{ subject to } \mathbf{z} \in \mathcal{S}, [0\ \ \mathrm{I}]\mathbf{z} = \mathbf{x}(t)\}$$

analytically. However the solution $\mathbf{c}^*(t)$ is not generally available in closed-form, and the implied control law cannot therefore be performed analytically in continuous-time. This section describes an alternative approach in which $\mathbf{c}(t)$ is adapted continuously (via $\dot{\mathbf{c}}(t)$) in the direction of the local cost gradient. Satisfaction of the constraint $\hat{u}_t \in \mathcal{U}$ is ensured by projecting the gradient of $J(t)$ into the subspace orthogonal to $\triangle_{\partial\mathcal{S}}^\perp(\mathbf{z})$ whenever $\mathbf{z} \in \partial\mathcal{S}$, thus forcing $\mathbf{z}$ to remain in $\mathcal{S}$.

The convergence of an adaptation law for $\mathbf{c}(t)$ and the stability of the resulting receding horizon control law can be determined from the time-variation of J. Using (5) the derivative of $J(t) = J(\mathbf{z}(t))$ can be expressed

$$\dot{J}(\mathbf{z}) = \frac{\partial}{\partial\tau} J(\Phi_\tau^{\mathbf{f}}(z))\Big|_{\tau=0} + \mathbf{d_c}J(\mathbf{z})(\dot{\mathbf{c}} - A_\phi\mathbf{c}) \tag{15}$$

(where $\mathbf{d_c} = [\partial/\partial c_1 \ \cdots \ \partial/\partial c_{n_c}]$). The following theorem uses the L^2-stability property of the prediction system (5) to derive a bound on $\partial J(\Phi^{\mathbf{f}}_\tau(\mathbf{z}))/\partial\tau\big|_{\tau=0}$.

Theorem 4.1. *The derivative along a trajectory of (5) of the cost J defined in Theorem 2.1 satisfies*

$$\frac{\partial}{\partial\tau} J(\Phi^{\mathbf{f}}_\tau(z))\bigg|_{\tau=0} \leq -\hat{e}_t^2(0) - \lambda\hat{u}_t^2(0). \tag{16}$$

Proof. From Theorem 2.1, $\partial J(\Phi^{\mathbf{f}}_\tau(\mathbf{z}))/\partial\tau$ is bounded for all $\tau \geq 0$ according to

$$\frac{\partial}{\partial\tau} J(\Phi^{\mathbf{f}}_\tau(z)) \leq -\hat{e}_t^2(\tau) - \lambda\hat{u}_t^2(\tau) - \hat{v}_t(\tau)[\gamma\hat{v}_t(\tau) - \hat{e}_t(\tau)].$$

Integrating over $[0,\tau)$ and using (7), we have $J(\Phi^{\mathbf{f}}_\tau(\mathbf{z})) - J(\mathbf{z}) \leq -\|\hat{e}_t\|^2_{2,\tau} - \lambda\|\hat{u}_t\|_{2,\tau}$. The bound (16) is then obtained from $\lim_{\tau\to 0}[J(\Phi^{\mathbf{f}}_\tau(\mathbf{z})) - J(\mathbf{z})]/\tau$. □

In the absence of constraints, the adaptation law

$$\dot{\mathbf{c}}(t) = A_\phi \mathbf{c}(t) - \gamma^c \left(\mathbf{d_c} J(\mathbf{z}(t))\right)^T \tag{17}$$

(where $\gamma_c > 0$ is an adaptive gain) therefore ensures that $\dot{J}(t) \leq 0$. Since J is positive definite and radially unbounded in $\mathbf{x}$, it follows that the origin $\mathbf{x} = 0$ of (1) is uniformly stable with the receding horizon control law $u(t) = g(\mathbf{z})$. In addition, the convergence of $e(t)$ and $u(t)$, and convergence of $\mathbf{c}(t)$ to a minimum point of J can be shown as follows.

Theorem 4.2. *If $\mathbf{c}(t)$ satisfies (17) and $\mathbf{z}(t) \in \mathcal{S}$ for all $t \geq 0$, then* $\lim_{t\to\infty} u(t) = 0$, $\lim_{t\to\infty} e(t) = 0$, *and* $\lim_{t\to\infty} \mathbf{d_c} J(\mathbf{z}(t)) = 0$.

Proof. In the absence of disturbances $\hat{u}_t(0) = u(t)$ and $\hat{e}_t(0) = e(t)$. From (15) and (16) we therefore have $\dot{J}(t) \leq -e^2(t) - \lambda u^2(t) - \gamma^c \mathbf{d_c} J(\mathbf{z}(t)) \left(\mathbf{d_c} J(\mathbf{z}(t))\right)^T$. The integral of this expression over $t \in [0,\infty)$ yields

$$\int_0^\infty [e^2(t) + \lambda u^2(t) + \gamma^c \mathbf{d_c} J(\mathbf{z}(t))(\mathbf{d_c} J(\mathbf{z}(t)))^T]\, dt \leq J(0) - \lim_{t\to\infty} J(t).$$

Since the RHS of this inequality is necessarily finite, and $e(t)$, $u(t)$, and $\mathbf{d_c} J(\mathbf{z}(t))$ are uniformly continuous, the convergence of $e(t)$, $u(t)$, and $\mathbf{d_c} J(\mathbf{z}(t))$ then follows from application of Barbalat's Lemma [10]. □

In the case of active constraints, the derivative $\dot{\mathbf{z}}_u(t)$ of the prediction system state corresponding to the unconstrained adaptation law (17) must be projected into $\partial\mathcal{S}$ to ensure that $\hat{u}_t \in \mathcal{U}$. To determine the form of the projection operator, let $\mathbf{z} \in \partial\mathcal{S}$ and define index sets $\mathcal{K}$ and $\mathcal{K}'$ according to

$$\triangle^\perp_{\partial\mathcal{S}}(\mathbf{z}) = \mathrm{span}\{(\mathbf{n}_k)_{k\in\mathcal{K}'}\}; \;\; \mathbf{n}_k^T \dot{\mathbf{z}}_u > 0 \;\; \forall k \in \mathcal{K} \subseteq \mathcal{K}'.$$

Also let the projection of $\dot{\mathbf{z}}_u$ into $\partial\mathcal{S}$ be $\Pi_\mathcal{K}\dot{\mathbf{z}}_u$. Since $\dot{\mathbf{x}}$ is independent of $\dot{\mathbf{c}}$, $\Pi_\mathcal{K}$ is of the form $[\Pi^T_{\mathcal{K},cx} \ [0 \ \mathrm{I}]^T]^T$ for some $\Pi_{\mathcal{K},cx} \in \mathbb{R}^{n_f \times n}$. Defining an orthonormal basis for span$\{(\mathbf{n}_k)_{k\in\mathcal{K}}\}$ as $\{(\mathbf{p}_k)_{k\in\mathcal{K}}\}$, a possible choice for $\Pi_\mathcal{K}$ is therefore

$$\Pi_\mathcal{K} = \mathrm{I} - \begin{bmatrix} \mathrm{I} \\ 0 \end{bmatrix} [\mathrm{I}\ \ 0] \sum_{k\in\mathcal{K}} \mathbf{p}_k\mathbf{p}_k^T, \tag{18}$$

and the resulting adaptation law is given by

$$\dot{\mathbf{c}}(t) = [\mathrm{I}\ \ 0]\Pi_\mathcal{K}\dot{\mathbf{z}}_u(t). \tag{19}$$

To show that the constrained parameter adaptation (19) results in a stable control law, first note that with (19) the second term in the RHS of (15) becomes

$$\mathbf{d_c}J(\mathbf{z})(\dot{\mathbf{c}} - A_\phi\mathbf{c}) = -\mathbf{d_c}J(\mathbf{z})\left[\sum_{k\in\mathcal{K}} \mathbf{p}_{k,c}\mathbf{p}_k^T\mathbf{f}(\mathbf{z}) + \gamma^c\Pi_{\mathcal{K},c}\left(\mathbf{d_c}J(\mathbf{z}(t))\right)^T\right]$$

where $\mathbf{p}_{k,c} = [\mathrm{I}\ 0]\mathbf{p}_k$ and $\Pi_{\mathcal{K},c} = \mathrm{I} - \sum_{k\in\mathcal{K}} \mathbf{p}_{k,c}\mathbf{p}_{k,c}^T$. Since $\mathbf{n}_k^T\mathbf{f}(\mathbf{z}) \le 0$ for all $k \in \mathcal{K}$ due to $\mathbf{z} \in \mathcal{S}$, $\mathbf{n}_k^T\dot{\mathbf{z}}_u > 0$ implies that $\gamma_c\mathbf{d_c}J(\mathbf{z})\mathbf{n}_{k,c} < \mathbf{n}_k^T\mathbf{f}(\mathbf{z}) \le 0$. It follows that $\mathbf{d_c}J(\mathbf{z})\sum_{k\in\mathcal{K}} \mathbf{p}_{k,c}\mathbf{p}_k^T\mathbf{f}(\mathbf{z}) \ge 0$ for all $k \in \mathcal{K}$, and therefore

$$\mathbf{d_c}J(\mathbf{z})[\mathrm{I}\ 0](\Pi_\mathcal{K}\dot{\mathbf{z}}_u - \mathbf{f}(\mathbf{z})) \le -\gamma^c\mathbf{d_c}J(\mathbf{z})\Pi_{\mathcal{K},c}\left(\mathbf{d_c}J(\mathbf{z})\right)^T. \tag{20}$$

This implies that $\dot{J}(t) \le 0$ for all $t \ge 0$, so that $\mathbf{x} = 0$ is uniformly stable. The following theorem shows in addition that $u(t)$ and $e(t)$ converge to zero, and that $\mathbf{c}(t)$ converges to a minimum of $J(\mathbf{z}(t))$ in the boundary of the admissible set.

Theorem 4.3. *If $\mathbf{c}(t)$ is adapted according to (19) for all $t \ge 0$, then $\lim_{t\to\infty} u(t) = 0$, $\lim_{t\to\infty} e(t) = 0$, and $\lim_{t\to\infty} \mathbf{d_c}J(\mathbf{z}(t))\Pi_\mathcal{K} = 0$.*

Proof. Combining (15), (16), and (20), in the absence of disturbances we have $\dot{J}(t) \le -e^2(t) - u^2(t) - \gamma^c\mathbf{d_c}J(\mathbf{z}(t))\Pi_{\mathcal{K},c}\left(\mathbf{d_c}J(\mathbf{z}(t))\right)^T$. The convergence of $e(t)$, $u(t)$, and $\mathbf{d_c}J(\mathbf{z}(t))\Pi_\mathcal{K}$ can therefore be shown analogously to the proof of Theorem 4.2. □

Remark 4.4. *The projection $\Pi_\mathcal{K}\dot{\mathbf{z}}_u$, with $\Pi_\mathcal{K}$ as defined in (18), does not violate inactive constraints. To show this let $\mathbf{n}_j = \mathbf{n}_{j,\mathcal{K}} + \mathbf{n}_{j,\mathcal{K}_\perp}$ for any $j \in \mathcal{K}' - \mathcal{K}$, where $\mathbf{n}_{j,\mathcal{K}} \in$ span$\{(\mathbf{n}_k)_{k\in\mathcal{K}}\}$ and $\mathbf{n}_k^T\mathbf{n}_{j,\mathcal{K}_\perp} = 0$ for all $k \in \mathcal{K}$, then $\mathbf{n}_{j,\mathcal{K}}^T\dot{\mathbf{z}}_u + \mathbf{n}_{j,\mathcal{K}_\perp}^T\dot{\mathbf{z}}_u \le 0$ due to $\mathbf{n}_j^T\dot{\mathbf{z}}_u \le 0$. But $\mathbf{n}_{j,\mathcal{K}}^T\dot{\mathbf{z}}_f > 0$, and $\mathbf{n}_{j,\mathcal{K}_\perp}^T\dot{\mathbf{z}}_u = \mathbf{n}_j^T\Pi_\mathcal{K}\dot{\mathbf{z}}_u$, which implies that $\mathbf{n}_j^T\Pi_\mathcal{K}\dot{\mathbf{z}}_u < 0$.*

5. Estimation of admissible set boundary

Implicit in the constrained adaptation law (19) is the assumption that $\triangle_{\partial\mathcal{S}}^\perp(\mathbf{z}(t))$ can be computed continuously. In general however, closed-form expressions for $\triangle_{\partial\mathcal{S}}^\perp(\mathbf{z}(t))$ are not available since the solutions for τ, $\Phi_\tau^\mathbf{f}(\mathbf{z})$ and $(\Phi_\tau^\mathbf{f})_*(\mathbf{z})$ in (14) corresponding to given $\mathbf{z} \in \partial\mathcal{S}$ cannot be obtained in closed-form. One approach to overcoming this problem is to approximate $\partial\mathcal{S}$, for example by interpolating over a set of discrete samples computed offline. However this is necessarily conservative and computationally expensive for large n.

Here we describe an alternative approach in which asymptotically convergent estimates $\boldsymbol{\xi}(t)$ of the points $\boldsymbol{\xi}^*(t) \in M$ of stationarity in distance from $\mathbf{z}(t)$ to M are computed online. We show that $\mathbf{z}(t)$ can be forced to remain in $\mathcal{S}$ by projecting $\dot{\mathbf{z}}_u(t)$ into the tangent space $\triangle_M(\boldsymbol{\xi}(t))$ (via an adaptation law of the form (19)) whenever a necessary condition for $\mathbf{z}(t) \in \partial\mathcal{S}$ based on $\boldsymbol{\xi}(t)$ is satisfied. In the limit as $\boldsymbol{\xi}(t) \to \boldsymbol{\xi}^*(t)$ this condition is equivalent to $\mathbf{z}(t) \in \partial\mathcal{S}$, and the asymptotic convergence properties of the constrained adaptation law of Section 4 are retained.

First consider the variation with $\mathbf{z}$ of a local minimum or maximum point of the distance $\|\boldsymbol{\zeta} - \mathbf{z}\|_2$, $\boldsymbol{\zeta} \in M$. Throughout this section we assume for convenience that $\boldsymbol{\xi}^*$ is a regular point of M (the treatment of singular points is analogous) and accordingly denote $\triangle_M^{\perp}(\boldsymbol{\xi}^*) = \mathbf{n}(\boldsymbol{\xi}^*)$, where $\mathbf{n}^T(\boldsymbol{\xi}^*) = \mathbf{d}(g \circ \Phi_\tau^{\mathbf{f}})(\boldsymbol{\xi}^*)$. By definition $\boldsymbol{\xi}^*(t)$ satisfies $[\mathrm{I} - \mathbf{n}(\boldsymbol{\xi}^*)\mathbf{n}^T(\boldsymbol{\xi}^*)/\|\mathbf{n}(\boldsymbol{\xi}^*)\|_2^2](\boldsymbol{\xi}^*(t) - \mathbf{z}(t)) = 0$ for all $t \geq 0$. Differentiating and solving for $\dot{\boldsymbol{\xi}}^*(t) \in \triangle_M(\boldsymbol{\xi}^*(t))$ we have

$$\begin{gathered}\dot{\boldsymbol{\xi}}^*(t) = \Pi_{\Gamma^*}\Gamma^{-1}(\mathbf{z}, \boldsymbol{\xi}^*)\dot{\mathbf{z}}(t) \\ \Gamma(\mathbf{z}, \boldsymbol{\xi}) = \mathrm{I} - \frac{\mathbf{n}^T(\boldsymbol{\xi})(\boldsymbol{\xi} - \mathbf{z})}{\|\mathbf{n}(\boldsymbol{\xi})\|_2^2}\frac{\partial \mathbf{n}}{\partial \boldsymbol{\xi}}(\boldsymbol{\xi}), \ \Pi_{\Gamma^*} = \mathrm{I} - \frac{\Gamma^{-1}(\mathbf{z}, \boldsymbol{\xi}^*)\mathbf{n}(\boldsymbol{\xi}^*)\mathbf{n}^T(\boldsymbol{\xi}^*)}{\mathbf{n}^T(\boldsymbol{\xi}^*)\Gamma^{-1}(\mathbf{z}, \boldsymbol{\xi}^*)\mathbf{n}(\boldsymbol{\xi}^*)}.\end{gathered} \tag{21}$$

By considering the variation of $\|\boldsymbol{\zeta} - \mathbf{z}\|_2$ with $\boldsymbol{\zeta}$ it is easily shown that, for all $\boldsymbol{\zeta}$ in a neighbourhood $N \subset M$ of $\boldsymbol{\xi}^*$: $\Gamma(\mathbf{z}, \boldsymbol{\zeta}) > 0$ if $\boldsymbol{\xi}^*$ is a minimum point of $\|\boldsymbol{\zeta} - \mathbf{z}\|_2$ in N, and $\Gamma(\mathbf{z}, \boldsymbol{\zeta}) < 0$ if $\boldsymbol{\xi}^*$ is a maximum point of $\|\boldsymbol{\zeta} - \mathbf{z}\|_2$ in N. From (21) it follows that $\boldsymbol{\xi}^*(t)$ is continuous whenever N contains a strict minimum or maximum point. The following theorem shows that in this case an adaptation law based on (21):

$$\begin{gathered}\dot{\boldsymbol{\xi}}(t) = -\gamma^{\mathbf{z}}\Pi(\boldsymbol{\xi}(t) - \mathbf{z}(t)) + \Pi_\Gamma\Gamma^{-1}(\mathbf{z}, \boldsymbol{\xi})\dot{\mathbf{z}}(t) \\ \Pi = \mathrm{I} - \frac{\mathbf{n}(\boldsymbol{\xi})\mathbf{n}^T(\boldsymbol{\xi})}{\|\mathbf{n}(\boldsymbol{\xi})\|_2^2}, \ \Pi_\Gamma = \mathrm{I} - \frac{\Gamma^{-1}(\mathbf{z}, \boldsymbol{\xi})\mathbf{n}(\boldsymbol{\xi})\mathbf{n}^T(\boldsymbol{\xi})}{\mathbf{n}^T(\boldsymbol{\xi})\Gamma^{-1}(\mathbf{z}, \boldsymbol{\xi})\mathbf{n}(\boldsymbol{\xi})}\end{gathered} \tag{22}$$

(where $\gamma^{\mathbf{z}}$ is an adaptive gain) results in asymptotic convergence of $\boldsymbol{\xi}(t)$ to $\boldsymbol{\xi}^*(t)$.

Theorem 5.1. *If $\boldsymbol{\xi}(t)$ satisfies (22) with $\Gamma(\mathbf{z}, \boldsymbol{\xi}) > 0$ and $\gamma^{\mathbf{z}} > 0$, or $\Gamma(\mathbf{z}, \boldsymbol{\xi}) < 0$ and $\gamma^{\mathbf{z}} < 0$, for all $t \geq 0$, then* $\lim_{t\to\infty} \Pi(\boldsymbol{\xi}(t) - \mathbf{z}(t)) = 0$.

Proof. With (22) we obtain

$$\frac{d}{dt}\tfrac{1}{2}\|\Pi(\boldsymbol{\xi}(t) - \mathbf{z}(t))\|_2^2 = -\gamma(\boldsymbol{\xi}(t) - \mathbf{z}(t))\Pi\Gamma(\mathbf{z}, \boldsymbol{\xi})\Pi(\boldsymbol{\xi}(t) - \mathbf{z}(t))$$

and the convergence of $\Pi(\boldsymbol{\xi}(t) - \mathbf{z}(t))$ follows by Lyapunov's direct method. □

Remark 5.2. *If $\Gamma(\mathbf{z}, \boldsymbol{\zeta})$ is singular for all $\boldsymbol{\zeta}$ in a neighbourhood $N \subset M$ of $\boldsymbol{\xi}^*$, asymptotic convergence of the estimate $\boldsymbol{\xi}(t)$ to a point in N can be obtained with a time-varying gain $\gamma^{\mathbf{z}}$ and Γ^{-1} replaced by a pseudo-inverse in (22). Note however that $\Gamma(\mathbf{z}, \boldsymbol{\xi}^*)$ is necessarily nonsingular if $\|\boldsymbol{\xi}^* - \mathbf{z}\|_2$ is sufficiently small.*

Estimates $\boldsymbol{\xi}(t)$ can be used to ensure satisfaction of the constraint $\mathbf{z}(t) \in \mathcal{S}$ via a test condition of the form $s(\mathbf{z}, \boldsymbol{\xi}) = 0$. Clearly s must have the property that $s(\mathbf{z}, \boldsymbol{\xi}) = 0$ whenever $\|\boldsymbol{\xi}^* - \mathbf{z}\|_2$ is sufficiently small. In addition, we require that $s(\mathbf{z}, \boldsymbol{\xi}^*) = 0$ if and only if $\mathbf{z} = \boldsymbol{\xi}^*$, so that the resulting control law is not conservative in the limit of convergence of $\boldsymbol{\xi}$ to $\boldsymbol{\xi}^*$. If s is defined as:

$$s(\mathbf{z}, \boldsymbol{\xi}) = (\boldsymbol{\xi} - \mathbf{z})^T \left[\frac{\mathbf{n}(\boldsymbol{\xi})\mathbf{n}^T(\boldsymbol{\xi})}{\|\mathbf{n}(\boldsymbol{\xi})\|_2^2} - \alpha \Pi \right] (\boldsymbol{\xi} - \mathbf{z}), \tag{23}$$

where $\alpha = 0$ if $\Gamma > \mathrm{I}$ and $\alpha \geq 1/\underline{\sigma}(\Gamma(\mathbf{z}, \boldsymbol{\xi})) - 1$ if $0 < \Gamma(\mathbf{z}, \boldsymbol{\xi}) \leq \mathrm{I}$ ($\underline{\sigma}(\Gamma)$ denotes the minimum singular value of Γ), then the second condition is satisfied since $s(\mathbf{z}, \boldsymbol{\xi}^*) = \|\boldsymbol{\xi}^* - \mathbf{z}\|_2^2$ for any choice of α.

Moreover $\alpha = 0$ results in $s(\mathbf{z}, \boldsymbol{\xi}) = 0$ whenever $\mathbf{n}^T(\boldsymbol{\xi} - \mathbf{z}) = 0$, and this condition is necessarily satisfied as $\mathbf{z}$ tends to $\boldsymbol{\xi}^*$ from the interior of $\mathcal{S}$ if $\Gamma > \mathrm{I}$, since $\Gamma > \mathrm{I}$ implies that a neighbourhood of $\boldsymbol{\xi}$ in M lies in the boundary of a convex set. Considering the variation of $s(\mathbf{z}, \boldsymbol{\xi})$ with $\mathbf{z}$ (for fixed $\boldsymbol{\xi}$, $\|\boldsymbol{\xi} - \mathbf{z}\|_2$) in the case that $0 < \Gamma \leq \mathrm{I}$, we have $s(\mathbf{z}, \boldsymbol{\xi}) \geq 0$ for all $\mathbf{z}$ such that $\boldsymbol{\xi} - \mathbf{z}$ lies in the direction of $\mathbf{n}(\boldsymbol{\xi})$, and $\alpha \geq 0$ implies that $s(\mathbf{z}, \boldsymbol{\xi})$ decreases monotonically with decreasing $\mathbf{n}^T(\boldsymbol{\xi})(\boldsymbol{\xi} - \mathbf{z})$. In addition, $(1 + \alpha)\Gamma \geq \mathrm{I}$ ensures that the surface defined by $s(\mathbf{z}, \boldsymbol{\xi}) = 0$ for fixed $\boldsymbol{\xi}$ is tangent to M at $\mathbf{z} = \boldsymbol{\xi}$. Note that it suffices to consider only the case of $\Gamma > 0$, since $\Gamma > 0$ can be assumed if $\|\boldsymbol{\xi}^* - \mathbf{z}\|_2$ is sufficiently small. The following theorem summarizes this discussion.

Theorem 5.3. *If $\mathbf{c}(t)$ satisfies (19) with $\mathbf{n}(\boldsymbol{\xi}^*(t))$ replaced by the estimate $\mathbf{n}(\boldsymbol{\xi}(t))$ in the definition of $\Pi_{\mathcal{K}}$, and $\boldsymbol{\xi}(t)$ is adapted according to*

$$\dot{\boldsymbol{\xi}}(t) = -\gamma^{\mathbf{z}} \Pi(\boldsymbol{\xi}(t) - \mathbf{z}(t)) + \dot{\mathbf{z}}(t) \tag{24}$$

whenever $s(\mathbf{z}, \boldsymbol{\xi}) = 0$, then $\mathbf{z}(t) \in \mathcal{S}$ for all $t \geq 0$, and the asymptotic convergence results of Theorem 4.3 hold.

Proof. The adaptation law (19) based on $\boldsymbol{\xi}(t)$ results in $\mathbf{n}^T(\boldsymbol{\xi}(t))\dot{\mathbf{z}}(t) = 0$. Differentiating (23) along a trajectory of (24), we therefore have

$$\dot{s}(\mathbf{z}, \boldsymbol{\xi}) = \gamma^{\mathbf{z}} (\boldsymbol{\xi} - z)^T \Pi[(1 + \alpha)\Gamma(\mathbf{z}, \boldsymbol{\xi}) - \mathrm{I}]\Pi(\boldsymbol{\xi} - \mathbf{z}) \geq 0.$$

Since $s(\mathbf{z}(t), \boldsymbol{\xi}(t)) = 0$ is necessary for $\mathbf{z}(t) \in \partial\mathcal{S}$, it follows that $\mathbf{z}(t)$ remains in $\mathcal{S}$ for all $t \geq 0$. Moreover, by the argument used in the proof of Theorem 5.1, the modified adaptation law (24) results in asymptotic convergence of $\boldsymbol{\xi}(t)$ to $\boldsymbol{\xi}^*(t)$, and since $s(\mathbf{z}(t), \boldsymbol{\xi}(t)) = 0$ is equivalent to $\mathbf{z} \in \partial\mathcal{S}$ in the limit of convergence, the constrained adaptation law of (19) is obtained asymptotically. □

6. Concluding remarks

A direct implementation of the continuous-time predictive control strategy outlined above would involve: (i) initializing a stationary point estimate $\boldsymbol{\xi}(t)$ at each stationary point of $\|\boldsymbol{\xi} - \mathbf{z}(0)\|_2$ for $\boldsymbol{\xi} \in M$, computed offline; (ii) updating online $\mathbf{c}(t)$ and $\boldsymbol{\xi}(t)$ according to (17) and (22) if $s(\mathbf{z}(t), \boldsymbol{\xi}(t)) > 0$; (iii) updating online $\mathbf{c}(t)$ and $\boldsymbol{\xi}(t)$ according to (19) and (24) if $s(\mathbf{z}(t), \boldsymbol{\xi}(t)) = 0$. This approach is clearly impracticable in the general case of an infinite number of stationary points in distance from $\mathbf{z}(t)$ to M. However, the asymptotic stability property of the prediction system (5) can be used to determine a bound T_{con} on the interval over which constraints are active as a function of $\mathbf{z}(t)$. A finite number of stationary points need only be computed if each estimate $\boldsymbol{\xi}(t)$ is then constrained to remain in the subset of M defined by $\bigcup_{\tau\in[0,T_{con}]} M \cap \partial G_\tau$.

Implementation on a digital computer would require some form of discretization. Since the control and parameter adaptation laws require few computational operations at each instant in time, this could be achieved using high sample rates and difference approximations to the derivatives of $\mathbf{c}(t)$ and $\boldsymbol{\xi}(t)$.

References

[1] S. S. Keerthi and E. G. Gilbert. Optimal infinite-horizon feedback laws for a class of constrained discrete-time systems. Stability and moving horizon approximations. *J. Optimiz. Theory Appl.*, 57(2):265–293, 1988.

[2] H. Michalska and D. Q. Mayne. Robust receding horizon control of constrained nonlinear systems. *IEEE Trans. Autom. Contr.*, 38(11):1623–1633, 1993.

[3] S. de Oliveira and M. Morari. Robust model predictive control for nonlinear systems. In *Proc. IEEE Conf. Dec. Contr.*, volume 4, pages 3561–3566, 1994.

[4] L. T. Biegler. Advances in nonlinear programming concepts for process control. *J. Proc. Contr.*, 1998. To appear.

[5] J. A. Rossiter, J. R. Gossner, and B. Kouvaritakis. Infinite horizon stable predictive control. *IEEE Trans. Autom. Contr.*, 41(10):1522–1527, 1996.

[6] B. Kouvaritakis, M. Cannon, and J. A. Rossiter. Recent developments in generalized predictive control for linear systems. *Int. J. Contr.*, 1998. In press.

[7] M. Cannon and B. Kouvaritakis. Infinite horizon predictive control of constrained continuous-time linear systems. Technical Report OUEL 2133/97, University of Oxford, 1997. Submitted to *Automatica*.

[8] D. W. Kammler. Approximation with sums of exponentials in $L^p[0, \infty)$. *J. Approx. Thy.*, 16:384–408, 1976.

[9] M. Vidyasagar. *Nonlinear Systems Analysis*. Prentice-Hall, Englewood Cliffs, NJ, second edition, 1993.

[10] M. Krstic, I. Kanellakopoulos, and P. Kokotovic. *Nonlinear and Adaptive Control Design*. Wiley-Interscience, New York, 1995.

Department of Engineering Science,
University of Oxford,
Parks Road
Oxford OX1 3PJ, UK
E-mail address: `mark.cannon@eng.ox.ac.uk`

Part II

Modelling and Computational Aspects in Nonlinear Predictive Control

Progress in Systems and Control Theory, Vol. 26

Efficient Solution of Dynamic Optimization and NMPC Problems

Lorenz T. Biegler

Abstract. Large scale optimization strategies have evolved considerably over the past two decades. Currently, continuous variable optimization problems (nonlinear programs) are solved on-line for steady state refinery models with several hundred thousand variables. Moreover, efficient NLP strategies have been developed for dynamic optimization problems. Still, to take the next step, on-line optimization of large dynamic chemical processes, a number of limitations and research challenges must be overcome.

Many of the advances in NLP algorithms have taken place by recognizing and exploiting the framework of Successive Quadratic Programming (SQP) algorithms. These are extensions of Newton type methods for converging to the solution of the KKT (optimality) conditions of the optimization problem. Because of this, fast convergence can be expected and a number of standard devices can be added to stabilize the algorithm to converge from poor starting points. Limitations of these Newton-based methods are also well-known. They experience difficulties in the presence of ill conditioning and extreme nonlinearities. Also, for optimization algorithms, nonconvexity can also lead to a number of difficulties and there is a need for software that allows exploitable structures for specific problem classes.

A number of innovations in algorithm design and problem formulation address these issues and greatly improve performance. As a result, very fast NLP algorithms can be derived for data reconciliation, parameter estimation, nonlinear model predictive control and dynamic optimization. Moreover, inequality constraints and variable bounds can be treated through advances in interior point strategies. These methods preserve the particular problem structure and scale well in performance for large-scale problems with many constraints.

Finally, the ability to solve nonlinear programs quickly also allows us to consider more challenging problems. These include the extension to solving nonconvex NLPs *globally* and the ability to assess the solution's tolerance to uncertainty by considering features of nonlinear programming sensitivity analysis and robust optimization through flexibility analysis and design.

1. Introduction

With increasing acceptance of model predictive control (MPC) as a general purpose strategy for process control, research and applications are proceeding to consider more complex nonlinear models. Two advantages of this strategy are the use of a general control law to handle MIMO systems directly and a framework to handle input and output constraints. Using linear dynamic plant models, MPC has become a standard tool with a wide variety of commercial applications and successes. Moreover, over the past five years, theoretical properties of MPC have been analyzed more deeply. Based on stabilty properties of LQG through Lyapunov analysis dating from the early 70s [28, 30, 27] the stability properties of MPC are now much better understood. In particular, Rawlings and coworkers (e.g., [42, 46, 45]) have identified and analyzed several MPC problem formulations (e.g., infinite horizon, endpoint constrained, dual mode) which inherit these stability properties.

Given the success of MPC with linear models, it is useful to question whether the more difficult task of incorporating nonlinear models is even worthwhile. While the former requires the solution of convex problems for each sampling time (usually quadratic programs (QPs) or linear programs (LPs)), nonlinear model predictive control (NMPC) requires the consideration (and at least a partial solution) of a nonconvex, nonlinear program (NLP). Difficulties with NMPC come from both theoretical and computational perspectives. As noted by Mayne [34], significant concerns are introduced by nonconvexity since nominal stability, which requires strong descent of the Lyapunov function, can no longer be guaranteed. A related concern is the expense and reliability of solving the NLP on-line.

Despite these concerns, NMPC are being applied in the chemical industry and these applications are viewed as essential for a number of process control problems. NMPC is needed especially for nonlinear, unsteady batch processes where a trajectory needs to be followed from the prediction of a nonlinear model. It is also needed for processes with wide swings in operation, beyond the ranges of a local linearization. This is especially true for processes operating at or near 'singular' points (e.g., sign changes in process gain and input multiplicities) that cannot be captured by linear controllers and where higher order information is needed.

A more important justification for nonlinear model predictive control is its position in the framework of on-line process optimization. Currently, on-line optimization is performed with steady-state nonlinear models and these pass setpoints and feedforward information to controllers represented by linear dynamic models. Fundamentally this leads to an inconsistency in the overall problem formulation and a potential conflict in the tasks of the controller and on-line optimizer. As a result, care needs to be applied in the formulation of constraints, both for the steady state model and the controller, and also in updating the steady state optima used as controller setpoints. Also implied in this process is that the timescales for the online optimization are much longer than for the controller.

A more natural approach would be to start with an overall, integrated, dynamic optimization problem and propose decompositions that fit the need of the process application. Such a problem could consider the influence of the scheduling and operating policies, the interactions of upstream or downstream processes and the local inputs which could include disturbances and changing performance requirements. At the "lowest" level, the resulting control strategy could still require the solution of nonlinear dynamic optimization problems and it is for the efficient execution of this task, that this paper is directed.

In the following sections, recent advances are covered for the solution of nonlinear dynamic optimization problems which have the characteristics of NMPC problems. This review is by no means complete but I hope to highlight some of the major computational trends and show their relevance to NMPC problem formulations. These become especially important as larger applications and problem formulations are considered.

This section concludes with a statement of the nonlinear programming problem derived from a nonlinear differential-algebraic (DAE) model. Section 2 continues with a Newton-based algorithm for the solution of nonlinear programs. Some general convergence properties of this approach will be treated along with the special case of NMPC problems. Section 3 then discusses sequential and simultaneous approaches and emphasizes the need for problem conditioning and the formulation of dichotomous problems. Section 4 discusses the exploitation of structure for simultaneous strategies as well as inequality constraint handling with emphasis on interior point strategies. With the availability of good local NLP solvers, section 5 briefly considers more challenging applications. In particular, this section deals with branch and bound strategies for global optimization as well as problem formulations for incorporating uncertainty into the NMPC algorithm. Finally, section 6 summarizes some of these points and gives directions for future work.

1.1. Problem definition

Consider a process model of the form:

$$\dot{x} = f(x, z, u, \delta; \theta) \tag{1}$$

$$c(x, z, u, \delta; \theta) = 0 \tag{2}$$

$$y = g(x; \theta) \tag{3}$$

with f, c and g twice continuously differentiable, where the state vector consists of the differential variables $x \in \mathbf{R}^{\mathbf{n_{sd}}}$, and algebraic variables $z \in \mathbf{R}^{\mathbf{n_{sa}}}$, the input vector is $u \in \mathbf{R}^{\mathbf{n_i}}$, δ is the disturbance vector, θ are adjustable model parameters, and the output vector is $y \in \mathbf{R}^{\mathbf{n_o}}$. Without loss of generality, assume that the algebraic constraints $c = 0$ can be formulated as index one and, as a result the algebraic equations, c, and state variables, z, can be eliminated. This is done for convenience of notation; direct elimination of algebraic equations is not required as all of the algorithms formulated below can accommodate them implicitly.

For most of this paper a perfect model is assumed and the state vector is available. Some relaxations to this assumption will be discussed in section 5. For a constant sampling time T it is convenient to define the discrete time operator corresponding to (1, 2) as $x_{i+1} = \chi(T; x_i, u_i)$. This operator can be evaluated explicitly through integration of the DAE model, or approximated and represented implicitly, for instance, using collocation on finite elements. Substitution of the discretized model leads to the following receding horizon problem formulation, at time i:

$$Min_{u_l,x_l,y_l} J = \sum_{l=i}^{i+p} h_i(y_{sp,l} - y_l, u_l - u_{r,l}) \tag{4}$$

$$s.t.\ x_{l+1} = \chi(x_l, u_l),\ \ y_l = g(x_l) \quad l = i, \ldots, i+p-1$$

$$u_l = u_{i+m-1},\ l = i+m, \ldots, i+p,\ \ u_a \le u_l \le u_b$$

$$y_p = g(x_p),\ \ x_a \le x_l \le x_b,\ \ y_a \le y_i \le y_b$$

where the terms in the objective function $h(\xi, \eta)$ are nonnegative, continuously differentiable and take the value zero iff ξ and η are zero. A typical example of this objective function is:

$$h(\xi, \eta) = (y_{sp,l} - y_l)^T Q_{yl}(y_{sp,l} - y_l) + (u_l - u_{r,l})^T Q_{ul}(u_l - u_{r,l}) \tag{5}$$

Here y_{sp} is the setpoint and u_r is a reference input compatible with this setpoint. Note that $\chi(x_l, u_l)$ is obtained through the evaluation of (1, 2) inside the sampling interval $t \in [t_l, t_{l+1}]$ only. To simplify the notation further, we define the concatenated vectors U, X and Y containing all values of the corresponding variables inside a predictive horizon: $U \equiv [u_i^T, u_{i+1}^T, \ldots, u_{i+m-1}^T]^T \in \mathbf{R}^{\mathbf{n_i m}}$, $X \equiv [x_i^T, x_{i+1}^T, \ldots, x_{i+p}^T]^T \in \mathbf{R}^{\mathbf{n_s p}}$, and $Y \equiv [y_i^T, y_{i+1}^T, \ldots, y_{i+p}^T]^T \in \mathbf{R}^{\mathbf{n_o p}}$. As a result, the NLP can be rewritten as:

$$Min_{U,X,Y} J(Y, U) \tag{6}$$

$$s.t.\ F(X, U) = 0,\ \ Y - G(X) = 0$$

$$U_a \le U \le U_b,\ \ X_a \le X \le X_b,\ \ Y_a \le Y \le Y_b$$

Note that this formulation is general enough to allow for the treatment of endpoint constraints or other devices to ensure stability of the nonlinear controller. It should be noted that (6) can now be solved using any NLP algorithm. In this paper we explore Newton-type approaches for this solution as these methods can be adapted to exploit the structure of the NMPC problem. These approaches have the advantage of simplicity along with well-known convergence properties. Moreover, unlike general purpose NLP solvers, they can be easily tailored to different applications, with significant benefits in performance. As a result, we briefly summarize the Successive Quadratic Programming (SQP) algorithm for (6) in the next section.

2. Basic SQP algorithm

Using SQP for problem (6), the following quadratic programming (QP) subproblem is created and solved at iteration k, and the solution, d, is used as a search direction to converge to the solution of (6):

$$\begin{aligned} Min_d \quad & \nabla_Y J(Y_k, U_k)^T d_Y + \nabla_U J(Y_k, U_k)^T d_U + 1/2 d^T B_k d \\ s.t. \quad & F(X_k, U_k) + \nabla_X F(X_k, U_k)^T d_X + \nabla_U F(X_k, U_k)^T d_U = 0 \\ & Y_k + d_Y - G(X_k) - \nabla_X G(X_k)^T d_X = 0 \\ & U_a \le U_k + d_U \le U_b, X_a \le X_k + d_X \le X_b \\ & Y_a \le Y_k + d_Y \le Y_b \end{aligned} \tag{7}$$

where $d^T = [d_U^T \ \ d_X^T \ \ d_Y^T]$ and B_k is the Hessian of the Lagrange function, $\nabla^2 L(U, X, Y, \lambda, \gamma) = \nabla^2 J + \Sigma_i \nabla^2 F_i \lambda_i - \Sigma_j \nabla^2 G_j \gamma_j$ or its approximation.

2.1. Summary of convergence properties

As with any Newton-type method, asymptotic convergence properties for SQP are also well-known (see e.g. [19]). If B_k is asymptotically equivalent to the actual Hessian at the optimum, $\nabla^2 L^*$ then the convergence rate is Q-quadratic, as is expected with any Newton method. However, $\nabla^2 L^*$ is rarely positive definite and this leads to difficulties in solving the QP, (7). Using a positive definite quasi-Newton approximation for B_k (common in many generic SQP solvers, such as NPSOL and VF13AD), leads only to (one- or two-step) Q-superlinear convergence. Finally, a less faithful approximation for B_k leads to only Q-linear convergence.

Similarly, a number of reliable stabilization algorithms are available to promote convergence from poor starting points [19]. First, it is well known that if the B_k matrix is uniformly positive definite in the reduced space of the active constraints, then d, the solution to (7), forms a strong descent direction for an exact penalty (or merit) function, the minimum of which corresponds to a solution of (6). Thus a stepsize along d can be found which is bounded away from zero and the convergence argument then proceeds in a straightforward way.

However, these properties do not *guarantee* convergence from poor starting points. First, if (6) is nonconvex, a solution to (7) cannot be guaranteed, as the region of linearized constraints may not be feasible. In this case, a trick that allows the SQP algorithm to continue is to introduce an additional scalar variable ξ and solve, for instance:

$$\begin{aligned} Min_{U,X,Y} \quad & J(Y, U) + M\xi \\ s.t. \quad & F(X, U)(1 - \xi) = 0 \\ & (Y - G(X))(1 - \xi) = 0 \\ U_a \le U \le U_b, \quad & X_a \le X \le X_b \\ Y_a \le Y \le Y_b, \quad & 0 \le \xi \le 1 \end{aligned} \tag{8}$$

where M is chosen suitably large to keep ξ as small as possible. The modified problem has a feasible point with the variables between their bounds and with $\xi = 1$. Under the above conditions, solution of (8) either satisfies the constraints of (6) or $\xi^* = 1$, at which point one is forced to try another starting point (or, better yet, re-examine the problem formulation for consistency).

Second, the SQP algorithm attracts to (first order) Kuhn-Tucker points that may not be a local minimum. In fact, simply tracking Kuhn Tucker points may even lead to a local maximum [34]. To ensure optimum solutions, however, one can examine the second order conditions at the converged point and follow directions of negative curvature if the point is not optimal. These directions ensure an immediate descent away from the converged point. Alternately, SQP methods have been developed that, under mild conditions, will attract to local minima. These usually require exact second derivatives as well as trust region strategies.

2.2. Convergence from poor starting points

Let's represent (6) in a briefer form as $Min\ J(w) \mid c(w) = 0,\ \ a \leq w \leq b$. with the resulting QP at iteration k as:

$$\begin{aligned} Min \quad & \nabla J(w_k)^T d + 1/2 d^T B d \\ s.t. \quad & c(w_k) + \nabla c(w_k)^T d = 0 \\ & a \leq w_k + d \leq b \end{aligned} \tag{9}$$

Because the QP keeps the sequence $\{w_k\}$ between bounds, a line search method can be developed that reduces a merit function, e.g., the ℓ_1 exact penalty function, $P(w) = J(w) + \beta\|c(w)\|_1$ along the solution vector d. Dividing the solution into range and null space components, $d = Y_k p_R + Z_k p_N$ where $A_k^T Z_k = 0$, A_k is the matrix of active (assumed linearly independent) constraint gradients, $Z_k^T B Z_k$ is positive definite and $[Y_k \ \ Z_k]$ is a square nonsingular matrix, it is straightforward to show the solution to (9) from:

$$(A_k^T Y_k) p_R = -\bar{c}(w_k) \tag{10}$$

$$(Z_k^T B Z_k) p_N = -(Z_k^T \nabla J(w_k) + Z_k^T B Y_k p_R) \tag{11}$$

Here the vector $\bar{c}^T = [c_+^T \ \ c_-^T \ \ \hat{c}^T]$ where c_+ refers to those elements of $c_j(w_k) \geq 0$, c_- refers to elements of $c_j(w_k) < 0$ and $\hat{c}$ are the residuals of the active bounds ($\hat{c}_j = (w_k - b)_j$ or ($\hat{c}_j = (w_k - a)_j$). Using this solution as a descent direction for the ℓ_1 penalty function leads to:

$$\begin{aligned} D_d P(w_k) &= \nabla J(w_k)^T (Z_k p_N + Y_k p_R) \\ &\quad + \beta \nabla c_+^T Y_k p_R - \beta \nabla c_-^T Y_k p_R && (12) \\ &= -\nabla J(w_k)^T Z_k (Z_k^T B Z_k)^{-1} (Z_k^T \nabla J(w_k) \\ &\quad + (Z_k^T B Y_k) p_R) + \nabla J(w_k)^T Y_k p_R - \beta \|c(w_k)\|_1 && (13) \\ &\leq -\kappa \| Z_k^T \nabla J(w_k), c(w_k) \| && (14) \end{aligned}$$

with positive κ bounded away from zero. The last two inequalities follow from (9) and by choosing β sufficiently large. Bounds on β can be derived from the constraint multipliers at the QP solution and a common device is to increase β as needed at each iteration. Since $\|Z^T \nabla J(w), c(w)\| = 0$ is a Kuhn-Tucker point, d is a direction of strong descent and, from a Taylor series argument (see [23, 7]) there exists a stepsize α bounded away from zero such that $w_{k+1} = w_k + \alpha d$ leads to a strictly decreasing sequence $\{P(w_k)\}$ until convergence.

However, this analysis assumes that $\hat{Q} = (A_k^T Y_k)$ or $\hat{Q} = (Z_k^T B Z_k)$ is well conditioned to avoid limiting points where $\hat{Q}\bar{p} = 0$. Ill-conditioning of either matrix can easily lead to failure of the line search strategy. Fortunately, these problems can be addressed through a trust region strategy, which requires an additional constraint in (9), of the form $\|d\|_2 \leq \Delta_c$. Here Δ_c is a positive scalar that represents the size of the current trust region. As with the line search algorithm, the stepsize Δ_c is chosen adaptively so that the new point $w_{k+1} = w_k + d$ reduces the merit function. However, unlike the line search method, changing Δ_c changes *both* the length and the direction of d.

Trust region strategies are well known for unconstrained minimization (see [15]) and a number of trust region algorithms have been developed for constrained minimization as well (see, e.g., [31]). For (9), the trust region is applied by considering the determination of p_R and p_N as separate subproblems. Each problem can be solved by updating a scalar μ that is used as a correction term with $(\hat{Q} + \mu I)$ in (10) or (11); choosing a larger μ corresponds to reducing the size of Δ_c.

However, this simple approach creates a convex problem that will not detect directions of negative curvature. Thus, for problems where $\hat{Q}$ has negative eigenvalues, the trust region problem must be considered more carefully. In particular, an inexpensive way to detect directions of negative curvature is to apply a preconditioned conjugate gradient algorithm to:

$$(Z_k^T B Z_k) p_N = -(Z^T \nabla J(w_k) + Z_k^T B Y_k p_R) \mid \|p_N\| \leq \Delta_c \tag{15}$$

Here a vector p_N will be found that either solves the linear system or is forced to the trust region bound [49, 12, 13]. In the latter case, p_N is a direction of negative curvature and the algorithm continues with an improved point.

Thus, line search strategies offer a cheap way to ensure convergence from poor starting points as long as $\hat{Q}$ is positive definite and well conditioned. Trust region strategies, on the other hand, can deal with indefinite and ill-conditioned $\hat{Q}$ but multiple subproblems may need to be solved at each iteration. Given these complementary features, a superior algorithm can be constructed that combines the trust region and line search algorithms. Here a trust region is chosen to ensure that the QP is well-conditioned and a line search is applied within the trust region to ensure the best steplength. This combined strategy leads to significantly better performance on both well and poorly conditioned problems [21, 51], as shown in the next example.

Example of trust region and line search methods
To illustrate the performance of the line search, trust region and the hybrid strategies, consider the following two similar examples. Choosing x_1 as the independent variable leads to a well conditioned matrix $A_k^T Y_k$ in the first example but to a poorly conditioned one in the second.

Number of Variables n	Line Search	SQP Iterations Trust Region	Combined
100	8 (8/0.09)	8 (8/0.11)	8 (8/0.11)
200	9 (10/0.18)	9 (9/0.16)	9 (9/0.17)
300	12 (16/0.31)	18 (18/0.31)	9 (11/0.29)
400	13 (20/0.43)	18 (18/0.45)	9 (10/0.33)
500	13 (20/0.48)	18 (18/0.54)	10 (11/0.45)
1000	14 (20/1.04)	18 (18/1.03)	10 (12/0.86)
100	11 (16/0.11)	145 (145/1.05)	14 (19/0.13)
200	51 (129/0.72)	46 (46/0.61)	26 (26/0.44)
300	51 (128/1.10)	103 (103/1.88)	16 (16/0.44)
400	51 (128/1.48)	66 (66/1.62)	16 (20/0.51)
500	51 (128/1.78)	74 (74/2.20)	16 (20/0.62)
1000	12 (18/0.87)	69 (69/4.14)	15 (15/1.16)

TABLE 1. Iterations (No. functions/CPU Seconds) for Examples 1 and 2.

Well conditioned example

$$\begin{array}{ll} Min & 1/2\sum_{i=1}^{n} x_i^2 \\ s.t. & c_j = x_1(x_{j+1}-1) - 10x_{j+1} = 0 \quad j = 1,\ldots,n-1 \\ & x_i^0 = 0.1 \quad x_j^* = 0 \quad i = 1,\ldots,n \end{array}$$

Poorly conditioned example

$$\begin{array}{ll} Min & 1/2\sum_{i=1}^{n} x_i^2 \\ s.t. & c_j = x_{j+1}(x_1-1) - 10x_1 = 0 \quad j = 1,\ldots,n-1 \\ & x_i^0 = 0.1 \quad x_i^* = 0 \quad i = 1,\ldots,n \end{array}$$

The number of SQP iterations required for convergence within a tolerance of 10^{-10} is given in Table 1, as a function of problem size, for three types of stabilization algorithms. Further details of this comparison are given in [51]. Note that the combined trust region and line search algorithms lead to improved performance on both types of problems. This is especially true in the second example where both line search and trust region methods need to work harder due to ill-conditioning.

2.3. Exploitation of structure for NMPC

Finally, we note that (6) has an exploitable structure that can accelerate the SQP algorithm. For instance, if we assume a regulatory problem where the setpoint is interior to the bound constraints, then the Kuhn Tucker conditions for this problem are:

$$\nabla_Y J + \gamma = 0 \tag{16}$$
$$\nabla_U J + \nabla_U F \lambda = 0 \tag{17}$$
$$\nabla_X F \lambda - \nabla_X G \gamma = 0 \tag{18}$$

From the first equation we can rewrite the relations as:

$$\nabla_U J + \nabla_U F \lambda = 0 \tag{19}$$
$$\nabla_X G \nabla_Y J + \nabla_X F \lambda = 0 \tag{20}$$

and since the evolution of the states is well defined, $\nabla_X F$ is square and nonsingular and $\lambda = -(\nabla_X F)^{-1} \nabla_X G \nabla_Y J$. Consequently, if the predicted outputs remain close to their setpoints, $\nabla_Y J = Q_Y(Y - Y_{sp}) \approx 0$ and $(\gamma, \lambda) \approx 0$ as well. As a result, we can simplify the evaluation of the B_k matrix in (7) since $\nabla^2 L(U, X, Y, \lambda, \gamma) \approx \nabla^2 J = Q$ where

$$Q = \begin{bmatrix} Q_U & 0 & 0 \\ 0 & 0 & 0 \\ 0 & 0 & Q_Y \end{bmatrix}.$$

Q is positive definite in the reduced space of the linearized equality constraints and setting $B_k = Q$ therefore leads to a convex QP with a unique solution.

This approach, termed the *Gauss-Newton property*, is well known for unconstrained least squares problems and has been demonstrated very successfully for constrained least squares problems [9]. In the case of zero residuals and nonlinear constraints, this simplification leads to a quadratic convergence rate. This rate becomes linear if projected contributions of the nonlinear constraints are nonzero, but in practice convergence is still quite good for problems with small residuals. Moreover for large residuals, variations of this approach also include hybrid strategies [19, 1, 2] which adapt and switch between quasi-Newton methods and the Gauss-Newton approximation for Q.

This informal survey of properties serves not only as background but also motivates the implementation of NLP algorithms for NMPC. In the next section we see that ill-conditioned problems (usually characterized by regions of instability in the nonlinear process model) can lead to failure of naive NLP implementations, but also can be handled through alternate approaches.

3. Sequential vs. simultaneous NLP strategies

Solution of (6) can be simplified by considering the forward structure of the equality constraints. Given the availability of efficient solvers for initial value DAEs (e.g., DASSL, DDASAC), it is straightforward to combine general purpose NLP software with these solvers to develop NMPC controllers. More efficient approaches can also be developed by considering the properties of the previous section and noting some parallels with linear MPC controllers.

3.1. Sequential approach

Because of the relation to convergence properties of SQP, the Gauss-Newton approximation leads to much faster NMPC algorithms than with general purpose NLP solvers. Moreover, for problems that converge to the setpoint (zero residuals), this leads to a Q-quadratic convergence rate. We first consider a sequential approach where the elimination of linearized equality constraints in (7) leads to a smaller inequality constrained QP entirely in d_U:

$$\begin{aligned} Min_{d_U} \quad & a_k^T d_U + 1/2 d_U^T (S^T Q_Y S + Q_U) d_U \\ s.t. \quad & U_a \le U_k + d_U \le U_b \\ & X_a \le X_k + r_k + R d_U \le X_b \\ & Y_a \le Y_k + s_k + S d_U \le Y_b \end{aligned} \tag{21}$$

Here $a_k = (\nabla_U J(Y_k, U_k) + S^T(\nabla_Y J(Y_k, U_k) + Q_Y s_k))$, $r_k = -\nabla_X F^{-T} F_k$, $R = -\nabla_X F^{-T} \nabla_U F^T$, $s_k = (G_k - Y_k) + \nabla_X G^T r_k$, and $S = \nabla_X G^T R$. These terms are derived from a forward elimination of the state equations which corresponds to an initial value problem (IVP) decomposition. This decomposition leads to the *Newton-type control algorithm* for nonlinear MPC described in our previous work [39] and is a nonlinear analog to the QDMC algorithm for linear models. Because the controller has a starting point based on the solution from the last sampling time, it usually requires few iterations of (21) to converge (6). In fact, a single solution of (21) is often sufficient to obtain good performance of the nonlinear controller [32].

While solution of (21) is efficient for nonlinear systems stable in the large, severe problems can be encountered if (21) is solved for locally unstable systems. This instability is encountered in two ways. First, integration of a DAE model (1,2) at an unstable point leads to unbounded profiles and immediate failure of the NLP algorithm. In certain process systems (e.g, closed loop systems with poorly tuned controllers, exothermic reactors) it may be difficult to find inputs that give bounded profiles except very close to the optimal solution [9, 8] and here the sequential approach is unsuitable for dynamic optimization.

The second problem occurs even if we obtain a bounded solution. This occurs in the forward decomposition of multiple shooting problems applied to (6). Here even though functions and gradients can be evaluated for the NLP solver, the influence of unstable modes leads to R and S matrices that are very poorly conditioned. For linear problems, this instability is treated by matrix factorizations that cancel the unstable modes (see [41]) or through a stabilizing reparameterization of the control, e.g., $u = Kx + v$. This reparameterization is often useful for nonlinear problems as well, but is difficult to apply to general nonlinear problems over the entire domain of inputs. Instead, stable solutions to nonlinear problems can be found through boundary value problem (BVP) formulations.

3.2. Simultaneous approach

To exploit the BVP formulation, we need to consider the larger system (7) where the state profiles remain bounded for integration of the predictive horizon and trust region methods can be applied. Here well conditioned formulations are ensured when the unstable forward modes are pinned down by final conditions. This complements the well-shown endpoint constraint formulations of [27, 42], where the constraint $x_{k+p} = 0$ is imposed in (6); it also applies to dual mode controllers [35] which confine the endpoint to a region around the setpoint.

"Pinning down" the unstable modes is characterized by the concept of dichotomy, where stable modes either in the forward or reverse direction are identified with their respective boundary conditions. In fact, for linear BVPs, de Hoog and Mattheij [24] note the equivalence of well-conditioned systems and dichotomous systems. Consequently, by ensuring well-conditioned systems, one can also ensure a dichotomous problem formulation. (Here, nonlinear BVPs are treated similarly through linearization about the solution [4]). To see how this works consider (9) and the linearization of the (now discretized) DAE model, $c(w_k) + \nabla c(w_k)^T d = 0$. By choosing independent (inputs) and dependent (state) w variables and partitioning $\nabla c(w_k)^T = [N \mid C]$ we effectively define the boundary conditions for this problem and solve for the state variables by using a well-conditioned C matrix as the state Jacobian. Necessary conditions for finding C include determination of a stable pivot sequence and rescaling the system, and these can be done with widely available linear algebra software. On problems with few degrees of freedom (e.g., NMPC problems with short input horizons), the partitioning can be combined with a reduced space SQP approach (with subproblems similar to (21)) and a reliable and efficient NLP strategy results, as shown in the following example.

BVP example with detection of dichotomy

To illustrate the selection of dichotomous models we consider a parameter estimation problem described in [14]. The model is an linear time varying (LTV) system and the optimization problem is given by:

$$Min \quad \sum_{i=1}^{n}(z_{1,m}(i) - z_1(i))^2 + \sum_{i=1}^{n}(z_{2,m}(i) - z_2(i))^2 \tag{22}$$

$$s.t. \quad \dot{z} = A(u,t)z + f(t)$$

$$z_j(0) = 1 \quad j = 1, \ldots, 5$$

where

$$A(u,t) \;=\; \begin{bmatrix} -u_1\tilde{c}_1 & 0 & \omega_1 + u_1\tilde{s}_1 & 0 & 0 \\ 0 & -u_2\tilde{c}_2 & 0 & \omega_2 + u_2\tilde{s}_2 & 0 \\ -\omega_1 + u_1\tilde{s}_1 & 0 & u_1\tilde{c}_1 & 0 & 0 \\ 0 & -\omega_2 + u_2\tilde{s}_2 & 0 & u_2\tilde{c}_2 & 0 \\ 0 & 0 & 0 & 0 & u_3 \end{bmatrix},$$

$\tilde{c}_k = \cos(2\omega_k t)$, $\tilde{s}_k = \sin(2\omega_k t)$, measured values are given by $z_{1,m}$ and $z_{2,m}$, and $f(t)$ is a nonlinear vector function chosen so that the solution of the ODE is $z_j(t) = e^t$, $j = 1, \ldots, 5$. The problem has three unstable modes and a bounded solution is obtained only if the parameters are chosen appropriately. The data are generated randomly around this solution with standard deviations σ at 0.05 and 0.10 and the parameters (u_1, u_2, u_3) initialized at 20 % above their true values. Here a straightforward partition with the parameters as the independent variables leads to unbounded solutions and immediate failure of the NLP algorithm; with a partition chosen automatically by the sparse matrix solver, MA28, convergence is obtained in 25 to 30 SQP iterations. Further details of this comparison are given in [14]. Moreover, this study also considers the dynamic optimization of unstable reactors and reactive separation systems. Current work also focuses on rapid element-by-element decompositions that also preserve dichotomy.

4. Exploiting structure of simultaneous approaches

In addition to exploiting the convergence properties of SQP and the conditioning of the QP subproblem, we also need to consider efficient strategies for the solution of (7). For structured QPs there is a tremendous difference in the performance of general purpose QP strategies vs. tailored methods. In the first part of this section we consider the case without variable bounds and then add these constraints in the second part. In particular, if we consider the Kuhn-Tucker matrix of (7), with inactive bounds, we have a system of order $(n_s + n_i + n_o)(p + 1)$. With a general purpose solver the computational effort is $O(\{(n_s+n_i+n_o)(p+1)\}^3)$. Considering a smaller input horizon and using either (21) or (7), where a decomposition is applied to solve the problem in the space of d_U, leads to computational effort that is proportional to $(n_i m)^3 + [(n_s + n_o)(p+1)]^\tau$, where $\tau = 3$ unless a sparse factorization is used for the linearized state equations. However, even this performance is unattractive for large systems.

4.1. Matrix structure of NMPC

To tailor the solution of these QPs we consider the structure of (6) in greater detail. Using equation (4) to define J we can reorder the Kuhn-Tucker matrix from (7) to form an almost block diagonal (ABD) system:

$$\begin{bmatrix}
Q_0 & A_0^T & C_0^T & 0 & & & & & & \\
A_0 & 0 & 0 & \bar{A} & & & & & & \\
C_0 & 0 & 0 & 0 & & & & & & \\
0 & \bar{A}^T & 0 & Q_1 & A_1^T & C_1^T & 0 & & & \\
 & & & A_1 & 0 & 0 & \bar{A} & & & \\
 & & & C_1 & 0 & 0 & 0 & & & \\
 & & & 0 & \bar{A}^T & 0 & \ddots & & & \\
 & & & & & & & Q_p & A_p^T & C_p^T \\
 & & & & & & & A_p & 0 & 0 \\
 & & & & & & & C_p & 0 & 0
\end{bmatrix} \tag{23}$$

where the matrices are defined as $\Phi_l = \frac{\partial\chi(x_l,u_l)}{\partial x_l}$, $\Gamma_l = \frac{\partial\chi(x_l,u_l)}{\partial u_l}$, $D_l = \frac{\partial g(x_l)}{\partial x_l}$, $A_l = [\Gamma_l^T \; \Phi_l^T \; 0]$, $\bar{A} = [0 \; -I \; 0]$ $C_l = [0 \; -D_l^T \; I]$ and

$$Q_l = \begin{bmatrix} Q_{ul} & 0 & 0 \\ 0 & 0 & 0 \\ 0 & 0 & Q_{yl} \end{bmatrix}$$

(In the last block, the constraints that involve A_p and x_p could be eliminated or invoked to deal with final time constraints.) This ABD system can be decomposed efficiently using a number of methods. First, a symmetric sparse solver such as MA27 or MA47 (Harwell Subroutine Library) can be used to decompose (23). This approach scales with a low power of the problem size (usually between one and two) and leads to stable decompositions. These routines simplify the analysis of the pivot sequence by exploiting symmetry. Similarly nonsymmetric sparse solvers such as MA28 or MA48, can also be used. These routines require more analysis of the pivot sequence but this is an offline cost. Moreover, our experience is that the factorization times with these solvers are less than with symmetric sparse solvers.

Second, Dunn and Bertsekas [17] represent the multipliers and control variables as affine functions of the the state variables. This Riccati-like approach leads to a backward and forward recursion for the multiplier and state variables, respectively. Albuquerque and Biegler [2] extended this approach to deal with DAE systems (1, 2) but this decomposition becomes ill-conditioned for unstable systems. Instead, the alternative SOLVEBLOK approach [10] can be applied to (23) if the system is unstable. However, neither decomposition exploits sparsity within each interval.

Given the scaling arguments for the above decomposition strategies, NMPC strategies can be solved with much faster performance than by using general purpose NLP solvers or even the modified linear MPC solver (21). On parameter and state estimation problems with the same structure as (23), performance improvements over general purpose NLP and QP solvers greater than three orders of magnitude were obtained [2]. Clearly the exploitation of this structure is often the key to getting performance improvements for NMPC. However, the examples in [2] include inequality constraints only on the model parameters. To deal with inequalities on the inputs and outputs, we also need to develop constraint handling strategies that preserve the structure of (23), as discussed next.

4.2. Constraint handling for NMPC

Inequality constraints for (7) can be treated either by active set strategies or interior point (IP) methods. A detailed discussion on the pros and cons of both methods for model predictive control is given by [52]. Active set strategies are most commonly used in general purpose QP solvers (such as QPSOL, MINOS, QPOPT, VE13AD, etc.), Here the active set of constraints is modified through rank one updates of the Kuhn-Tucker matrix (23). One disadvantage of active set strategies is that they add a combinatorial element to the identification of the active set and the solution of the QP, and this leads to a large increase in computational effort with problem size. Performance characteristics of the interior point method, on the other hand, seem to be insensitive to problem size. For instance, with a small NMPC example [3] the number of active set iterations for MINOS and QPSOL increases polynomially to over 1500 for 1000 constraints while the IP method remains at a constant 6 iterations.

To illustrate the application of IP methods, we reconsider the QP subproblem (7) where we combine all of the linearized equality constraints into $w_k + W_k d = 0$ and the variable bounds into inequality constraints: $h_k + H_k d \leq 0$. Introducing slack variables s into these constraints (and suppressing the subscript k) leads to a quadratic program written as:

$$\begin{aligned} min \quad & a^T d + \frac{1}{2} d^T Q d \\ s.t. \quad & w + Wd = 0 \\ & h + Hd + s = 0, \quad s \geq 0 \end{aligned} \tag{24}$$

where $a^T = [\nabla_U J(Y_k, U_k)^T,\ 0,\ \nabla_Y J(Y_k, U_k)^T]$,

$$W = \begin{bmatrix} \nabla_U F(X_k, U_k)^T & \nabla_X F(X_k, U_k)^T & 0 \\ 0 & \nabla_X G(X_k)^T & -I \end{bmatrix}$$

and h, H are defined to represent the bounds on U, X, Y and their gradients, respectively. The optimality conditions of (24) are given by (25)-(28).

$$Qd + W^T\lambda + H^T\nu + a = 0 \tag{25}$$

$$w + Wd = 0 \tag{26}$$

$$h + Hd + s = 0 \tag{27}$$

$$SVe = 0, \quad (s, \nu) \geq 0 \tag{28}$$

with $e^T = [1, 1, \ldots, 1]$, and λ and ν as the multipliers of the equalities and inequalities, respectively. Also, $S = diag\{s\}$ and $V = diag\{\nu\}$. Note that this system is nonlinear because of the complementarity conditions given in (28).

To solve this system reliably, a *central path* is introduced, which keeps the iterates biased towards the interior of the nonnegative region $(s, \nu) > 0$, and also forces all the complementarity pairs to converge at a similar rate. Here the *complementarity gap* μ is a measure of the average value of the products $s_i \nu_i$ and is defined as $\mu = s^T \nu / n$. To keep the iterates in the interior, (28) is replaced with $SVe = \sigma\mu e$ where $\sigma \in [0, 1]$ is a parameter. Setting $\sigma = 1$ corresponds to a pure centering direction, *ie.* moving all the pairwise products to the current average value μ, while $\sigma = 0$ gives a standard Newton step. Various *path following methods* differ in the way the search direction is modified and σ is chosen. An excellent survey and analysis of these methods is given in [53].

The most widely used approach for selecting σ to solve IP equations is the Mehrotra predictor-corrector algorithm [37]. Each Mehrotra iterate solves an affine scaling *predictor* step (29) and a *corrector* step (30), where r_1, r_2, r_3 are the residuals of (25)-(27) at the current point. The predictor step helps to reduce the complementarity gap while centering is accomplished through the corrector step, which also incorporates a second order correction term in (30)

$$\begin{bmatrix} Q & W^T & H^T & 0 \\ W & 0 & 0 & 0 \\ H & 0 & 0 & I \\ 0 & 0 & S & V \end{bmatrix} \begin{bmatrix} d^p \\ \Delta\lambda^p \\ \Delta\nu^p \\ \Delta s^p \end{bmatrix} = - \begin{bmatrix} r_1 \\ r_2 \\ r_3 \\ SVe \end{bmatrix} \tag{29}$$

$$\begin{bmatrix} Q & W^T & H^T & 0 \\ W & 0 & 0 & 0 \\ H & 0 & 0 & I \\ 0 & 0 & S & V \end{bmatrix} \begin{bmatrix} d^c \\ \Delta\lambda^c \\ \Delta\nu^c \\ \Delta s^c \end{bmatrix} = \begin{bmatrix} 0 \\ 0 \\ 0 \\ \sigma\mu e - \Delta S_p \Delta V_p e \end{bmatrix} \tag{30}$$

For the solution of linear programs, IP methods have been developed into powerful, commercial grade solvers. These have also been extended to the solution of quadratic programs and convex NLPs in general. For example, Rao et al. [41] developed an efficient solution strategy that applies the Mehrotra predictor-corrector to linear MPC problems. Here a Riccati transformation (with a linear relation between the d and λ) was used to decompose the matrices in (29)-(30).

While the number of IP iterations is not a function of problem size, the cost per IP iteration is generally much larger than simple rank one updates to active sets; as a result IP methods are not a universal successor to active set methods. However, (29)-(30) can be rearranged along the lines of the ABD structure in (23) so for NMPC problems, where a large number of active constraints are expected, significant benefits can be obtained with IP approaches. To illustrate, the next example presents a comparison of this IP-based NMPC solver for the reactor portion of the Tennessee Eastman Challenge problem [16].

NMPC example based on Tennessee Eastman problem

This example deals with control of the Tennessee Eastman reactor, a two phase, jacketed CSTR, with the reaction dynamics occurring in the vapor phase. The DAE model given in [3] consists of 8 differential and 28 algebraic equations and two control profiles. Because this problem is open loop unstable, it must be handled with a simultaneous formulation. The nonlinear system was solved with a full space SQP method with the QP subproblems solved using an IP method that exploits the ABD structure. Here we highlight the performance characteristics of this IP-based NLP strategy, using sparse symmetric solvers MA27 and MA47, and the nonsymmetric solver MA28 from Harwell. In this initial solution of the nonlinear optimal control problem the initial point was chosen close to the steady state values (as is typical with NMPC applications). As a result relatively few SQP iterations were required.

Performance results using a Mehrotra predictor-corrector implementation and ISQP are given in Table 2. This table also gives information about the problem size (length of output horizon n_y, the number of primal variables, equality constraints, inequality constraints), the number of active inequalities at the solution

n_y	Var/Eq/Ineq/Act	$\frac{Time}{MA28}$	$\frac{Time}{MA27}$	Newt/PC/ISQP
10	1100/1081/140/24	1.3	4.39	2/16/17
20	2200/2161/280/64	14.7	22.05	4/39/38
40	4400/4321/560/144	31.7	41.31	3/31/36
80	8800/8641/1120/304	91.3	104.94	3/33/36
160	17600/17281/2240/624	263.2	286.7	3/31/30
320	35200/34561/4480/1264	691.7	836.9	3/33/32

TABLE 2. Problem size, CPU seconds and iterations using MA28 and MA27 with Mehrotra predictor-corrector and SQP implementations

as well as total CPU seconds (DEC Alpha/400) for solving the nonlinear optimal control problem. For all of the simulations the number of Newton iterations remains the same and the number of interior point iterations for each simulation only varied by one or two. The results show that the MA28 implementations were faster than the MA27 versions and the MA47 solver was much slower [3].

Note that the number of IP iterations remains independent of the problem size. In fact the simulation examples reveal that approximately 10 interior point iterations are required per QP. The benefit of using interior point techniques becomes especially evident when we consider the output horizon of 320 with 1264 active constraints; an active set method would add these constraints one at a time.

5. Beyond solution of NLPs

Given the rapid advances in the efficient solution of NLPs that arise from NMPC problems, there are a number of additional challenges that can be reconsidered. In this section we consider two open questions related to the implementation of NMPC. The first question relates to robust stability properties for nonlinear systems. Robust stability for linear MPC has been tackled by a number of efficient strategies [29, 5, 11]. However, the analog to NMPC is still difficult. The second question relates to the stability of NMPC controllers [34]. In the solution of NMPC problems, one may attract to an infeasible point with local methods. Thus it is interesting to see if we can explore a more reliable *global* algorithm, and at what cost. These two topics are briefly surveyed below.

5.1. Dealing with uncertainty in NMPC

So far we have only dealt with the nominal system for (1, 2). For on-line implementations, the question of robustness to uncertain inputs and model mismatch clearly needs to be addressed. To deal with the development of stable NMPC controllers, we revisit the basic stability result developed by [27] for the perfect model case with state feedback. Here the imposition of a final time constraint ($x_{i+p} = 0$) for (4) leads to a solution with the following properties:

$$J^*(x_i) = \sum_{l=i}^{i+p-1} h(x_l, u_l) + \underbrace{h(x_{i+p}, u_{i+p})}_{=0} \tag{31}$$

Therefore the shifted optimal solution $\{x_{i+1}, u_{i+1}, x_{i+2}, u_{i+2}, \dots, x_{i+p}, u_{i+p}, 0, 0\}$, is a feasible solution for the problem (4) at time index $i+1$. Thus the difference between $J^*(x_i)$ and the value of the objective function at $i+1$ evaluated with these state and input sequences, $J(x_{i+1})$, gives

$$J^*(x_i) - J^*(x_{i+1}) \geq J^*(x_i) - J(x_{i+1}) = h(x_i, u_i) \tag{32}$$

Taking the sum of the differences given by (32) over N we obtain

$$J^*(x_1) - J^*(x_{N+1}) = \sum_{i=1}^{N} (J^*(x_i) - J^*(x_{i+1})) \geq \sum_{i=1}^{N} h(x_i, u_i) \tag{33}$$

Because the sequence $\{J^*(x_i)\}$ is decreasing, then as $N \to \infty$, $h(x_i, u_i) \to 0 \Rightarrow x_i \to 0$.

To include the presence of model mismatch, let $\bar{x}_{i+1}$, be the state prediction given by the solution of (4) at time i for time index $i+1$. To account for the existence of mismatch, consider the difference between $J^*(x_i)$ and $J^*(x_{i+1})$ by adding and subtracting $J^*(\bar{x}_{i+1})$, that is

$$J^*(x_i) - J^*(x_{i+1}) = (J^*(x_i) - J^*(\bar{x}_{i+1})) - (J^*(x_{i+1}) - J^*(\bar{x}_{i+1})) \tag{34}$$

The term $J^*(x_i) - J^*(\bar{x}_{i+1})$ represents the difference between the optimal objective functions at time indices i and $i+1$ for the model prediction and satisfies the inequality (32). It follows that

$$J^*(x_i) - J^*(x_{i+1}) \geq h(x_i, u_i) - (J^*(x_{i+1}) - J^*(\bar{x}_{i+1})) \tag{35}$$

To ensure the sequence $\{J^*(x_i)\}$ is decreasing, we need the quantity (35) to be positive and bounded by a positive function $W(\|x_i\|)$ of class $\mathcal{K}_\infty$ defined in [27], such that

$$h(x_i, u_i) - (J^*(x_{i+1}) - J^*(\bar{x}_{i+1})) \geq W(\|x_i\|), \tag{36}$$

with $W(\|x_i\|) \to 0$ as $\|x_i\| \to 0$, for all i, $i \geq 0$. Therefore, a sufficient robust stability condition is that, for all i, the mismatch term $J^*(x_{i+1}) - J^*(\bar{x}_{i+1})$ is sufficiently less than $h(x_i, u_i)$. Under this condition, by taking the sum of the differences over N time indices, we obtain

$$J^*(x_1) - J^*(x_{N+1}) = \sum_{i=1}^{N} (J^*(x_i) - J^*(x_{i+1})) \geq \sum_{i=1}^{N} W(\|x_i\|) \tag{37}$$

Then as $N \to \infty$, $\|x_N\| \to 0$ and robust stability is obtained.

The inequality (36) helps us consider two questions for robust stability. If we consider $J(x_{i+1}, \theta)$ for all possible values of uncertainty or model mismatch we can add the constraints:

$$J(x_{i+1}, \theta) - J^*(\bar{x}_{i+1}) \leq h(x_i, u_i) \quad \forall \theta \tag{38}$$

to (4). Provided that this extended problem has a solution, robust stability for NMPC is then guaranteed. Badgwell [5] considered this formulation for linear and nonlinear MPC problems, and demonstrated the superiority of robust control on a number of process examples. For the linear case, the addition of (38) leads to a convex programming problem; an efficient interior point strategy was also developed to solve this problem [26].

Instead of imposing constraint (38) directly, one can use

$$J^*(x_{i+1}, \theta) - J^*(\bar{x}_{i+1}) \leq h(x_i, u_i)$$

to assess *how much* uncertainty (θ) can be tolerated by the on-line solution of (4). In this case, by bounding the model mismatch by a function $W_m(\|x_i\|)$, of class $\mathcal{K}_\infty$, one can assess the sensitivity of $J^*(x_{i+1}, \theta)$ through an off-line NLP sensitivity analysis [44]. By differentiating the Kuhn Tucker conditions for (4) one can evaluate how much the state and control variables will change as a function of θ, and correspondingly in $J^*(x_{i+1}, \theta)$. Moreover, from the norms on the Kuhn-Tucker matrices, bounds on $(\Delta x_l, \Delta u_l, \Delta J^*)$ can be derived which are related to the uncertainties θ. Applying the inequality (36) then yields the level of uncertainty that can be tolerated by (4) for robust stability. Details of this derivation along with small examples are given in [44].

Finally, given an uncertainty description for model mismatch and unmeasured disturbances, robust optimization techniques can also be applied to the NMPC problem [38, 40] through the application of flexibility analysis and the solution of multiperiod optimization problems. Under certain conditions, these formulations guarantee the feasibility of the controller over the entire range of the uncertainty description. While these have been applied to small, off-line examples, the results have shown greatly improved performance over the nominal case. Further research in this area as well as specialization to individual applications will also lead to rigorous and efficient ways to design robust NMPC controllers.

5.2. Global optimization algorithms

A sufficient condition for nominal stability of NMPC controllers is the global solution of the nonconvex problem (6). The difficulty of this task is problem dependent and the intriguing question is what happens when the NLP algorithm 'gets stuck' now and then. Here relaxations to the global solutions have been developed and analyzed [46] and also a well implemented special purpose NMPC implementation may find this occurence to be rare. Nevertheless, it is useful to consider how global optimization algorithms work and how they can help.

Global solution of nonconvex problems is clearly a more expensive task. Instead of attracting to a stationary point, these algorithms require a complete enumeration of the search space. Virtually all of this algorithms do this via lower and upper bounding strategies; efficiency of these algorthms then depends on how problem structure can be exploited. The state of the art in global optimization algorithms is reviewed in [25]. Progress in the development and application of global methods for process systems engineering is covered in recent work by Floudas [20] and Grossmann [22].

Perhaps the most general and popular strategy for global optimization is a spatial branch and bound approach. Given a closed, convex region in the search space, a feasible point or locally optimal solution to (6) serves as an upper bound to the global minimum. A convex relaxation of this problem is then set up and solved in this region and this leads to a lower bound to (6). The branch and bound strategy continues by examining upper and lower bounds on active regions and eliminating regions whose lower bound is worse than the current upper bound. The remaining regions are then partitioned, new bounds are evaluated and this process continues until all regions are explored and either the lower and upper bounds coincide, or no feasible point can be found.

Clearly the key features to this algorithm are a) the efficiency of the search and partitioning strategies, b) the tightness of the lower bound obtained by the convex relaxation, c) the tightness of the upper bound, and d) the cost of obtaining these bounds. Fortunately, considerable research has gone into advancing the first two points. Efficient partitioning strategies have been developed and analyzed according to the general framework proposed in [25]. These strategies are often problem specific, but there also are a number of general purpose strategies for the efficient partitioning of search spaces [18]. Moreover, tight convex envelopes within objective functions and inequality constraints (equalities can be treated as doubly bounded inequalites) have been developed for virtually all nonconvex terms and these can be expanded in a constructive way to deal with complicated problem formulations. Here, a clever implementation has been developed [47] that automates the generation of convex lower bounding problems and applies an efficient spatial branch and bound algorithm. This approach has been demonstrated on a variety of challenging examples in process synthesis.

Global algorithms for transfer function identification have also been developed [48] and these are somewhat related in structure to NMPC problems. For this application, the efficiency solving the lower and upper bounding problems is the key factor for the success of the algorithm. An example that describes this approach is shown next.

Spatial branch and bound example for global optimization

Consider the problem of identifying the parameters of a discrete rational transfer function model of a system which can be written as:

$$y(t) = \sum_{i=1}^{na} a_i(y(t-i) - v(t-i)) + \sum_{j=1}^{nb} b_j u(t-j) + v(t)$$

Here $y(t)$, $u(t)$, and $v(t)$ represent the output, input, and the disturbance signals respectively with the sequences of $u(t)$ and $v(t)$ uncorrelated and $u(t)$ persistently excited. We assume that the inputs and disturbances are given and all of the model outputs, y, and the parameter estimates are estimated by formulating the least squares problem:

$$Min \quad J = \sum_{t=1}^{Ndata} (y(t) - \hat{y}(t))^T \, (y(t) - \hat{y}(t))$$

$$s.t. \quad y(t) = \sum_{i=1}^{na} a_i(y(t-i) - v(t-i)) + \sum_{j=1}^{nb} b_j u(t-j) + v(t) \tag{39}$$

The motivation for solving this NLP directly is rooted in the consistency result of output error formulations outlined in [33] and its theoretical advantages over equation error and instrumental variable formulations. However, this NLP formulation has nonconvex bilinear terms in (39) and these introduce the existence of local solutions.

In the spatial branch and bound strategy [25] the nonconvex bilinear terms are replaced by a set of convex inequalities known as McCormick underestimators [36] and given by:

$$a_i y(t-i) \geq a_i^U y(t-i) + y(t-i)^U a_i - a_i^U y(t-i)^U \tag{40}$$

$$a_i y(t-i) \geq a_i^L y(t-i) + y(t-i)^L a_i - a_i^L y(t-i)^L \tag{41}$$

$$a_i y(t-i) \leq a_i^L y(t-i) + y(t-i)^U a_i - a_i^L y(t-i)^U \tag{42}$$

$$a_i y(t-i) \leq a_i^U y(t-i) + y(t-i)^L a_i - a_i^U y(t-i)^L \tag{43}$$

The resulting relaxed convex problem is given by:

$$Min \quad J = \sum_{t=1}^{Ndata} (y(t) - \hat{y}(t))^T \, (y(t) - \hat{y}(t)) \tag{44}$$

$$\begin{aligned} s.t. \quad y(t) &= \sum_{i=1}^{na} (w_i - a_i v(t-i)) + \sum_{j=1}^{nb} b_j u(t-j) + v(t) \\ w_i &\geq a_i^U y(t-i) + y(t-i)^U a_i - a_i^U y(t-i)^U \\ w_i &\geq a_i^L y(t-i) + y(t-i)^L a_i - a_i^L y(t-i)^L \\ w_i &\leq a_i^L y(t-i) + y(t-i)^U a_i - a_i^L y(t-i)^U \\ w_i &\leq a_i^U y(t-i) + y(t-i)^L a_i - a_i^U y(t-i)^L \end{aligned}$$

and this provides a lower bound to the objective within the partitioned region, represented by bounds on the coefficients, a_i^L, a_i^U. With this relaxation, the approximation error for the bilinear terms goes to zero as either of the upper and lower variable bounds converge. Consequently, successive partitioning of the variables leads to a monotonically increasing sequence of lower bounds. Since feasible upper bound solutions are readily available, the method also provides a decreasing

Problem Size				Total	Nodes
Ndata	Var/Eq/Ineq	a	b	Time (Min)	Solved
20	40/20/80	0.896	0.413	1.8	181
40	80/40/160	0.896	0.392	1.0	69
80	160/80/320	0.899	0.391	1.7	67
160	320/160/640	0.900	0.393	3.1	61
320	640/320/1280	0.899	0.397	7.7	61
640	1280/640/2560	0.900	0.398	18.0	61
1280	2560/1280/5120	0.899	0.401	32.6	47

TABLE 3. Transfer Function Identification Results

sequence of upper bounds and convergence of the upper and lower bounds yields the global solution to the original nonconvex optimization problem. Moreover the structure of (45) is the same as (7) and this problem can be solved efficiently by the approach in the previous section.

To demonstrate this approach, we consider a first order plant model with $y(0) = 0, a = 0.9, b = 0.4$ and an uncertainty description given by:

$$\begin{aligned} v(t) &= -0.8v(t-1) - 0.8e(t-1) + e(t) \qquad (45) \\ &\bar{E}\{u(t)\} = 0, \bar{E}\{u^2(t)\} = 1 \\ &\bar{E}\{e(t)\} = 0, \bar{E}\{e^2(t)\} = 0.333 \\ &\bar{E}\{v^2(t)\} = 2.70 \end{aligned}$$

For the implementation of our approach, the noise bound was chosen as $\sigma_y = 10$, and the parameter bounds on a were $a^L = -1$ and $a^U = 1$. This choice of parameter bounds includes all possible stable transfer function models. In Table 3, the results show that the output error formulation correctly identifies the parameters despite the noisy signal and leads to very quick convergence of the parameters to their correct values. In contrast, neither the equation error (solved by recursive least squares) nor the instrumental variable approaches obtained converged parameter values even for Ndata = 1280. Solution times for this global optimization approach (with up to 1282 variables) are also reasonable.

6. Conclusions and future work

Optimization formulations have always been the key to describing nontrivial process control problems. In previous years, however, difficulties and inefficiency in solving large nonlinear programs have caused control researchers to shy away from direct solution of these optimization formulations. Instead, clever problem reformulations have been proposed in the hope of avoiding the optimization problem altogether.

This paper reviews a number of recent developments and properties for the solution of NLPs for dynamic optimization problems. Included in this summary is an informal description of local and global convergence properties as well as the Gauss-Newton property, which can be applied to many NMPC problems. These lead to fast, large-scale optimization algorithms tailored to the NMPC problem. Moreover, a number of on-line applications have already been considered in industry that make use of these NLP formulations.

Dealing with large, complex NMPC problems also involves the exploitation of structure for nonlinear dynamic problems. Here the focus is on simultaneous problem formulations, as they allow reliable consideration of locally unstable and ill-conditioned systems. The NMPC problem with many degrees of freedom and application of the Gauss Newton properties can be solved efficiently with a full space SQP algorithm and a variety of efficient decomposition schemes can be applied to the Kuhn-Tucker matrix. Moreover, interior point techniques can be applied to handle bound constraints in the QP subproblem and these preserve the original structure of the Kuhn-Tucker matrix as well. It should be noted that IP strategies have been very successful for the solution of convex NLPs and these include a number of control applications dealing with linear matrix inequalities and the design of robust controllers [29, 11]. In addition, IP strategies have also been developed for the direct solution of nonconvex NLPs in [12, 13].

As a result, efficient solution of NMPC problems also allows us to consider more difficult tasks, including the global solution of NLPs through spatial branch and bound algorithms. Early applications of this approach on small to moderate sized dynamic systems have demonstrated the feasibility of these approaches.

Also, as noted in a discussion by Rawlings, the (possibly nonconvex) MPC problem is solved on-line in lieu of a more difficult off-line feedback control problem (e.g., through the Hamilton-Jacobi-Bellman partial differential equations). On the other hand, numerical solutions to the HJB equations have recently been developed for general nonlinear systems without constraints [6], using Galerkin approximations. While this numerical approach is expensive for large problems and constraints have not yet been incorporated, the off-line character of these problems may make this feedback approach an appealing alternative to NMPC.

Finally, future research will lead to many more applications of (offline and online) dynamic optimization for process control. With advances in problem formulations, solution algorithms at various levels, convergence properties, computer software and hardware, it is clear that much larger and more complex nonlinear systems will be considered. Coupled with the development of improved NLP algorithms, NMPC strategies will be helped by novel problem formulations. These can include greater incorporation of convex formulations and the development of lower bounding strategies if global algorithms need to be applied. Also, the application of NLP tools and decompositions will be useful in the development of robust NMPC controllers.

Acknowledgements
The author is grateful to Vipin Gopal, Lino Santos, George Staus and to the anonymous reviewers for helpful input and suggestions. Financial support from the National Science Foundation (CTS-9729075) and Air Products and Chemicals is gratefully acknowledged.

References

[1] Albuquerque, A. and L. T. Biegler, "Decomposition algorithms for on-line estimation with nonlinear models," Comp. Chem. Engr., 19, p. 1031 (1995)

[2] Albuquerque, A. and L. T. Biegler, *Decomposition algorithms for on-line estimation with nonlinear DAE models,* Comp. Chem. Engr., 21, p. 283 (1997)

[3] Albuquerque, J., V. Gopal, G. Staus, L. T. Biegler, and B. E. Ydstie, "Interior Point SQP Strategies For Structured Process Optimization Problems" Process Syste ms Engineering '97

[4] Ascher, U. M., R.M.M. Mattheij and R.D. Russel, *Numerical Solution of Boundary Value Problems for Ordinary Differential Equations,* Prentice Hall Englewood Cliffs,NJ. (1988)

[5] Badgwell, T., "Robust model predictive control for nonlinear plants," presented at Annual AIChE meeting, Chicago, IL (1996)

[6] Beard, R. W., G. N. Saridis, and J. T. Wen, "Galerkin Approximations of the Generalized Hamilton-Jacobi-Bellman Equation," Automatica, 33, 12, P. 2159 (1997)

[7] Biegler, L. T., J. Nocedal and C. Schmid, *A Reduced Hessian Method for Large-Scale Constrained Optimization,* SIAM J. Optimization, 5, p. 314 (1995)

[8] Biegler, L. T., Advances in Nonlinear Programming Concepts for Process Control," *J. Proc. Control*, to appear (1998)

[9] Bock, H. G., "Recent advances in parameter identification techniques," in P. Deuflhard and E. Hairer (eds.), *Numerical Treatment of Inverse Problems*, Birkhäuser, Heidelberg (1983)

[10] deBoor, C. W., and R. Weiss, "SOLVEBLOK: A package for solving ABD linear ssytems," *ACM TOMS*, 6 ,p. 80 (1980)

[11] S. Boyd, C. Crusius, and A. Hansson, Control Applications of Nonlinear Convex Programming. *J. Process Control*, to appear, 1997

[12] Byrd, R., J-C. Gilbert and J. Nocedal, "A trust region method based on interior point methods for nonlinear programming," Technical Report OTC 96/02, Northwestern University (1996)

[13] Byrd, R., M. E. Hribar and J. Nocedal, "An interior point algorithm for large-scale nonlinear programming," Technical Report OTC 97/05, Northwestern University (1997)

[14] Cervantes, A. and L. T. Biegler, "Large-Scale DAE Optimization Using a Simultaneous NLP Formulation," *AIChE J.*, to appear (1998)

[15] Dennis, J. and R. Schnabel, *Numerical Methods for Unconstrained Optimization and Nonlinear Equations*, SIAM, Philadelphia (1996)

[16] Downs, J.J. and E.F. Vogel, *A Plant-Wide Industrial Process Control Problem,* Computers Chem. Eng., Vol 17(3), p. 245–255 (1993)

[17] Dunn, J. C. and D. Bertsekas, "Efficient dynamic programming implementations of Newton's method for unconstrained optimal control problems," *J. Opt. Theo. Appl.*, 62, p. 23 (1989)

[18] Epperly, T. G. W., and E. N. Pistikopoulos, "A reduced space branch and bound algorithm for global optimization," J. Global Optimization, 11, p. 287 (1997)

[19] Fletcher, R., *Practical Methods for Optimization*, Wiley, Chichester (1987)

[20] Floudas, C. A., "Deterministic Global Optimization in Design, Control, and Computational Chemistry," in *Large Scale Optimization with Applications*, Volume 93: Part II: Optimal Design and Control, IMA Volumes in Mathematics and Applications, Springer Verlag, New York (1997)

[21] Gopal, V., PhD thesis, Carnegie Mellon University, Pittsburgh, PA, 1997.

[22] Grossmann, I. E. (ed.) *Global Optimization in Engineering Design*, Kluwer, Netherlands (1996)

[23] S-P. Han. "A globally convergent method for nonlinear programming." *JOTA*, 22(3):297–310, 1977.

[24] deHoog, R., and R. M. Mattheij, "On the conditioning of multipoint and integral boundary value problems," *SIAM J. Math. Anal.*, 20, 1, p. 200 (1989)

[25] Horst, R. and H. Tuy, *Global Optimization* Springer Verlag, Berlin (1996)

[26] Kassmann, D. E., and T. A. Badgwell, "Interior Point Methods in Robust Model Predictive Control," presented at Annual AIChE Meeting, Los Angeles, CA (1997)

[27] Keerthi, S.S., E.G. Gilbert, "Optimal Infinite-Horizon Feedback Laws for General Class of Constrained Discrete-Time Systems: Stability and Moving-Horizon Approximations", *IEEE Trans. Auto. Cont.*, **57**(2), 265–293 (1988)

[28] Kleinman, D. L., "An Easy Way to Stabilize a Linear Constant System," IEEE Trans. Aut. Cont., AC-15, p. 692 (1970)

[29] Kothare, M., V. Balakrishnan and M. Morari, "Robust Constrained Model Predictive Control using Linear Matrix Inequalities," *Automatica*, 32, 10, p. 1361 (1996)

[30] Kwon, W. H. and A. E. Pearson, "On Feadback Stabilization of Time Varying Discrete Linear Systems," IEEE Trans. Aut. Cont., AC-23, p. 838 (1977)

[31] Lalee, M., J. Nocedal and T. Platenga, "On the implementation of an algorithm for large-scale equality constrained optimization," *SIAM J. Opt.*, to appear (1998)

[32] Li, W-C. and L. T. Biegler, "Multistep, Newton-type Control Strategies for Constrained Nonlinear Processes," *Chemical Engineering Research and Design*, **67**, p. 562 (1989)

[33] Ljung, L., (1987), *System Identification, Theory for the User*, Prentice Hall, Inc., Englewood Cliffs, New Jersey.

[34] Mayne, D. Q., , "Nonlinear model predictive conrol: an assessment," Proceedings of CPC-V, (1996)

[35] Michalska, H., D.Q. Mayne (1993). "Robust Receding Horizon Control of Constrained Nonlinear Systems", *IEEE Trans. Auto. Cont.*, **38**(11), 1623–1633.

[36] McCormick, G.P.,"Computability of Global Solutions to Factorable Nonconvex Programs: Part I: Convex underestimator problems", *Math. Prog.*, Vol. 10, p.147 (1976)

[37] Mehrotra, S., *Quadratic convergence in a primal dual method,* Math. Oper. Res., 15, p. 741 (1993)

[38] Mohideen, M. J., PhD Thesis, Imperial College, London (1996)

[39] Oliveira, N.M.C., L.T. Biegler "Newton-type Algorithms for Nonlinear Process Control. Algorithm and Stability Results", *Automatica*, 31, 2, p. 281 (1995)

[40] Pistikopoulos, E., "Uncertainty in Process Design and Operations," *Comp. Chem. Engr.*, 19, p. S553 (1995)

[41] Rao, C., J. B. Rawlings and S. Wright, "Application of Interior Point Methods to Model Predictive Control," *J. Opt. Theo. Applics.*, to appear (1998)

[42] Rawlings, J.B., K.R. Muske (1993). "The Stability of Constrained Receding Horizon Control", *IEEE Trans. Auto. Cont.*, **38**(10), 1512–1516.

[43] Santos, L.O., N. Oliveira and L.T. Biegler, "Reliable and Efficient Optimization Strategies for Nonlinear Model Predictive Control," Proc. Fourth IFAC Symposium on Dynamics and Control of Chemical Reactors, Distillation Columns and Batch Processes (DYCORD '95), p. 33 (1995).

[44] L. O. Santos and L. T. Biegler, "A Tool to Analyze Robust Stability for Model Predictive Controllers," *J. Process Control*, to appear (1998)

[45] Scockaert, P. and J. B. Rawlings, "Optimization formulations for model predictive control," Proceedings of IMA Workshop on Large Scale Optimization (1997)

[46] Scockaert, P., D. Q. Mayne and J. B. Rawlings, "Suboptimal model predictive control," submitted for publication (1997)

[47] Smith, E. M. B., and C. Pantelides, "Global Optimisation of Nonconvex MINLPs," *Comp. Chem. Engr.*, 21, p. S791–S796 (1997)

[48] Staus, G. H., L. T. Biegler and B. E. Ydstie, "Global Optimization for Identification," submitted for publication (1998)

[49] Steihaug, T., "The conjugate gradient method and trust regions in large scale optimization," *SIAM J. Num. Anal.*, 20, p. 626 (1983)

[50] Tanartkit, P., L.T. Biegler, "Stable Decomposition for Dynamic Optimization", *I& EC Research*, 34, p. 1253 (1995)

[51] Ternet, D., PhD thesis, Carnegie Mellon University, Pittsburgh, PA, 1997.

[52] Wright, S., "Applying New Optimization Algorithms to Model Predictive Control," Proceedings of CPC-V, (1996)

[53] Wright, S., *Primal-Dual Interior Point Methods*, SIAM, Philadelphia (1997)

Department of Chemical Engineering
Carnegie Mellon University
Pittsburgh, PA USA-15213
E-mail address: biegler@cmu.edu

Progress in Systems and Control Theory, Vol. 26

A Direct Multiple Shooting Method for Real-time Optimization of Nonlinear DAE Processes

H. G. Bock, M. M. Diehl, D. B. Leineweber, J. P. Schlöder

Abstract. The direct multiple shooting method has long been known as a fast off-line optimization method in ODE and DAE (e.g. [B81, BP84, P81]).

Some factors crucial for its fast performance are briefly reviewed, such as structure exploiting quadratic programming, partitioned high rank updates, use of boundary and algebraic consistency conditions and invariants for gradient and Hessian reduction techniques, and parallel solution approaches (e.g. [L96, L97a, SBS98]).

Recently, this approach has been successfully adapted to the specific requirements of real-time optimization [L97b]. Special strategies have been developed to effectively minimize the on-line computational effort. They use precalculated information as far as possible – e.g. Hessians, gradients and QP presolves for nominal or neighbouring situations – to minimize response time in case of perturbations. In typical real-time problems they have proven much faster than fast off-line strategies. Comparisons for a fed-batch fermentation process show a drastic drop of computational costs.

NMPC may be interpreted as a special real-time optimization problem. Ways to exploit the inherent structure of NMPC by a suitable shift-and-embed strategy are described. Preliminary numerical results show excellent performance for a benchmark CSTR problem.

1. Introduction

On-line control based on nonlinear dynamic models and real time application of optimization techniques has received more and more attention over the past decade, e.g. in chemical engineering [GPM89, ER90, ER92, BR91]. Among the advantages of this approach are the flexibility provided in formulating the objective and the process model, the capability to directly handle equality and inequality constraints, and the possibility to treat disturbances fast.

One important precondition, however, is the availability of reliable and efficient numerical optimal control algorithms.

In this paper, we sketch the *direct* multiple shooting approach to the solution of such optimal control problems, discuss the adaptation of this approach to the requirements of real-time optimization and demonstrate the excellent performance of the algorithm for a bioengineering process. Extensions and modifications for the application in NMPC as another special real-time optimization problem are presented.

Direct optimal control methods reformulate the original infinite dimensional optimization problem adaptively as a sequence of finite nonlinear programming (NLP) problems by a parametrization of the controls and states. Such a method is called a *simultaneous* solution strategy, if the NLPs are solved by an infeasible point method such as sequential quadratic programming (SQP) or generalized Gauss-Newton, because then the discretized model equations have to be satisfied only for the final solution, and therefore, optimization and simulation proceed simultaneously.

Our simultaneous solution method, the *direct multiple shooting method*, originally introduced by Bock and Plitt [BP84, P81], leads to a large, but specially structured NLP problem which is solved very efficiently by a tailored SQP technique, taking advantage of the inherent problem structure in several respects, e.g. the generation of gradients, the approximation of the block diagonal Hessian or the fast structure exploiting solution of the quadratic programming (QP) subproblems. The original approach for ordinary differential equation (ODE) models has later been extended to differential algebraic equation (DAE) models [L97a, L96, BES88, L98]. Since an optimization boundary value problem is solved, we speak of a multiple shooting based *boundary value problem (BVP) approach*. Multiple shooting has the following characteristics, which are important in the context of large-scale process optimization:

- Advanced, fully adaptive DAE solvers are employed to calculate the function values and derivatives. Therefore, our technique can be combined with popular dynamic process simulators and modeling platforms.
- Since the integrations are decoupled on different multiple shooting intervals, the method is well suited for parallel computation.
- The approach allows a natural treatment of multistage problems, control and path constraints as well as multipoint boundary conditions.
- Multiple shooting has been shown to be considerably more stable and efficient than single shooting for the solution of optimization boundary value problems.

Especially the last item may deserve some explanation. It is well known that an optimization boundary value problem may be well-conditioned even though the initial value problem for the underlying dynamic process is ill-conditioned, i.e. the differential equation has unstable modes. If a single shooting approach is used, the algorithm inherits this ill-conditioning. Its round-off and discretization errors are propagated according to the – bad – condition number of the initial value problem, rather than the – good – condition number of the original optimization boundary value problem: hence the algorithm is numerically unstable. This effect is avoided by the multiple shooting approach which has even been able to solve optimization problems connected with highly chaotic systems [BBB92, KBS93]. A detailed numerical stability analysis for the case of parameter estimation is given in [B87].

The efficiency of the approach, which has been observed in many practical applications, has several reasons. One of the most important is the possibility to include information about the approximate behaviour of the state trajectory (which is often well known) into the initial guess for the iterative solution procedure; this can damp the influence of poor initial guesses for the controls (which are usually much less known).

The original direct multiple shooting code MUSCOD, implemented by Plitt in 1981 [P81], has since then undergone several revisions. Our latest, modular implementation MUSCOD2 (Leineweber [L97a, L97b, L98]) is able to treat very general multistage control problems and is used for the numerical investigations in this paper. Other recent developments of direct multiple shooting algorithms are described, e.g. by Tanartkit and Biegler [TB96] and Petzold *et al.* [PR97].

The paper is organized as follows:

- In Chapter 2 we introduce a general class of optimal control problems that can be treated by the current implementation of the direct multiple shooting method.
- In Chapter 3, the reformulation of infinite dimensional control problems as finite dimensional NLP problems is described.
- In Chapter 4, we sketch an SQP method specially tailored to the solution of such highly structured NLP problems.
- The use of our techniques in the context of real-time optimization is discussed in Chapter 5. The optimization of a disturbed batch process in real-time is presented as a numerical example.
- In Chapter 6 we specialize these real-time considerations to nonlinear model predictive control. A strategy to increase the efficiency of the optimizing procedure is presented, and preliminary numerical results for the control of a continuous stirred tank reactor are given.

2. Multistage optimal control problems

Many dynamic process optimization problems of practical relevance can be expressed as multistage optimal control problems in DAE. We consider the following class of M-stage optimal control problems, where the time horizon of interest $[t_0, t_M]$ is divided into M subintervals. The aim is to minimize a generalized Mayer-type objective function which is the sum of pointwise defined scalar functions ϕ_i.

$$\min_{x_i,\, z_i,\, u_i,\, p,\, t_i} \sum_{i=0}^{M-1} \phi_i(x_i(t_{i+1}), z_i(t_{i+1}), p, t_{i+1}) \tag{1}$$

subject to the model stages $i = 0, 1, \dots, M-1$ modeled by DAE

$$\left.\begin{array}{rcl} B_i(\cdot)\,\dot{x}_i(t) & = & f_i(x_i(t), z_i(t), u_i(t), p, t) \\ 0 & = & g_i(x_i(t), z_i(t), u_i(t), p, t) \end{array}\right\}, \quad t \in [t_i, t_{i+1}], \tag{2}$$

the control and path constraints

$$h_i(x_i(t), z_i(t), u_i(t), p, t) \geq 0, \quad t \in [t_i, t_{i+1}], \quad i = 0, 1, \dots, M-1, \tag{3}$$

the stage transition conditions

$$x_{i+1}(t_{i+1}) = c_i(x_i(t_{i+1}), z_i(t_{i+1}), p), \quad i = 0, 1, \dots, M-2, \tag{4}$$

and the multipoint boundary conditions

$$\sum_{i=0}^{M-1} \left(r_i^s(x_i(t_i), z_i(t_i), p, t_i) + r_i^e(x_i(t_{i+1}), z_i(t_{i+1}), p, t_{i+1}) \right) \left\{ \begin{array}{c} = \\ \geq \end{array} \right\} 0. \tag{5}$$

Matrix $B(\cdot)$ is assumed to be invertible so that the quasilinear-implicit nonlinear DAE (2) is of semi-explicit type. The control vector on stage i is denoted by $u_i(t)$, the differential and algebraic states by $x_i(t)$ and $z_i(t)$ ($t \in [t_i, t_{i+1}]$). The vector p contains global model parameters. The dynamic model equations and in particular the dimensions may change from one stage to the next; the coupling is provided by the stage transition conditions (4). Continuous control and path constraints (3) and general, linearly coupled multipoint boundary conditions (5) are included. All problem functions are assumed to be twice continuously differentiable.

Throughout this paper we will assume that the DAE is of *index one*. Higher index problems can be treated by index reduction. However, it is then advisable both for efficiency and stability, to take the resulting invariants into account. Effective ways to achieve this have been introduced by Schulz *et al.* [SBS98].

3. Parametrization of the optimal control problem

To reformulate the infinite problem of the previous section as a finite dimensional NLP problem, we parametrize both, its controls and its states. The latter one is the essential step to obtain the multiple shooting formulation of the problem.

3.1. Control discretization

We first approximate the control functions by suitable parametrizations. On each interval $[t_i, t_{i+1}]$ we employ a time transformation

$$\theta_i(\tau; t_i, d_i) = t_i + \tau d_i, \quad t_{i+1} = t_i + d_i, \quad \tau \in [0, 1] \tag{6}$$

and choose a multiple shooting grid

$$0 = \tau_{i,0} < \tau_{i,1} < \cdots < \tau_{i,m_i} = 1 \tag{7}$$

such that $\theta_i(\tau_{i,0}) = t_i$ and $\theta_i(\tau_{i,m_i}) = t_{i+1}$.

Then we use a *piecewise* representation $\hat{u}_i$ of the control functions u_i on each subinterval

$$\hat{u}_i(\tau) \stackrel{\text{def}}{=} \varphi_{ij}(\tau, q_{ij}), \quad \tau \in I_{ij} = [\tau_{ij}, \tau_{i,j+1}], \quad j = 0, 1, \ldots, m_i - 1, \tag{8}$$

The functions φ_{ij} are given functions with local support, typically polynomials (e.g., constant, linear, or cubic), parametrized by the local parameters q_{ij}.

The piecewise control representation need not be continuous at the inner grid points τ_{ij}, $j = 1, 2, \ldots, m_i - 1$. Continuity may be enforced by additional conditions

$$\varphi_{ij}(\tau_{i,j+1}, q_{ij}) - \varphi_{i,j+1}(\tau_{i,j+1}, q_{i,j+1}) \;=\; 0, \quad j = 0, 1, \ldots, m_i - 2, \tag{9}$$

similarly higher-order continuity. Our implementation provides great flexibility in choosing suitable control approximations – the parametrization type and possible continuity conditions can be specified separately for each control function on each model stage.

3.2. State parametrization by multiple shooting

In a second step the DAE are parametrized by *multiple shooting* (as described, e.g., in [BES88] and [SBS98]). We introduce additional optimization variables s^x_{ij}, s^z_{ij} which serve as initial values for differential and algebraic variables $x_{ij}(\tau_{ij}) = s^x_{ij}$ and $z_{ij}(\tau_{ij}) = s^z_{ij}$ for the following set of independent initial value problems (IVPs) on each of the multiple shooting subintervals I_{ij}.

$$\left.\begin{aligned} B_i(\cdot)\, dx_{ij}/d\tau &= f_i(x_{ij}, z_{ij}, \varphi_{ij}(\tau, q_{ij}), p, \theta_i(\tau; t_i, d_i))\, d_i \\ 0 &= g_i(x_{ij}, z_{ij}, \varphi_{ij}(\tau, q_{ij}), p, \theta_i(\tau; t_i, d_i)) \\ &\quad - \alpha_{ij}(\tau)\, g_i(s^x_{ij}, s^z_{ij}, \varphi_{ij}(\tau_{ij}, q_{ij}), p, \theta_i(\tau_{ij}; t_i, d_i)) \end{aligned}\right\} \quad \tau \in I_{ij}, \tag{10}$$

We denote the differential components of the solution of (10) on interval I_{ij} by $x_{ij}(\tau; s^x_{ij}, s^z_{ij}, q_{ij}, p, t_i, d_i)$, indicating the dependency on initial values, local control and global system parameters, as well as on the variables describing the time transformation.

By including the continuity conditions at the end of each interval

$$x_{ij}(\tau_{i,j+1}; s^x_{ij}, s^z_{ij}, q_{ij}, p, t_i, d_i) - s^x_{i,j+1} \;=\; 0, \quad j = 0, 1, \ldots, m_i - 1 \tag{11}$$

and the consistency conditions at its beginning

$$g_i(s^x_{ij}, s^z_{ij}, \varphi_{ij}(\tau_{ij}, q_{ij}), p, \theta_i(\tau_{ij}; t_i, d_i)) \;=\; 0, \quad j = 0, 1, \ldots, m_i \tag{12}$$

as equality constraints in the NLP problem, we ensure that the solution will satisfy the original DAE (2). It should be stressed that the multiple shooting formulation is an equivalent transformation, and no discretization of the continuous problem. It is only through the *numerical* integration of (10) that discretization errors are introduced into the DAE solution.

Note that the equations (10) are a *relaxed* version of the original DAE (2). The term subtracted in the algebraic equations allows for inconsistent initial values s^x_{ij} and s^z_{ij}.

The scalar factor α_{ij} may e.g. be given by

$$\alpha_{ij}(\tau) = \exp\left(-\beta \frac{\tau - \tau_{ij}}{\tau_{i,j+1} - \tau_{ij}}\right), \quad \beta \geq 0, \text{ constant}, \tag{13}$$

in order to damp the relaxation term, or any other suitable function which satisfies $\alpha_{ij}(\tau_{ij}) = 1$.

In view of efficiency, the possibility to allow for inconsistent initial values is an important feature of our BVP approach.

A more classical alternative would of course be to omit the relaxation term from (10), to drop s^z_{ij} from the set of optimization variables and to determine the values of s^z_{ij} by solving the consistency conditions (12) at the beginning of each integration. This would result in a feasible point strategy for conditions (12). The frequent underlying iterations at each discontinuity required by this alternative, however, have been found to become very expensive for larger process models, even in the case of pure simulation problems (see e.g. [M96, KM97]). Therefore, we deal with consistency only at the level of the optimization problem, i.e. require it only at the solution. A detailed numerical comparison of both approaches is in preparation.

Another advantage of the relaxation approach is that it easily generalizes to higher index problems, as shown in [SBS98]. This contrasts with the feasible point approach, for which it is not straightforward to satisfy the invariants generated by index reduction.

Note that, if the unrelaxed DAE with consistent initial values is solvable without singular or impasse points, then so is the relaxed DAE in a neighbourhood of these values. Indeed, numerical experience indicates that our optimization algorithm, when started with consistent data satisfying conditions (12), remains close to the constraint manifold throughout the SQP iterations.

Advanced, fully adaptive DAE solvers can be employed to compute

$$x_{ij}(\tau_{i,j+1}; s^x_{ij}, s^z_{ij}, q_{ij}, p, t_i, d_i)$$

and the required sensitivities. For MUSCOD2 several integrators are available, such as the stiff DAE integrator DAESOL [Ba94, Ba97] or the integrator library MBSSIM [vS97, A93, W96] for mechanical multibody systems. They efficiently treat relaxed DAE, including model discontinuities. First and second order derivatives can be generated in a fast and error controlled way, following the principle of internal numerical differentiation (for details, see e.g. [B81, B83, B87, SWS96, SBS98]).

It has already been explained that the multiple shooting approach provides considerably enhanced stability and efficiency, compared to methods based on single shooting. Furthermore, the integrations – including sensitivity generation – are independent on each multiple shooting interval and can hence be performed in parallel. Since the integrations are usually by far the most expensive part of the computation, significant speed-ups can be obtained. Efficient parallel algorithms for the solution of optimization boundary value problems by direct multiple shooting, including load balancing strategies, were developed by Gallitzendörfer and coauthors [G96, GB94, L97a]. The reported numerical results demonstrate good scalability for moderate processor numbers on distributed memory multiprocessor systems based on message passing, like workstation clusters or MPP machines.

3.3. The finite constrained optimization problem

The parametrization of problem (1)–(5) using multiple shooting and piecewise control representation leads to the following structured NLP problem :

$$\min_{s^x_{ij},\, s^z_{ij},\, q_{ij},\, p,\, t_i,\, d_i} \sum_{i=0}^{M-1} \phi_i(s^x_{i,m_i},\, s^z_{i,m_i},\, p,\, \theta_i(\tau_{i,m_i}; t_i, d_i)) \tag{14}$$

subject to continuity and consistency conditions at each node on each model stage $i = 0, 1, \dots, M-1$

$$\begin{array}{rcll} x_{ij}(\tau_{i,j+1}; s^x_{ij}, s^z_{ij}, q_{ij}, p, t_i, d_i) - s^x_{i,j+1} & = & 0, & j = 0, 1, \dots, m_i - 1 \\ g_i(s^x_{ij}, s^z_{ij}, \varphi_{ij}(\tau_{ij}, q_{ij}), p, \theta_i(\tau_{ij}; t_i, d_i)) & = & 0, & j = 0, 1, \dots, m_i \end{array} \tag{15}$$

discretized control and state constraints

$$h_i(s^x_{ij}, s^z_{ij}, \varphi_{ij}(\tau_{ij}, q_{ij}), p, \theta_i(\tau_{ij}; t_i, d_i)) \geq 0, \quad j = 0, 1, \dots, m_i, \tag{16}$$

stage transition conditions

$$s^x_{i+1,0} = c_i(s^x_{i,m_i}, s^z_{i,m_i}, p), \quad i = 0, 1, \dots, M-2, \tag{17}$$

and multipoint constraints

$$\sum_{i=0}^{M-1} \left(r^s_i(s^x_{i,0}, s^z_{i,0}, p, \theta_i(\tau_{i,0}; t_i, d_i)) + r^e_i(s^x_{i,m_i}, s^z_{i,m_i}, p, \theta_i(\tau_{i,m_i}; t_i, d_i)) \right) \left\{ \begin{array}{c} = \\ \geq \end{array} \right\} 0. \tag{18}$$

Lower and upper bounds may be specified for each of the NLP variables.

It can be shown by the Weierstrass Theorem that an optimal solution of the *finite* constrained optimization problem exists under mild conditions, which include the existence of a feasible solution, compactness of feasible optimization variables and a growth condition on the dynamics.

4. Blending multiple shooting and SQP

We will briefly discuss some of the basic ideas underlying our special SQP method which is tailored to the multiple shooting parametrization, such as limited-memory block updates for the Hessian, a line search globalization, and a condensing strategy for the efficient solution of the QP subproblems. For notational convenience, we write the structured NLP problem (14)–(18) in the abbreviated form

$$\min_w F(w) \quad \text{subject to} \quad \left\{ \begin{array}{rcl} G(w) & = & 0 \\ H(w) & \geq & 0, \end{array} \right. \tag{19}$$

where w contains all the state variables and control parameters, as well as the model parameters. The discretized dynamic model is included in the equality constraints $G(w) = 0$. Starting from an initial guess w^0, an SQP method for the solution of (19) iterates

$$w^{k+1} = w^k + \alpha^k \Delta w^k, \quad k = 0, 1, \dots , \tag{20}$$

where $\alpha^k \in]0, 1]$ is a relaxation factor, and the search direction Δw^k is the solution the quadratic programming (QP) subproblem

$$\min_{\Delta w \in \Omega^k} \; \nabla F(w^k)^T \, \Delta w + \frac{1}{2} \, \Delta w^T \, A^k \, \Delta w \tag{21}$$

subject to

$$G(w^k) + \nabla G(w^k)^T \, \Delta w = 0, \qquad H(w^k) + \nabla H(w^k)^T \, \Delta w \geq 0.$$

A^k denotes an approximation of the Hessian of the *Lagrangian function* L,

$$L(w, \lambda, \mu) \;=\; F(w) - \lambda^T G(w) - \mu^T H(w), \tag{22}$$

where λ and μ are the Lagrange multipliers.

By appropriate strategies for the choice of the relaxation factor α^k or the trust region Ω^k or both the convergence can be globalized. (For a detailed description of globalisation strategies available in the latest version of MUSCOD2 the reader is referred to Leineweber [L98])

4.1. Exploiting the structure of the problem

It turns out, that the NLP problem and the resulting QP problems have a special, very desirable structure, which is due to our deliberate choice of state and control parametrizations.

The Lagrangian function is *partially separable*, i.e., it can be written in the form

$$L(w, \lambda, \mu) \;=\; \sum_{i=0}^{M-1} \sum_{j=0}^{m_i} L_{ij}(w_{ij}, \lambda, \mu), \tag{23}$$

where w_{ij} are the components of the primal variables w which are effective on interval (i, j) only. This separation is possible if we simply interpret the parameters p, t_i, d_i as piecewise constant continuous controls.

As a consequence, the Hessian of L has a *block diagonal structure* with blocks $\nabla^2_{w_{ij}} L_{ij}(w_{ij}, \lambda, \mu)$. Similarly, the multiple shooting parametrization introduces a characteristic *block sparse structure* of the Jacobian matrices $\nabla G(w)^T$ and $\nabla H(w)^T$. The resulting structure of the corresponding Karush-Kuhn-Tucker matrix is sketched in Figure 1.

It is of crucial importance for performance and numerical stability of the direct multiple shooting method that these structures of (19) are fully exploited. Therefore several special algorithms have been developed, some of which are summarized in the following

- Partitioned high rank updates preserve the block diagonal structure of the Hessian and speed up local convergence [P81, BP84].

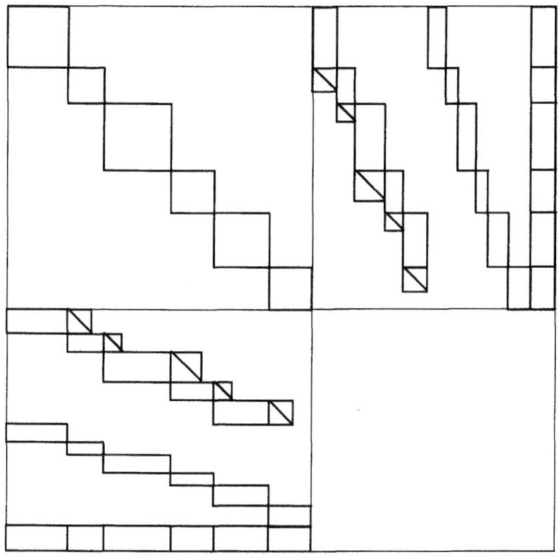

FIGURE 1. Structure of the symmetric indefinite Karush-Kuhn-Tucker matrix. The left upper part is the block diagonal Hessian. Each Hessian block corresponds to the optimization variables effective on one multiple shooting interval: differential and algebraic states and controls. The left lower part shows first the block sparse structure of the Jacobian of continuity conditions. The squares with diagonals represent identity matrices(derivatives of (11) w.r.t. $s^x_{i,j+1}$). Second, the (local) consistency conditions and decoupled interior point conditions are depicted, and finally the linearly coupled interior point conditions at the bottom.

- Special recursive QP solvers are used for problem (21) that exploit the block sparse structure shown in Figure 1. Both active set strategies and interior point methods are available for the treatment of large systems of inequality constraints [P81, BP84, L96, St95].
- Leineweber [L98] developed a reduction technique for DAE systems with a large share of algebraic variables. He exploits the linearized algebraic consistency conditions for a reduction in variable space, so that only reduced gradients and Hessian blocks need to be calculated, which correspond to the differential variables only (cf. [S188, BES88])

In the next subsection we will explain the first aspect in more detail.

4.2. High rank updates and other Hessian approximations

How can the block diagonal structure of the Hessian be preserved and exploited by the choice of the approximation A^k? Several approaches are possible and have been been realized within direct multiple shooting.

The first approach, which corresponds to a Newton method, would be to use the "exact" Hessian blocks, i.e. a numerically computed approximation. This version is recommended if the computation of the Hessian is cheap or does not have to be considered for other reasons. This is the case in the real time optimization algorithm presented in the next chapter, where the Hessian can be computed and stored in advance. The use of the "exact" Hessian has excellent local convergence properties. Since the Hessian can become indefinite far from the optimal solution, a globalisation based on trust regions is needed.

Since in many cases the cost for the Hessian computation is prohibitive, and moreover positive definite approximations allow for relatively simple line search globalisations, secant update formulae have been popular for nearly four decades – this second approach is the standard in our off-line direct multiple shooting method. Clearly, the application of any of the standard rank-two updates would immediately destroy the Hessian block structure. Therefore, Bock and Plitt have proposed a generalized high-rank update procedure [P81, BP84]: each diagonal block of A^k is updated *separately* using the corresponding components of the step δ^k,

$$\delta^k = \alpha^k \Delta w^k,$$

and the Lagrangian gradient difference γ^k,

$$\gamma^k = \nabla_w L(w^{k+1}, \lambda^{k+1}, \mu^{k+1}) - \nabla_w L(w^k, \lambda^{k+1}, \mu^{k+1}).$$

A given update formula

$$A^{k+1} = A^k + U(A^k,\, \delta^k,\, \gamma^k), \tag{24}$$

then generalizes to

$$A^{k+1} = A^k + \sum_{i=0}^{M-1} \sum_{j=0}^{m_i} U(P_{ij} A^k P_{ij},\, P_{ij}\delta^k,\, P_{ij}\gamma^k), \tag{25}$$

with corresponding projections P_{ij}. This update preserves the block diagonal structure of the matrices A^k. Extensive numerical tests have shown that the asymptotic convergence rate is considerably improved by this high-rank update procedure. A similar partitioned update procedure for unconstrained optimization of partially separable functions has been proposed by Griewank and Toint [GT82]. As the default update we employ a modified BFGS formula which keeps the Hessian approximations positive definite (Powell [P78]). Performance is improved by a limited-memory strategy.

A third approach to obtain a cheap Hessian approximation – the constrained Gauss-Newton method – is recommended in the special case of a least squares type cost function $F(w) = \frac{1}{2}\|C(w)\|_2^2$.

The matrix $\nabla_w C^T \nabla_w C$ is already available from the gradient computation. It provides an excellent approximation of

$$\nabla_w^2 L = \nabla_w C^T \nabla_w C + \nabla_w^2 C C - \nabla_w^2 G \lambda - \nabla_w^2 H \mu$$

if the residual $C(w)$ of the cost function is sufficiently small at the solution, and consequently also the adjoint variables (The latter can easily be seen, e.g., for the equality constrained case, where $\lambda = \nabla_w G(w)^\dagger \nabla_w C(w) C(w)$). In this case, all terms involving second order derivatives of cost function and constraints can be omitted.

The Gauss-Newton approximation obviously preserves the block diagonal structure of the Hessian. For a detailed analysis of local and global convergence properties we refer to [B81, B87].

An alternative to SQP methods that use approximations of the full Hessian are so called *reduced* or *partially reduced* SQP methods, which use linearizations of all or some of the equality constraints of the NLP problem to project gradients and/or Hessian to the corresponding null space. This approach was originally introduced by Gabay [G82]. An analysis of partially reduced variants for discretized control problems is given by Schulz [S98].

Leineweber [L98] shows how the the consistency conditions can be used to eliminate the algebraic variables both from gradient computation and high rank Hessian updates for the direct multiple shooting approach discussed in this paper. He demonstrates enormous savings for applications in which the number of algebraic states exceeds by far the number of differential ones.

For the case of Gauss-Newton methods such reduction techniques were introduced by Schlöder [Sl88] for the ODE case, and generalized in [BES88] and [SBS98] to DAE of index one and higher.

5. Real-time optimal control problems

Open loop solutions as determined by the off-line methods of the previous section are of limited use in many control applications where disturbances call for feedback corrections.

After disturbances the precalculated nominal controls may cause violations of state constraints and boundary conditions and lead to a significant loss of process performance. For small perturbations a linearized neighbouring feedback control might be sufficient. In the general case, however, it would be highly desirable to be able to calculate on-line an optimal solution for the disturbed situation which satisfies all restrictions (estimates of the current state given).

More generally one may understand a real-time optimal control problem as one, the specifications of which (such as problem functions or problem data) become only available while the process is in progress. For large-scale systems, or short process time scales, the calculations required for a full optimization are time-critical, and even the fastest off-line optimization methods may not deliver the optimal control fast enough. Hence special optimization methods for real-time problems have to be developed. We would like to stress two main differences that distinguish our real-time approaches from off-line methods. First, one observes that

TABLE 1. *Overview of Real-Time Strategies.*

Strategy	Precalculated Information Used
COLD	same initial guesses as used for *off-line* computation of reference solution (for comparison only)
COOL	state, control, and parameter values of reference solution as initial guesses
WARM	like COOL, plus gradients and function values, as well as final BFGS approximation of Hessian from off-line computation, QP presolves
HOT	like WARM, but precalculated "exact" Hessian of reference solution

any calculations performed off-line must be counted differently w.r.t. the algorithmic complexity and efficiency. Therefore we propose a real-time approach that utilizes one or more off-line calculated scenarios (reference solutions) on which the solution method is based, e.g., that serve as a starting point for an iteration method. Second, at least for time-critical situations, we propose a dovetailed coupling of the real-time progress of the process with the progress of the optimization procedure.

As a realization of this concept we will now discuss how direct multiple shooting can be adapted to the specific requirements of real-time optimization. We describe strategies which exploit knowledge of an off-line computed optimal solution as reference trajectory in order to greatly reduce the total computational cost.

5.1. Fast multiple shooting for real-time optimization

We have developed specialized *real-time strategies* for direct multiple shooting which use precalculated information about a reference solution to a varying degree (see the overview in Table 1). The reference solution is chosen to be the off-line calculated optimal solution for an undisturbed process. The "COLD" strategy corresponds to the off-line method (i.e., no use at all is made of information about the reference solution) and has been included for comparison only. "COOL" uses the reference values of all optimization variables as initial guesses but no other information. "WARM", in addition, uses the final BFGS approximation of the Hessian obtained during the off-line computation of the reference solution,

"HOT" uses the precalculated "exact" Hessian at the reference solution, up to discretization errors. Observe that this calculation is performed *off-line* and does not enter the on-line computational effort.

In all cases, the functions and gradients are evaluated on each major iteration, as usual in SQP-type methods. For WARM and HOT, however, they are calculated off-line for the first iteration, as well as major parts of the quadratic program. COLD, COOL, and WARM employ standard modified BFGS block updates for the Hessian and a line search globalization.

Various possibilities exist for the realization of HOT. In the variant tested in table 3, the Hessian is kept *fixed* (no updates, in contrast to the method presented in section 6.4), and – deliberately – a *full-step* method is employed. In this case the first iteration corresponds to a full Newton step and all others to simplified Newton steps.

Note that multiple shooting is well suited for these strategies, because it allows to directly include the reference state values at *all* discretization points as initial guesses for the real-time optimization.

Remark: the use of terms like "cold", "cool" etc. stems from our group's jargon about starting strategies for continuation methods and should not be confused with the use of such terms by other authors.

5.2. Application to a fed-batch fermentation process

In order to test and compare the above real-time strategies in a realistic setting, we have performed an real-time optimization study for a fed-batch fermentation process for the production of Nikkomycin (an antibiotics) as described by King and coauthors [HKG90, KBG90, K94, KW95]. The filamentous organism *Streptomyces tendae* is grown in a synthetic medium containing glucose, ammonia, phosphate, and trace elements. The fermentor is well-stirred, and temperature and pH are controlled (27°C, pH 6). During fermentation, the substrates glucose, ammonia, and phosphate can be added to the reactor at defined rates. The process is terminated when a certain maximum broth volume is reached.

The objective is to maximize the final amount of product by piecewise constant feeding profiles for the substrates. This nonlinear ODE model involves 12 states (masses of biomass fractions, substrates, and product; activity, volume), 3 controls (feed rates of glucose, ammonia, and phosphate). The nonlinearities arise from generalized Michaelis-Menten enzyme kinetics. Apart from bounds on all states and controls, the control problem contains 5 nonlinear path constraints on concentrations of substrates and total biomass. The batch time is free. The initial states are prescribed.

First, an optimal solution is calculated off-line using 16 equidistant multiple shooting nodes. The optimal final product mass is 71.32 g and the optimal batch time 183.7 hr. Second, a complete failure of the ammonia feed pump for 85.8 hr $\leq t \leq$ 98.0 hr is assumed which leads to a deviation from the nominal trajectory. If the control profiles of the nominal solution were used one would observe a performance loss of 13.8% – only 61.45 g of product could be recovered.

Hence, a real-time optimization of the process is desirable, from the new disturbed initial state at $t = 98.0$ hr to a new terminal point. For the computed new optimal solution, a final product mass of 68.67 g is obtained at a slightly increased batch time of 191.5 hr. I.e., the performance loss is reduced to 3.7%. Table 2 provides a comparison of the nominal, disturbed, and optimized solutions in terms of process performance.

For this real-time optimization, the four strategies described above are compared, using the unperturbed optimal solution as the reference. The results shown

TABLE 2. *Comparison of Numerical Solutions.*

Solution	Objective	
nominal	71.32	(100%)
disturbed	61.45	(86.2%)
reoptimized	68.67	(96.3%)

TABLE 3. *Comparison of Real-Time Strategies.*

Strategy	Iterations	CPU Time
COLD	37	64.9 s
COOL	13	19.3 s
WARM	8	12.3 s
HOT	4	6.8 s

in Table 3 indicate that especially the use of precalculated exact Hessians in HOT leads to very significant speed-ups. Compared to the unmodified off-line strategy, the total CPU time is reduced by a factor of almost ten (see Leineweber *et al.* [L97b]).

For the numerical integration and sensitivity generation we used a Runge-Kutta-Fehlberg 4/5 integrator equipped with internal numerical differentiation. The integration tolerance was 10^{-7}. All calculations were performed on an SGI Indy workstation (MIPS R4000 processor, 100 MHz speed).

Remark: We have chosen a relatively small single stage ODE problem since it was at hand and provides an interesting real world background. However, we would like to point out that the implementation of the described methods in our optimization package MUSCOD2 is indeed capable of treating (real-time) optimization problems for large scale multistage DAE systems.

5.3. Interpretation of the numerical results

Obviously, for a problem with time scales as generous as in this application (hours), even our unmodified off-line method in COLD or COOL version would be sufficiently fast to be used for real-time optimization.

But what can we conclude for systems of this size where the optimization is time-critical?

- The first iterate for WARM and HOT takes less than a millisecond since all gradient and Hessian information is precomputed and the QP solution needs only a few operations to compute the reoptimized control parameters after the trajectory disturbance is known (due to a presolve procedure). Hence, the control can "immediately" respond to a disturbance. Note that the first step for HOT is a full Newton step. In the special case where no pure state inequality constraints are present, it is easy to show that this step is equivalent to the result of a so called "linearized neighbouring feedback control" (which takes all linearized constraints into account). Such methods for control constrained problems were described, e.g by Krämer-Eis et al. [Kr85, KrB87]. Whereas these methods are also able to adapt to changes in the switching structure, they only satisfy feasibility and optimality conditions up to first order.

- Any further full iteration step takes less than 2 seconds. This means, that not only a new control iterate but also *a new reference trajectory* including *new gradients* is available for subsequent use.
- This time can be reduced by a substantial factor if "near-optimality" is sufficient and feasibility is the dominant aim. In this case we can avoid the computation of new gradients and large parts of the QP – whereas function evaluations are still necessary for the SQP iterations.

In the numerical results above, the optimization was iterated to convergence (residual of KKT-System less than $1.0 \cdot 10^{-6}$). However, if the computing time per iteration is not negligible compared to the time scale of the process, it is advisable to adapt the QP problem solved in each iteration to the real-time progress of the process. Therefore, we propose to dovetail the SQP iterations with the dynamics of the process, i.e. to predict the initial states that the process will have reached in real-time after having performed the actual iteration (e.g. by on-line simulation of the process), and to use these initial states and the corresponding remaining interval as input for this very iteration. The authors claim that this procedure in combination with HOT is contractive under mild conditions, if the first initial states for the real-time optimization are sufficiently close to the reference trajectory. Therefore, if the iteration time tends to zero, the iterates converge linearly to the optimal solution on the remaining intervals for the respective initial states. The contraction rate grows linearly with the initial perturbation.

Notice that this real-time optimization technique gives much more than classical linearized neighbouring feedback control, as mentioned above, since each such iteration improves both, feasibility and optimality.

Recall that all time critical components of this algorithm can be very effectively parallelized if its performance should be insufficient for a practical application. Hence for a system of the above complexity process durations in the range of seconds should be tractable by real-time direct multiple shooting.

6. Efficient multiple shooting for NMPC

Next we will describe how problems of NMPC can be efficiently treated by direct multiple shooting. Since NMPC is closely related to the real-time optimization problem of the previous section, we expect that the techniques described there will also prove useful in this context.

6.1. Formulation of a NMPC optimization problem

A rather general NMPC Problem with sampling interval δ, prediction horizon T_p comprising N predicted sampling steps ($T_p = N\delta$) and control horizon $T_c = M\delta$ ($M \leq N$) may be formulated as follows:

$$\min_u \int_{t_0}^{t_0+T_p} F(x(t), z(t), u(t), t)\, dt \quad + \quad E(x(t_0+T_p), z(t_0+T_p), t_0+T_p) \tag{26}$$

subject to the DAE

$$B(\cdot)\dot{x}(t) = f(x(t), z(t), u(t), t) \tag{27}$$

$$0 = g(x(t), z(t), u(t), t) \tag{28}$$

with initial value

$$x(t_0) = x_0 \tag{29}$$

the control and path constraints

$$h(x(t), z(t), u(t), t) \geq 0 \tag{30}$$

and possible restrictions on the state at the endpoint

$$r(x(t_0+T_p), z(t_0+T_p), t_0+T_p) \left\{ \begin{array}{c} = \\ \geq \end{array} \right\} 0. \tag{31}$$

The controls are held constant on each sampling interval:

$$u(t) = u_i \quad \forall t \in [t_i, t_{i+1}] \quad t_i = t_0 + i\delta \quad i = 0, 1, \ldots N-1 \tag{32}$$

and are fixed after the Mth interval:

$$u_j = u_{M-1} \quad \text{for} \quad M \leq j \leq N \tag{33}$$

Note that more generally than (32) any control parametrization with local support can be treated within our framework.

6.2. Direct multiple shooting for NMPC

NMPC problems can be solved by direct multiple shooting in the following way. With the sampling time δ being a natural choice for the interval length, we introduce the node values s_i^x and s_i^z (both combined: s_i) of differential and algebraic states at times t_i and get the following NLP problem, to which we will refer as P in the sequel:

$$\min_{u,s} \sum_{i=0}^{N-1} F_i(s_i^x, s_i^z, u_i) \quad + \quad E(s_N^x, s_N^z, t_N) \tag{34}$$

subject to the initial value constraint

$$s_0^x = x_0, \tag{35}$$

the continuity conditions

$$s_{i+1}^x = x_i(t_{i+1}) \qquad i = 0, 1, \ldots N-1, \tag{36}$$

and the consistency conditions

$$0 = g(s_i^x, s_i^z, u_i, t_i) \qquad i = 0, 1, \ldots N, \tag{37}$$

where

$$F_i(\cdot) = \int_{t_i}^{t_i+\delta} F(x_i(t), z_i(t), u_i, t)\, dt \tag{38}$$

and $x_i(t), z_i(t)$ are the solutions of the relaxed decoupled initial value problems on $[t_i, t_{i+1}]$:

$$B(\cdot)\dot{x}_i(t) = f(x_i(t), z_i(t), u_i, t) \tag{39}$$

$$0 = g(x_i(t), z_i(t), u_i, t) - \alpha_i(t) g(s_i^x, s_i^z, u_i, t_i) \tag{40}$$

$$x_i(t_i) = s_i^x \tag{41}$$

$$z_i(t_i) = s_i^z \tag{42}$$

where α_i is a scalar damping factor used for the relaxation of the DAE, e.g. given by eq. (13). The control and path constraints have to be discretized as

$$h(s_i^x, s_i^z, u_i, t_i) \geq 0 \qquad i = 0, 1, \ldots N - 1, \tag{43}$$

and the restrictions at the endpoint become

$$h(s_N^x, s_N^z, u_{N-1}, t_N) \geq 0 \tag{44}$$

$$r(s_N^x, s_N^z, t_N) \left\{ \begin{array}{c} = \\ \geq \end{array} \right\} 0. \tag{45}$$

Again,

$$u_j = u_{M-1} \quad \text{for} \quad M \leq j \leq N, \tag{46}$$

which can be implemented efficiently via a two model-stage approach: one for the control horizon, and a second one for the remaining part of the prediction horizon, for which the constant control then is a global parameter.

6.3. Enhancing the efficiency of the algorithm

The similarity between the optimization problems at two consecutive time steps can be exploited to increase the efficiency of the optimization algorithm. There are various ways to accomplish this. Basic aim is to minimize on-line computing effort (evaluation of functions, gradients and the Hessian; QP solution) and to find the best available initial guesses for the trajectory and the controls. We just describe one particular strategy which computes the first iterate for the next time step with only a minimum amount of additional linear algebra – and without any computation of DAE solutions or gradients.

To change from the solution of problem P at a time t_0 to the next problem P' at time $t_0' = t_0 + \delta$, we use the fact that the first $N{-}1$ multiple shooting stages of P' coincide with the last $N{-}1$ stages of P. Hence, the solution of P can be used as an initial guess for P', after a shift:

$$\left.\begin{array}{lcl} u'_i & := & u_{i+1} \\ s'_i & := & s_{i+1} \end{array}\right\} \quad \text{for} \quad i = 0, 1, \ldots N-2 \tag{47}$$

Note that in our approach the initial guess for s'_0 is **not** given by the new measured state x'_0: instead, the value x'_0 enters the problem only implicitly through the initial condition (35)

$$(s_0^x)' = x'_0,$$

which can be considered a special linear embedding of the perturbation by an initially infeasible constraint. The violation of this constraint is corrected in the first SQP iteration due to its linearity.

The initial guesses for the new final interval $[t'_{N-1}, t'_N]$ are not specified, yet. A possible choice is $u'_{N-1} := u_{N-1}$, $s'_{N-1} := s_{N-1}$ and $s'_N := s_N$, i.e., we simply duplicate initial value, trajectory, gradients and Hessian of interval $[t_{N-1}, t_N]$ and accept that the continuity condition (36) is violated initially at t'_{N-1}.

As the Hessian is block diagonal, it can be recycled by a shift. The old Hessian

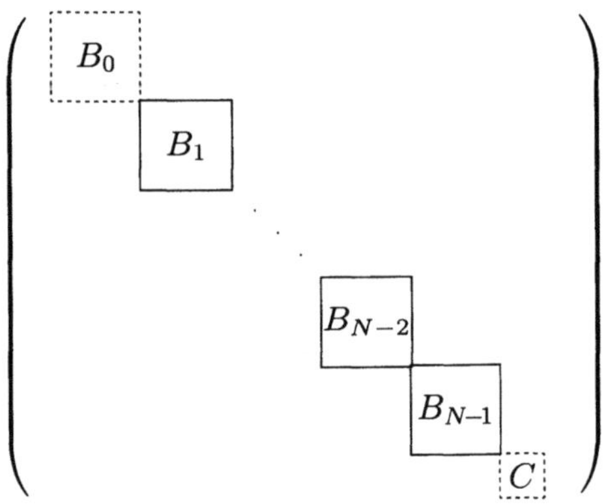

is replaced by:

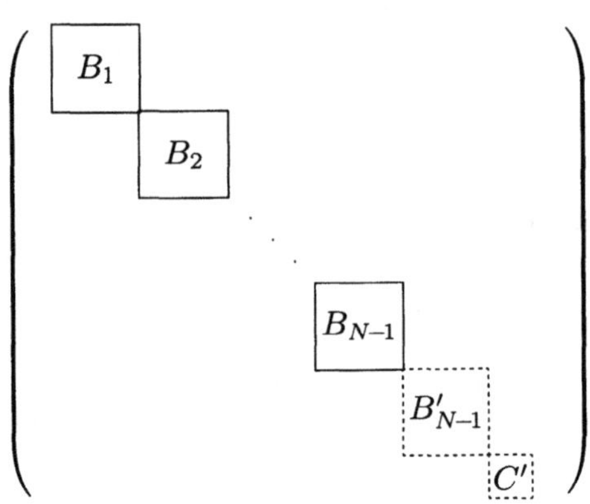

The blocks of the last interval, B'_{N-1}, and of the endpoint, C', are initialized by the canonical choice $B'_{N-1} := B_{N-1}$ and $C' := C$.

So all information necessary for the first QP subproblem is given without any further computational effort – the first SQP iteration needs no function evaluations at all, and only a little extra linear algebra is necessary to get the first correction to the problem variables if intermediate results from the last QP solution are suitably stored.

6.4. Preliminary numerical results

To test our algorithm we have implemented the control of a continuous stirred tank reactor (CSTR) presented as a benchmark example for NMPC in [CKA95]. This system – originally introduced in [KE93] – models the production of cyclopentenol from cyclopentadiene in a CSTR with a cooling jacket; it is described by ODE and shows a sign change in the steady state gain at the setpoint. It has four states: the concentrations c_A and c_B of educt and product, the tank temperature ϑ and the cooling jacket temperature ϑ_K. The two controls are feed flow and heat removal.

The control aim is to steer the system into a steady state, the setpoint, which is optimal with respect to yield. Its state values are: $c_A = 2.34\,\mathrm{mol\,l^{-1}}$, $c_B = 1.29\,\mathrm{mol\,l^{-1}}$, $\vartheta = 114.2\,^\circ\mathrm{C}$, $\vartheta_K = 112.9\,^\circ\mathrm{C}$.

We use the controller setup proposed by Chen in [Ch97]("Controller C/C", Chapter 5, pp. 61-66). The controller has a sample time of 20 seconds and predicts the control moves for 22 sample intervals, i.e. $T_p = T_c = 440$ sec.. The integral objective term $F(x(t), u(t))$ is the weighted quadratic deviation of states and controls from the setpoint. Stability is ensured by an end point constraint $r(x(t_0+T_p)) \geq 0$ forcing the final state to lie in a region around the setpoint, and by a quadratic penalty term $E(x(t_0 + T_p))$ penalizing a remaining final state deviation.

We consider the following test case: for a given perturbed state the system has to be driven back to the setpoint. There are no further disturbances in the process.

We have carried out control simulations for the same four perturbed initial states treated by Chen [Ch97]. Each run simulates a period of 2000 seconds and thus comprises 100 control steps, i.e. 100 consecutive NMPC optimization problems iterated to convergence.

TABLE 4. *Elapsed CPU times for different initial disturbances and a simulation time of 2000 seconds – i.e. 100 consecutive optimization problems iterated to convergence*

Initial state				Elapsed CPU time
c_A $[\frac{\mathrm{mol}}{\mathrm{l}}]$	c_B $[\frac{\mathrm{mol}}{\mathrm{l}}]$	ϑ [°C]	ϑ_K [°C]	Contr. C/C
0.0	0.0	85.0	85.0	69.3 sec.
5.1	2.5	134.2	132.9	49.9 sec.
0.0	2.5	85.0	132.9	56.1 sec.
5.1	0.0	134.2	85.0	48.1 sec.

For the first NMPC step in a simulation run the optimizer is initialized with gradients and Hessian evaluated at the setpoint trajectory. In all following steps the information available from the previous optimization step is re-used for the first QP solution, using the strategy proposed in this section.

The CPU times for the full control simulation runs are listed in Table 4. The maximum time for one optimization step is 6 seconds, spent on 14 SQP iterations. Mostly only one SQP iteration is needed, each of which takes less than half a second.

As the time scale for the calculation of a fully optimized trajectory iterated to convergence is already well below the 20 seconds between two control moves, the preliminary numerical results suggest that the proposed direct multiple shooting method is well suited for the use in NMPC.

Dovetailing of the iterations with the real time of the process, as described in the previous chapter, would extend the applicability of the approach.

For the numerical integration and sensitivity generation we used a Runge-Kutta-Fehlberg 4/5 integrator equipped with internal numerical differentiation. The integration tolerance was 10^{-6}. The simulations were performed on an SGI Indy workstation (MIPS R4000 processor, 100 MHz speed).

7. Conclusions

We have presented the direct multiple shooting method for the optimization of nonlinear processes described by DAE. It features excellent stability and efficiency properties, as well as general and flexible applicability. Being a very fast method already in its off-line version due to structure exploitation, it can be further accelerated for real-time optimization. Key features are the extensive use of precalculated off-line information and the possible dovetailing of the iterative solution with the process dynamics for both of which the direct multiple shooting approach is particularly apt. In addition, multiple shooting is well suited for parallelisation. The performance of the real-time algorithm is demonstrated on a 12 state biofermentation process.

Finally, it is shown how these remarkable features can be carried over to moving horizon NMPC problems by a shift-and-embed strategy for the Kuhn-Tucker-System. Encouraging preliminary numerical results for a benchmark CSTR control problem prove the direct multiple shooting approach to be a very promising candidate for the use in nonlinear model predictive control.

Acknowledgements

The authors gratefully acknowledge fruitful discussions with R. W. Longman and B. Alexander. We would like to thank R. Findeisen for pointing out to us the CSTR benchmark problem. We are further indebted to the anonymous referees for their suggestions and comments that helped to improve the paper.

References

[A93] Andrzejewski, Th., H. G. Bock, E. Eich, and R. von Schwerin, Recent Advances in the Numerical Integration of Multibody Systems. In W. Schiehlen (ed.) *Advanced Multibody System Dynamics – Simulation and Software Tool,* 127–151, Kluwer Acad. Publ., Dordrecht, NL (1993).

[Ba94] Bauer, I., Numerische Behandlung Differentiell-Algebraischer Gleichungen mit Anwendungen in der Chemie. *Diploma thesis*, University of Augsburg (1994).

[Ba97] Bauer, I., F. Finocchi, W. J. Duschl, H.-P. Gail, and J. P. Schlöder, Simulation of chemical reactions and dust destruction in protoplanetary accretion discs. *Astron. Astrophys.* **317**, 273–289 (1997).

[BBB92] Baake, E., M. Baake, H. G. Bock and K. Briggs. Fitting Ordinary Differential Equations to Chaotic Data. Phys. Rev. A, 45 (1992).

[KBS93] Kallrath, J., H. G. Bock and J. P. Schlöder, Least Squares Parameter Estimation in Chaotic Differential Equations. Celestial Mechanics and Dynamical Astronomy, 56 (1993).

[BR91] Biegler, L. T., J. B. Rawlings, *Optimization Approaches to Nonlinear Model Predictive Control*, in Y. Arkun, W. H. Ray (eds.), *Chemical Process Control – CPC IV*, 543–571, The CACHE Corp., Austin, Texas (1991).

[B81] Bock H.G., "Numerical treatment of inverse problems in chemical reaction kinetics", *Modelling of chemical reaction systems, eds. Ebert, Deuflhard, Jäger, Springer series Chemical Physics*, **18**, 102–125, (1981).

[B83] Bock, H. G., Recent advances in parameter identification techniques for O.D.E. In P. Deuflhard, and E. Hairer (eds.), *Numerical Treatment of Inverse Problems in Differential and Integral Equations*. Birkhäuser, Boston (1983).

[BP84] Bock, H.G., Plitt K.J., "A multiple shooting algorithm for direct solution of optimal control problems", *Proc. 9th IFAC World Congress Budapest, Pergamon Press*, (1984).

[B87] Bock, H. G., Randwertproblemmethoden zur Parameteridentifizierung in Systemen nichtlinearer Differentialgleichungen. *Bonner Mathematische Schriften 183*, University of Bonn (1987).

[BES88] Bock, H. G., E. Eich, J. P. Schlöder, *Numerical Solution of Constrained Least Squares Boundary Value Problems in Differential-Algebraic Equations*, in K. Strehmel (ed.), *Numerical Treatment of Differential Equations*, Teubner, Leipzig (1988).

[Ch82] Chamberlain, R. M., C. Lemaréchal, H. C. Pedersen, and M. J. D. Powell, The watchdog technique for forcing convergence in algorithms for constrained optimization. *Math. Progr. Study* **16**, 1–17 (1982).

[Ch97] Chen H., *Stability and Robustness Considerations in Nonlinear Model Predictive Control.*, Fortschr.-Ber. VDI-Series 8, No. 674, VDI Verlag, Düsseldorf, D (1997).

[CKA95] Chen H., A. Kremling and F. Allgöwer, Nonlinear Model Predictive Control of a Benchmark CSTR. *Proc. 3rd European Control Conference ECC'95, Rome/Italy* 3247–3252, Rome, I (1995).

[ER90] Eaton, J. W., J. B. Rawlings, *Feedback Control of Chemical Processes Using On-Line Optimization Techniques*, Comput. Chem. Engng 14, p. 469 (1990).

[ER92] Eaton, J. W., J. B. Rawlings, *Model Predictive Control of Chemical Processes*, Chem. Engng Sci. 47, p. 705 (1992).

[G82] Gabay, D., Reduced quasi-Newton methods with feasibility improvement. *Math. Progr. Study* **16**, 18–44 (1982).

[GPM89] García, C. E., D. M. Prett, M. Morari, *Model Predictive Control: Theory and Practice – a Survey*, Automatica 25, p. 335 (1989).

[GB94] Gallitzendörfer, J. V., and H. G. Bock, Parallel algorithms for optimization boundary value problems in DAE. In H. Langendörfer (ed.), *Praxisorientierte Parallelverarbeitung*. Hanser, München (1994).

[G96] Gallitzendörfer, J. V., Parallele Algorithmen für Optimierungsrandwertprobleme. *Ph.D. thesis*, University of Heidelberg (1996).

[GT82] Griewank, A., and P. L. Toint, Partitioned variable metric updates for large structured optimization problems. *Numer. Math.* **39**, 119–137, 1982.

[HKG90] Hege-Treskatis, D., R. King, E. D. Gilles, *Nikkomycin Production of* Streptomyces tendae *During Continuous Fermentations in a Defined Medium*, in D. Behrens (ed.), *DECHEMA Biotechnology Conferences, Vol. 4*, VCH Verlagsgesellschaft, Weinheim (1990).

[KE93] Klatt K.-U. and S. Engell, Rührkesselreaktor mit Parallel- und Folgereaktion, in *VDI-Bericht Nr. 1026* pp. 101–108, VDI Verlag, Düsseldorf, D (1993).

[KW95] Kendlbacher, Th., W. Waldraff, G. Breuel, R. Biener, R. King, E. D. Gilles, *Application of Model-Predictive Control to Fermentations with* Streptomyces, in R. D. Schmid (ed.), *Biochemical Engineering 3 – Stuttgart*, E. Kurz & Co, Stuttgart (1995).

[KBG90] King, R., T. K. Buschulte, H. Yang, E. D. Gilles, *Mathematical Models of Filamentously Growing Microorganisms*, in D. Behrens (ed.), *DECHEMA Biotechnology Conferences, Vol. 4*, VCH Verlagsgesellschaft, Weinheim (1990).

[K94] King, R., *Mathematische Modelle mycelförmig wachsender Mikroorganismen*, Fortschrittberichte VDI, Reihe 17, Nr. 103, VDI-Verlag, Düsseldorf (1994).

[Kr85] Krämer-Eis, P., Ein Mehrzielverfahren zur numerischen Berechnung optimaler Feedback-Steuerungen bei beschränkten nichtlinearen Steuerungsproblemen. *Bonner Mathematische Schriften 166*, University of Bonn (1985).

[KrB87] Krämer-Eis, P., and H. G. Bock, Numerical Treatment of State and Control Constraints in the Computation of Feedback Laws for Nonlinear Control Problems. In P. Deuflhard et al. (eds.), *Large Scale Scientific Computing,* pp. 287-306, Birkhäuser, Boston (1987).

[KM97] Kröner, A., W. Marquardt, and E. D. Gilles, Getting around consistent initialization of DAE systems? *Comput. Chem. Engng* **21**, 145–158 (1997).

[L96] Leineweber D.B., "The theory of MUSCOD in a nutshell", *IWR preprint 96-19*, (1996).

[L97a] Leineweber D.B., Bock H.G., Schlöder, J.P., Gallitzendörfer J.V., Schäfer A., Jansohn P.,"A boundary value problem approach to the optimization of chemical processes described by DAE models", submitted to *Comput. Chem. Eng.* (1997).

[L97b] Leineweber D.B., Bock H.G., Schlöder, J.P., "Fast direct methods for real-time optimization of chemical processes", *Proc. 15th IMACS World Congress on Scientific Computation, Modelling and Applied Mathematics Berlin, Wissenschaft- und Technik-Verlag Berlin*, (1997).

[L98] Leineweber, D. B., Efficient reduced SQP methods for the optimization of chemical processes described by large sparse DAE models. *Ph.D. thesis*, University of Heidelberg (1998).

[M96] Majer, C., and E. D. Gilles, Erweiterung einer Simulationsumgebung für verfahrenstechnische Prozesse zur Parameterschätzung, Versuchs- und Trajektorienplanung. *Chemie Ingenieur Technik* **68**, 1066–1067 (1996).

[P81] Plitt K.J., "Ein superlinear konvergentes Mehrzielverfahren zur direkten Berechnung beschränkter optimaler Steuerungen", Diplomarbeit, Bonn (1981)

[P78] Powell, M. J. D., A fast algorithm for nonlinearly constrained optimization calculations. In G. A. Watson (ed.), *Numerical Analysis, Dundee 1977*. (Lecture Notes in Mathematics 630.) Springer, Berlin (1978).

[PR97] Petzold, L., J. B. Rosen, P. E. Gill, L. O. Jay and K. Park, Numerical optimal control of parabolic PDEs using DASOPT. In Biegler, Coleman, Conn, Santosa (eds.): *Large Scale Optimization with Applications, Part II*, pp. 271–299, Springer (1997).

[S98] Schulz V.H., Solving discretized optimization problems by partially reduced SQP methods. *Comput. Visual. Sci.*, **1**, 83–96, (1998).

[Sl88] Schlöder, J. P., Numerische Methoden zur Behandlung hochdimensionaler Aufgaben der Parameteridentifizierung. *Bonner Mathematische Schriften 187*, University of Bonn (1988).

[SBS98] Schulz V.H., Bock H.G., Steinbach M.C., "Exploiting invariants in the numerical solution of multipoint boundary value problems for DAEs", *SIAM J. Sci. Comp.*, **19**, 440–467, (1998).

[vS97] Schwerin, R. von, Numerical methods, algorithms, and software for higher index differential-algebraic equations in multibody system simulation. *Ph.D. thesis*, University of Heidelberg (1997).

[St95] Steinbach, M. C., Fast recursive SQP methods for large-scale optimal control problems. *Ph.D. thesis*, University of Heidelberg (1995).

[SWS96] Schwerin, R. von, M. Winckler and V. H. Schulz, Parameter Estimation in Discontinuous Descriptor Models, In D. Bestle and W. Schiehlen (eds.): *IUTAM Symposium on Optimization of Mechanical Systems 1995*, pp. 269–276, Kluwer Academic Publishers, Dordrecht, NL (1996).

[TB96] Tanartkit, P., and L. T. Biegler, A nested, simultaneous approach for dynamic optimization problems – I. *Comput. Chem. Engng* **20**, 735–741 (1996).

[W96] Winckler, M., R. von Schwerin, Simulating Vehicle Systems with Discontinuous Effects using MBSSIM, in *Proc. of ICIAM 95, vol. 1 of Special Issue of ZAMM* 587-588, Akademie-Verlag, Berlin, (1996).

IWR, Universität Heidelberg
Im Neuenheimer Feld 368
D-69120 Heidelberg
E-mail address: scicom@iwr.uni-heidelberg.de

Progress in Systems and Control Theory, Vol. 26

Modeling and Identification for Nonlinear Model Predictive Control: Requirements, Current Status and Future Research Needs

Jay H. Lee

Abstract. In this paper, we present a critical look into modeling / identification strategies for nonlinear model predictive control applications. General modeling requirements for nonlinear model predictive control applications are discussed. Then, strengths and weaknesses of fundamental modeling as well as various empirical modeling methods are examined in light of these requirements. A promising new paradigm of nonlinear subspace identification is put forth. We conclude the paper with a discussion of future research needs.

1. Introduction

Nonlinear model predictive control (NLMPC), despite the wide publicity and the intensive research efforts it attracted over the past few decades, is still being perceived as an academic concept rather than a practicable control strategy. Rarely do we hear reports about successful uses of NLMPC in solving problems of any practical significance. In stark contrast, linear model predictive control has long established its reputation as a powerful and broadly-applicable tool among the practicing engineers, compiling an impressive track record along the way and leaving a permanent mark on the industrial control practice [16, 29, 36]. Its industrial impact has been nothing short of revolutionary, restructuring the way engineers view and practice control. In attempting to explain the apparent disparity, we can bring to the table a garden variety of issues that are at play. Asked to choose the most important, however, most of us will, without hesitation, point the arrow to one common factor: *the inability to construct a nonlinear model on a reliable and consistent basis.* This brings into focus what is probably the most urgent and pressing question in the realm of nonlinear model predictive control research today:

> How do we best use available information to build models that are most useful for designing and implementing a nonlinear model predictive controller?

This article is about bringing a perspective to the above question, examining how close we are to providing a useful answer to it, and where and how future research efforts may best be spent in quest for such an answer. We begin our inquiry at the very end of the road map we wish to construct: What attributes do we require of a model in NLMPC applications? This establishes a set of criteria upon which the effectiveness of various modeling strategies available at present can be judged; without a well-defined end objective, any attempt to discuss the workings and failings of a particular modeling strategy would be futile. Understanding the end objective has an added effect of bringing forth future research needs and promising directions: What are the missing components that will enable us to complete the road map to the promised land?

Our scrutiny into the current state of the art begins with a critical look at fundamental modeling, examining what exactly is required to build and use fundamental models in the context of NLMPC and how realistic these requirements are. This will point us to a number of open research issues as well as the need for an alternative (or complementary) framework that enables us to take advantage of information hidden in plant data, an often abundant yet grossly under-utilized resource for modeling. The question of whether and to what extent science is able to present us with such a framework at current time is investigated next. Common features among the popular nonlinear empirical modeling strategies are identified and their strengths and weaknesses are examined, again all from the standpoint of how they fit the criteria established earlier. In addition to the well-studied strategies, we put forth a promising new paradigm inspired by linear subspace identification. We end the paper by contemplating a possibility of marrying the two seemingly incompatibles, fundamental modeling and empirical modeling. After all, one is likely to have access to both types of information, *a priori* and data. An ideal vision then is that of an engineer intertwining the very different ways of modeling to obtain a best possible model for given information.

The intended goal of this paper is to bring a fresh perspective to the area and invigorate / diversify the discussion. Given this goal, the style that I have chosen for the paper is one of an informal assessment rather than of a rigorous scientific deliberation or a comprehensive review. Hence, an emphasis is given to the future over the past, and to intrinsic merits / drawbacks over quantitative details. This also means that no attempt was made to survey all the available results in the area or to compile a complete list references and give proper credits to the contributors. For these, there already are several excellent survey / review articles and books [4, 30, 31, 45]. Finally, the paper contains, though necessarily, a number of speculative and subjective statements that not everyone may agree. I am also sure that many of the points brought up in this paper have already been published or deliberated by the researchers in this area. An apology is due for not being able to credit all of them.

2. General requirements in modeling for NLMPC

Before rushing into the discussion of which modeling techniques are effective and which are not, it would serve us well to invest some time into examining what we require a model to do in a typical NLMPC application. Often is the case that a discussion on comparative merits of various modeling techniques winds up in confusion due to a lack of clearly defined criteria.

The mere term "predictive control" makes clear what the model is supposed to do: Models here are to be used for *prediction*, that is, to predict future system behavior. However, what is not always made sufficiently clear is that the prediction is to be done *in the face of both known and unknown changes*. Known changes here refer to those in the manipulated variables and measured disturbances; unknowns are unmeasured disturbances and system noise. To keep track of the unknown effects, we are most likely to use an observer of some sort and update the system states before we proceed with a predictive control calculation. For the design of an observer, it is best that statistical information about these effects be provided in the form of a stochastic model.

I bring up this point because identification of unknown input effects is seldom emphasized in the nonlinear system identification literature [31]. This can probably be traced to linear system identification where the usual practice is to model the effects of known inputs rigorously (for example, through several plant tests) but then add to this model some disturbance model that is arbitrarily chosen or derived based on some heuristics. However, extending this practice to nonlinear system modeling / identification would fundamentally be wrong as we cannot separate the two effects for nonlinear systems in general; these effects are inherently coupled. In addition, we must require that any *general* modeling technique be capable of yielding models leading to cross-channel measurement updates if needed. Such a capability is important in many practical applications (e.g., those requiring inferential control), as it allows us to predict the behavior of those variables that are not on-line measurable, measurable with significant delays, or whose measurements are occasionally missing, based on more readily available measurements.

Another important, yet often overlooked point is that models are to be used *in a closed-loop environment*. Because systems' input output patterns can change dramatically after closing a loop, it is entirely possible that a model that provided wonderful predictions in open loop lose that capability once a controller is designed and loop is closed. While it may be possible to establish patterns of disturbances and system noises before a closed-loop implementation, either from prior knowledge or from available open-loop plant data, it is generally very difficult to do the same for manipulated inputs whose patterns will depend on, among many things, the controller. Hence, it is desirable that model predictions for manipulated input effects be accurate in a global sense, that is, accurate for all the possible input patterns. On the other hand, closing a loop should not alter the disturbance behavior. Hence, a model that provides accurate predictions of disturbance effects in open loop should equally be effective in a closed-loop environment. Of course, the validity of this statement is lost to some degree, when one considers the effect of model errors.

In summary:

> In nonlinear model predictive control, a model should enable us to predict accurately effects of both known and unknown changes on the system (output) behavior, possibly in a closed-loop setting.

3. Fundamental modeling

Let us first examine the advantages and disadvantages associated with building a nonlinear model predictive controller by fundamental principles. A major point of attraction is that a model obtained on the basis of fundamental principles would be globally valid and therefore well-suited to the tasks of optimization and control, which can require extrapolation beyond the range of data used to fit the model. On the other hand, a main drawback is the difficulty associated with obtaining a fundamental model. Another obstacle is that rigorous modeling often yields a model of very high order, complicating estimator / controller designs and associated on-line computation.

3.1. Difficulties associated with modeling

The difficulty associated with deriving necessary system equations for NLMPC through fundamental modeling is larger than often perceived. Following are the typical tasks involved in building a fundamental model that meets the requirements we laid out previously.

- *Gather Necessary Information To Perform Unsteady-State Mass / Energy Balances:* In order to obtain the governing differential equations, one has to define the dynamic states and gather the necessary kinetic, thermodynamic and transport information to perform the respective mass and energy balances. Given the usual lack of these information, this alone can be a daunting task.

  The governing equations will be of the form,

$$\frac{dx}{dt} = f(x, u, p) \tag{1}$$

  In the above, u represents the known inputs (manipulated inputs and measured disturbances) and p the unknown inputs (disturbance variables and varying parameters).

- *Relate Output Variables to State Variables:* For estimation and control, it is necessary to derive output equations in the form of

$$y = g(x, p) + \nu \tag{2}$$

  or, more descriptively,

$$\begin{aligned} y_c &= g_c(x, p) + \nu_c \\ y_m &= g_m(x, p) + \nu_m \end{aligned} \tag{3}$$

where y_c are variables to be controlled and y_m are those that are measured. ν_c and ν_m represent measurement noises.

Most output variables can be related to state variables in a transparent manner. On the other hand, there can be occasions when it is not obvious how to relate some of the output variables, e.g., those defining product qualities, to state variables. One may even find that a relation between model states and these variables is not well-defined because the model states had been chosen "parsimoniously" in an effort to keep the system order reasonable [42].

- *Estimate Parameters:* Values for many important model parameters may not be available *a priori* and may have to be estimated using input / output data. The level of difficulty for this task is primarily determined by the number of equations, the number of parameters, and the amount of available data. In most cases, a nonconvex optimization is needed [41].
- *Assign Statistical Models to Disturbance Variables / Time-Varying Parameters and Initial States:* Most popular estimation techniques [21, 40] require information about temporal and spatial patterns (i.e., auto- and cross-correlation) of the unknown input vector p. These information are best given in the form of a stochastic system model. In addition, for batch processes or other transient processes, the covariance matrix for initial states needs to be specified. The current practice is to choose them in an *ad hoc* manner, with popular choices being independent random steps for p, white noises with covariance matrices of a scalar-times-identity for state and measurement noises, and a "large" initial covariance matrix. This arbitrariness means lower efficiency of estimation [12]. In some case, it may be difficult even to defin p as disturbance sources and time-varying parameters may not be completely known.

  There are estimation techniques developed entirely in a deterministic setting and do not require statistical information [23, 46]. For these techniques, the main goal is to achieve the property of asymptotic convergence, that is, convergence of estimates to true values. The absence of statistical consideration means that these techniques do not *directly* exploit spatial and temporal correlation that exist, and consequently, the transient performance could suffer. This is especially true when the state dimension is significantly higher than that for the measured output.

3.2. Difficulties associated with model complexity

Many chemical processes are distributed parameter systems or are composed of a large number of subsystems (e.g., trays in a distillation column). Rigorous application of fundamental principles to modeling would yield hundreds or possibly thousands of equations, some of which could be algebraic equations. Since most estimation and NLMPC techniques require on-line solutions to nonlinear programs (NLP) [40, 3, 37], or at least nonlinear integration, Jacobian matrix calculation, *etc.* [21, 17, 25], it is of practical importance to keep the model order low.

A major stumbling block in obtaining a low-order model through fundamental modeling is the apparent *lack of a suitable model order reduction technique* applicable to *general* nonlinear systems. Most techniques proposed thus far are based on linear projections, obtained either by applying a linear model reduction technique to the linearized system or by applying the Principal Component Analysis (PCA) and other related methods to simulation data [1]. In addition to these, there are techniques that exploit certain qualitative attributes of dynamic system behavior, that are inherent to specific process types [28, 2].

In terms of building a model order reduction technique that is specifically geared toward control applications, no consensus has been reached on what quantitative objective should be used, as the issue of how errors from a model reduction ultimately affect closed-loop performance is not well understood at this point.

4. Empirical modeling

When the information necessary to build a fundamental model is lacking or renders a model that is too complex for an on-line use, empirical modeling is a viable alternative. However, one must be aware of the limitations and potential dangers that come with using an empirical model for NLMPC. Hence, we begin the section with a discussion on this particular point.

4.1. What can we expect from an empirical model?

Empirical modeling is a process of transforming available input output data into an input output relation that can be used to predict future trends, particularly those of the outputs, *when patterns same or similar to those contained in the model building data set emerge.* A caveat is that empirical modeling is inductive by nature and therefore nothing conclusive can be said about the extrapolation capability of any empirical model. On the other hand, optimization and control generally require accurate extrapolations from a model. Hence, one must exercise a caution in using an empirical model for predictive control. One option is to restrict the search space so that a solution be found within the space covered by the model building data, but this space is often difficult to define and can be overly restrictive in the context of nonlinear dynamical systems.

However, an effective empirical modeling method can still be a very useful tool in designing an effective NLMPC strategy. Below are some possible scenarios in which empirical models may be useful:

- Use an empirical model only on a passive basis, that is, merely to predict future output trends in an open-loop environment. The predictions can be used to make manual control adjustments.
- Complement a fundamental model with an empirical model. As mentioned, it may be difficult to model the disturbance and noise effects completely at a fundamental level. In this case, an effective strategy may be to model

the residual effects separately with an empirical model. The model can subsequently be fused with the fundamental model.

- Use an empirical model to perform optimal control computation. The robustness problem may be handled in two different ways. First, the search can be constrained to the range of available data [9]. Second, the model, both its parameters and structure, can continually be updated on the basis of incoming data. The two are not mutually exclusive and can be integrated in a synergistic way. For example, one may constrain the control moves to lie within the data space, but allow some restrictive probing (sending of experimental control moves) in order to generate fresh information that can be used to reduce the performance-bottlenecking portion of uncertainty.

4.2. Popular empirical model structures

For a comprehensive look into various empirical model structures, there are several excellent surveys [4, 31, 45] and books [34, 47] that provide all the details as well as extensive lists of references. Here, we resist the temptation of duplicating them. Instead, we attempt to identify some common features among the popular model structures and seek a unification.

Most nonlinear empirical models studied in the literature fit into the following form:

$$y(k) = F(\phi(k), \theta) + \varepsilon(k) \tag{4}$$

ϕ and y are all real vectors of dimensions n_ϕ and n_y respectively. F is a function that maps ϕ to the output vector y. The transformation is usually estimated in a parameterized form (in terms of the vector denoted by θ). $\varepsilon(k)$ is the residual error sequence, and is often modeled as a random sequence with certain statistical properties (e.g., an independent, identically distributed sequence).

Before a parameter estimation method can be applied, the user must first make the following choices:

- *Input Vector:* One has to decide which variables to include in ϕ as well as how many. This is critical as different choices lead to models with fundamentally different characteristics (see [31] for a detailed discussion). For instance, choosing

$$\begin{aligned}\phi(k) = [u^T(k-1), u^T(k-2), \quad & \ldots, u^T(k-m_u), \\ \varepsilon^T(k-1), \quad & \ldots, \varepsilon^T(k-m_\varepsilon)]^T\end{aligned} \tag{5}$$

leads to a nonlinear counterpart to the linear finite impulse response model, referred to as the Nonlinear Moving Average model with eXogenous input (NMAX) in the literature. For the convenience of parameter estimation, the terms, $\varepsilon(k-1), \ldots, \varepsilon(k-m_\varepsilon)$, are often excluded from the above [31, 32]. However, with these terms taken out, it is not possible to express a coupling between input effects and unmeasured disturbance effects. In addition, NMAXs cannot produce multiple steady states or subharmonic generation behavior [31]. On the other hand, choosing

$$\phi(k) = [y^T(k-1), \ldots, y^T(k-m_y), \tag{6}$$
$$u^T(k-1), u^T(k-2), \ldots, u^T(k-m_u)]^T \tag{7}$$

gives a Nonlinear Auto Regressive model with eXogenous input (NARX), which can exhibit both multiple steady state and subharmonic generation phenomena. In addition, despite the absence of residual error terms in the input vector, the presence of autoregressive terms makes it possible to introduce a nonlinear coupling between known input effects and unknown input effects. For this, the NARX structure is fundamentally less restrictive than the NMAX structure. For the most generality, one could choose

$$\phi(k) = [y^T(k-1), \ldots, y^T(k-m_y), \\ u^T(k-1), u^T(k-2), \ldots, u^T(k-m_u), \varepsilon^T(k-1), \ldots, \varepsilon^T(k-m_\varepsilon)]^T \tag{8}$$

which results in a Nonlinear AutoRegressive Moving Average model with eXogenous input (NARMAX).

Note that these definitions are consistent with those found in the standard linear system identification literature [26] and are different from those found in some notable nonlinear system identification papers (For example, the NARX structure as defined above is often referred to as the NARMAX model in the literature [31, 19].

Once the model type is decided, the number of lagged inputs / outputs $(m_y, m_u, m_\varepsilon)$ should be determined. Available methods for determining these parameters include: the False Nearest Neighborhood (FNN) method [38], Akaike Information Criteria (AIC) and its related concepts [31, 34, 47], and cross-validation [31, 47]. The FNN method has the advantage over the others that the order selection can be carried out independently of finding the nonlinear operator F.

- *Parameterization of F:* F may be parameterized as a series sum of *a priori* chosen basis functions (e.g., polynomials, radial basis functions, wavelets):

$$y(k) = \sum_{i=1}^{M} b_i f_i(\phi(k)) \tag{9}$$

For example, expanding NMAX with polynomial bases yields Volterra models that have been closely studied [7, 32, 27]. Basis functions can be classified into global functions (e.g., polynomials) and local functions (e.g., radial bases and wavelets).

One can also opt for a nonlinear parameterization based on adaptive basis functions

$$y(k) = \sum_{i=1}^{M} f_i(\phi(k), \theta_i) \tag{10}$$

or a neural network.

In some cases, F is expressed as a weighted sum of several models [22]:

$$y(k) = \sum_{j=1}^{r} w_j(\psi(k))F_j\left(\phi_j(k), \theta_j\right) + \varepsilon(k) \tag{11}$$

$w_j(\psi(k))$ is the weighting function for the j_{th} model given in terms of certain relevant variables (called "scheduling variables") which can be same as those in the regressor.

In addition to the type, the number of basis functions (or the number of hidden nodes in the case of a neural network) also has to be decided.

The parameter values are typically decided based on some optimization criterion (e.g., prediction error minimization). If the basis functions were fixed kinds yielding a linear-in-parameter structure and ϕ did not contain any unknown term (e.g., residuals), one would get a linear regression problem that can be solved quite efficiently. However, use of adaptable basis functions as well as neural networks would require a nonconvex optimization. The same is true when ϕ contains unknown residual terms (as in the NARMAX model), although the best bet is to obtain a suboptimal solution through pseudo-linear regression.

4.3. Methods for reducing the dimensionality

Many general empirical model structures, especially those employing a linear parameterization lead to a large nubmer of parameters to be estimated. An excessive number of unknown coefficients lead to an ill-conditioned estimation problem causing numerical difficulties and high sensitivity to noise and other identification errors. The following are some of the suggested remedies for fighting the dimensionality problem:

- *Lowering of the Regressor Dimension Through Input Projection:* In some applications, F is identifed as two maps, P and $\hat{F}$, for convenience:

$$F(\phi(k), \theta) \triangleq \hat{F}(\underbrace{P(\phi, \beta)}_{\hat{\phi}(k)}, \hat{\theta}) \tag{12}$$

  P is a projection operator that maps the regressor ϕ to a reduced regressor $\hat{\phi}$ of substantially lower order. In most cases, this transformation is not chosen *a priori* but is determined on the basis of certain statistics computed using the data. Proposed methods are mostly extensions of the Principal Component Analysis and Partial Least Squares [20, 35] that have been found to be effective in dealing with the same problem in linear modeling [39]. For instance, in the PLS Neural Network proposed by Qin and McAvoy, one dimensional *linear* projections are decided one at a time based on the residuals resulting after fitting a neural network to the projected data. Methods to obtain nonlinear projections have also been proposed, but the computational complexities associated with these extensions have made it very difficult to apply them to realistic problems [24, 14].

- *Constraining Structures:* Another way to fight the dimensionality of the estimation problem is to impose some restrictions on the general model structure so that fewer parameters need to be estimated. For instance, for a linear-in-parameter model structures, one gets a "nonsaturated" model by setting some of the coefficients to zero *a priori*. An example of such a model is the following restricted form of the NMAX structure [32]:

$$y(k) = \sum_{i=1}^{m_u} F_i(u(k-i)) + \varepsilon(k) \tag{13}$$

The Hammerstein model belongs to the above class. A similar simplification of the NARX model is

$$y(k) = \sum_{i=1}^{m_y} F_i(y(k-i)) + \sum_{i=1}^{m_u} G_i(u(k-i)) + \varepsilon(k) \tag{14}$$

The above is referred to as a Nonlinear Additive Auto Regressive model with eXogenous input (NAARX) in Pearson and Ogunnaike [31].

At present time, it is not well understood what is lost by imposing limitations of above sorts to the general structures. Because of this, it is often difficult in practice to choose a restricted structure that is meaningful for a given process.

- *Stepwise Regression:* This method is applicable to linear-in-parameter structures. In the stepwise regression [5], one starts with a simple "nonsaturated" structure. The model structure gradually is made to evolve to a proper one by the process of evaluating the excluded bases in terms of their importance to the goodness of fit and adding the most important ones while dropping those that are no longer important.
- *Biased Regression:* For structures with a linear parameterization (and ϕ containing no unknown terms), one gets a linear regression problem. In this case, various biased regression methods like Principal Component Regression and Partial Least Squares can be used to obtain estimates that are more robust with respect to (near-)colinearities in the data.

4.4. Assessment: Limitations and shortcomings of current methods

Despite the variety of structures and remedies for the dimensionality problem, fitting a nonlinear empirical model to given data in practice is considered to be a very difficult task. It is also a problem that has not been resolved satisfactorily from a theoretical standpoint. The following are some of the potential shortcomings of the currently available methods:

- Identifying the proper structure can be a laborious task requiring iteration between choosing a particular structure and then finding the best parameter values. This is particularly true in the case of multivariable systems.
- One is likely to end up with a model of very high order despite the available methods for deciding the system order. Again, this is particularly true in

the case of multivariable and / or *stochastic* systems. For example, we know that, in the linear case, if we restrict the regressor to contain lagged outputs and inputs only (such as in the ARX structure), we need in general an infinite order regressor in order to model a state-space stochastic system of finite order. For the same reason, the parameter estimation problem is likely to be very ill-conditioned, requiring special regression methods (*e.g.*, the PLS).

- Depending on the structure, parameter estimation through prediction error minimization leads to a nonconvex optimization. For some model structures, it is difficult even to write the model in a predictor form, which is needed to formulate prediction error minimization. For instance, this would be the case if the input vector ϕ includes terms involving ε.
- Since MPC needs good multi-step prediction, it makes more sense to optimize the model parameters based on multi-step prediction errors rather than single-step prediction errors. This difference is not of any significance in the linear system identification where one can achieve the effect of a multi-step prediction error minimization by performing a single-step error minimization with appropriate prefiltering [26]. Doing the same should be more difficult for nonlinear systems, however.

These points motivate an alternative method which is discussed next.

4.5. Subspace identification

While the idea of directly identifying a nonlinear state-space model (which bears a strong structural resemblance to fundamental models) strikes us as being quite attractive, it is seldom practiced due to the difficulties that stem from the unavailability of data for system states. Even in the linear case, the traditional approach requires a canonical parameterization followed by a nonlinear parameter estimation [26]. In the nonlinear case, it has been proposed to use internally recurrent neural networks, which unfortunately are known to be very difficult to train [50].

In the linear system identification side, an alternative approach called subspace identification has been attracting much attention lately [49]. In this approach, one first constructs state data from input output data. Construction of state data is based on the idea that, from an input output system's perspective, states can be defined as any (but preferably minimal) memory that stores information about past inputs and outputs needed to predict future output behavior. Once state data become available, it is relatively straightforward to fit a state-space model. In this section, we explore the possibility of extending this idea to nonlinear system identification. In order to gain necessary insights, we first review the main ideas in linear subspace identification.

4.5.1. SUBSPACE IDENTIFICATION FOR LINEAR SYSTEMS. Suppose underlying is linear state-space system

$$\begin{aligned} x(k+1) &= Ax(k) + Bu(k) + w(k) \\ y(k) &= Cx(k) + \nu(k) \end{aligned} \tag{15}$$

where $w(k)$ and $\nu(k)$ are white noise sequences. The usual assumptions of controllability and observability as well as stationarity are in effect here.

The following Kalman filter equation is equivalent to the above system in an input-output sense and is called innovation representation of (15):

$$\begin{aligned} \hat{x}_{\infty}(k+1) &= A\hat{x}_{\infty}(k) + Bu(k) + K_{\infty}\varepsilon_{\infty}(k) \\ y(k) &= C\hat{x}_{\infty}(k) + \varepsilon_{\infty}(k) \end{aligned} \tag{16}$$

In the above, $\hat{x}_{\infty}$'s represent consecutive state estimates from a *steady-state* Kalman filter, K_{∞} the Kalman gain, and $\varepsilon_{\infty}(k)$ the corresponding innovation sequence, which is white. The system equation to be identified is

$$\begin{aligned} \hat{x}_{\bar{n}+1}(k+1) &= A\hat{x}_{\bar{n}}(k) + Bu(k) + K_{\bar{n}}\varepsilon_{\bar{n}}(k) \\ y(k) &= C\hat{x}_{\bar{n}}(k) + \varepsilon_{\bar{n}}(k) \end{aligned} \tag{17}$$

where $\hat{x}_{\bar{n}+1}(k+1)$ and $\hat{x}_{\bar{n}}(k)$ represent two consecutive estimates from a *non-steady-state* Kalman filter that was started at time $k-\bar{n}$ with the initial estimate of $\hat{x}_0(k-\bar{n}) = 0$ and the covariance matrix chosen as the open-loop steady-state covariance. The subscripts $[\cdot]_{\bar{n}+1}$ and $[\cdot]_{\bar{n}}$ are added to denote the number of time steps the Kalman filter had been running. $\bar{n}$ should be chosen larger than state dimension n_x (*i.e.*, choose one larger than an upper-bound of n_x if it is not known *a priori*).

(17) serves as an approximation to (16). Note that system matrices for the deterministic parts of (17) (i.e., A, B, C) are same as those appearing in (16). The same is not true for the stochastic parts, but with a reasonably large $\bar{n}$, $K_{\bar{n}}, \mathrm{Cov}\{\varepsilon_{\bar{n}}(k)\})$ should be close to $K_{\infty}, \mathrm{Cov}\{\varepsilon_{\infty}(k)\})$. The goal is to identify system matrices, $(A, B, C, K_{\bar{n}}, \mathrm{Cov}\{\varepsilon_{\bar{n}}(k)\})$, within a state coordinate transformation.

As a first step toward this goal, we note that

$$\begin{bmatrix} y(k) \\ y(k+1) \\ \vdots \\ \vdots \\ y(k+\bar{n}-1) \end{bmatrix} = \begin{bmatrix} C \\ CA \\ CA^2 \\ \vdots \\ CA^{\bar{n}-1} \end{bmatrix} \hat{x}_{\bar{n}}(k) + \begin{bmatrix} 0 & 0 & \cdots & 0 \\ CB & 0 & \ddots & 0 \\ CAB & CB & \ddots & \vdots \\ \vdots & \ddots & \ddots & \vdots \\ CA^{\bar{n}-2}B & \cdots & \cdots & CB \end{bmatrix}$$

$$\times \begin{bmatrix} u(k) \\ u(k+1) \\ \vdots \\ u(k+\bar{n}-2) \end{bmatrix} + \begin{bmatrix} e_{\bar{n}}(k|k-1) \\ e_{\bar{n}}(k+1|k-1) \\ \vdots \\ \vdots \\ e_{\bar{n}}(k+\bar{n}-1|k-1) \end{bmatrix} \tag{18}$$

$e_{\bar{n}}(k+i|k-1)$ represents prediction error. For the convenience of notation, let us denote the above as

$$\mathcal{Y}^{0+}(k) = \Gamma_o\hat{x}_{\bar{n}}(k) + L_3\mathcal{U}^{0+}(k) + \mathcal{E}_{\bar{n}}^{0+}(k|k-1) \tag{19}$$

It can be easily shown by propagating the Kalman filter equation that $\hat{x}_{\bar{n}}(k)$ is simply a linear function of $y(k-1), \ldots, y(k-\bar{n})$ and $u(k-1), \ldots, u(k-\bar{n})$. Hence, let us express

$$\hat{x}_{\bar{n}}(k) = \begin{bmatrix} H_1 & H_2 \end{bmatrix} \begin{bmatrix} \mathcal{Y}^-(k) \\ \mathcal{U}^-(k) \end{bmatrix} \tag{20}$$

where H_1 and H_2 are functions of system matrices as well as covariance matrices (including the initial covariance matrix chosen for the nonstationary Kalman filter). Combining (19) and (20), we obtain

$$\mathcal{Y}^{0+}(k) = \underbrace{\Gamma_o \begin{bmatrix} H_1 & H_2 \end{bmatrix}}_{\begin{bmatrix} L_1 & L_2 \end{bmatrix}} \begin{bmatrix} \mathcal{Y}^-(k) \\ \mathcal{U}^-(k) \end{bmatrix} + L_3 \mathcal{U}^{0+}(k) + \mathcal{E}_{\bar{n}}^{0+}(k|k-1) \tag{21}$$

$\mathcal{E}_{\bar{n}}^{0+}(k|k-1)$ is orthogonal – in a statistical sense – to $\mathcal{Y}^-(k)$, $\mathcal{U}^-(k)$ and $\mathcal{U}^{0+}(k)$, given that the input sequence $u(k)$ is white.[1] Therefore, the least squares regression gives an unbiased, consistent estimate for $[L_1 \ \ L_2]$.

It can be easily verified that, since $[L_1 \ L_2] = \Gamma_o[H_1 \ H_2]$, if the system is controllable / observable, $[L_1 \ L_2]$ has rank of n_x. Hence, one can obtain the system order simply by examining its rank. This map also defines the state space, but we are free to choose any coordinate axes within the state space (recall that system matrices can only be identified within a state coordinate transformation). Suppose that the SVD of the matrix is given as

$$\begin{bmatrix} L_1 & L_2 \end{bmatrix} = \begin{bmatrix} U_1 & U_2 \end{bmatrix} \begin{bmatrix} \Sigma & 0 \\ 0 & 0 \end{bmatrix} \begin{bmatrix} V_1^T \\ V_2^T \end{bmatrix} \tag{22}$$

Then the extended observability matrix Γ_o can be chosen as $U_1 P$ where P is any nonsingular (square) matrix. To obtain a balanced system, we may choose $\Gamma_o = U_1 \Sigma^{1/2}$. This fixes the state coordinate axes and

$$\begin{aligned} \hat{x}_{\bar{n}}(k) &= \Gamma_o^\dagger \begin{bmatrix} L_1 & L_2 \end{bmatrix} \begin{bmatrix} \mathcal{Y}^-(k) \\ \mathcal{U}^-(k) \end{bmatrix} && (23) \\ &= \Sigma^{1/2} V_1^T \begin{bmatrix} \mathcal{Y}^-(k) \\ \mathcal{U}^-(k) \end{bmatrix} && (24) \end{aligned}$$

Using the above equation, we can create data for the Kalman estimate $\hat{x}_{\bar{n}}(k)$ from input-output data.

In practice, one would be given only a finite data set and hence have $[\hat{L}_1 \ \ \hat{L}_2]$, an estimate for $[L_1 \ \ L_2]$, whose rank can be significantly higher than n_x because of various identification errors. A logical way to choose the system order and define the state space is to use the SVD of the matrix:

$$\begin{bmatrix} \hat{L}_1 & \hat{L}_2 \end{bmatrix} \begin{bmatrix} \mathcal{Y}^-(\bar{n}+1) & \cdots & \cdots & \mathcal{Y}^-(N+1-\bar{n}) \\ \mathcal{U}^-(\bar{n}+1) & \cdots & \cdots & \mathcal{U}^-(N+1-\bar{n}) \end{bmatrix} \tag{25}$$

[1]The assumption of white input can be relaxed by adopting a different initialization of the Kalman filter. See Van Overschee and De Moor [49] for details.

Based on the above SVD, Γ_o can be defined in a similar manner as before. We assumed here that input output data for N sample steps were available for identification. The system order can be determined by examining its singular values and finding a large gap between two successive values – a somewhat subjective decision. The state data matrix can be obtained by

$$\begin{aligned} \left[\ \hat{x}_{\bar{n}}(\bar{n}+1) \quad \cdots \quad \cdots \quad \hat{x}_{\bar{n}}(N+1-\bar{n})\ \right] &= \Gamma_o^{\dagger} \left[\ \hat{L}_1 \quad \hat{L}_2\ \right] \\ &\left[\begin{array}{cccc} \mathcal{Y}^-(\bar{n}+1) & \cdots & \cdots & \mathcal{Y}^-(N+1-\bar{n}) \\ \mathcal{U}^-(\bar{n}+1) & \cdots & \cdots & \mathcal{U}^-(N+1-\bar{n}) \end{array} \right] \end{aligned} \tag{26}$$

From the viewpoint of model reduction, choosing the reduced state space this way amounts to minimizing the squared sum of prediction error vector $\mathcal{E}_{\bar{n}}^{0+}(k|k-1), k = \bar{n}+1, \ldots, N+1-\bar{n}$ for given data.

Next, we create data for $\hat{x}_{\bar{n}+1}(k+1)$ in a similar manner. An important, but often overlooked point is that one cannot simply time-shift $\hat{x}_{\bar{n}}(k)$ to obtain $\hat{x}_{\bar{n}+1}(k+1)$. This way, one would have $\hat{x}_{\bar{n}}(k+1)$ that represents an estimate from a Kalman filter started at a different time (*i.e.*, at $k-\bar{n}+1$ instead of $k-\bar{n}$). Rather, one has to follow the same procedure as before with $\mathcal{Y}^-(k+1)$ and $\mathcal{U}^-(k+1)$ replaced by $\bar{\mathcal{Y}}^-(k+1)$ and $\bar{\mathcal{U}}^-(k+1)$ that include $y(k)$ through $y(k-\bar{n})$ and $u(k)$ through $u(k-\bar{n})$ respectively. The state vector can be created by using

$$\begin{aligned} \left[\ \hat{x}_{\bar{n}+1}(\bar{n}+2) \quad \cdots \quad \cdots \quad \hat{x}_{\bar{n}+1}(N+2-\bar{n})\ \right] &= \bar{\Gamma}_o^{\dagger} \left[\ \bar{L}_1 \quad \bar{L}_2\ \right] \\ &\left[\begin{array}{cccc} \bar{\mathcal{Y}}^-(\bar{n}+2) & \cdots & \cdots & \bar{\mathcal{Y}}^-(N+2-\bar{n}) \\ \bar{\mathcal{U}}^-(\bar{n}+2) & \cdots & \cdots & \bar{\mathcal{U}}^-(N+2-\bar{n}) \end{array} \right] \end{aligned} \tag{27}$$

where $\bar{\Gamma}_o$ is the same Γ_o used to create $\hat{x}_{\bar{n}}$ but with the last n_y rows eliminated and $[\bar{L}_1 \quad \bar{L}_2]$ is the matrix identified using the least squares as before.

Once data for $\hat{x}_{\bar{n}}(k)$ and $\hat{x}_{\bar{n}+1}(k+1)$ are created according to the above procedure, system matrices A, B, and C can be estimated using the linear least squares method. In addition, the covariance matrix for $\varepsilon_{\bar{n}}(k)$ and $K_{\bar{n}}$ can be estimated on the basis of the residuals from the least squares.

Advantages of the subspace identification method over the traditional method are several-fold:

- Structure identification is automatic. It avoids having to try models of different orders or the need for a special parameterization that requires difficult-to-obtain prior information like the observability indices.
- It yields a stochastic system model in the form of a state estimator without any nonlinear optimization, which is needed, for instance, to identify an ARMAX model.
- Model reduction is already built into the identification.

4.5.2. Extension to nonlinear systems. In keeping with the spirit of linear subspace identification, we can attempt to identify nonlinear systems in the form of following nonlinear state estimator:

$$\begin{array}{rcl}\hat{x}_{\infty}(k+1) & = & f(\hat{x}_{\infty}(k), u(k), \varepsilon_{\infty}(k)) \\ y(k) & = & g(\hat{x}_{\infty}(k)) + \varepsilon_{\infty}(k)\end{array} \tag{28}$$

In the above, $\hat{x}_{\infty}(k)$ represents the state estimate at time k based on infininte-length past data. The idea of identifying a nonlinear model directly in the state estimator form is very attractive since most on-line applications including MPC require a state estimator and it is not at all straightforward to design an effective nonlinear state estimator starting from a general system description.

As in the N4SID, in order to make the problem of state data construction tractable, we approximate the above steady-state estimator with the following non-steady-state version:

$$\begin{array}{rcl}\hat{x}_{\bar{n}+1}(k+1) & = & f_{\bar{n}}(\hat{x}_{\bar{n}}(k), u(k), \varepsilon_{\bar{n}}(k)) \\ y(k) & = & g(\hat{x}_{\bar{n}}(k)) + \varepsilon_{\bar{n}}(k)\end{array} \tag{29}$$

Here, $\hat{x}_{\bar{n}+1}(k+1)$ and $\hat{x}_{\bar{n}}(k)$ are state estimates resulting from running a non-steady-state estimator being started at time $k - \bar{n}$.

A multi-step output prediction equation based on $\hat{x}_{\bar{n}}(k)$ can be developed by running the model forward in time (with the future innovations set to zero) and will be of the form,

$$\mathcal{Y}^{0+}(k) = \mathcal{F}(\hat{x}_{\bar{n}}(k), \mathcal{U}^{0+}(k)) + \mathcal{E}_{\bar{n}}^{0+}(k|k-1) \tag{30}$$

where $\mathcal{E}_{\bar{n}}^{0+}(k|k-1)$ is the prediction error vector. Note that, this time, $\hat{x}_{\bar{n}}(k)$ is some *nonlinear* function of $\mathcal{Y}^{-}(k)$ and $\mathcal{U}^{-}(k)$, which we denote by

$$\hat{x}_{\bar{n}}(k) = \mathcal{P}(\mathcal{Y}^{-}(k), \mathcal{U}^{-}(k)) \tag{31}$$

Following the same logic, we can choose the state dimension and fit the maps $\mathcal{F}$ and $\mathcal{P}$ – equivalents of $[\Gamma_o L_3]$ and $[H_1 H_2]$ – so that $\mathcal{E}_{\bar{n}}^{0+}(k|k-1)$ is minimized in the mean square sense. Even though we cannot offer any rigorous theoretical basis at this time (*e.g.*, the convergence of $\mathcal{F}$ and $\mathcal{P}$ to some equivalents of original system parameters), it is certainly sensible from a practical standpoint given the fact that the main role of model state is to store the information about past inputs and outputs that are necessary for future output prediction.

Of course, a major additional complication over the linear case is that there does not appear to be a straightforward way to extract out appropriate system dimension n_x and maps $\mathcal{P}$ and $\mathcal{F}$ after creating the overall map

$$\mathcal{F}\left(\mathcal{P}(\mathcal{Y}^{-}(k), \mathcal{U}^{-}(k)), \mathcal{U}^{0+}(k)\right)$$

representing the nonlinear equivalent of oblique projection. Rather, $\mathcal{P}$ and $\mathcal{F}$ need to be optimized simultaneously. The state dimension must be chosen by increasing the output dimension of $\mathcal{P}$ one at a time and examining the incremental reduction in the prediction error, attainable after the optimization of $\mathcal{F}$ and $\mathcal{P}$.

The above seems to be a formidable task, but there are some promising directions:

- Neural networks can be used to create maps $\mathcal{P}$ and $\mathcal{F}$. This is demonstrated in Figure 1.

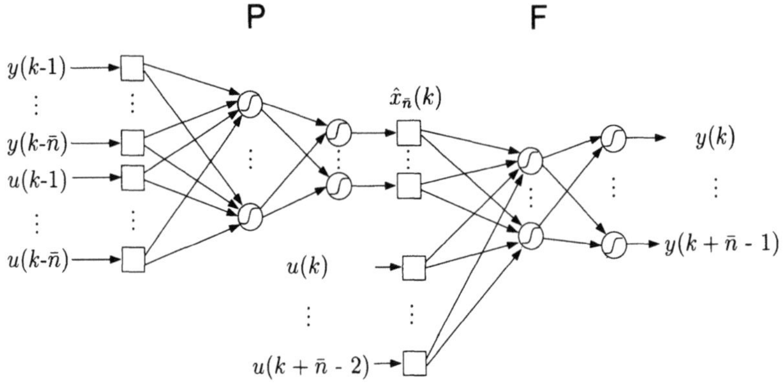

FIGURE 1. Creating states from input-output data using a neural network with 3 hidden layers

The structure bears much resemblance to the *auto-associative neural network* proposed by Kramer [24] for nonlinear PCA. The main difference is that the above network does not try to map the input vectors to themselves with a limited number of nonlinear modes (which would amount to orthogonal projection). Rather it tries to find modes that contain the most relevant information about future outputs – as in oblique projection. A drawback is that the above neural net is known to be very difficult to train due to the presence of three hidden layers.

- In order to avoid the formidable task of training the above five-layer network, we can create a set of nonlinear basis functions of past data $\mathcal{Y}^-(k), \mathcal{U}^-(k)$ *a priori* and model $\hat{x}_{\bar{n}}(k)$ as a linear combination of these nonlinear functions. The network to train is shown in Figure 2.

A potential problem is the exploding dimensionality in creating a sufficiently complete set of nonlinear basis functions for a very large dimensional vector. For this, one may want to employ a linear projection first to extract out significant linear modes before moving onto the nonlinear projection (which will help further reduce the state dimension). Of course, the linear projection should be chosen conservatively so that there occurs no significant loss of information relevant for predicting $\mathcal{Y}^{0+}(k)$.

In this case, the projection operator $\mathcal{P}$ plays a role similar to "trimming" of the basis functions that was discussed earlier in the context of input-output time-series modeling. However, there is an important difference from the trimming methods discussed in the literature [5]. Here, the projection is decided based on the objective of minimizing *multi-step* prediction errors. In the context of time-series modeling, the basis functions are usually trimmed or combined on the basis of the *single-step* prediction error. Obviously, the former is more appropriate for defining the states for a state-space model, which circumvents the need to store all the past output and input values.

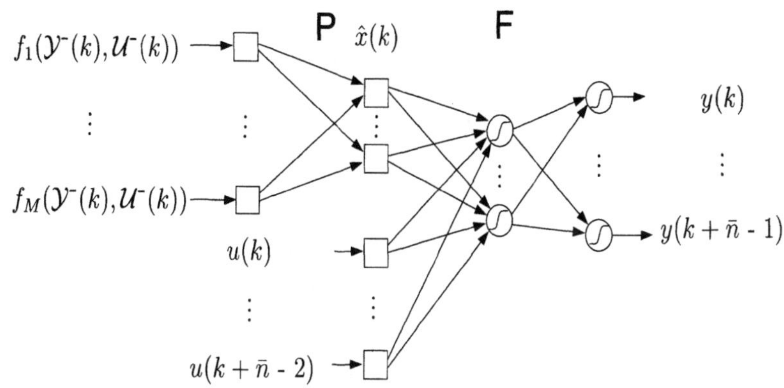

FIGURE 2. Creating states from input-output data using a conventional neural network

Once the maps are created and the optimal order determined, we can use the map $\mathcal{P}$ for the chosen order to obtain data for the state vector $\hat{x}_{\bar{n}}(k)$. We can also apply the method in a similar manner to derive the data for $\hat{x}_{\bar{n}+1}(k+1)$.

Once these data are obtained, the nonlinear observer equations can be constructed. For this, we first obtain the map g through nonlinear regression. This gives the residual sequence $\varepsilon_{\bar{n}}(k)$. We can then find the map $f_{\bar{n}}$ through another nonlinear regression. Note that, even if we use neural networks to perform the required oblique projection, this does not mandate us to build the state-space model using neural networks as well. While neural networks certainly represent good choices for the nonlinear regression, other choices are possible. For instance, fitting hinging hyperplanes would yield a piece-wise linear state-space model which has many attractive features for nonlinear model predictive control implementation [8].

4.5.3. ADVANTAGES OF THE SUBSPACE APPROACH. While the idea is still at its infancy, its attractions are clear:

- Multiple input, multiple output (MIMO) problems are not any more difficult – not in a fundamental sense. As is the case with linear systems, there is no need to choose the system order or some canonical parameterization *a priori* in order to ensure identifiability.
- The conditioning problem would be less severe here since a model with only a limited number of modes are fitted to the output vector of *multiple*

time steps. Special regression methods like the PLS or PCR are not likely to be needed unless one chooses excessively high number of modes.

- It yields the model in the form of nonlinear state estimator, which can directly be used in nonlinear model predictive control [37, 29]. This is a significant advantage because a nonlinear state estimator minimizing multi-step prediction errors is very difficult to design in general. Note that, even though $\hat{x}_{\bar{n}}$ is created as a function of finite-dimensional memory vectors $\mathcal{Y}^-$ and $\mathcal{U}^-$, because the states are updated recursively in a real-time implementation, model states could actually possess infinite-dimensional memories. In other words, model states updated according to the estimator equation would be, in general, a function not of a fixed number of lagged inputs and outputs but of all the inputs and outputs that had been registered up to that time. This is a feature that cannot be achieved with a NARX model [19], for instance.
- Multi-step prediction equations relating $\hat{x}(k)$ to $\mathcal{Y}^{0+}(k|k-1)$ are directly provided by the map $\mathcal{F}$. The multi-step prediction can also be constructed indirectly using the constructed state-space equations; however, this may not provide as good a prediction since these equations are not fitted to minimize the multi-step prediction error directly. Since any empirical nonlinear model would carry with it some errors, prediction models obtained from a direct fitting should be preferred.

As a final note, I want to point out that the subspace methods, both linear and nonlinear, can be used to identify models for batch processes as well as continuous processes. For batch processes, k would be the *batch index* instead of time index. Hence, the states are to be defined as memories about the past batches relevant for predicting the current and future batches. In addition, the input and output vectors, $u(k)$ and $y(k)$ will contain the input and output values for an entire batch rather than at a specific time instant. Since the inputs for the k_{th} batch should affect the outputs for the same batch, one would have a direct feed-through term (a nonzero D matrix) even though the causality calls for a specific structure (a lower triangular D). For details and discussions on how to handle the high dimensionality, readers are referred to [43].

5. Future research issues

I would like to close this paper by pointing out what I believe to be important open research issues in this area. First, I hope to have conveyed a view that we need different approaches in this area. I do not believe a single approach or method will be sufficient to solve a large majority of industrial problems. There is no point in arguing over relative merits of fundamental modeling vs. empirical modeling; the two are complementary rather than mutually exclusive. While we should strive for a variety of approaches to the problem, it will be helpful to base our future endeavors on an understanding of the same basic set of requirements for the modeling (even though, at the end, some of our efforts may lead to an approach satisfying only a partial set of such requirements). This way, the diversity will not be a source of distraction and confusion, but an opportunity for an engineer to create a more flexible, synergistic solution.

5.1. Issues in fundamental modeling

In the arena of fundamental modeling, one of the most pressing issue appears to be *model order reduction.* Fundamental principles behind most chemical processes are quite complex and describing them with requisite details can lead to hundreds or sometimes thousands of equations. For NLMPC, on the other hand, not only are such complexities unwarranted, but are also unnecessary. It should be possible to describe important, relevant input-output dynamics with a much smaller number of equations. In performing a model order reduction, our goal should be to obtain a low order model that results in a good prediction performance under the intended closed-loop condition. We should strive to find ultimately an order reduction method that specifically addresses this end goal. However, at this point of time, even methods based on some meaningful open-loop criteria would be of interest.

In the field of fluid dynamics, it is a popular practice to reduce the order of simulation models by generating a large set of data and performing an orthogonal projection (*e.g.*, PCA) on the data [1, 44]. This approach has not received much attention in the control field, probably because the model order reduction problem for the linear case could be solved without the aid of data. However, this may be an effective way to tackle the more difficult nonlinear model order reduction problem. If data are to be used as the main vehicle for model order reduction, the distinction between the problem of model order reduction and that of system identification gets blurred, the only major difference being data coming from simulation rather than an actual experiment. In this context, the subspace identification method seems to be quite appealing, particularly since data for states can readily be made available in simulation. In addition, one may have to iterate between model reduction and closed-loop simiulation in order to generate meaningful data.

Another interesting problem in this area is how to augment / transform a fundamental model into a form suitable for an estimator design. It was mentioned in the paper that design of an efficient estimator requires an accurate *stochastic description* of system disturbances and noises. How one determines the structure and the covariance matrices of a stochastic system based on plant data is an open question. In fact, it would be desirable to have a method that transforms fundamental model equations directly into observer equations based on plant data.

5.2. Issues in empirical modeling

In the empirical modeling side, the most fundamental and important problem seems to be that of *data generation.* It is fundamental in the sense that nature and accuracy of any empirical model is intrinsically linked to the data used for model-building. It is important in view of the fact that, for most nonlinear systems, it is not known how to excite the system to obtain data that are "persistently exciting" [26], even if one were allowed to conduct the experiment in any wanted way for an infinite time period. The only general results the author is aware of are those specific to the second-order Volterra systems [33, 32, 31].

While the persistency of excitation is an issue of theoretical importance, it is a practical challenge to find ways to make identification experiments *efficient*, that is, to generate data that are rich in the information most relevant to the intended application (*i.e.*, data that would distribute model error in an optimal manner for control). It is clear that, when the intended application is closed-loop control, good identification experiments would attempt to create an actual closed-loop condition under which the model will be used. However, this condition depends on the controller, which in turn depends on the model yet to be found. This is the main dilemma in designing control-oriented system identification experiments, and the only solution appears to be an *iterative design*, where designs are continuously improved based on collected information. In this regard, the framework discussed in [10] or that in [13] may be of some interest. In [10], for instance, the model and the uncertainty descriptions are derived at the end of each segment of an experiment and this information along with control requirements are then used to design test signals for the remaining segments so that the model resulting after the experiment yields the best closed-loop performance. This philosophy can be applied to both linear and nonlinear system identification, but unlike in the linear case, it is expected to be much more difficult to derive computational methods that can be implemented.

In the area of model fitting, *reduction of dimensionality* is an issue that deserves further attention. Existing methods, such as the AIC and their variants, do not appear to be adequate, especially in the multivariable systems' context. In addition, the issue of identifying the stochastic part of a system has largely been ignored. It would be the best if the model can be identified in a form conducive to straightforward estimator design. For the both issues, the concept of *nonlinear subspace identification* introduced in this paper holds much promise. More research is needed to explore its merits and to develop it into a practicable method.

Because any empirical model is likely to be a gross approximation of underlying dynamics, the issue of *model uncertainty* cannot be ignored or treated heuristically. In general, both structural and parametric uncertainties should be significant. Uncertainty of a model is mainly determined by amount and distribution of data; hence, the issue is inherently linked with that of experiment design. On the flip side, through the controller design, we do exercise some influence over how uncertainty actually plays out in the closed loop. For instance, we can design the controller or restrict the control computation such that uncertain parts of system dynamics are not unduly excited to cause undesirable oscillations or an instability. This problem falls under the umbrella of *robust control*. In order to implement the idea, however, we need a method to capture uncertainty on a quantitative basis [11, 18].

Another way to reduce the amount of uncertainty is to continue to improve the model on-line on the basis of incoming signals. This leads to the concept of *adaptive control*, which, I believe, is better motivated in the context of nonlinear model based control. It is best that on-line model estimation be combined with intelligently designed on-line identification experiments. This is logical in view of the fact that a smart decision maker tends to experiment with new strategies in order to improve his / her decision-making skills but without being overly aggressive. Closely related to this is the concept of dual control that was brought forth by Feldbaum in the 60's [15] but still remains as an unresolved issue, even for linear systems. While a rigorous theoretical treatment may not be possible, I believe we can still workout interesting, creative strategies that will also be effective in practice. However, many questions need to be answered. For example, when and how should we revive the parameter estimation? What about adding more basis functions? How much and what kind of test signals should be sent at each time step? In light of these questions, it seems logical that we look at the problems of experiment design, model fitting and performance monitoring / diagnosis as related problems. It also seems to be a logical place to combine system identification / control theories with artificial intelligence tools.

5.3. Issues in integrating fundamental modeling and empirical modeling

My long-term vision of modeling is that of engineers using the two approaches in a complementary manner rather than relying exclusively on one or the other. This makes sense because in practice one would have both fundamental knowledge and plant data, which, by themselves without the other, may not be sufficient for good modeling. In such cases, the success of modeling effort hinges upon the ability to combine the two types of information in a creative, synergistic way.

The integration can occur at several different levels.

- A fundamental model may be used strictly for simulation in order to gain insights into the experiment design and model structure selection. Since simulators impose virtually no restriction in terms of the amount and the type of data that can be collected, it is ideal for designing suitable identification experiments that will draw necessary information without jeopardizing the stability of an on-going operation. In addition, it will offer some indications as to how much data need to collected and what kind of closed-loop performance one can expect ultimately. Even in the case that one has a full confidence in a fundamental model including its parameter values, empirical modeling may offer a way to obtain a simpler model that is more amenable to on-line use.
- A typical situation would be that one can write down the basic equations representing mass and energy balances, but misses some critical information like kinetics and thermodynamic relationships. In such a case, one may embed black boxes into a fundamental model. The key issue is how to train these black boxes or estimate relevant parameters based on input, output data from the plant. The resulting model may still have only

a limited extrapolation capability and lead to problems in a closed-loop setting; however, the problem is likely to be less severe.

- One can also bring into empirical modeling some constraints that arise from fundamental process understandings. This approach is referred to as *grey box identification* in the system identification literature [48, 6]. While it may not be possible to find constraints that are universally applicable, it would certainly be worthwhile to examine what types of constraints would be most relevant for specific classes of processes. This type of understandings, in fact, will be a key to identifying a reliable model with a limited amount of data and in the face of adversities like nonlinearity, disturbances, *etc.*
- In certain instances, low-order model structures can be derived on the basis of an understanding about certain intrinsic characteristics of system dynamics. The traveling wave model applicable to high-purity distillation columns and pack-bed reactors is a good example [28, 2]. More research along this line should yield fruitful results.
- Even when a complete set of governing equations can be written out and all the parameters can be estimated accurately, design of a predictive controller may be hampered by the fact that it is not possible to identify all the sources of disturbances and their statistical characteristics. In this case, residuals (prediction errors from the fundamental model) may be modeled with an empirical model. This is particularly attractive in view of the fact that feedback will only affect the manipulated inputs whose effects are predicted by the fundamental model, the part of the combined model that should have a good extrapolation capability. Residuals representing the effect of unmeasured disturbances and system noise can be modeled effectively using historical data. In other words, closing a new loop is likely to have little (or less significant) effect on their distribution. One would need stochastic modeling for this; again, nonlinear subspace identification seems ideal.

Acknowledgment

The author gratefully acknowledges the support of the National Science Foundation NYI program (grant CTS-9357827), DuPont and Aspen Technology.

References

[1] H. Aling, R. L. Kosut, A. Emami-Naeini, and J. L. Ebert. Nonlinear model reduction with application to rapid thermal processing. In *Proceedings of the 35th Conference on Decision and Control*, pages 4305–4310, Kobe, Japan, 1996.

[2] L. S. Balasubramhanya and F. J. Doyle III. Nonlinear control of a high purity distillation column using a traveling wave model. *AIChE Journal*, 43:703–714, 1997.

[3] B. W. Bequette. Nonlinear control of chemical processes: A review. *Ind. Eng. Chem. Res.*, 30:1391–1413, 1991.

[4] S. A. Billings. Identification of nonlinear systems-a survey. In *IEE Proc.*, volume 127, Pt. D–6, pages 272–285, 1980.

[5] S. A. Billings and W. S. F. Voon. A prediction-error and stepwise-regression estimation algorithm for non-linear systems. *Int. J. Control*, 44(3):803–822, 1986.

[6] T. Bohlin. A case study of grey box identification. *Automatica*, 30(2):307–318, 1994.

[7] Boyd S. and Chua L. O. Fading memory and the problem of approximating nonlinear operators with volterra series. *IEEE Tras. Circuits and Systems*, 32:1150–1161, 1985.

[8] Y. Chikkula, J. H. Lee, and B. Ogunnaike. Dynamically scheduled model predictive control of nonlinear processes using hinging hyperplane models. *AIChE Journal*, to appear, 1998.

[9] Y. Chikkula, J. H. Lee, and B. Ogunnaike. Robust model predictive control of nonlinear systems using input-output models. In *Proceedings of American Control Conf.*, pages 2205–2209, Seattle, WA, 1995.

[10] B. Cooley and J. H. Lee. Control-relevant experiment design for multivariable systems. *Automatica*, under review, 1997.

[11] B. Cooley and J. H. Lee. Integrated identification and robust control. *Proceedings of IFAC ADCHEM Conference*, Banff, Canada, to appear in *Journal of Process Control*, 1997.

[12] A. K. Datta, J. H. Lee, and A. Krishnagopalan. Reducing batch-to-batch variability of pulp qulaity through model-based estimation. *Pulp and Paper Canada*, 98(4):46–49, 1997.

[13] F. De Bruyne, B. D. O Anderson, M. Gevers and N. Linard. Robust model predictive control of nonlinear systems using input-output models. In *Proceedings of American Control Conf.*, pages 2205–2209, Seattle, WA, 1995.

[14] D. Dong and T. J. McAvoy. Nonlinear principal component analysis – based on principal curves and neural networks. *Computers and Chemical Engineering*, 20(1):65–78, 1996.

[15] A. A. Feldbaum. *Optimal Control Theory.* Academic Press, New York, NY, 1965.

[16] C. E. Garcia, D. M. Prett, and M. Morari. Model predictive control: theory and practice – a survey. *Automatica*, 25(3):335–348, 1989.

[17] C.E. Garcia. Quadratic dynamic matrix control of nonlinear processes- an application to a batch reaction process. In *AIChE Annual Meeting*, 1984.

[18] M. Gevers. Connecting identification and robust control: a new challenge. In *9th IFAC Symposium on Identification and System Parameter Estimation*, Budapest, Vol 1. Pp. 1–10, 1991.

[19] E. Hernandez and Y. Arkun. Control of nonlinear systems using polynomial ARMA models. *AICHE Journal*, 39(3):446–460, 1993.

[20] T. R. Holcomb and M. Morari. PLS / neural network. *Computers and Chemical Engineering*, 16(4):393–411, 1992.

[21] A. H. Jazwinski. *Stochastic Processes and Filtering Theory*. Academic Press, New York, NY, 1970.

[22] T. A. Johansen and B. A. Foss. Identification of nonlinear system structure and parameters using regime decomposition. *Automatica*, 31(2):321–326, 1995.

[23] J. C. Kantor. A finite dimensional nonlinear observer for an exothermic stirred-tank reactor. *Chem. Eng. Sci.*, 44(7):1503–1510, 1989.

[24] M. A. Kramer. Nonlinear principal component analysis using autoassociative neural networks. *AIChE J.*, 37(2):233–243, 1991.

[25] J. H. Lee and N. L. Ricker. Extended Kalman filter based model predictive control. *Ind. Eng. Chem. Res.*, pages 1530–1541, 1994.

[26] L. Ljung. *System Identification: Theory for the User*. Prentice-Hall Information and System Sciences Series, Englewood Cliffs, NY, 1987.

[27] B. Maner and F. J. Doyle III. Simulated polymerization reactor control using autoregressive plus volterra-based mpc. *AIChE Journal*, 43:1763–1784, 1997.

[28] W. Marquart. Wave propagation in counter-current separtion processes and its significance to model reduction. *Automatisierungstechnik*, 35(4):156–162, 1987.

[29] M. Morari and J. H. Lee. Model predictive control: Past, present and future. *presented at PSE97/ESCAPE5, Computers and Chemical Engineering.*, page to appear, 1997.

[30] R. K. Pearson. Nonlinear input / output modeling. *Journal of Process Control*, 5(4):197–211, 1995.

[31] R. K. Pearson and B. A. Ogunnaike. *Nonlinear Process Control, Henson, M. A. and Seborg, D. E. (eds.)*, chapter Nonlinear Process Identification. Prentice Hall PTR, 1997.

[32] R. K. Pearson, B. A. Ogunnaike, and F. J. Doyle III. Identification of structurally constrained second-order volterra models. *IEEE Trans. Signal Processing*, 44:2837–2846, 1996.

[33] R. K. Pearson, T. A. Ogunnaike, and F. J. Doyle III. Identification of nonlinear input/output models using non-gaussian input sequences. In *Proc. American Control Conference*, pages 1465–1469, San Francisco, 1993.

[34] M. B. Priestley. *Non-linear and Non-Stationary Time Series Analysis*. Academic Press, San Diego, CA, 1988.

[35] S. J. Qin and T. J. McAvoy. Nonlinear PLS modeling using neural networks. *Computers and Chemical Engineering*, 16(4):379–391, 1992.

[36] S. Joe Qin and T.A. Badgwell. An overview of industrial predictive control technology. In *Chemical Process Control – V, Assessment and New Directions for Research*, Tahoe City, CA, January 1996.

[37] J. B. Rawlings, E. S. Meadows, and K. R. Muske. Nonlinear model predictive control: a tutorial and survey. In *Proc. of ADCHEM*, Kyoto, Japan, 1994.

[38] C. Rhodes and M. Morari. Determining the model order of nonlinear input/output systems. In *AIChE Journal*, volume 44 (1), pages 151–163, 1998.

[39] N. L. Ricker. Use of biased least squares estimators for parameters in discrete time pulse responses. *Industrial and Engineering Chemistry Research*, 27(2):343–350, 1988.

[40] D. Robertson, J. H. Lee, and J. Rawlings. A moving horizon based approach for least squares state estimation. *AIChE Journal*, 42:2209–2224, 1996.

[41] D. G. Robertson and J. H. Lee. A constrained least squares formulation of state estimation. *Journal of Process Control*, 5(4):291–299, 1995.

[42] D. G. Robertson, S. A. Russel, J. H. Lee, and B. Ogunnaike. Control of a batch condensation polymerization reactor. In *Proceedings of American Control Conf.*, pages 4453–4457, Seattle, WA, 1995.

[43] S. A. Russel, P. Kesavan, J. H. Lee, and B. Ogunnaike. Recursive data-based prediction and control of product quality for batch and semi-batch processes applied to a Nylon 6,6 autoclave. *97' AIChE Annual Meeting*, Los Angeles, CA, *AIChE Journal*, submitted, 1998.

[44] L. Sirovich. *New Perspectives in Turbulence, Sirovich, L. (eds.)*, chapter Empirical Eigenfunctions and Low-Dimensional Systems. Springer-Verlag, 1991.

[45] J. Sjoberg, Q. Zhang, L. Ljung, A. Benveniste, B. Delyon' et al. Nonlinear black-box modeling in system identification: a unified overview *Automatica*, 31(12):1691–1724, 1995.

[46] M. Soroush. Nonlinear state-observer design with application to reactors. *Chemical Engineering Science*, 39:387–404, 1993.

[47] H. Tong. *Non-linear Time Series.* Oxford University Press, New York, 1990.

[48] H. J. A. F Tulleken. Grey-box modeling and identification using physical knowledge and bayesian techniques. *Automatica*, 29(2):285–308, 1993.

[49] P. Van Overschee and B. de Moor. N4SID: Subspace algorithms for the identification of combined deterministic stochastic systems. *Automatica*, 30(1):75–93, 1994.

[50] R. J. Williams. *Neural Networks for Control, Miller, W. T., Sutton, R. S., and Werbos, P. J. (eds.)*, chapter Adaptive State Representation and Estimation Using Recurrent Connectionist Networks. The MIT Press, Cambridge, MA, 1990.

School of Chemical Engineering,
Purdue University,
CHME,
West Lafayette, IN 47907-1283
E-mail address: `jhl@ecn.purdue.edu`

Progress in Systems and Control Theory, Vol. 26

Structural Concepts for Optimization Based Control of Transient Processes

Achim Helbig, Olaf Abel, Wolfgang Marquardt

Abstract. This paper treats optimal operation of transient processes, characterized by intentionally time-varying variable profiles. Based on the generic formulation of a reconciliation and a control problem, specific structural concepts for the realization of optimization based control are discussed. In particular, a direct and several decomposition approaches are presented. Applicability of the different approaches is treated with respect to the time scales of the process, the disturbances and the available process models. A number of illustrative examples from the optimization of semi-batch reactors are given.

1. Introduction

In the operation of chemical processes stationary and transient phases can be distinguished. In the former case, the goal is to keep the process at one single steady state. In the latter case, all process variables are intentionally time-varying and the process is called *transient*. Examples include the start-up, the shut-down, and grade changes of continuously operated plants. Furthermore, processes exist which are permanently operated in transient phases such as batch and periodically forced continuous processes.

In order to achieve an economic optimal operation under a set of given restrictions, such as a specific amount of material to be produced or certain quality specifications, operational degrees of freedom have to be fixed by designing a control system. For stationary processes, this comprises the determination of an optimal operating point and the regulation at the corresponding setpoints in order to reject disturbances. For transient processes, the control system has to drive the process along a trajectory in the state space.

Many of the problems arising in this context are naturally formulated in terms of an optimization problem. Therefore, optimization based control techniques offer great potential. The flexibility to formulate an appropriate cost function and the ability to handle constraints on state and manipulated process variables contribute to a feedback control system which operates the process as close as possible to the true economic optimum even in the presence of changing environmental or market conditions and disturbances. For continuous processes, which are predominantly operated at steady states, this concept has already been realized in the form of real-time optimization (RTO) schemes [20], often used in combination with model predictive controllers (MPC) [30] driving the processes to the calculated setpoints.

This paper discusses structural concepts for optimization based control of transient processes. Special emphasis is on the proper reflection of the different relevant time scales in the control concept. A generic formulation for the description of optimal operations is given in the next section. Different structural concepts to realize optimal operation are discussed in the subsequent Sections 3 and 4.

2. Operational objectives and problem setting

The goal of optimal process operation is to maximize profit. In order to reach this goal the planning and scheduling level in the decision hierarchy associates some economic cost function Φ_i to each operational phase i of duration $\Delta_i = [t_{0,i}, t_{f,i}]$. The cost function has to be minimized subject to production constraints $\boldsymbol{m}_i$ and additional process constraints $\boldsymbol{g}_i$ fixed by the planning and scheduling level of the decision hierarchy, in the sequel referred to as decision maker. While $\boldsymbol{m}_i$ together with Φ_i reflect the overall operational objectives, $\boldsymbol{g}_i$ concatenates those constraints which have to be enforced during any possible operation. Examples for $\boldsymbol{m}_i$ include product capacity and quality, $\boldsymbol{g}_i$ comprises for instance existing inventory limits. The endpoint $t_{f,i}$ of Δ_i depends on the operational objective and may cover the duration of a stationary or periodic phase of a continuous process. For transient phases of a continuous process due to grade changes, $t_{f,i}$ corresponds to the time where a new steady state is reached, and for batch processes $t_{f,i}$ equals the final batch time.

If a perfect model of the process exists, if the initial state $\boldsymbol{x}_{0,i}$ at the beginning of Δ_i is known exactly and if the process is not disturbed, the associated optimal trajectory for the operational degrees of freedom $\boldsymbol{u}_i$ can be determined entirely off-line through the solution of an optimal control problem for any operational phase Δ_i. For notational simplicity the index i will be omitted in the sequel. The optimization problem can then be formulated as:

$$\min_{\boldsymbol{u}} \Phi(\boldsymbol{x}, \boldsymbol{u}, t_0, t_f) \tag{1}$$

$$\begin{aligned} \text{s.t.} \quad 0 &= \boldsymbol{f}(\dot{\boldsymbol{x}}, \boldsymbol{x}, \boldsymbol{u}, \boldsymbol{d}) \,, \quad \boldsymbol{x}(t_0) = \boldsymbol{x}_0 \,, \\ \boldsymbol{y} &= \boldsymbol{h}(\boldsymbol{x}, \boldsymbol{u}, \boldsymbol{d}) \,, \\ 0 &\geq \boldsymbol{g}(\boldsymbol{x}, \boldsymbol{u}, \boldsymbol{d}) \,, \\ 0 &\geq \boldsymbol{m}(\boldsymbol{x}, \boldsymbol{u}, \boldsymbol{d}) \,, \\ & t \in [t_0, t_f] \,. \end{aligned}$$

Here, $\boldsymbol{x}$ denotes the system state, $\boldsymbol{u}$ free operational variables, and $\boldsymbol{d}$ given parameters. $\boldsymbol{f}$ contains the process model consisting of a set of differential and algebraic equations, $\boldsymbol{h}$ maps the system state to the outputs $\boldsymbol{y}$. $\boldsymbol{g}$ and $\boldsymbol{m}$ contain the path constraints to be enforced. Additionally, $\boldsymbol{m}$ can comprise point and periodicity constraints. Problem (1) may be solved by sequential or simultaneous techniques for the direct optimization of dynamic systems [5, 39, 40].

In all practical situations the above assumptions are not fulfilled. This is mainly due to parametric or unstructured model uncertainty and time-varying exogenous inputs, now covered by an unknown time-varying disturbance vector $\boldsymbol{d}(t)$. In the presence of disturbances, it is necessary to introduce feedback of available measurements $\boldsymbol{\eta}$ of the system outputs $\boldsymbol{y}$ and/or of exogenous inputs $\boldsymbol{u}$ and $\boldsymbol{d}$. The arising optimal output feedback control problem can naturally be separated into a dynamic data and model reconciliation problem (2) and an optimal control problem (3). Whilst the optimal control problem is formulated on a prediction horizon $[t_c, t_f]$, the reconciliation problem incorporates process measurements over some horizon $[t_r, t_c]$ in the past. In general, the following dual problems have to be solved at any current time $t_c \in \Delta$:

$$\min_{\boldsymbol{x}_{r,0},\boldsymbol{d}_r} \Phi_r(\boldsymbol{y}_r, \boldsymbol{\eta}, \boldsymbol{x}_{r,0}, \boldsymbol{d}_r, t_r, t_c) \tag{2}$$

$$\begin{aligned} \text{s.t.}\quad 0 &= \boldsymbol{f}(\dot{\boldsymbol{x}}_r, \boldsymbol{x}_r, \boldsymbol{u}_r, \boldsymbol{d}_r)\,, \\ & \boldsymbol{x}_r(t_r) = \boldsymbol{x}_{r,0}\,, \\ \boldsymbol{y}_r &= \boldsymbol{h}(\boldsymbol{x}_r, \boldsymbol{u}_r, \boldsymbol{d}_r)\,, \\ \boldsymbol{u}_r &= \mathcal{U}[\boldsymbol{u}_c(\cdot)]\,, \\ 0 &\geq \boldsymbol{g}_r(\boldsymbol{x}_r, \boldsymbol{u}_r, \boldsymbol{d}_r)\,, \\ & t \in [t_r, t_c]\,, \end{aligned}$$

$$\min_{\boldsymbol{u}_c} \Phi_c(\boldsymbol{x}_c, \boldsymbol{u}_c, t_c, t_f) \tag{3}$$

$$\begin{aligned} \text{s.t.}\quad 0 &= \boldsymbol{f}(\dot{\boldsymbol{x}}_c, \boldsymbol{x}_c, \boldsymbol{u}_c, \boldsymbol{d}_c)\,, \\ & \boldsymbol{x}_c(t_c) = \boldsymbol{x}_r(t_c)\,, \\ \boldsymbol{y}_c &= \boldsymbol{h}(\boldsymbol{x}_c, \boldsymbol{u}_c, \boldsymbol{d}_c)\,, \\ \boldsymbol{d}_c &= \mathcal{D}[\boldsymbol{d}_r(\cdot)]\,, \\ 0 &\geq \boldsymbol{g}_c(\boldsymbol{x}_c, \boldsymbol{u}_c, \boldsymbol{d}_c)\,, \\ 0 &\geq \boldsymbol{m}(\boldsymbol{x}_c, \boldsymbol{u}_c, \boldsymbol{d}_c)\,, \\ & t \in [t_c, t_f]\,. \end{aligned}$$

The indices r and c characterize the reconciliation and the optimal control problem, respectively. Both problems are subject to process specific constraints $\boldsymbol{g}_r$ and $\boldsymbol{g}_c$. In many practical situations additional levels of complexity have to be added to problem (3). These include integer variables among the operational degrees of freedom $\boldsymbol{u}_c$ as well as hybrid discrete-continuous models [1, 4]. The problems (2) and (3) are coupled. Previously calculated manipulated variables $\boldsymbol{u}_c$ are used in (2) as indicated by the operator $\mathcal{U}$ which takes the entire $\boldsymbol{u}_c$ signal on $[t_r, t_c]$ as input. The initial condition $\boldsymbol{x}_c(t_c)$ of (3) is determined through (2). Furthermore, disturbances $\boldsymbol{d}_r$ estimated over the reconciliation horizon $[t_r, t_c]$ provide the basis for disturbance prediction in the optimization model. The operator $\mathcal{D}$ reflects a generalized disturbance predictor. The prediction may be based on extrapolation or on simple disturbance models as discussed e.g. in [32]. For applicability it has to be assumed that enough information can be extracted from the measured data in order to enable a proper reconciliation.

Optimization concepts which rely on real-time solution of problems (2) and (3) on the controller sampling frequency are called direct approaches and will be discussed in the next section. The application of this computationally demanding and for many problems yet intractable approach is, however, not necessary in all cases, depending on the nature of the disturbances $\boldsymbol{d}(t)$. Specifically, the following cases are of relevance:

(D1) $\boldsymbol{d}(t) = \boldsymbol{d}_0 + \Delta\boldsymbol{d}(t)$ with expectation value $E\{\Delta\boldsymbol{d}(t)\} = 0$. In this case it is assumed that a constant part $\boldsymbol{d}_0$ and high frequency zero mean disturbance variations $\Delta\boldsymbol{d}(t)$ can be distinguished.

(D2) $\boldsymbol{d}(t) = \boldsymbol{d}_0(t) + \Delta\boldsymbol{d}(t)$ with $E\{\Delta\boldsymbol{d}(t)\} = 0$. Here, the disturbance can be separated into a slow trend (or drift) $\boldsymbol{d}_0(t)$ and added high frequency zero mean disturbance variations.

(D3) No special structure in $\boldsymbol{d}(t)$ is assumed.

In cases (D1) and (D2), decomposition approaches can be applied to optimization based control of transient processes since the time scales of the disturbances and the optimization problem are sufficiently separated. Case (D2) has already been proposed for the treatment of stationary processes [23]. Appropriate concepts of different complexity will be discussed in Section 4. In case (D3), decomposition is questionable and a direct approach as discussed next is more favorable.

3. Direct approach

In direct approaches, a single optimizing feedback control layer exists in the decision hierarchy between the decision maker and the process with base control, as shown in Fig. 1. Problems (2) and (3) are solved in real time in each sampling interval $\delta_c = t_{j+1} - t_j$ of the feedback system. Since no separation of time-scales is applied, the controller has to carry out economic optimization exploiting the low frequency part of the disturbances $\boldsymbol{d}(t)$ whilst simultaneously rejecting the high frequencies (noise) contained in $\boldsymbol{d}(t)$. Different realizations of the direct approach can be distinguished on the duration of the prediction horizon used for the optimal control problem. Approaches which extend prediction to the endpoint t_f of Δ, as formulated in (3), can be distinguished from schemes with fixed prediction horizon $[t_c, t_c + H_c\delta_c]$, with $H_c\delta_c \ll \Delta$. In all cases a suitable algorithm for real-time dynamic data and model reconciliation or at least state and parameter estimation has to be applied [6, 12].

Assuming that a process model with long range prediction qualities is available, direct approaches can be realized by repeated real-time solution of problems (2) and (3). The economic objective function Φ_c is used directly on the control level here, and the enforcement of endpoint constraints is straightforward. For batch processes methods including parameter update schemes and elements from stochastic optimization have been developed [34, 38]. These so called on-line re-optimization schemes could also be interpreted as shrinking horizon MPC. Problem (3) can alternatively be tackled by so-called optimal feedback control [27], where a mapping from the system state to optimal manipulated variables is analytically derived first. This (possibly very complex) state-input law is then evaluated on-line. The main drawback of this approach is that the switching structure of active constraints has to be known a priori.

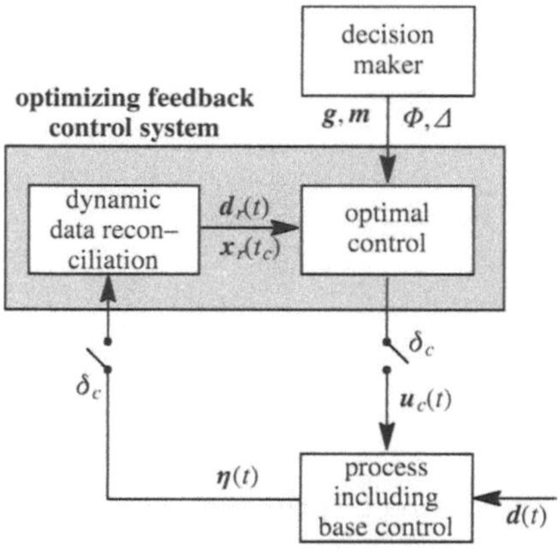

FIGURE 1. Direct optimization approach.

Direct approaches based on fixed prediction horizons are naturally formulated in the framework of MPC. In order to be able to achieve the true closed loop optimum on Δ through optimization on shorter horizons $H_c\delta_c$, both the cost function Φ_c and possible endpoint constraints contained in $\boldsymbol{m}$ have to be suitably translated onto the short horizon. No performance loss is encountered if an "instantaneous" equivalent Φ_H for Φ_c can be defined on $H_c\delta_c$ and if no endpoint constraints are imposed on the problem [38]. In all other cases, direct approaches with fixed prediction horizons are either sub-optimal (see [14] for an example) or totally inapplicable.

Even if an appropriate function Φ_H exists and no endpoints constraints have to be enforced the viability [3] of the system with respect to the path constraints $\boldsymbol{g}_c$ has to be ensured. Otherwise, deadlock situations may occur where control actions lead to infeasibilities of the optimization problems to be solved at later times. The prevention of deadlocks requires, in parallel to the computation of the nominal manipulated variables, the analysis of worst case scenarios on a sufficiently long horizon $H_\xi\delta_c$, with H_ξ not necessarily equal to H_c.

This may be illustrated by an example from the MPC-based on-line optimization of exothermic semi-batch reactors. In this example safety relevant variables of a strongly exothermic and autocatalytic two-phase semi-batch reactor such as pressure and temperature must be guaranteed to stay below pre-specified limits even in case of a cooling system failure. This defines the worst case situation for a possible deadlock in the reactor safety constraints. The method proposed in [2] explicitly accounts for failure scenarios by using a hybrid discrete-continuous process model within the MPC. The process behavior for each type of failure ξ is observed on specific failure horizons $H_\xi\delta_c$. In the considered example, the feed flowrate to the reactor is to be determined such that time minimal operation is achieved. This overall economic objective can be transformed to the maximization of the feed addition rate on the current time window.

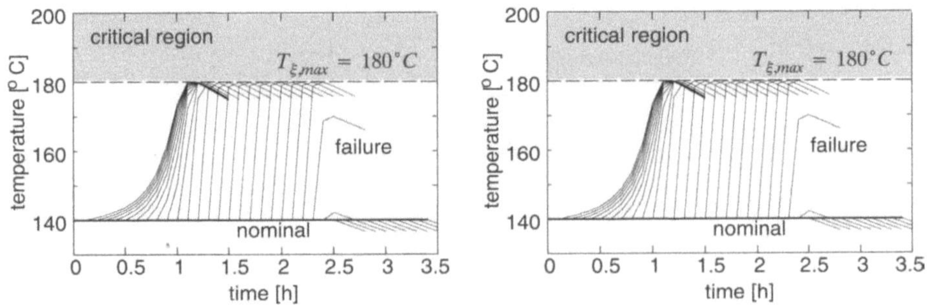

FIGURE 2. MPC-based direct optimizing control of a semi-batch polymerization reactor with incorporation of safety scenarios.

Some of the obtained results are shown in Figure 2. A reduction of the feed period from $6.5h$ for conventional operation to $2.4h$ is achieved. Simultaneously, the safety constraints on pressure and temperature are enforced. For the reason of brevity, Fig. 2 (right) only depicts the temperature evolution in case of a failure for every prediction horizon. As can be seen, the corresponding constraint is the active limitation preventing further optimization. For the example treated, an optimizing MPC scheme not analyzing the worst case scenario indeed drives the process into a deadlock [2].

If deadlock situations can be excluded in advance, short prediction horizons suffice for process optimization. This offers the possibility to use crude process models without long range prediction capabilities within MPC. These models have to be updated at every sampling interval leading to a nonlinear adaptive predictive control scheme. Clearly the method used for the adaptation of the model is crucial for the success.

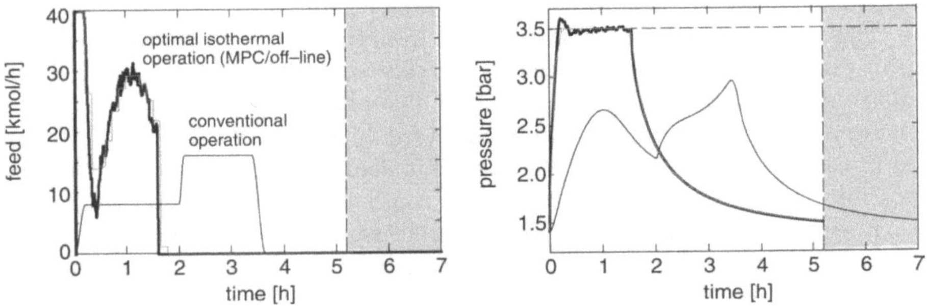

FIGURE 3. MPC-based direct optimizing control of a semi-batch ethoxilation reactor with calorimetric process model.

An illustrating example, taken from [14], considers a semi-batch ethoxilation reactor which fulfills the rather restrictive assumptions for the applicability of the MPC based on-line optimization scheme with short prediction horizon. The feed flowrate to the reactor is again to be determined such that time minimal operation is achieved. The case studied treats an isothermal vapor-liquid reactor with a pressure limit as path constraint. The results achieved with an MPC algorithm containing only a calorimetric process model without any knowledge of either the reaction or the heat and mass transfer kinetics are shown in Fig. 3.

Compared to the conventional operation, which is plotted in thin lines in Fig. 3, a time reduction of nearly two hours is achieved (shaded area) without violating the pressure constraint. Fig. 3 also depicts the optimal solution calculated off-line (gray line) using the very detailed nominal model. The differences between this trajectory and the MPC solution with a calorimetric model are remarkably small.

4. Decomposition approaches

The unifying characteristic of decomposition approaches for the operation of transient processes is the separation of problem (3) into the generation of an open-loop optimal trajectory and a feedback control problem. The major assumption underlying this approach is a significant difference between dominant time scales of the disturbances and of the optimal trajectory. This assumption may be naturally fulfilled with high quality or it has to be postulated if a direct approach is computationally intractable. Decomposition approaches always require the availability of a dynamic model with appropriate endpoint prediction qualities. In the following section, a simple decomposition approach will be discussed first. A short discussion of possible extensions follows in Section 4.2 before a time scale decomposition approach is treated in Section 4.3.

4.1. Simple decomposition approach

In the simple decomposition approach it is assumed that $\boldsymbol{d}(t)$ can be well described by a constant part $\boldsymbol{d}_0$ with added zero-mean high frequency noise (case (D1), see above). In this case, rejection of $\Delta\boldsymbol{d}(t)$ on the feedback control level is sufficient in order to operate close to the true plant optimum. The calculation of an optimal trajectory for Δ through solution of (3) can be carried out entirely off-line before the transient phase starts. From the optimal solution, the desired trajectories $\boldsymbol{x}_d(t)$, $\boldsymbol{y}_d(t)$, and $\boldsymbol{u}_d(t)$ are passed to the feedback controller which has to track $\boldsymbol{x}_d(t)$ or $\boldsymbol{y}_d(t)$ using $\boldsymbol{u}_d(t)$ in a feedforward manner. The decomposition approach results in a transparent structure (Fig. 4) and circumvents the computationally demanding problem of real-time solution of problems (2) and (3).

For stationary processes the simple decomposition approach is currently industrial standard. Based on a disturbance value $\boldsymbol{d}_0$, the stationary counterparts of problems (2) and (3) are solved once. Information passed to the underlying feedback controller for disturbance rejection comprises (constant) setpoint values

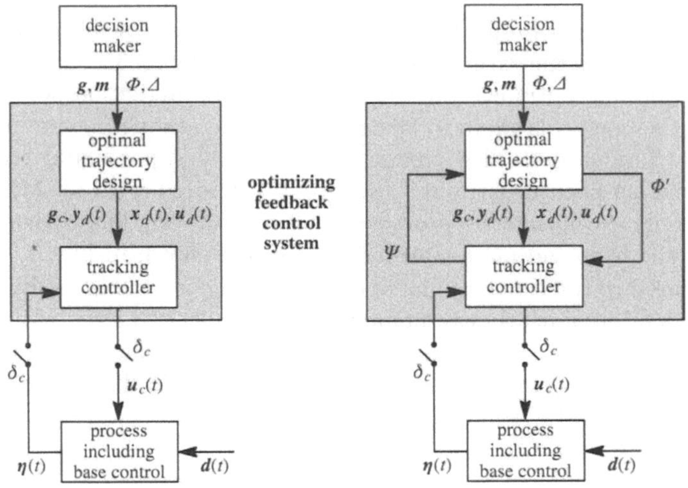

FIGURE 4. Simple (left) and extended (right) decomposition approaches.

$\boldsymbol{y}_s$, optimal stationary values for the manipulated variables $\boldsymbol{u}_s$, and a set of relevant constraints $\boldsymbol{g}_s$ with the index s indicating the optimal stationary solution. Since the range of operation is restricted to a small regime around the selected operating point $\boldsymbol{y}_s, \boldsymbol{x}_s, \boldsymbol{u}_s$, the use of linear dynamic models and the reduction to output feedback seems appropriate on the feedback control level. This is done in standard linear MPC, which at the same time enforces the constraints $\boldsymbol{g}_s$.

In the field of transient processes the simple decomposition approach has found significant interest in the framework of optimal design of batch or periodic processes [1, 8, 36, 37]. Although the number of reported cases is still small, a high industrial application potential can be assigned to this approach since excessive real-time computations are avoided.

4.1.1. OPTIMAL TRAJECTORY DESIGN Two main requirements have to be fulfilled in the design of optimal trajectories. First, all operational constraints $(\boldsymbol{g}_c, \boldsymbol{m})$ have to be enforced under all possible variations of the process. This leads to the notion of flexible design [10]. Second, the "difficulty" of the problem posed for the feedback controller has to be limited. Consequently, controllability aspects have to be included in the design step. Ultimately, the entire process including the corresponding equipment should be designed such that the resulting optimal trajectory is both easy to control and insensitive to model errors.

The rigorous consideration of the flexibility requirements may add additional complexity to the generic problem formulation (3) and may render the overall problem stochastic and/or hybrid (discrete-continuous). For the treatment of stochastic elements the reader is referred to [16, 38]. Hybrid systems arise e.g. from the rigorous inclusion of process dynamics after failures which is an important aspect in the design of inherently safe trajectories [1]. Generally, a set Ξ of possible failure modes ξ has to be considered which may be entered at any time during nominal operation. Their introduction leads to the notion of scenario-integrated dynamic optimization necessitating the extension of (3) by the following constraints:

$$\begin{aligned} 0 &= \boldsymbol{f}_\xi(\dot{\boldsymbol{x}}_\xi(\tau_\xi), \boldsymbol{x}_\xi(\tau_\xi), \boldsymbol{u}_\xi(\tau_\xi), \boldsymbol{d}_\xi(\tau_\xi), \boldsymbol{p}_\xi), \\ \boldsymbol{x}_\xi(\tau_\xi = t) &= \boldsymbol{h}_\xi(\boldsymbol{x}_d(t), \xi, t) , \\ 0 &\geq \boldsymbol{g}_\xi(\boldsymbol{x}_\xi(\tau_\xi), \boldsymbol{y}_\xi(\tau_\xi), \boldsymbol{u}_\xi(\tau_\xi)) , \\ & \xi \in \Xi, \quad \tau_\xi \in [t, t+\tau_{\xi,f}], \; . \end{aligned} \tag{4}$$

Here, $\boldsymbol{f}_\xi$ and $\boldsymbol{g}_\xi$ represent process model and constraints for each failure scenario ξ. At any possible time instant for the different failures, $\boldsymbol{h}_\xi$ maps the nominal state to the initial condition of the failure model. It is worth noting that scenario-integrated optimization requires problem analysis along multiple time axes t, τ_ξ.

In order to give an example for the scenario-integrated trajectory design the semi-batch polymerization reactor example from [1] is considered. The aim is to derive the time minimal reactor temperature setpoint and feeding policy for a highly exothermic two-phase reactor. For safety reasons the reactor temperature is not allowed to exceed $423\,K$ even if the cooling system fails. Problem (3) under consideration of (4) is solved for this single failure mode assuming that the feed is switched off at the moment the failure occurs. The derived optimal trajectories which also satisfy additional path and endpoint constraints are shown in Fig. 5. The nominal solution is shown in bold lines whilst thin lines indicate the system behavior after failures occurring at different times over the batch. The importance of the failure considerations directly follows from the fact that the safety constraint is active over the entire batch.

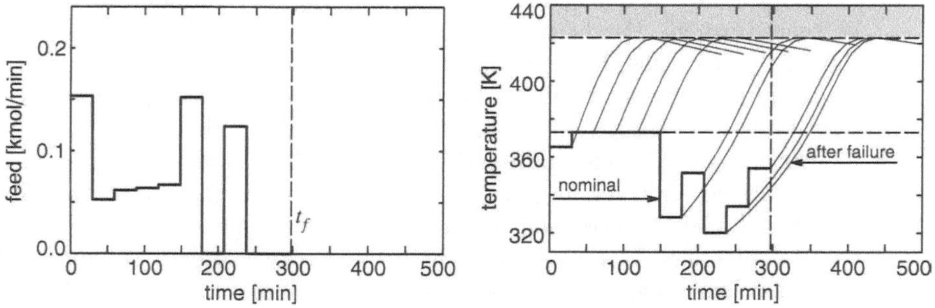

FIGURE 5. Scenario-integrated trajectory optimization for a semi-batch polymerization reactor.

The integration of controllability restrictions in trajectory design leads to the notion of control-integrated design of transient processes which to a large extent remains an unsolved problem. A system is called (input-output) controllable if a controller exists which meets some specified performance criteria. Mature techniques which extend the controller independent analysis of invertibility limitations carried out for process models linearized at one single operating point [22, 35] to nonlinear and transient systems are lacking. Currently, no rigorous means exists to directly translate invertibility requirements or controller specific limitations into constraints of the trajectory design problem. Taking a heuristic approach, anticipated controller performance problems can be addressed through bounding of suitable functions of inputs and outputs or through appropriate (simple) parameterizations of the setpoint trajectory $\boldsymbol{y}(t)$. Another possibility requires an explicit representation of the controller. The defining equations can be used to augment the process model $\boldsymbol{f}$. The setpoints can then be calculated directly for the nominal closed loop system obeying controller limitations as shown in [21] for stationary processes.

4.1.2. TRACKING OF OPTIMAL TRAJECTORIES The task of the tracking controller is to keep the process as close as possible to the calculated optimal trajectory suppressing process disturbances. The specific constraints $\boldsymbol{g}_d$ to be obeyed by the controller are also specified by the trajectory design step. Two fundamentally different approaches to this task can be distinguished on the basis of the amount of knowledge on the optimal trajectory exploited in the tracking controller.

The first class of approaches uses only the desired output trajectory $\boldsymbol{y}_d$. The calculation of appropriate input variables $\boldsymbol{u}_c$ is based on an output error $\Delta\boldsymbol{y}_c$ defined as difference between measured or calculated outputs and the desired outputs. The controllers belonging to this class can further be separated into the following subclasses:

- Approaches exploiting a full nonlinear process model. Nonlinear MPC may belong to this subclass. Applications can mainly be found in the area of batch processes [11, 25]. MPC for periodically forced chemical reactors is treated in [26]. Furthermore, geometric control techniques must be assigned to this class. In principle, they rely on the transformation of the tracking problem to a regulation problem. Inversion based approaches can than be used to design an appropriate controller [9, 19, 41].
- Approaches using models of reduced complexity [15]. These may be specifically derived for tracking purposes, possibly on the basis of an available detailed model. A prominent example for this subclass are linear model based controllers. However, since the desired trajectories $\boldsymbol{x}_d(t)$ and $\boldsymbol{y}_d(t)$ may pass large regions of the state space where the influence of the process nonlinearities can be very high, the performance achievable with one (single) linear model may not be sufficient [18].
- Model free solutions. The properly tuned PID controller certainly belongs to the most frequently used algorithms in this subclass.

In the second class of controllers the trajectories of $\boldsymbol{x}_d$ and $\boldsymbol{u}_d$ in addition to $\boldsymbol{y}_d$ are actively used for the determination of $\boldsymbol{u}_c$. The task of the tracking controller can in principle be derived from (3) through a change of the objective function and an insertion of the desired optimal trajectory (index d) into the process model. Then, the problem can be entirely formulated in deviation variables:

$$\min_{\Delta \boldsymbol{u}_c(t)} \|\Delta \boldsymbol{y}_c\|_Q + \epsilon \|\Delta \boldsymbol{u}_c\|_P \tag{5}$$

$$\begin{aligned}
\text{s.t.}\quad 0 &= \boldsymbol{f}(\dot{\boldsymbol{x}}_d + \Delta\dot{\boldsymbol{x}}_c, \boldsymbol{x}_d + \Delta\boldsymbol{x}_c, \boldsymbol{u}_d + \Delta\boldsymbol{u}_c, \boldsymbol{d}_{c,0} + \Delta\boldsymbol{d}_c)\ ,\\
&\quad \boldsymbol{x}_c(t_c) = \boldsymbol{x}_r(t_c)\ ,\\
\Delta\boldsymbol{y}_c &= \boldsymbol{h}(\boldsymbol{x}_d + \Delta\boldsymbol{x}_c, \boldsymbol{u}_d + \Delta\boldsymbol{u}_c, \boldsymbol{d}_{c,0} + \Delta\boldsymbol{d}_c) - \boldsymbol{y}_d\ ,\\
\Delta\boldsymbol{d}_c &= \mathcal{D}(\boldsymbol{d}_r(\tau))\ ,\quad \tau \in [t_r, t_c]\ ,\\
0 &\geq \boldsymbol{g}_d(\boldsymbol{x}_d + \Delta\boldsymbol{x}_c, \boldsymbol{u}_d + \Delta\boldsymbol{u}_c, \boldsymbol{d}_{c,0} + \Delta\boldsymbol{d}_c)\ ,\\
&\quad t \in [t_c, t_{c,f}]\ .
\end{aligned}$$

Here, $t_{c,f}$ is the considered time horizon for the tracking problem. Additionally, an analogous dual problem derived from (2) can be stated. Techniques specifically tailored to the solution of problem (5) include:

- Model predictive control. This scheme still offers the advantage of being able to enforce the constraints $\boldsymbol{g}_d$. If the deviations from the nominal trajectory are sufficiently small, even a treatment of (5) in a linear setting is possible. Application of linear MPC [24, 31] to the problem derived from linearization of $\boldsymbol{f}$ and $\boldsymbol{h}$ around the optimal trajectory is certainly an approach worthwhile studying.
- Calculation of the best linear time-variant corrector. Applying the method of neighboring extremals [29], an optimal state feedback law for the optimized trajectory can be derived entirely off-line on the basis of the linearized necessary conditions for the trajectory design problem (3). The concept can be extended to the on-line interpolation between a set of optimal trajectories calculated for some model parameter variations. The major restrictions of this method are due to a limited deviation region ("controllability region" [29]) and problems associated to determination of switching points for the active constraints.

In the past approaches using only the desired process outputs $\boldsymbol{y}_d$ have found considerably more attention both in academia and industrial practice. This is certainly caused by the fact that the trajectories of $\boldsymbol{x}_d$ and $\boldsymbol{u}_d$ are often not available. In a decomposition approach, as discussed here, both trajectories are however known directly as a result of the trajectory design step and future research effort is desirable to investigate control schemes exploiting this knowledge.

4.2. Extended decomposition approach

As shown in the right part of Fig. 4, two possibilities exist in order to enhance the performance of the simple decomposition approach under the assumption (D1). The first mechanism acts on the trajectory design level and is based on the assessment of the feedback controller performance. Appropriate measures Ψ with respect to control error, bound violations, and control energy might be evaluated on-line. They can be defined e.g. on the basis of cost functions evaluated within the controller or during its design. Poor performance with respect to these measures may necessitate changes in the constraints to be enforced and consequently also in the trajectories to be tracked. Therefore, controller dependent constraints, such as discussed in Section 4.1.1, may be altered on a heuristic basis. The trajectory design problem is then resolved and an updated setpoint trajectory is passed to the feedback controller. The updating procedure can be executed several times within Δ.

The second possibility concerns the feedback controller. The requirement for approximately perfect tracking of a setpoint trajectory $\boldsymbol{y}_d(t)$ parameterized over time might be relaxed. After a disturbance has caused a deviation from the originally optimal path $\boldsymbol{x}_d(t)$ at some time instant t_c, it is of interest to force the system to the "closest" state $\boldsymbol{x}_d^*(t_c)$ on $\boldsymbol{x}_d(t)$. For periodic processes this concept is known and an optimal controller to bring the state back to the optimal periodic orbit has been developed in [41]. Generally $\boldsymbol{x}_d^*(t_c)$ differs from $\boldsymbol{x}_d(t_c)$ and the relation

$$\boldsymbol{x}_d^*(t_c) = \boldsymbol{x}_d(t_c + \vartheta) \tag{6}$$

with $\vartheta \neq 0$ holds. The time shift ϑ becomes a degree of freedom in the controller. The same idea can be applied in terms of the system outputs $\boldsymbol{y}_c$. The feedback controller has to minimize the modified objective function $\Phi^{‘}$ in this case

$$\min_{\Delta u_c(t),\vartheta} \|\boldsymbol{y}_c(t) - \boldsymbol{y}_d^*(t+\vartheta)\|_Q + \epsilon \|\Delta \boldsymbol{u}_c(t)\|_P. \tag{7}$$

Generalizations to other parameterizations of the desired trajectory on the basis of a priori knowledge may be possible. In principle, any monotonic process variable (e.g. conversion) can be used to replace time [13]. Similar ideas have already been applied for the monitoring and analysis of unequal length batch processes [33] but, to the authors' knowledge, they have not yet been used for control purposes.

4.3. Time scale decomposition approach

The decomposition approaches discussed so far rely on the assumption (D1) that the existing disturbances $\boldsymbol{d}(t)$ can be separated into a constant part $\boldsymbol{d}_0$ and a zero mean high frequency part $\Delta\boldsymbol{d}(t)$. Often this assumption is not justified due to slowly time-varying non-zero mean disturbances or drifting model parameters. Then it is unavoidable to provide some kind of feedback not only from the process to the tracking controller but also from the process to the trajectory design level. Assumption (D2) might provide a possible basis for such an adaptation scheme. Here, the disturbances are separated into a slow drift $\boldsymbol{d}_0(t)$ and a high frequency part $\Delta\boldsymbol{d}(t)$. An update of the slowly varying part $\boldsymbol{d}_0(t)$ can be performed on a time interval Δ_c which is larger than the sampling time δ_c but still significantly smaller than the transient phase Δ.

In the optimization based control of stationary processes such a scheme is state of the art in the industrial practice and is commonly referred to as real-time optimization (RTO) [20]. The goal is to keep the process at an optimal steady state under varying parameters [23]. Thus, the stationary phase Δ is divided into a number of subintervals Δ_s, and on each subinterval, RTO treats the disturbances as $\boldsymbol{d}_s(t) = \boldsymbol{d}_{s,0} + \Delta \boldsymbol{d}_s(t)$ with $E\{\Delta \boldsymbol{d}_s(t)\} = 0$. As shown in Fig. 6, a new optimal steady state is calculated and passed to the control system as soon as deviations in $\boldsymbol{d}_{s,0}$ are determined which lead to a significant performance deterioration. In order to be consistent with the model assumptions the parameter update step can only be executed if the process is at or at least close to a steady state. Therefore, the scheme is only applicable if the subintervals Δ_s are significantly longer than the dominant time constant of the closed loop system.

The concept of RTO can be seen as the simplest case of a time scale separation of the disturbances between a short time scale for the feedback control level and an infinitely long time scale associated with the update of the steady state model. An extension of this concept to the case of transient processes has to replace the infinite time scale by a long but finite one. Postulating the validity of assumption (D2), a distinction of time scales is performed by a time scale separator analyzing the process measurements $\boldsymbol{\eta}(t)$. Both parts $\boldsymbol{\eta}_0(t)$ and $\Delta\boldsymbol{\eta}(t)$ of $\boldsymbol{\eta}(t)$ are dealt with separately in different model update schemes. The economic optimization problem is resolved as significant changes in $\boldsymbol{d}_0(t)$ are detected. Fig. 7 shows the corresponding scheme.

Although there are currently no mature realizations of this concept available, necessary characteristics and some ideas can be highlighted here. One aspect concerns the choice of appropriate models for feedback control and trajectory optimization related to the chosen time scale separation. It is desirable to use reduced models derived from $\boldsymbol{f}, \boldsymbol{h}$ in (1) each providing a sufficient resolution on the corresponding scale. One way to achieve this could be to project the original model in (1) from a high order infinite-dimensional vector function space into a lower order

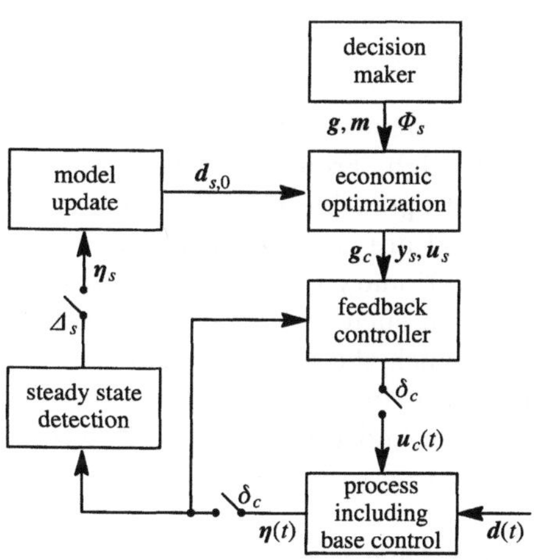

FIGURE 6. RTO for stationary processes.

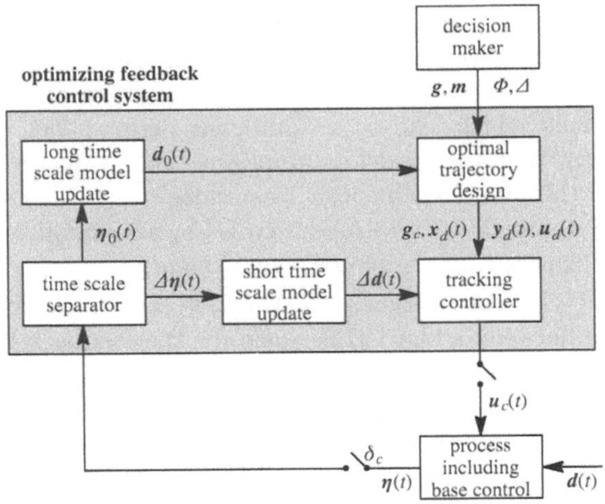

FIGURE 7. Time scale decomposition for transient processes.

infinite-dimensional vector function space for feedback control and into a finite-dimensional vector function space for the trajectory design level. For the projection of the available information into a reduced state space several (nonlinear) model reduction techniques may be used. Furthermore, a reduced model can be obtained through identification of input-output models [28]. For the necessary projection on the trajectory design level, solutions are currently lacking. Possibilities may include Wavelet-Galerkin [6] or multi-grid [7] methods leading to models which support only a restricted dynamic behavior of the state and output trajectories.

5. Conclusions

The probability that the significance of transient process operations will dramatically grow in the future is high. From a practical point of view, the design of a control system for optimal operation on time-varying trajectories poses a major challenge. What is needed are concepts which extend the successful ideas of real-time optimization for stationary processes to transient processes. Several approaches for the realization of such concepts have been discussed in this paper. Clearly, the time scales on which the process operates and on which disturbances enter the process play a key role in the design of appropriate control systems. The control approach to be chosen and the models to be used on the various control levels have to reflect the existing differences and may therefore be developed on multiple time scales, too.

Only if a separation of time scales is impossible a direct approach to the optimal operation of transient processes will be applied. In order to enhance the transparency, approaches decomposing the overall task into the trajectory design and the feedback control level will generally be preferred. However, due to disturbances and uncertainties, simple schemes without the repetitive solution of the design problem will suffer from considerable performance and profit loss for realistic cases. Consequently, solutions will be preferred which provide some interaction between the process and the trajectory design level (e.g. model update) on the one hand, and between the trajectory design and the feedback control level on the other hand. For the latter part, the different approaches can be distinguished by the extent of economics consideration on the control level. A paradigm which offers the flexibility to define various objective functions, the possibility to incorporate constraints, and which allows the use of all kind of process models can be very useful at this point. Therefore, the importance of nonlinear MPC in the optimal operation of transient processes may even become higher than it is already today in the control of their stationary counterparts.

Acknowledgment

Fruitful discussions with T. Backx and O.H. Bosgra are greatfully acknowledged. This research was partly supported by the German BMBF under grant FKZ 3A731140497.

References

[1] O. Abel, A. Helbig, and W. Marquardt. Optimization approaches to control-integrated design of industrial batch reactors. In R. Berber and C. Kravaris, editors, *Nonlinear Model Based Process Control*, NATO-ASI Series, pages 513–551, Dordrecht, The Netherlands, 1998. Kluwer Academic Publishers.

[2] O. Abel and W. Marquardt. A model predictive control scheme for safe and optimal operation of exothermic semibatch reactors. In C. Georgakis, editor, *Proceedings of IFAC Symposium DYCOPS-5*, pages 761–766, Corfu, Greece, 8–10 June 1998.

[3] J.-P. Aubin. *Viability Theory.* Systems & Control: Foundations & Applications. Birkhäuser, Boston, 1991.

[4] P.I. Barton, R.J. Allgor, and S. Galan. Dynamic optimization in a discontinuous world. *Ind. Eng. Chem. Res.*, 37(3):966–981, 1998.

[5] L. T. Biegler. Advances in nonlinear programming concepts for process control. *J. Proc. Cont.*, 8(5/6):301–311, 1998.

[6] T. Binder, L. Blank, W. Dahmen, and W. Marquardt. Towards multiscale dynamic data reconciliation. In R. Berber and C. Kravaris, editors, *Nonlinear Model Based Process Control*, NATO-ASI Series, pages 623–665, Dordrecht, The Netherlands, 1998. Kluwer Academic Publishers.

[7] J.H. Bramble. *Multigrid Methods.* Number 294 in Pitman Research Notes in Mathematics. Longman, London, 1993.

[8] J.-S. Chang, J.-S. Hsu, and Y.-T. Sung. Trajectory tracking of an optimizing path in a batch reactor: Experimantal study. *Ind. Eng. Chem. Res.*, 35(7):2247–2260, 1996.

[9] S. Desavia, D. Chen, and B. Paden. Nonlinear inversion-based output tracking. *IEEE Trans. Auto. Cont.*, 41(7):930–942, 1996.

[10] V.D. Dimitriadis and E.N. Pistikopoulos. Flexibility analysis of dynamic systems. *Ind. Eng. Chem. Res.*, 34:4451–4462, 1995.

[11] B. Ettedgui, M.-V. LeLann, M. Cabassud, and G. Casamatta. Nonlinear model predictive control of semi-batch chemical reactors. In *Proc. of the 3rd European Control Conference ECC'95*, pages 3123–3127, Rome, Italy, 1995.

[12] A. Gelb. *Applied Optimal Estimation.* MIT Press, Cambridge, 1974.

[13] L.M. Gugliotta, J.R. Leiza, M. Arotçarena, P.D. Armitage, and J.M. Asua. Copolymer composistion control in unseed emulsion polymerization using calorimetric data. *Ind. Eng. Chem. Res.*, 34(11):3899–3906, 1995.

[14] A. Helbig, O. Abel, and W. Marquardt. Model predictive control for on-line optimization of semibatch reactors. In *Proceedings of the American Control Conference ACC'98*, volume 3, pages 1695–1699, Philadelphia, USA, 24–26 June 1998.

[15] A. Helbig, O. Abel, A. M'hamdi, and W. Marquardt. Analysis and nonlinear model predictive control of the Chylla-Haase benchmark problem. In *Proc. of CONTROL'96*, pages 1172–1177, Exeter, UK, 2–5 September 1996.

[16] R.P. Isaacs. *Differential Games.* Krieger, New York, 3. edition, 1975.

[17] A. Isidori. *Nonlinear Control Systems: An Introduction.* Springer-Verlag, Berlin, 1995.

[18] T.A. Johansen and B.A. Foss. Operating regime based process modeling and identification. *Comput. Chem. Eng.*, 21(2):159–176, 1996.

[19] P. Lu. Tracking control of nonlinear systems with bounded controls and control rates. *Automatica*, 33(6):1199–1202, 1997.

[20] T.E. Marlin and A.N. Hrymak. Real-time operations optimization of continuous processes. In J.C. Kantor, C.E. Garcia, and B. Carnahan, editors, *Proc. of Fifth International Conference on Chemical Process Control*, volume 93 of *AIChE Symposium Series No. 316*, pages 156–164. AIChE and CACHE, 1997.

[21] M.J. Mohideen, J.D. Perkins, and E.N. Pistikopoulos. Optimal design of dynamic systems under uncertainty. *AIChE J.*, 42(8):2251–2272, 1996.

[22] M. Morari. Design of resilient processing plants – III. A general framework for the assessment of dynamic resilience. *Chem. Eng. Sci.*, 38(11):1881–1891, 1983.

[23] M. Morari, Y. Arkun, and G. Stephanopoulos. Studies in the synthesis of control structures for chemical processes. Part I: Formulation of the control problem. Process decomposition and the classification of the control tasks. Analysis of the optimizing control structures. *AIChE J.*, 26(2):220–232, 1980.

[24] K.R. Muske and J.B. Rawlings. Model predictive control with linear models. *AIChE J.*, 39(2):262–287, 1993.

[25] Z. Nagy and Ş. Agachi. Model predictive control of a PVC batch reactor. *Comput. Chem. Eng.*, 21(6):571–591, 1997.

[26] F. Özgülşen, S.J. Kendra, and A. Çinar. Nonlinear predictive control of periodically forced chemical reactors. *AIChE J.*, 39(4):589–598, 1993.

[27] S. Palanki, C. Kravaris, and H.Y. Wang. Synthesis of state feedback laws for end-point optimization in batch reactors. *Chem. Eng. Sci.*, 48(1):86–97, 1993.

[28] R.K. Pearson and B.A. Ogunnaike. Nonlinear process identification. In M.A. Henson and D.E. Seborg, editors, *Nonlinear Process Control*, pages 11–110. Prentice-Hall, 1997.

[29] H.J. Pesch. A practical guide to the solution of real-life optimal control problems. *Control and Cybernetics*, 23(1/2):7–60, 1994.

[30] S.J. Qin and T.A. Badgwell. An overview of industrial model predictive control technology. In J.C. Kantor, C.E. Garcia, and B. Carnahan, editors, *Proc. of Fifth International Conference on Chemical Process Control*, volume 93 of *AIChE Symposium Series No. 316*, pages 232–256. AIChE and CACHE, 1997.

[31] J.B. Rawlings, E.S. Meadows, and K.R. Muske. Nonlinear model predictive control: A tutorial and survey. In *Preprints of the IFAC Symposium ADCHEM'94*, pages 203–214, Kyoto, Japan, 25–27 May 1994.

[32] N.L. Ricker. Model predictive control: State of the art. In Y. Arkun and W.H. Ray, editors, *Chemical Process Control - CPC IV*, CACHE, pages 271–296. Elsevier, 1991.

[33] S.G. Rothwell, E.B. Martin, and A.J. Morris. Comparison of methods for handling unequal length batches. In C. Georgakis, editor, *Proceedings of IFAC Symposium DYCOPS-5*, pages 66–71, Corfu, Greece, 8–10 June 1998.

[34] D. Ruppen. *A Contribution to the Implementation of Adaptive Optimal Operation for Discontinuous Chemical Reactors.* PhD thesis, ETH, No. 10634, Zürich, Switzerland, 1994.

[35] S. Skogestad and I. Postlethwaite. *Multivariable Feedback Control.* John Wiley & Sons, Chicester, England, 1996.

[36] M. Soroush and C. Kravaris. Optimal design and operation of batch reactors. 1. Theoretical framework. *Ind. Eng. Chem. Res.*, 32:866–881, 1993.

[37] M. Soroush and C. Kravaris. Optimal design and operation of batch reactors. 2. A case study. *Ind. Eng. Chem. Res.*, 32:882–893, 1993.

[38] P. Terwiesch, M. Agarwal, and D.W.T. Rippin. Batch unit optimization with imperfect modelling: A survey. *J. Proc. Cont.*, 4(4):238–258, 1994.

[39] V.S. Vassiliadis, R.W.H. Sargent, and C.C. Pantelides. Solution of a class of multistage dynamic optimization problems. 1. Problems without path constraints. *Ind. Eng. Chem. Res.*, 33(9):2111–2122, 1994.

[40] V.S. Vassiliadis, R.W.H. Sargent, and C.C. Pantelides. Solution of a class of multistage dynamic optimization problems. 2. Problems with path constraints. *Ind. Eng. Chem. Res.*, 33(9):2123–2133, 1994.

[41] Q. Wang, J.L. Speyer, and L.D. Dewell. Regulators for optimal period processes. *IEEE Trans. Auto. Cont.*, 40(10):1767–1778, 1995.

Lehrstuhl für Prozesstechnik
RWTH Aachen
Turmstr. 46
D–52056 Aachen, Germany
E-mail address: {helbig,abel,marquardt}@lfpt.rwth-aachen.de

Progress in Systems and Control Theory, Vol. 26

Efficient Nonlinear Modeling Using Wavelet Compression

Michael Nikolaou and Diwakar Mantha

Abstract. We demonstrate how to use wavelets for the reparametrization of second-order Volterra models in terms of a substantially smaller number of coefficients. The resulting structure retains several of the advantages of the Volterra structure, while being parsimonious, thus making feasible the identification of Volterra models from experimental data. A simulation study on a polymerization reactor elucidates the proposed approach.

1. Introduction

The key element in any model predictive control (MPC) scheme is the plant model. Conventional MPC schemes employ linear models to approximate the dynamic behavior of a plant. Linear models may be inadequate in modeling highly nonlinear plants, as is the case with several chemical plants. Consequently, a nonlinear model structure is essential for capturing the dynamics of such plants.

The use of a general nonlinear model for the identification and analysis of any nonlinear plant would be too cumbersome to be practical, simply because the behavior of plants that are not linear, hence nonlinear, can vary widely. Among several nonlinear model structures ([19]; [7]), polynomial Volterra models have been attractive in modeling stable nonlinear plants for MPC purposes. Due to their structural similarity to linear finite-impulse-response (FIR) models, Volterra models have several advantages as well as disadvantages that are also typical for linear FIR models. In fact, a linear FIR model is a first-order finite Volterra series model. The simplest nonlinear extension of the linear FIR model class is the second-order Volterra model class. An advantage of Volterra models relatively to other nonlinear models is that the parameters of a Volterra model appear linearly, keeping their least-squares identification problem linear. In addition, similarly to linear FIR models, least-squares identification of Volterra modes through open-loop experiments involving random plant inputs produces unbiased parameter estimates, even if colored noise enters the identified plant. In practice, however, as with linear FIR models, the number of coefficients required to approximate the dynamic behavior of an actual plant by a Volterra series model for the purpose of MPC is quite large (typically thousands of coefficients, corresponding to a few tens of past time-steps) particularly if the plant is sampled relatively frequently. This fact makes the identification of Volterra models from experimental data practically impossible for most chemical plants, because of inordinately high requirements on data and experimentation time.

One solution to the above practical difficulty is to reparametrize a Volterra model in terms of considerably fewer coefficients, without destroying the model's structure. Using this idea, Nikolaou and Vuthandam [12] demonstrated that wavelets and their time-frequency localization properties can be exploited to generate parsimonious reparametrizations of linear FIR models by compressing the underlying FIR kernel. The present work is an extension of that idea to the case of second-order discrete-time Volterra models. The two-dimensional discrete wavelet transform (2-D DWT) is employed. The central idea of this article is that certain prior knowledge about the plant, based on either first-principles or empirical data, can be translated into a sparseness pattern of the 2-D DWT of the model to be identified. According to the proposed approach, the original second-order Volterra kernel is first transformed into its 2-D DWT. A priori information is then used to discard a large number of coefficients by setting them identically equal to zero. The resulting parsimonious model retains the same structure as the original model, is simpler to identify, and can be easily converted to the original form by employing the two-dimensional inverse discrete wavelet transform (2-D IDWT). The set of coefficients of the 2-D DWT of the Volterra kernel that are set to zero may be refined in an iterative manner. The compression methodology used above is related to that employed in image processing using wavelet transform coding [4].

The preceding ideas are illustrated in Figures 1 and 2. Figure 1 depicts a typical first-order (i.e., linear) discrete Volterra kernel. The number of coefficients $\{h_j\}_{j=1,2,\ldots}$ is dictated by the fixed frequency used to sample the continuous-time plant output. But, while virtually all coefficients of the linear kernel $\{h_j\}_{j=1,2,\ldots}$ need to be identified for small values of j, only a few coefficients need to be identified for large j. Figure 2 similarly demonstrates the general nature of a typical second-order (i.e., quadratic) discrete Volterra kernel, and again it is obvious that not all coefficients $\{g_{ij}\}_{i,j=1,2,\ldots}$ need to be identified. Values of h_j and g_{ij} not explicitly identified can be constructed through appropriate interpolation. One question, then, is how to select which terms of the sequences $\{h_j\}_{j=1,2,\ldots}$ and $\{g_{ij}\}_{i,j=1,2,\ldots}$ to identify explicitly and how to construct terms not explicitly identified. In fact, as shown in the next section, the true problem in discrete Volterra model identification is that of estimating the kernels $\{h_j\}_{j=1,2,\ldots}$ and $\{g_{ij}\}_{i,j=1,2,\ldots}$ through reconstruction of corresponding *continuous functions* from *sampled* values of those functions. Clearly, techniques more sophisticated than uniform sampling can be used for the sampling of those continuous functions.

Methodologies using wavelets for identification have been proposed in the literature in various contexts. Tsatsanis and Giannakis [23] proposed an algorithm for the identification of time-varying autoregressive (AR) and autoregressive moving-average (ARMA) models using wavelets. Pati et al.[14] examined their

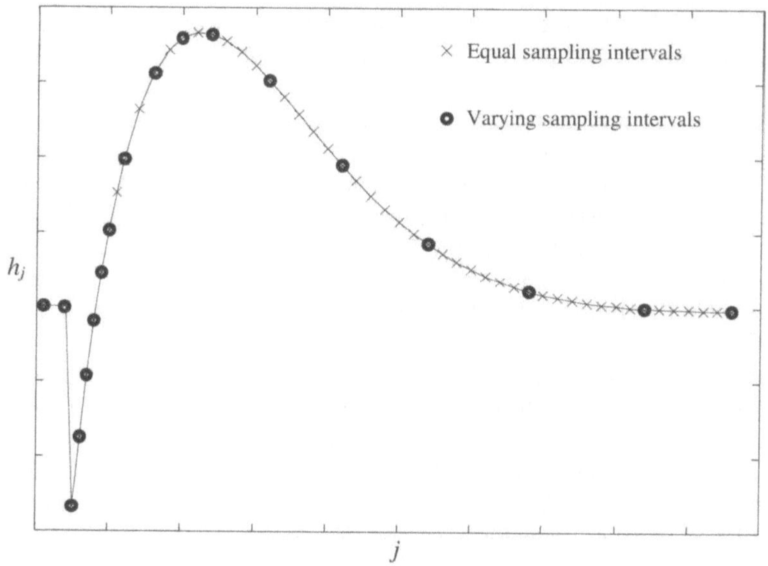

FIGURE 1. Typical impulse response (first-order Volterra kernel).

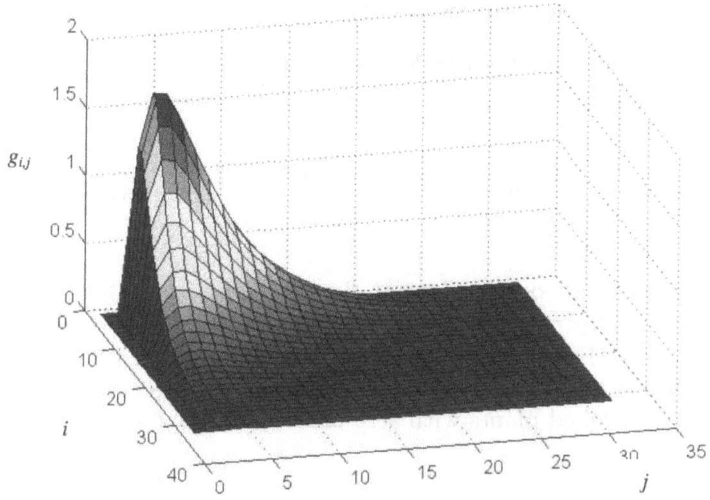

FIGURE 2. Typical second-order discrete Volterra kernel.

method for model-order reductions of linear stable systems using wavelet approximations. Sureshbabu and Farrell [21] studied a wavelet-based system-identification method for nonlinear systems. A method for the denoising of input-output identification data using wavelet-based prefiltering was presented by Palavajjhala et al. [13]. Wavelets were applied by Tsatsanis and Giannakis [22] in detecting transient plant disturbances and jumps. Development of parsimonious Volterra models by imposing structural restrictions were discussed by Pearson et al. [15]. Doyle et al. [5] proposed analysis, synthesis, and design procedures for nonlinear model-based control employing Volterra models. Ling and Rivera [8] presented a control-relevant model reduction technique of Volterra models using nonlinear internal model control (IMC) as a design basis.

The rest of this article is organized as follows. First, we discuss the structure of the second-order Volterra model for discrete-time systems, and explain how continuous-time one- and two-dimensional functions underlie that structure. Next, we present a brief overview of identification methods for such models. Then, we introduce the DWT and IDWT in one and two dimensions. We subsequently present the use of the wavelet transform in the development of parsimonious Volterra models and the resulting identification methodology. To illustrate the proposed methodology, we present a simulation study on a polymerization reactor. We then summarize our results and present directions for future research.

2. Identification of second-order discrete Volterra models

2.1. Continuous- and discrete-time Volterra models

Consider a continuous-time, stable, causal, plant modeled by the second-order Volterra model

$$y(t) = y_1(t) + y_2(t) \hat{=} \int_0^\infty h(r)u(t-r)dr + \int_0^\infty \int_0^\infty g(r,s)u(t-r)u(t-s)drds \tag{1}$$

where $y(t)$ is the output of the plant at time t, $u(t)$ is the input to the plant at time t, and $h(r)$ and $g(r,s)$ are the first-order and second-order Volterra kernels corresponding to $y_1(t)$ and $y_2(t)$, respectively [18]. If the plant is sampled uniformly every T time units, with $u(r) = u(t_{j-1})$ constant for $r \in [t_{j-1}, t_j)$, as is the case with computer-controlled plants with zero-order hold, the output at time $t_k = kT$ can be written as

$$y_1(t_k) = \sum_{j=1}^{\infty} h_j u(t_{k-j}) \approx \sum_{j=1}^{m_1} h_j u(t_{k-j}) \tag{2}$$

$$y_2(t_k) = \sum_{i=1}^{\infty}\sum_{j=1}^{\infty} g_{ij} u(t_{k-i}) u(t_{k-j}) \approx \sum_{i=1}^{m_2}\sum_{j=1}^{m_2} g_{ij} u(t_{k-i}) u(t_{k-j}) \tag{3}$$

The coefficients h_j and g_{ij} of the discrete-time first- and second-order kernels can be easily shown to be

$$h_j \hat{=} \int_{t_j}^{t_{j+1}} h(r)dr = \int_{jT}^{(j+1)T} h(r)dr, \quad j = 1, 2, \ldots \tag{4}$$

$$g_{ij} \hat{=} \int_{t_j}^{t_{j+1}} \int_{t_j}^{t_{j+1}} g(r,s)drds = \int_{jT}^{(j+1)T} \int_{t_j}^{t_{j+1}} g(rs)drds, \quad i, j = 1, 2, \ldots \tag{5}$$

respectively. As equations (4) and (5) imply, the identification of the model in eqns. (2) and (3) entails the identification of the corresponding integrals of the continuous-time kernels $h(r)$ and $g(r,s)$, which, in turn, can be approximated as $h(r) \approx \sum_{l=1}^{m_h} c_l^h v_l^h(r)$ and $g(r,s) \approx \sum_{l=1}^{m_g} c_l^g v_l^g(r,s)$ where $\{v_l^h(r)\}_{l=1}^{m_h}$ and $\{v_l^g(r)\}_{l=1}^{m_g}$ are sets of basis functions for the subspaces of functions approximating h and g, respectively. Substituting the above equations into equations (4) and (5) yields

$$h_j \approx \sum_{l=1}^{m_h} c_l^h \int_{jT}^{(j+1)T} v_l^h(r)dr = \sum_{l=1}^{m_h} c_l^h w_{jl}^h \Longrightarrow$$

$$\mathbf{h} \hat{=} \begin{bmatrix} h_1 \\ \vdots \\ h_{m_1} \end{bmatrix} = \begin{bmatrix} w_{11}^h & \cdots & w_{1m_h}^h \\ \vdots & \ddots & \vdots \\ w_{m_1 1}^h & \cdots & w_{m_1 m_h}^h \end{bmatrix} \begin{bmatrix} c_1^h \\ \vdots \\ c_{m_h}^h \end{bmatrix} = \sum_{l=1}^{m_h} c_l^h \mathbf{w}_l^h \tag{6}$$

and

$$g_{ij} \approx \sum_{l=1}^{m_g} c_l^g \int_{iT}^{(i+1)T} \int_{jT}^{(j+1)T} v_l^g(r,s)drds = \sum_{l=1}^{m_g} c_l^g w_{ijl}^g \Longrightarrow$$

$$\mathbf{G} \hat{=} \begin{bmatrix} g_{11} & \cdots & g_{1m_2} \\ \vdots & \ddots & \vdots \\ g_{m_2 1} & \cdots & g_{m_2 m_2} \end{bmatrix} = \sum_{l=1}^{m_g} c_l^g \mathbf{W}_l^g = [\ \mathbf{W}_1^g \cdots \mathbf{W}_{m_g}^g\ ] \left(\begin{bmatrix} c_1^g \\ \vdots \\ c_{m_g}^g \end{bmatrix} \otimes \mathbf{I}_{m_2} \right) \tag{7}$$

where the Kronecker product symbol $\otimes$ is defined as

$$\mathbf{A} \otimes \mathbf{B} \hat{=} \begin{bmatrix} a_{11}\mathbf{B} & \cdots & a_{1q}\mathbf{B} \\ \vdots & \ddots & \vdots \\ a_{p1}\mathbf{B} & \cdots & a_{pq}\mathbf{B} \end{bmatrix} \tag{8}$$

[1]. The above equations make it clear that identification of the Volterra model of equation (1) can be transformed into identification of the vectors $[c_1^h \cdots c_{m_h}^h]^T$ and $[c_1^g \cdots c_{m_h}^g]^T$. The advantage of this transformation is that the basis functions $\{v_l^h(r)\}_{l=1}^{m_h}$ and $\{v_l^g(r)\}_{l=1}^{m_g}$ can be virtually always chosen in a way such that $m_h \ll m_1$ and $m_g \ll m_2$. Indeed, while the values of m_1 and m_2 are determined by

the continuous-time system's sampling period T (which, in turn, should be small enough to prevent aliasing of the plant output y), the values of m_h and m_g are not. In fact, as Figures 1 and 2 show, both the first- and second-order Volterra kernels exhibit a similar pattern, namely substantial excusrsions for small values of kernel indices, but smooth tail ends. Hence, both kernels need not be sampled uniformly, but rather "more densely" loosely speaking, close to the origin and "less densely" towards their tail ends.

The above discussion naturally leads to the problem of how to select the basis functions and. Certainly, there are several alternatives. In the sequel we propose an approach relying on the DWT. Our motivation for using wavelets is based on their time-frequency localization properties, which have been extensively exploited by the signal compression research community.

In the sequel, whenever k is an integer, $x(k)$ will denote the value of the variable x at time kT.

2.2. Identification of Volterra models

2.2.1. MODEL PARAMETRIZATION. Assume the following discrete convolution structure for the system to be identified:

$$y(t) = y_1(t) + y_2(t) + \varepsilon\,(k) \tag{9}$$

$$y_1(k) \hat{=} \sum_{j=1}^{m_1} h_j u(k-j) = \mathbf{u}_1(k)^T \mathbf{h} \tag{10}$$

$$y_2(k) \hat{=} \sum_{i=1}^{m_2} \sum_{j=1}^{m_2} g_{ij} u(k-i) u(k-j) = \mathbf{u}_2(k)^T \mathbf{G} \mathbf{u}_2(k) \tag{11}$$

where $\mathbf{u}_1(k) \hat{=} [u(k-1) \cdots u(k-m_1)]^T$, $\mathbf{u}_2(k) \hat{=} [u(k-1) \cdots u(k-m_2)]^T$ and $\varepsilon(k)$ is output-additive random noise with zero mean. Usually $m_1 = m_2$, although this is not necessary for the following analysis. In the double summation of equation (11), we can assume, without loss of generality, that the kernel coefficient matrix $\mathbf{G}$ is symmetric. This means that out of the m_2^2 second-order coefficients, only $M_2 \hat{=} m_2(m_2+1)/2$ coefficients are distinguishable. For notational consistency, we also define $M_1 \hat{=} m_1$ to be the number of distinct linear coefficients.

The vector $vec_{lt}(\mathbf{G})$ (Appendix A) contains all the distinct elements of $\mathbf{G}$, i.e., all elements from the lower-triangular part of $\mathbf{G}$, and equation (11) may be written (as shown in Appendix B) as

$$y_2(k) = \mathbf{u}_{2\mathrm{lt}\Omega}^T(k) \mathbf{g}_{\mathrm{lt}} \tag{12}$$

where $\mathbf{u}_{2\mathrm{lt}\Omega}(k) \hat{=} \Omega_{m_2} \mathrm{vec}_{\mathrm{lt}}[\mathbf{U}_2(k)]$, $\mathbf{U}_2(k) \hat{=} \mathbf{u}_2(k)\mathbf{u}_2(k)^T$, $\mathbf{g}_{\mathrm{lt}} \hat{=} \mathrm{vec}_{\mathrm{lt}}(\mathbf{G})$, and

$$\Omega_{m_2} \hat{=} diag(\underbrace{1,2,2,\ldots,2}_{m_2 \text{ elements}} \quad \underbrace{1,2,2,\ldots,2}_{m_2-1 \text{ elements}} \quad \cdots \quad \underbrace{1}_{1 \text{ element}}) \in \Re^{M_2 \times M_2}. \tag{13}$$

Hence, the entire Volterra model of equation (9) can be parametrized in the standard linear regression form as

$$y(k) = \phi(k)^T\theta + \varepsilon(k) \tag{14}$$

where $\phi(k) \hat{=} \begin{bmatrix} \mathbf{u}_1(k) \\ \mathbf{u}_{2\mathrm{lt}\Omega}(k) \end{bmatrix}$ $\theta \hat{=} \begin{bmatrix} \mathbf{h} \\ \mathbf{g}_{\mathrm{lt}} \end{bmatrix}$.

2.2.2. IDENTIFICATION OF VOLTERRA MODELS USING ORDINARY LEAST SQUARES (OLS). The model equations (14) for input-output data for N observations can be written in matrix form as

$$\mathbf{y} = \Phi^T\theta + \varepsilon \tag{15}$$

where $\mathbf{y}\hat{=}[y(k-N+1)\cdots y(k)]^T$, $\Phi\hat{=}[\phi(k-N+1)\cdots\phi(k)]^T$, $\varepsilon\hat{=}[\phi(k-N+1)\cdots\phi(k)]^T$. The standard ordinary least-squares (OLS) estimation method solves the minimization problem

$$\min_{\mathbf{h},\mathbf{g}_{\mathrm{lt}}} \sum_{k=1}^{N} \left(y(k) - \sum_{i=1}^{m_1} h_i u(k-i) - \sum_{i=1}^{m_2}\sum_{j=1}^{m_2} g_{ij} u(k-i)u(k-j) \right)^2 =$$

$$= \min_{\theta} ||\mathbf{y} - \Phi^T\theta||_2^2 \tag{16}$$

to provide the OLS estimate $\hat{\theta}$ of the parameter vector θ, given by

$$\Phi\Phi^T\hat{\theta} = \Phi\mathbf{y} \tag{17}$$

DECOUPLING OF 1ST AND 2ND ORDER KERNEL IDENTIFICATION. The OLS objective, equation (16), may be written explicitly in terms of the Volterra kernels $\mathbf{h}$ and $\mathbf{g}_{\mathrm{lt}}$ as

$$\min_{\mathbf{h},\mathbf{g}_{\mathrm{lt}}} \left\| \mathbf{y} - \begin{bmatrix} \Phi_1^T & \Phi_2^T \end{bmatrix} \begin{bmatrix} \mathbf{h} \\ \mathbf{g}_{\mathrm{lt}} \end{bmatrix} \right\|_2^2 \Longrightarrow \begin{bmatrix} \frac{\Phi_1\Phi_1^T}{N} & \frac{\Phi_1\Phi_2^T}{N} \\ \frac{\Phi_2\Phi_1^T}{N} & \frac{\Phi_2\Phi_2^T}{N} \end{bmatrix} \begin{bmatrix} \hat{\mathbf{h}} \\ \hat{\mathbf{g}}_{\mathrm{lt}} \end{bmatrix} = \begin{bmatrix} \frac{\Phi_1}{N} \\ \frac{\Phi_2}{N} \end{bmatrix} \mathbf{y} \tag{18}$$

If the random inputs $u(k)$ are independent and identically distributed (i.i.d.) following a symmetric distribution with zero mean, then their odd-order moments are zero. Consequently, for large values of N, the terms $\Phi_1\Phi_2^T/N$ and $\Phi_2\Phi_1^T/N$ in equation (18) have entries $\frac{1}{N}\sum_k u(k-i)u(k-j)u(k-l) \approx Eu(k-i)u(k-j)u(k-l)$ which are approximately zero, where "E" is the expectation operator. Thus, the estimation of the linear and quadratic kernels are decoupled (Pearson et al. (1992)), i.e., $\Phi_1\Phi_1^T\hat{\mathbf{h}} = \Phi_1\mathbf{y}$ and $\Phi_2\Phi_2^T\hat{\mathbf{g}}_{\mathrm{lt}} = \Phi_2\mathbf{y}$.

UNBIASED ESTIMATES. Multiplication of equation (9) by $u(k-l)$ and by $u(k-l_1)u(k-l_2)$, where the inputs $u(k)$ are as above, application of the expectation operator on both sides of the equation, and use of ergodicity, shows (after straightforward calculations) that the estimates of $\mathbf{h}$ and $\mathbf{g}_{\mathrm{lt}}$ in equation (14) are unbiased, even if the zero-mean noise term $\varepsilon(k)$ is colored.

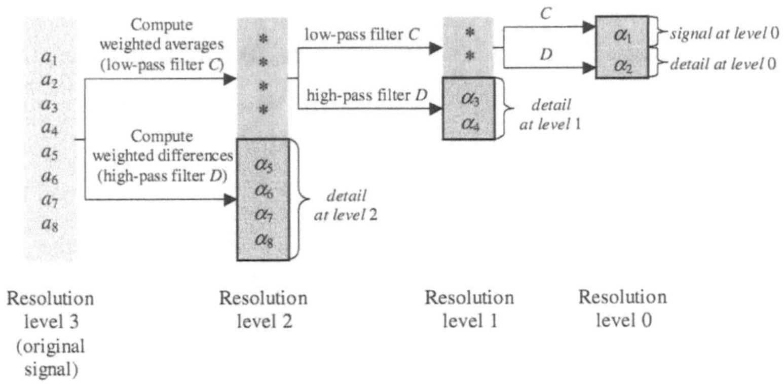

FIGURE 3. Schematic illustration of computing the 1-D DWT.

3. Discrete wavelet transform (DWT)

3.1. One-dimensional DWT

The (one-dimensional) DWT of a vector is the result of a linear transformation that generates a new vector of dimension equal to that of the original vector. The process of this transformation is also called decomposition, and can be achieved efficiently ($O(n)$ computations) by employing Mallat's multiresolution analysis (MRA) algorithm [9]. The key elements required for computation of the DWT are two discrete filters, the decomposition lowpass filter C and the decomposition highpass filter D that are associated with the *primal* wavelet. For computing the inverse DWT (IDWT), we need the corresponding reconstruction lowpass filter $\tilde{C}$ and reconstruction highpass filter $\tilde{D}$ associated with the *dual* wavelet. The two wavelets are biorthogonal [20].

Figure 3 schematically illustrates the process of taking the DWT. Let $\mathbf{a}$ be a vector in $\Re^{2^r}$ whose DWT is to be computed. This vector is designated to be the *approximation* at resolution level r, denoted $\mathbf{a}^r$. Filtering this vector with the lowpass filter C will result in a vector that is of the same dimension (assuming a suitable method of signal extension), and downsampling (i.e., removing alternate elements from) this vector gives the vector $\mathbf{a}^{r-1}$ of half the length, i.e., of dimension $2^{r-1} \times 1$. This vector is called the approximation at resolution level $r-1$. Similarly, filtering ar with the *highpass* filter D and then downsampling gives the vector $\mathbf{b}^{r-1}$ that is half the length of $\mathbf{a}^r$, and is called the *detail* at level $r-1$. The detail $\mathbf{b}^{r-1}$ is kept aside, and the approximation $\mathbf{a}^{r-1}$ is again filtered and downsampled to yield two vectors $\mathbf{a}^{r-2}$ and $\mathbf{b}^{r-2}$ that are half the length of $\mathbf{a}^{r-1}$. These are the approximation and detail, respectively, at resolution level $r-2$. Proceeding in this manner down to resolution level 0, the DWT vector, $\alpha \in \Re^{2^r}$, with respect to the primal wavelet, is the aggregate

$$\alpha \hat{=} \begin{bmatrix} (\mathbf{a}^0)^T & (\mathbf{b}^0)^T & (\mathbf{b}^1)^T & \cdots & (\mathbf{b}^{r-1})^T \end{bmatrix}^T \tag{19}$$

It is of the same dimension as the original vector **a**, and consists of the approximation at the lowest resolution level, the detail at that level, and all the details at higher levels. The vectors $\mathbf{a}^j$ and $\mathbf{b}^j$ of dimension $2^j \times 1$ $(0 \leq j \leq r-1)$ can be obtained recursively using the linear transformation through a $2^j \times 2^j$ matrix $\mathbf{F}_j$. Each step of this recursive algorithm is called the single-step discrete wavelet decomposition:

$$\left[(\mathbf{a}^{j-1})^T \quad (\mathbf{b}^{j-1})^T\right]^T = \mathbf{F}_j \mathbf{a}^j, \quad j = 1, 2, \ldots, r; \quad \mathbf{a}^r \hat{=} \mathbf{a} \tag{20}$$

$$\mathbf{F}_j \hat{=} \left[\mathbf{C}_j^T \quad \mathbf{D}_j^T\right]^T \tag{21}$$

where $\mathbf{C}_j$ and $\mathbf{D}_j$ are the $2^{j-1} \times 2^j$ matrices constructed to filter $\mathbf{a}^j$ by the lowpass filter C and the highpass filter D, respectively, and then downsample. For given filters C and D, the filter matrices can be constructed in several ways to filter a finite-length signal, depending on the various methods of signal extension, such as periodic, symmetric or constant extension [20]. In this work, we use constant extension. As an example, if C and D have four coefficients each (as in the case of the quadratic-spline (QS) wavelet), $\mathbf{C}_j$ is constructed as

$$\mathbf{C}_j = \begin{bmatrix} c_0 + c_1 & c_2 & c_3 & & & & \\ & c_0 & c_1 & c_2 & c_3 & & \\ & & & c_0 & c_1 & c_2 & c_3 \\ & & & \ddots & \ddots & \ddots & \\ & & & & c_0 & c_1 & c_2 + c_3 \end{bmatrix} \tag{22}$$

and $\mathbf{D}_j$ is constructed similarly, with the d_j in place of c_j.

The simplest filters correspond to the Haar wavelet, which has the advantage of orthogonality (self-duality), i.e., $\tilde{C} = C$ and $\tilde{D} = D$, but it has poor filtering characteristics. The QS wavelet has better interpolating properties (it is smoother). The filters for the QS wavelet used in this work are

$$(c_0 \quad c_1 \quad c_2 \quad c_3) = \frac{1}{2\sqrt{2}}(-1 \quad 3 \quad 3 \quad -1) \tag{23}$$

$$(\tilde{c}_0 \quad \tilde{c}_1 \quad \tilde{c}_2 \quad \tilde{c}_3) = \frac{1}{2\sqrt{2}}(1 \quad 3 \quad 3 \quad 1) \tag{24}$$

$$(d_0 \quad d_1 \quad d_2 \quad d_3) = \frac{1}{2\sqrt{2}}(-1 \quad 3 \quad -3 \quad 1) \tag{25}$$

$$\left(\tilde{d}_0 \quad \tilde{d}_1 \quad \tilde{d}_2 \quad \tilde{d}_3\right) = \frac{1}{2\sqrt{2}}(1 \quad 3 \quad -3 \quad -1) \tag{26}$$

The DWT of $\mathbf{a}$, with respect to (w.r.t.) the primal wavelet, is then given by the linear transformation

$$\alpha = \mathbf{W}_1 \mathbf{a} \tag{27}$$

where $\mathbf{W}_1 \hat{=} \mathbf{N}_1 \mathbf{N}_2 \cdots \mathbf{N}_r$, $\mathbf{N}_j \hat{=} \begin{bmatrix} \mathbf{F}_j & \mathbf{0} \\ \mathbf{0} & \mathbf{I} \end{bmatrix} \in \Re^{2^r \times 2^r}$, $j = 1, 2, \ldots, r$. The DWT of $\mathbf{a}$ with respect to the dual wavelet can also be obtained in a similar manner as

$$\tilde{\alpha} = \tilde{\mathbf{W}}_1 \mathbf{a} \tag{28}$$

Note that the biorthogonality property of the wavelets leads to the equality

$$\tilde{\mathbf{F}}_j^T \mathbf{F}_j = \mathbf{I}_{2^j}, \quad j = 1, 2, \ldots, r \Rightarrow \tilde{\mathbf{W}}_1^T \mathbf{W}_1 = \mathbf{I}_{2^r} \tag{29}$$

The IDWT is the process of reconstructing $\mathbf{a}$ from α (or $\tilde{\alpha}$), and the key recursion here comes from the inverse of relation (20) using equation (29):

$$\mathbf{a}^j = \begin{bmatrix} \tilde{\mathbf{C}}_j^T & \tilde{\mathbf{D}}_j^T \end{bmatrix} \begin{bmatrix} \mathbf{a}^{j-1} \\ \mathbf{b}^{j-1} \end{bmatrix} \tag{30}$$

That is, the single-step discrete wavelet reconstruction involves upsampling followed by filtering with the reconstruction filters [20]. Hence, $\mathbf{a}$ is reconstructed as

$$\mathbf{a} = \tilde{\mathbf{W}}_1^T \alpha = \mathbf{W}_1^T \tilde{\alpha} \tag{31}$$

Note that the matrix inverses $\mathbf{W}_1^{-1}$ or $\tilde{\mathbf{W}}_1^{-1}$ need not be computed numerically. The IDWT also uses a numerically efficient recursive algorithm ($O(n)$ computations).

3.2. Two-dimensional DWT

The 2-D DWT of a matrix is the result of a transformation that generates a new matrix of dimensions equal to those of the original matrix. The transformation is such that each element of the 2-D DWT is a linear combination of the elements of the original matrix. In the following, we only consider the class of 2-D wavelets that are *biorthogonal* and are a *separable* MRA approximation of $\Re^2$ [9]. Biorthogonality is used to avoid inefficient numerical inversion of the DWT transformation matrix ($\mathbf{W}_1$ or $\mathbf{W}_2$) during the computation of the IDWT (equations (31) and (38)). Separability has two advantages that lead to easier analysis. First, 2-D filtering with the wavelet filters may be achieved as two successive 1-D filterings (equation (32)). Second, the 2-D DWT of the symmetric kernel $\mathbf{G}$ is also symmetric, so that the unknown parameters to be identified are in the same form in both the kernel and its 2-D DWT. A brief outline of the 2-D DWT is given next.

Let $\mathbf{L}$ be a square matrix of dimension $2^r \times 2^r$, whose 2-D DWT is to be computed. This matrix is designated to be at resolution level r, and it is sought to obtain the approximations and details at the lower resolution levels. The matrices

$\mathbf{L}^j$, $\mathbf{M}_H^j$, $\mathbf{M}_V^j$ and $\mathbf{M}_D^j$ of dimension $2^j \times 2^j$ are called respectively the 'approximation', 'horizontal-detail', 'vertical-detail' and 'diagonal-detail' coefficients at resolution level j $(0 \leq j \leq r-1)$, and are obtained recursively by the relation

$$\left[\begin{array}{c|c} \mathbf{L}^{j-1} & \mathbf{M}_V^{j-1} \\ \hline \mathbf{M}_H^{j-1} & \mathbf{M}_D^{j-1} \end{array}\right] = \mathbf{F}_j \mathbf{L}^j \mathbf{F}_j^T = \left[\mathbf{F}_j (\mathbf{F}_j \mathbf{L}^j)^T\right]^T, \quad j = 1, 2, \ldots r; \quad \mathbf{L}^r \hat{=} \mathbf{L} \tag{32}$$

where the matrix $\mathbf{F}_j$ is as in equation (21). This is the one-step discrete wavelet decomposition in the 2-D case, where we use the approximation matrix at some resolution level j, $\mathbf{L}^j$, to obtain the approximation matrix and the three types of detail matrices at the next lower resolution level $j-1$. From equation (32), we see that we first take the single-step 1-D decomposition (equation (20)) of all columns of $\mathbf{L}^j$, transpose the result, again take the 1-D decomposition, and finally transpose again. The approximation matrix is the result of lowpass filtering in the vertical direction (along the columns of the matrix) followed by lowpass filtering in the horizontal direction (along the rows). This is denoted 'low-low'. Similarly, the horizontal details come from 'high-low', vertical details from 'low-high' and diagonal details from 'high-high' filtering. Proceeding down to the resolution level 0, the 2-D DWT of $\mathbf{L}$ with respect to the primal wavelets is the aggregate matrix

$$\Lambda \hat{=} \left[\begin{array}{c|c} \begin{array}{c|c|c} \begin{array}{c|c} \mathbf{L}^0 & \mathbf{M}_V^0 \\ \hline \mathbf{M}_H^0 & \mathbf{M}_D^0 \end{array} & \mathbf{M}_V^1 & \cdots \\ \hline \mathbf{M}_H^1 & \mathbf{M}_D^1 & \\ \hline \vdots & & \ddots \end{array} & \mathbf{M}_V^{r-1} \\ \hline \mathbf{M}_H^{r-1} & \mathbf{M}_D^{r-1} \end{array}\right] \tag{33}$$

Again, the 2-D DWT consists of the approximation matrix at the lowest resolution level, 0, and the three types of details at this and all the higher levels. The elements in Λ can be obtained using the 2-D DWT transformation matrix $\mathbf{W}_2$ of dimension $2^{2r} \times 2^{2r}$, analogous to $\mathbf{W}_1$ in the 1-D case and constructed in [10], as

$$\mathrm{vec}(\Lambda) = \mathbf{W}_2 \mathrm{vec}(\mathbf{L}) \tag{34}$$

where the vec function is defined [11] as

$$\mathrm{vec}(\mathbf{A}) \hat{=} \begin{bmatrix} \mathbf{A}_{\bullet 1} \\ \vdots \\ \mathbf{A}_{\bullet 1} \end{bmatrix} \in \Re^{pq \times 1} \tag{35}$$

and $\{\mathbf{A}_{\bullet j}\}_{j=1}^{q}$ are the columns of the matrix $\mathbf{A}$. Similarly, the elements of the 2-D DWT with respect to the dual wavelets are given by

$$\mathrm{vec}(\tilde{\Lambda}) = \tilde{\mathbf{W}}_2 \mathrm{vec}(\mathbf{L}) \tag{36}$$

and again it can be shown [10] that biorthogonality of the wavelets leads to

$$\tilde{\mathbf{W}}_2^T \mathbf{W}_2 = \mathbf{I}_{2^{2r}} \tag{37}$$

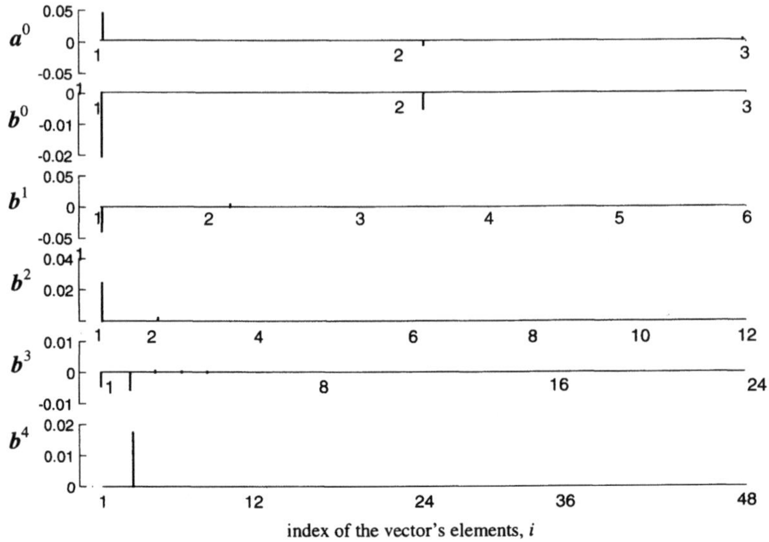

FIGURE 4. The 1-D DWT of the FIR kernel in Figure 1.

As a corollary [10] we note that, owing to the separability condition mentioned above, whenever $\mathbf{L}$ is symmetric Λ is also symmetric. With the use of biorthogonality, the 2-D IDWT of Λ (or $\tilde{\Lambda}$) is computed as

$$\text{vec}(\mathbf{L}) = \tilde{\mathbf{W}}_2^T \text{vec}(\Lambda) = \mathbf{W}_2^T \text{vec}(\tilde{\Lambda}) \tag{38}$$

again obviating the numerical computation of the matrix inverses $\mathbf{W}_2^{-1}$ or $\tilde{\mathbf{W}}_2^{-1}$.

4. Compression of a discrete Volterra model using the DWT

4.1. Compression of Volterra models

4.1.1. FIRST-ORDER KERNEL. Assume that the memory length of the linear Volterra kernel $\mathbf{h}$ is an integral power of 2, i.e., $m_1 = 2^{r_1}$. Using the biorthogonality condition (29), equation (10) may be reparametrized in terms of the DWT as

$$y_1(k) = \mathbf{u}_1(k)^T \mathbf{h} = [\tilde{\mathbf{W}}_1 \mathbf{u}_1(k)]^T \mathbf{W}_1 \mathbf{h} = \tilde{v}_1^T(k)\eta \tag{39}$$

where $\tilde{v}_1(k) \hat{=} \tilde{\mathbf{W}}_1 \mathbf{u}_1(k)$, $\eta \hat{=} \mathbf{W}_1 \mathbf{h}$. That is, η is the DWT of the linear kernel $\mathbf{h}$ w.r.t. the primal wavelet and $\tilde{v}_1(k)$ is the DWT of the input vector $\mathbf{u}_1(k)$ w.r.t. the dual wavelet. Figure 4 shows a typical DWT of an FIR kernel.

Model compression (approximation) using wavelets first involves determining the DWT coefficients of the kernel. Of the obtained DWT coefficients, only the coefficients with magnitude above a certain threshold value are retained, the rest being discarded (assumed to be zero) thus creating a new vector

$$\eta_c = \mathbf{P}_{c1}\eta \tag{40}$$

of dimension $M_{c1} < M_1$, where the matrix $\mathbf{P}_{c1}$ of dimensions $M_{c1} \times M_1$ selects the desired elements of the vector η. Good compression can be attained if a large number of DWT coefficients are negligible, i.e. $M_{c1} \ll M_1$. The term y_1 can then be approximated as

$$y_1(k) \approx \tilde{v}_{1c}(k)^T \eta_c \tag{41}$$

where $\tilde{v}_{1c}(k) \hat{=} \mathbf{P}_{c1}\tilde{v}_1(k)$. If the compressed DWT η_c is known, then the original linear kernel $\mathbf{h}$ can be approximated as

$$\mathbf{h} \approx \mathbf{h}_c = \tilde{\mathbf{W}}_1^T \mathbf{P}_{c1}^T \eta_c = \mathbf{X}_1 \eta_c \tag{42}$$

where $\mathbf{X}_1 = \tilde{\mathbf{W}}_1^T \mathbf{P}_{c1}^T$. The compressed kernel $\mathbf{h}_c$ is of the same dimension as the uncompressed kernel $\mathbf{h}$. Due to the binary-tree structure of the MRA algorithm, the removal of a single detail coefficient from η (when obtaining η_c) at a resolution level j, $0 \le j \le r-1$, will have an effect on $O(2^{r-j})$ elements of $\mathbf{h}$. Discarding neighboring coefficients from η will affect overlapping sets of elements of $\mathbf{h}$. Since $\mathbf{h}$ typically does not exhibit oscillatory behavior in its tail end, η will have many coefficients close to the highest resolution level ($j \approx r-1$) that are approximately equal to zero. Therefore, removal of these coefficients from η and reconstruction of $\mathbf{h}_c$ from η_c will (i) compress $\mathbf{h}$ significantly, because the number of coefficients at each resolution level j of η is 2^j, which implies that higher resolution levels contain exponentially more coefficients, and (ii) have a localized effect on $\mathbf{h}$. For example, for the QS wavelet used in this work, the effect of removing a coefficient from η at resolution level $r-1$ will affect no more than 4 coefficients of $\mathbf{h}$, which is relatively very small when r is large.

4.1.2. SECOND-ORDER KERNEL. The quadratic kernel may be treated similarly, but the algebra is more involved and presented here only briefly, due to space constraints. Details can be found in [10]. Suppose that the 2-D DWT of the kernel $\mathbf{G}$ is Γ. Since $\mathbf{G}$ is symmetric, Γ is also symmetric. Define the vector-valued function 'vec$_{\text{lt}}$' of an $(m \times m)$ square matrix $\mathbf{Q}$ as the $(m(m+1)/2 \times 1)$ vector

$$\text{vec}_{\text{lt}}(\mathbf{Q}) \hat{=} \left[\underbrace{q_{11} \quad \cdots \quad q_{m1}} \quad \underbrace{q_{22} \quad \cdots \quad q_{m2}} \quad \cdots \quad \underbrace{q_{mm}} \right]^T \tag{43}$$

obtained by taking the lower triangular part of $\mathbf{Q}$ and stacking the columns (of decreasing length) in a vector. Therefore, the vector

$$\gamma_{\text{lt}} \hat{=} \text{vec}_{\text{lt}}(\Gamma) \tag{44}$$

contains all the distinct elements of Γ, and the output of the quadratic kernel (equation (11)) can be shown to be

$$y_2(k) = \tilde{v}_{2\text{lt}\Omega}^T(k)\gamma_{\text{lt}} \tag{45}$$

where the vector $\tilde{v}^T_{2\mathrm{lt}\Omega}(k)$ is a corresponding transformation of the input vector $\mathbf{u}_2(k)$, that involves the 2-D DWT w.r.t. the dual wavelet. Again, when only $M_{c2} \ll M_2$ elements of the total of M_2 elements of γ_{lt} are significant, we can compress γ_{lt} to the vector

$$\gamma_{\mathrm{ltc}} = \mathbf{P}_{c2}\gamma_{\mathrm{lt}} \tag{46}$$

analogously to the linear-kernel case, and correspondingly compress the transformed input vector to the vector

$$\tilde{v}^T_{2\mathrm{lt}\Omega c}(k) = \mathbf{P}_{c2}\tilde{v}^T_{2\mathrm{lt}\Omega}(k) \tag{47}$$

The quadratic term of the output is now approximated as

$$y_2(k) \approx [\tilde{v}^T_{2\mathrm{lt}\Omega c}(k)]^T \gamma_{\mathrm{ltc}} \tag{48}$$

Again, given the compressed 2-D DWT γ_{ltc}, we can reconstruct the discrete-time quadratic kernel $\mathbf{G}_c$ using the transformation

$$\mathrm{vec}(\mathbf{G}_c) = \mathbf{X}_2\gamma_{\mathrm{ltc}} \tag{49}$$

where the matrix $\mathbf{X}_2$ involves a 2-D IDWT [10].

4.2. Identification using wavelet compression

Assume that a number of coefficients of the DWTs η and γ_{lt} of the kernels $\mathbf{h}$ and $\mathbf{g}_{\mathrm{lt}}$, respectively, to be identified are set to be identically zero, as discussed below. Then, using equations (4.5) and (4.11), the full Volterra model of equation (2.8) can be approximated as

$$y(k) = y_1(k) + y_2(k) + \varepsilon(k) \approx \phi_c^T(k)\theta_c + \varepsilon(k) \tag{50}$$

where $\phi_c(k) \hat{=} \begin{bmatrix} \tilde{v}_{1c}(k) \\ \tilde{v}^T_{2\mathrm{lt}\Omega c}(k) \end{bmatrix}$ and $\theta_c \hat{=} \begin{bmatrix} \eta_c \\ \gamma_{\mathrm{ltc}} \end{bmatrix}$. This model has similar structure to equation (14). However, instead of identifying the vector θ of length $(M_1 + M_2)$ in the time domain, we now identify a (much) shorter vector of length $(M_{c1} + M_{c2})$ in the wavelet domain. With N observations, using the definitions of y and ε in equation (15), we can write the outputs as

$$\mathbf{y} = \Phi_c^T\theta_c + \varepsilon \tag{51}$$

where $\Phi_c \hat{=} \begin{bmatrix} \phi_c(k-N+1) & \cdots & \phi_c(k) \end{bmatrix}$. The compressed discrete-time domain kernels can be reconstructed from the standard least-squares estimate of the parameter vector θ_c as

$$\begin{bmatrix} \hat{\mathbf{h}}_c \\ \mathrm{vec}(\hat{\mathbf{G}}_c) \end{bmatrix} = \begin{bmatrix} \mathbf{X}_1\hat{\eta}_c \\ \mathbf{X}_2\hat{\gamma}_{\mathrm{ltc}} \end{bmatrix} = \left[\begin{array}{c|c} \mathbf{X}_1 & \mathbf{0} \\ \hline \mathbf{0} & \mathbf{X}_2 \end{array}\right]\hat{\theta}_c \tag{52}$$

The selection of the coefficients in η and γ_{lt} that are set to zero is crucial for good identification of the Volterra model. If too many coefficients are set to zero, then the approximate Volterra model will be inaccurate. On the other hand, if too few coefficients are removed, the resulting estimator will have high variance because of overparametrization. Consequently, the coefficients that must be set to zero have to be selected iteratively, using thresholding and standard goodness-of-fit

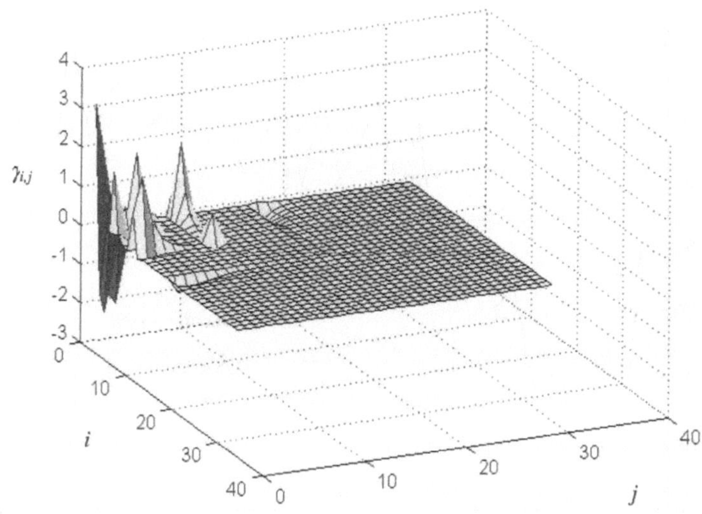

FIGURE 5. The 2D-DWT matrix Γ of the discrete second-degree Volterra kernel $\mathbf{G}$ in Figure 2.

criteria at each iteration ([12]; [19]). Prior knowledge about the dynamic behavior of the modeled plant can provide useful guidelines as to which coefficients to set to zero at the start of the iterative procedure. Indeed, as Figures 4 and 5 illustrate, a large number of coefficients corresponding to the tail ends of the kernels are virtually zero. To define the tail ends of the kernel in terms of corresponding indices of $\mathbf{h}$ and $\mathbf{G}$, one can use information from either first-principles modeling or from empirical modeling at a different operating point.

5. Case study

In this section we consider the application of the proposed wavelet compression to the identification of a second-order discrete Volterra model for a polymerization reaction taking place in a jacketed CSTR. The example is adapted from [5]. The development of the original physical model can be obtained from [2], [3], and [16]. The reaction under consideration is the free-radical polymerization of methylmethacrylate with azo-bis-isobutyronitrile as initiator and toluene as solvent.

Following the development presented in [5], we define the state variables $x_1 \hat{=} C_m$, $x_2 \hat{=} C_I$, $x_3 \hat{=} D_O$, and $x_4 = D_I$ where C_m is the monomer concentration, C_I is the initiator concentration, and D_I/D_O is the number-average molecular weight. The volumetric flow rate of the initiator ($u' = F_I$) and the number-average molecular weight ($y' = D_I/D_O$) are taken as the input and output variables,

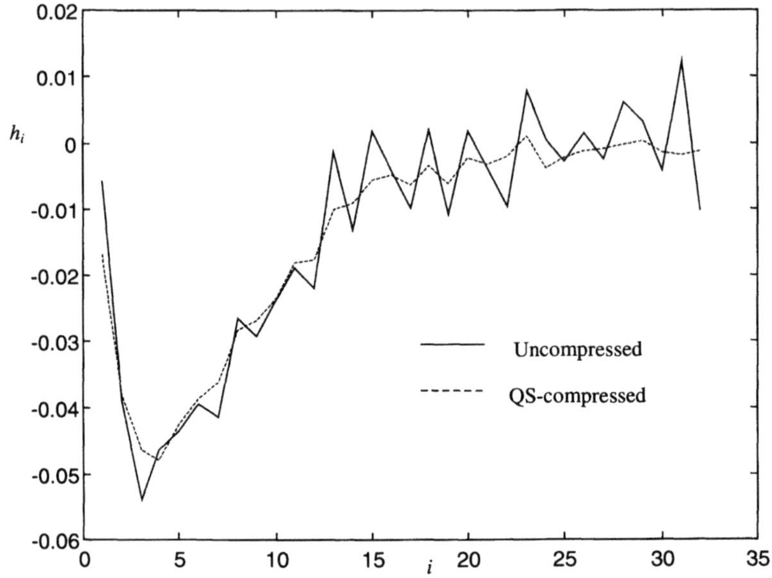

FIGURE 6. First-order discrete Volterra kernels ($m_1 = 32$, $T = 1/32hr$) identified without and with QS-wavelet compression. 18/32 DWT coefficients retained.

respectively, whose relationship is sought to be modeled as a second-order discrete Volterra model. The operating conditions chosen were $x_{10} = 5.50677$, $x_{20} = 0.132906$, $x_{30} = 0.0019752$, $x_{40} = 49.3818$, $u' = 0.016783$, and $y'_0 = 25000.5$. Under the assumptions mentioned in [5], the model then takes the form $\dot{x}_1 = 10(6 - x_1) - 2.4568x_1\sqrt{x_2}$, $\dot{x}_2 = 80u' - 10.1022x_2$, $\dot{x}_3 = 0.00241121x_1\sqrt{x_2} + 0.112191x_2 - 10x_3$, $\dot{x}_4 = 245.978x_1\sqrt{x_2} - 10x_4$, $y' = \frac{x_4}{x_3}$. The input u' and output y' are normalized as $u = (u' - u'_0)/u'_0$ and $y = (y' - y'_0)/y'_0$ so that these normalized variables are zero at the operating condition, have roughly equivalent ranges and are dimensionless. The discrete-time input/output data for the reactor are simulated using this first-principles model. The input is a sequence of independent random signals, uniformly distributed between -1 and +1 and implemented every $T = 1/32hr$. The continuous output is sampled with the same sampling period. We ran the simulation from $t = 0$ to $40hr$, thus obtaining 1280 input-output data points. The output was corrupted by additive Gaussian white noise with zero mean and variance 0.00035. The first 640 points were used for identification, and the next 640 points were used for cross validation of the identified model. With equal kernel memories $m_1 = m_2 = 32$, the full model requires $M_1 \hat{=} m_1 = 32$ linear kernel parameters and $M_2 \hat{=} m_2(m_2+1) = 528$ quadratic kernel parameters to be determined (a total of 560 parameters).

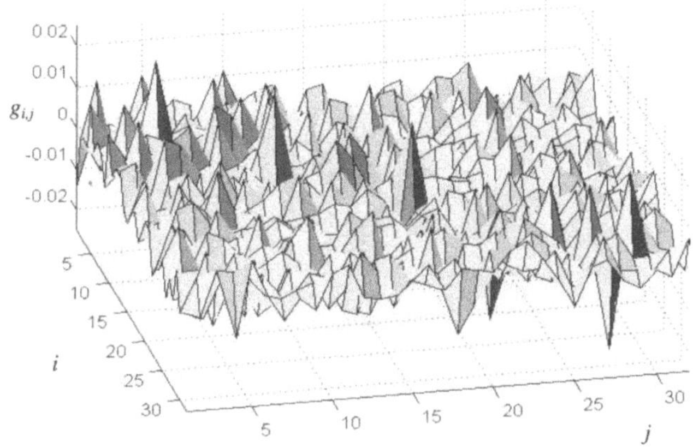

FIGURE 7. Second-order discrete Volterra kernel ($m_1 = m_2 = 32$, $T = 1/32hr$) identified without compression. 528 parameters identified from 640 data points.

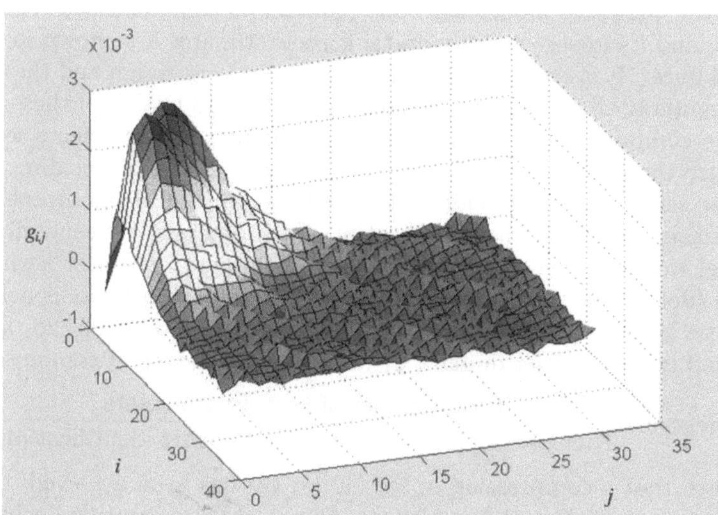

FIGURE 8. Second-order discrete Volterra kernel ($m_1 = m_2 = 32$, $T = 1/32hr$) identified without compression. 528 parameters identified from 1280 data points.

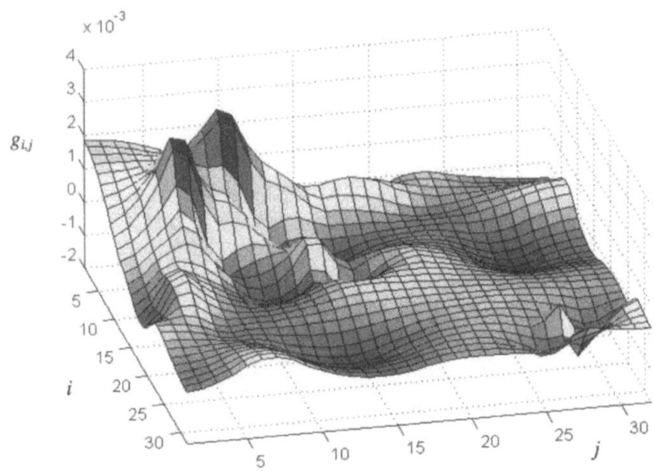

FIGURE 9. Second-order discrete Volterra kernel ($m_1 = m_2 = 32$, $T = 1/32hr$) identified with QS-wavelet compression. 28/528 coefficients in 2-D DWT retained.

First, the complete model with 560 parameters was identified from the 640 data points, and its linear- and quadratic-kernel estimates are shown in Figures 6 and 7 (solid lines). It is observed that the kernels are very rough and the quadratic kernel is essentially shapeless, as expected. Next, the QS DWTs of these identified kernels were computed, and the numerically insignificant coefficients were set to zero by a hard thresholding process (see, for example, [12]), thus yielding a starting point for the set of those coefficients of η_c and γ_{ltc} which are significant and need to be identified. Only 18 out of a total of 32 parameters corresponding to the linear kernel were retained, and the resulting identified first-order kernel, shown in Figure 6 (dotted), is much smoother. More striking smoothness is achieved for the quadratic kernel (Figure 7c), where only 28 out of a total of 528 parameters are estimated in the wavelet domain. If we define the degree of compression as

$$\text{Compression}=\frac{\text{Total number of kernel parameters}}{\text{Number of parameters retained during identification}} \tag{53}$$

it can be seen that a compression of 560/46 (≈ 12) has been achieved.

The uncompressed quadratic Volterra kernel was also identified using all 1280 data points of the simulation experiment, and the result is plotted in Figure 8. This kernel exhibits the two peaks that were captured by the wavelet-compressed kernel (Figure 9).

The uncompressed and compressed Volterra models developed from the first 640 data points were used to generate predictions of the plant output corresponding to the validation input data set (not used in identification), for comparison with

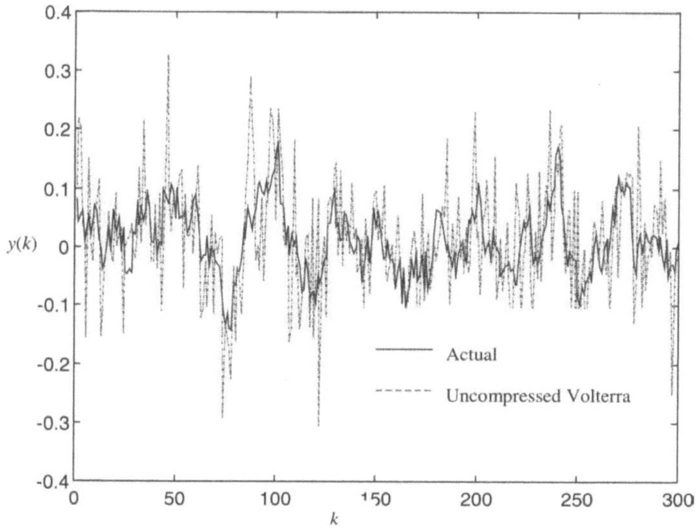

FIGURE 10. Outputs ($T = 1/32hr$) of the actual plant and the uncompressed Volterra model (in Figures 6 and 7) for the validation data set.

the actual plant (i.e., the first-principles model) output. The compressed model follows the actual plant very closely, as shown in Figure 11, while the uncompressed model is glaringly poor, as expected (Figure 10). The residuals (difference between actual and predicted output) for the compressed and the uncompressed models are plotted in Figure 8c, and the contrast is obvious. The reason that the uncompressed model does so poorly is, of course, the overfitting of data, i.e. too many parameters identified from a limited data set.

6. Conclusions and future work

In this article we proposed a new methodology for the identification of discrete second-order Volterra models using wavelet compression. On the basis of closeness of fit to the true plant model, we have demonstrated that the proposed model compression methodology (with the QS-wavelet basis) can reduce the requirements on experimental data for identification by at least an order of magnitude in comparison to the original Volterra model structure. One of the attractive features of the proposed approach is that qualitative prior knowledge about a plant model can be easily translated into quantitative information that can be utilized in parsimonious identification.

Future work can include the following. (i) Evaluate the proposed methodology by thoroughly comparing it with other nonlinear model structures (see, for example, [6]; [24]; [19]). (ii) Test the proposed methodology using various model

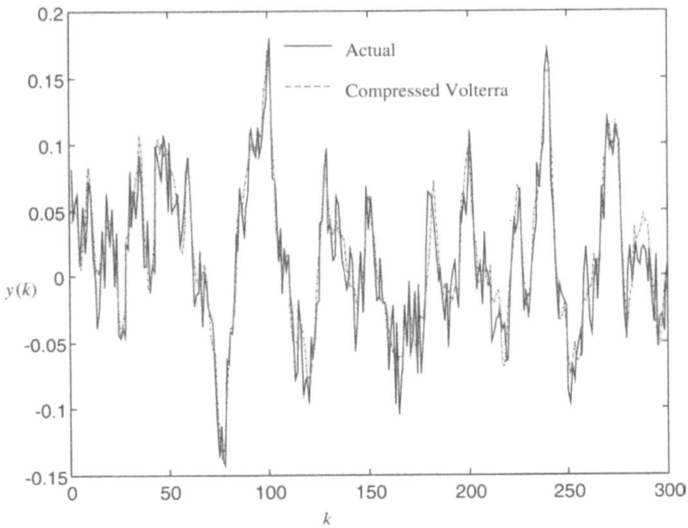

FIGURE 11. Outputs ($T = 1/32hr$) of the actual plant and the QS-wavelet-compressed Volterra model (in Figures 6 and 9) for the validation data set.

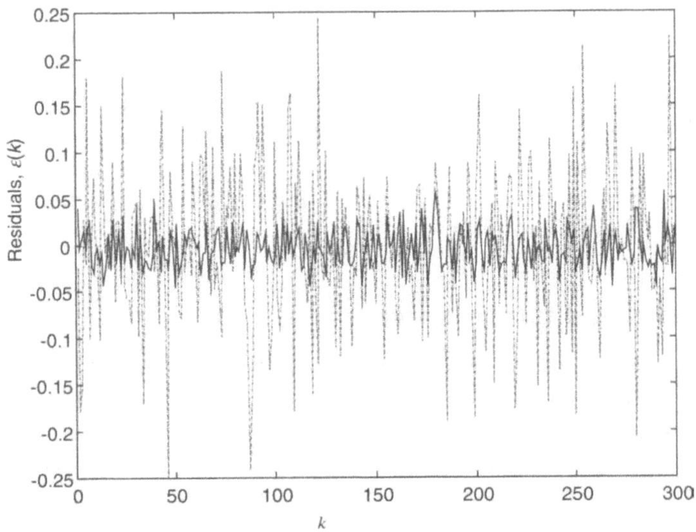

FIGURE 12. Comparison of the output-prediction residuals ($T = 1/32hr$) obtained using the compressed model (solid) and the uncompressed model (dotted).

validation procedures (iii) Refine schemata for retaining or rejecting DWT coefficients in the compressed model (iv) Explore and compare the compression potential of different wavelet bases (v) Study the effects of compressed model identification on closed-loop properties of MPC.

References

[1] Brewer, J. W., *Kronecker products and matrix calculus in system theory,* IEEE Trans. Circuits and Systems, CAS-25(9) (1978), 772–781.

[2] Congalidis, J. J. Richards and W. Ray, *Feedforward and feedback control of a copolymerization reactor,* AIChE J., 35 (1989), 891–907.

[3] Daoutadis, P., M. Soroush and C. Kravaris, *Feedforward/feedback control of multivariable nonlinear processes,* AIChE J., 36 (1990), 1471–1484.

[4] DeVore, R.A., B. Jawerth and B.J. Lucifer, *Image Compression Through Wavelet Transform Coding,* IEEE Transactions on Information Theory, 38(2) (1992), 719–746.

[5] Doyle, F. J. III, B. A. Ogunnaike and R. K. Pearson, *Nonlinear model-based control using second-order Volterra models,* Automatica, 31(5) (1995), 697–714.

[6] Kurth and Rake, *Identification of Nonlinear Systems with Reduced Volterra Series,* SYSID 94 Preprints, 10th IFAC Symposium on System Identification (1994).

[7] Juditsky, A., H. Hjalmarsson, A. Benveniste, B. Delyon, L. Ljung, J. Sjöberg, and Q. Zhang, *Nonlinear Black-box Models in System Identification: Mathematical Foundations,* Automatica, 31, 12 (1995), 1725–1750.

[8] Ling, W. M. and D. E. Rivera, *Control relevant model reduction of Volterra series models,* J. Proc. Cont., 8(2) (1998), 79–88.

[9] Mallat, S.G., *A Theory for Multiresolution Signal Decomposition: The Wavelet Representation,* IEEE Transactions on Pattern Analysis and Machine Intelligence, 11(7) (1989), 674–693.

[10] Mantha, D., *Parsimonious Identification of Discrete Volterra Models using Wavelet Compression,* M.S. Thesis, Department of Chemical Engineering, University of Houston (1998).

[11] Neudecker, H., *Some theorems on matrix differentiation with special reference to Kronecker matrix products,* J. Amer. Stat. Assoc., 64 (1969), 953–963.

[12] Nikolaou, M. and P. Vuthandam, *FIR Model Identification: Parsimony Through Kernel Compression with Wavelets,* AIChE J., 44(1) (1998), 141–150.

[13] Palavajjhala, S., R. L. Motard, and B. Joseph, *Process Identification using Discrete Wavelet Transforms: Design of Prefilters,* AIChE J., 42(3) (1996), 777–790.

[14] Pati, Y. C., R. Rezaiifar, P. S. Krishnaprasad, and W. P. Dayawansa, *A Fast Recursive Algorithm for System Identification and Model Reduction using Rational Wavelets,* Proc. of the 27th Asilomar Conference on Signals, Systems, and Computers, Pacific Grove, CA, 1 (1993), 35–39.

[15] Pearson, T., B. Ogunnaike and F. Doyle III, *Identification of discrete convolution models for nonlinear processes,* Proc. AIChE Annual Meeting (1992).

[16] Ray, W., *On the mathematical modeling of polymerization reactors,* J. Macromol. Sci.-Rev. Macromol. Chem., C8, 1–56 (1972).

[17] Rugh, W., Nonlinear System Theory – The Volterra/Wiener Approach, The Johns Hopkins University Press, Baltimore (1981).

[18] Schetzen, M., The Volterra and Wiener Theories of Nonlinear Systems, Wiley, New York (1980).

[19] Sjöberg, J., Q. Zhang, L. Ljung, A. Benveniste, B. Delyon, P.-Y. Glorennec, H. Hjalmarsson, and A. Juditsky, *Nonlinear Black-box Modeling in System Identification: a Unified Overview,* Automatica, 31, 12 (1995), 1691–1724.

[20] Strang, G. and T. Nguyen, Wavelets and Filter Banks, Wellesley, Cambridge, MA (1996).

[21] Sureshbabu, N., and J. A. Farrell, *Wavelet Based System Identification for Nonlinear Control Applications,* Proc. of 10th IEEE International Symposium on Intelligent Control, Monterey, CA (1995), 236–241.

[22] Tsatsanis, M. K., and G. B. Giannakis, *A Basis Expansion Approach for Detecting Transient Plant Disturbances and Jumps,* Proc. 33rd Conf. on Dec. and Cont., Lake Buena Vista, FL (1994), 3412–3417.

[23] Tsatsanis, M. K., and G. B. Giannakis, *Time-varying System Identification and Model Reduction using Wavelets,* IEEE Transactions on Signal Processing, SP-41 (1993), 3512–23.

[24] Wellers, and Rake, *Nonlinear Model Predictive Control of Simulated Chemical Processes using Volterra-Series,* in Nonlinear Model Based Process Control, R. Berber (Ed.), NATO ASI Series, Kluwer (1998).

Chemical Engineering Dept.
University of Houston
Houston, TX 77204-4792
USA
Tel: +1 713 743 4309, Fax: +1 713 743 4323
E-mail address: `nikolaou@uh.edu`

Progress in Systems and Control Theory, Vol. 26

Iterative Active-set Method for Efficient On-line MPC Design

Luigi Chisci[1], J. Anthony Rossiter[2] and Giovanni Zappa[1]

Abstract. This paper presents an efficient computational technique for the solution of the linear quadratic control problem for a discrete-time system subject to parallelotopic constraints in the control space. The problem solution is carried out in a state-space framework and makes use of both the *Active Set Method* and *Dynamic Programming*. It is shown how the optimal solution can be updated after inclusion or removal of an active constraint by a simple dynamic programming procedure requiring in the order of kn^2 operations, n being the system order and k the time at which the constraint is included or removed. The resulting fast algorithm offers a promising solution to alleviate the on-line computational burden of MPC design.

1. Problem formulation

Consider a discrete-time LTI system

$$x_{k+1} = A\, x_k + B\, \delta u_k \tag{1}$$
$$z_k = \delta u_k + E x_k + d \tag{2}$$

where: $k \in \mathbb{Z} \triangleq \{0, 1, 2, \dots\}$, $x_k \in \mathbb{R}^n$ is the state vector, $z_k \in \mathbb{R}^m$ a vector subject to the pointwise-in-time constraints

$$\| z_k \|_\infty \leq 1, \ \forall k \geq 0 \tag{3}$$

and, as is normal in predictive control, the system incorporates an integrator i.e. the manipulated variable $\delta u_k \triangleq u_k - u_{k-1} \in \mathbb{R}^m$ is the input increment. To measure control performance, the following quadratic cost is adopted

$$J(x_0, \delta u) \triangleq \sum_{k=0}^{\infty} x_k^T Q x_k + \delta u_k^T R \delta u_k \tag{4}$$

with $Q = Q^T \geq 0$ and $R = R^T > 0$. It is assumed that $\left(A, B, Q^{\frac{1}{2}}\right)$ is stabilizable and detectable. Next, in order to get a finite-dimensional optimization problem and also recover optimal LQ performance, whenever this is compatible with the constraints (3), the cost (4) is minimized subject to the terminal constraints

$$\delta u_k = K x_k, \ \forall k \geq N \tag{5}$$

where $N \geq 1$ is the control horizon and K is the LQ-optimal infinite-horizon feedback gain

$$K = -(R + B^T PB)^{-1} B^T PA, \tag{6}$$

P being the unique nonnegative symmetric solution of the Algebraic Riccati Equation

$$P = A^T PA - A^T PB(R + B^T PB)^{-1} B^T PA + Q. \tag{7}$$

Remark 1.1. *The ∞-norm bound constraints (3), with z_k defined by (2), encompass typical practical limitations taken into account in control system design such as, for instance,* amplitude *or* rate actuator *(input) constraints, or* amplitude *or* rate *output constraints. Notice that (2) and (3) allow to consider separately* input amplitude/ input rate/ output amplitude / output rate *constraints, but rules out the possibility to consider them simultaneously. Another efficient algorithmic technique, based on the* Mixed Weights Least-Squares *approach* [1], *has been proposed in* [2] *to overcome these constraint restrictions. It is worth pointing out, however, that whenever a single type of constraints (e.g. saturation constraints) is present (which is the case in several applications) a more efficient algorithmic technique can be obtained using the* Active Set Method *as shown hereafter. Previous work in this direction has been carried out in* [3] *following a polynomial approach; here, instead, a state-space approach is pursued.*

Recent work [4] has shown that if the problem (1)–(5) is feasible for a given state-space region of the initial condition, then there exists a sufficiently large N such that the infinite-horizon constraints (3) can be replaced by the finite-horizon ones

$$\|z_k\|_\infty \leq 1, \qquad k = 0, 1, \ldots, N-1. \tag{8}$$

Summing up, the subsequent part of the paper will concentrate on efficient solution of the optimal control problem CLQ_N defined by minimization w.r.t. δu_k, $0 \leq k \leq N-1$, of the cost (4) for the system (1)–(2) subject to constraints (5) and (8). For numerical and computational convenience, we follow the approach pursued in [5] whereby the perturbations with respect to the unconstrained LQ optimal control law

$$c_k \triangleq \delta u_k - K x_k \tag{9}$$

are used in place of δu_k as degrees of freedom in the optimization. Clearly the constraints (5) become $c_k = 0, \forall k \geq N$, and it is easy to see that, in the variables c_k, the problem is re-expressed as follows

$$\min_{c_0, c_1, \ldots, c_{N-1}} \left\{ x_0^T P x_0 + \sum_{k=0}^{N-1} c_k^T \overline{R} c_k \right\} \text{ subject to } \left\{ \begin{array}{l} \|z_k\|_\infty \leq 1, 0 \leq k < N \\ x_{k+1} = \overline{A} x_k + B c_k, \; k \geq 0 \\ z_k = c_k + \overline{E} x_k + d, \; k \geq 0 \end{array} \right\} \tag{10}$$

where $\overline{A} \triangleq A + BK$, $\overline{E} \triangleq E + K$ and $\overline{R} \triangleq R + B^T PB$. In fact, under the constraints (5), the cost (4) becomes

$$\begin{aligned} J(x_0, \delta u) &= x_N^T P x_N + \sum_{k=0}^{N-1} \left[x_k^T Q x_k + \delta u_k^T R \delta u_k\right] \\ &= (Ax_{N-1} + B\delta u_{N-1})^T P (Ax_{N-1} + B\delta u_{N-1}) + x_{N-1}^T Q x_{N-1} \\ &\quad + \delta u_{N-1}^T R \delta u_{N-1} + \sum_{k=0}^{N-2} \left[x_k^T Q x_k + \delta u_k^T R \delta u_k\right] \\ &= x_{N-1}^T (A^T PA - A^T PB\overline{R}^{-1} B^T PA + Q) x_{N-1} + (\delta u_{N-1} - Kx_{N-1})^T \\ &\quad \overline{R}(\delta u_{N-1} - Kx_{N-1}) + \sum_{k=0}^{N-2} \left[x_k^T Q x_k + \delta u_k^T R \delta u_k\right] \\ &= x_{N-1}^T P x_{N-1} + c_{N-1}^T \overline{R} c_{N-1} + \sum_{k=0}^{N-2} \left[x_k^T Q x_k + \delta u_k^T R \delta u_k\right] \end{aligned}$$

so that, iterating the above procedure, (10) is obtained.

Notice that each $z_k = z_k(c)$ depends affinely on the variabless c_i, $i = 0, 1, \ldots, k$. Therefore the set defined by (8) represents a parallelotope $\mathcal{P}$ in the c-variable (control) space $\mathbb{R}^{Nm}$. Hence, from a geometrical point of view, the above QP optimization problem amounts to finding, in the control space, the tangency point between the family of ellipsoids $\mathcal{E}(\gamma) \triangleq \{c : J(x_0, c) \leq \gamma\}$, centered in the origin, and the parallelotope $\mathcal{P}$, provided that the unconstrained optimal control law $c_k = 0$ $(k = 0, 1, \ldots, N-1)$ is infeasible.

2. Active set method

QP problems are usually solved by the so called *Active Set Method* (ASM), briefly described hereafter [6]. For simplicity of exposition we shall consider the single-input case, i.e. $m = 1$ and $\delta u_k, z_k, c_k \in \mathbb{R}$. The vector $a = [a_0, a_1, \ldots, a_{N-1}]^T$, $a_i \in \{-1, 0, 1\}$, is used to characterize the active constraints as follows:

$$a_i = \begin{cases} 0, & \text{if no constraint on } c_i \text{ is active} \\ +1, & \text{if } z_i = 1 \text{ i.e. } c_i = -\overline{E}x_i - d + 1 \\ -1, & \text{if } z_i = -1 \text{ i.e. } c_i = -\overline{E}x_i - d - 1 \end{cases} \tag{11}$$

Moreover, the set

$$\mathcal{A}(a) \triangleq \{i : a_i \neq 0\}.$$

will be called the *active set.*

ASM Algorithm

Initialization

Select a and a feasible solution $\hat{c} \in \mathcal{P}$, satisfying $z_k(\hat{c}) = a_k$ for $k \in \mathcal{A}$. A possible choice, corresponding to the center of $\mathcal{P}$, is obviously $\hat{c}$ such that $z_k(\hat{c}) = 0$ for $k = 0, 1, \ldots, N-1$ and, accordingly, $a = 0$ and $\mathcal{A}(a) = \emptyset$.

Generic step

1. Solve the optimization problem
$$\min_{c} J(x_0, c) \tag{12}$$
subject to $z_k = a_k$, $k \in \mathcal{A}$, and let c^* be the minimizer.
2. Check the constraints
$$-1 \leq z_k \leq 1 \text{ for } k \notin \mathcal{A} \tag{13}$$
If they are all satisfied set $\hat{c} = c^*$ and go to 4.
3. Compute
$$\hat{\mu} = \max\{\mu \mid \hat{c} + \mu(c^* - \hat{c}) \text{ satisfies } (13)\}$$
and set
$$\hat{c} = \hat{c} + \hat{\mu}(c^* - \hat{c})$$
Moreover let $q \notin \mathcal{A}$ be the index for which $|z_q(\hat{c})| = 1$, set $a_q = sign(z_q)$, $\mathcal{A} = \mathcal{A} \cup \{q\}$ and go to 1.
4. Compute, for $k \in \mathcal{A}$, the KKT (Karush-Kuhn-Tucker) multipliers λ_k defined by
$$\frac{\partial J}{\partial c}(\hat{c}) \quad = \quad - \sum_{k \in \mathcal{A}} \lambda_k \; a_k \; \frac{\partial z_k}{\partial c}$$
where z_k is the constrained variable. If $\lambda_k > 0$ for all $k \in \mathcal{A}$, $\hat{c}$ is the solution; otherwise let
$$q \quad = \quad \arg \min_{k \in \mathcal{A}} \lambda_k .$$
set $a_q = 0, \mathcal{A} = \mathcal{A} \backslash \{q\}$ and go to 1. □

Notice that the linear manifold $\{c \mid z_k(c) = a_k, k \in \mathcal{A}\}$, over which the minimization of (12) is carried out, corresponds to a face of the parallelotope $\mathcal{P}$. It can be easily shown that the ASM algorithm visits each face at most once. Hence the algorithm converges to the optimal solution in a finite number of steps which, however, can be exponential in N (3^N in the worst case).

The most demanding part of the ASM algorithm is step 1, namely the solution of a sequence of equality-constrained QP subproblems in which each subproblem differs from the previous one either by the inclusion or by the deletion of a single constraint. We will focus on this part of the algorithm in the next section.

3. Dynamic programming approach

It is well known that in the unconstrained case *Dynamic Programming* (DP) is by far the most convenient way to compute the LQ_N solution whenever the control horizon N is much larger than the plant order n. In fact using DP allows to get a computational load $O(Nn^2)$, which can further be reduced to $O(Nn)$ for a time-invariant system by exploiting *fast* algorithms and canonical representations

[7]. On the other hand, not using DP, at least $O(N^2)$ operations are required. Obviously, since ASM amounts to a sequence of equality constrained QP subproblems, DP can be applied to solving each subproblem. However, in this way, each subproblem is solved separately and the computational complexity of each ASM iteration is $O(Nn^2)$. In fact, due to the presence of active constraints, the problem is no longer time-invariant and fast-algorithms cannot be exploited. Hereafter it will actually be shown that the solution of each subproblem can be obtained by updating the previous one in a very simple way, requiring in the worst case only $O(Nn^2)$ operations for generic state-space representations and $O(Nn)$ for canonical representations.

Let us conjecture that, on the linear manifold $\{c : z_i(c) = a_i, i \in \mathcal{A}\}$, the cost $J_\mathcal{A}(x_0, c)$ is expressed by

$$J_\mathcal{A}(x_0, c) \;=\; V(x_0) \;+\; \sum_{\substack{i=0\\ i\notin\mathcal{A}}}^{N-1} \varrho_i(c_i - F_i x_i - g_i)^2 \tag{14}$$

Since, for $i \notin \mathcal{A}$, c_i is unconstrained, the optimal control law is given by

$$c_i \;=\; \begin{cases} F_i x_i + g_i, & i \notin \mathcal{A} \quad (a_i = 0) \\ -\overline{E} x_i - d + a_i, & i \in \mathcal{A} \quad (a_i \neq 0). \end{cases} \tag{15}$$

The conjecture (14) is certainly true for $\mathcal{A} = \emptyset$ $(a = 0)$; in such a case we have from (10)

$$\varrho_i = \overline{R},\; F_i = 0,\; g_i = 0,\; \forall i;\; V(x_0) = x_0^T P x_0.$$

It will be shown constructively that (14) actually holds true for an arbitrary active set $\mathcal{A}$.

The objective is to show how to efficiently update $\{\varrho_i, F_i, g_i : 0 \leq i \leq N-1, i \notin \mathcal{A}\}$ whenever

(a): the constraint

$$c_k \;=\; -\overline{E} x_k - d + \beta_k, \beta_k \in \{-1, 1\} \tag{16}$$

is added to the active set;

(b): the constraint (16) is removed from the active set.

First of all notice that both (a) and (b) leave unchanged the cost terms (14) and the control law (15) for $i > k$. Hence the problem is only to update (ϱ_i, F_i, G_i) for $i < k$. Secondly, it's important to stress that for any active constraint there is the necessity to keep track of all the relevant information needed to possibly make it inactive at a subsequent time. This is actually the most critical and tricky part of the updating procedure, for which a clever and efficient solution will be derived. It is worth pointing out that it is essential in the derivation that the constraints are of the form (2), (8) which guarantees that there can be at most one active constraint for each control degree-of-freedom c_k. The proposed update procedure will make use of the following simple algebraic results, which can be checked almost by inspection.

Lemma 3.1. *Let $\alpha, \beta, c \in \mathbb{R}$; $F^T, G^T, x \in \mathbb{R}^n$. Then,*

$$J(x,c) \triangleq \alpha(c - Fx)^2 + \beta(c - Gx)^2 = \tilde{\alpha}(c - \widetilde{F}x)^2 + \tilde{\beta}(\widetilde{G}x)^2 \tag{17}$$

with

$$\tilde{\alpha} = \alpha + \beta,\ \tilde{\beta} = \frac{\alpha\beta}{\alpha+\beta},\ \widetilde{F} = \frac{\alpha F + \beta G}{\alpha + \beta},\ \widetilde{G} = F - G. \tag{18}$$

Remark 3.2. *If the objective is to minimize $J(x,c)$ w.r.t. c, with x given, Lemma 3.1 yields both the optimal argument $c^*(x) = \widetilde{F}x$ and the optimal cost $J^*(x) \triangleq J(x, c^*(x)) = \tilde{\beta}(\widetilde{G}x)^2$.*

Lemma 3.3. *Let $J(x,c) \triangleq \alpha(c - Fx)^2$; $\alpha, c \in \mathbb{R}$; $F^T, x, G^T \in \mathbb{R}^n$. Then, imposing $c = Gx$, one gets*

$$J(x,c) = \alpha(\widetilde{G}x)^2 \tag{19}$$

with $\widetilde{G} = F - G$.

Remark 3.4. *Lemma 3.3 shows how the active constraint $c = Gx$ affects the cost $J(x,c) = \alpha(c - Fx)^2$, in particular $J(x, Gx) = \alpha(\widetilde{G}x)^2$ is the penalty incurred, with respect to the optimal choice for c, in introducing the constraint.*

Consider a linear constraint

$$c = Gx \tag{20}$$

and a set of quadratic cost terms

$$\varrho_i(c - F_i x)^2 \qquad i = 1, 2, \ldots \tag{21}$$

to which the linear constraint is applied thus producing

$$\varrho_i((F_i - G)x)^2 \qquad i = 1, 2, \ldots \tag{22}$$

When the constraint (20) is removed, the current cost must be updated by adding all the cost terms (21) and subtracting all the terms (22). The total cost update is easily accomplished by means of the following result.

Lemma 3.5. *Let $c, \varrho_i \in \mathbb{R}$; $F_i^T, G^T, x \in \mathbb{R}^n$. Then,*

$$\sum_i \varrho_i(c - F_i x)^2 - \sum_i \varrho_i((F_i - G)x)^2 = \varrho(c - Fx)^2 - \varrho((F - G)x)^2 \tag{23}$$

with

$$\varrho = \sum_i \varrho_i, \quad F = \frac{\sum_i \varrho_i F_i}{\varrho}. \tag{24}$$

Remark 3.6. *Formula (23) tells us that in order to remove the active constraint (20), one needs to accumulate all cost terms in (21) into a single one* $\varrho(c - Fx)^2$ *by (24) and add it to the current cost; then apply the constraint (20) to* $\varrho(c - Fx)^2$ *thereby producing* $\varrho((F - G)x)^2$*, and subtract the latter from the current cost. Hence, all that is needed to possibly disactivate the constraint (16) is to store* ϱ *and* F.

3.1. Constraint inclusion

Making use of the above lemmas, we shall consider constraint addition first. Assume that the constraint (16) must be added. Let $\mathcal{A}$ be the active set, with associated vector a, before constraint addition and $\mathcal{A}' \triangleq \mathcal{A} \cup \{k\}$ with associated vector a' such that

$$a'_i = a_i \text{ for } i \neq k \text{ and } a'_k = \beta_k.$$

Further, let us rearrange the terms of the cost (14) according to

$$J_{\mathcal{A}}(x_0, c) = J_{\mathcal{A}}(k+1 : N-1) + J_k(x_k, c_k) + J_{\mathcal{A}}(0 : k-1) + V_{\mathcal{A}}(x_0) \qquad (25)$$

where

$$\begin{aligned} J_i(x_i, c_i) &\triangleq \varrho_i(c_i - F_i x_i - g_i)^2 \\ J_{\mathcal{A}}(r : s) &\triangleq \sum_{\substack{i = r \\ i \notin \mathcal{A}}}^{s} J_i(x_i, c_i) \\ V_{\mathcal{A}}(x_0) &\triangleq x_0^T P_{\mathcal{A}} x_0 + 2x_0^T p_{\mathcal{A}} + \varrho_{\mathcal{A}}. \end{aligned} \qquad (26)$$

The objective is to find $J_{\mathcal{A}'}(x_0, c)$. The idea is to use DP, that is to proceed backward recursively over the control horizon $[0, k)$, updating the costs $J_i(x_i, c_i)$ and the optimal controls $c_i = F_i x_i + g_i$ for $i = k-1$ down to 0. First of all note that, by (16) and using Lemma 3.3, the term $J_k(x_k, c_k)$ in (25) is manipulated as follows

$$J_k(x_k, c_k) = J_k(x_k, -\overline{E}x_k + a'_k - d) = \widetilde{J}_k(x_k) \triangleq \tilde{\varrho}_k \left(\widetilde{F}_k x_k + \tilde{g}_k\right)^2 \qquad (27)$$

with

$$\tilde{\varrho}_k = \varrho_k, \widetilde{F}_k = \overline{E} + F_k, \tilde{g}_k = d + g_k - a'_k. \qquad (28)$$

Recall that (27) represents the penalty incurred in activating the constraint (16), which of course affects the choice of c_i for $0 \leq i \leq k-1$. Exploiting closed-loop dynamics

$$\begin{aligned} \widetilde{J}(x_k) &= \tilde{\varrho}_k(\widetilde{F}_k B c_{k-1} + \widetilde{F}_k \overline{A} x_{k-1} + \tilde{g}_k)^2 \\ &= \tilde{\varrho}_k(\widetilde{F}_k B)^2 \left(c_{k-1} + \frac{\widetilde{F}_k \overline{A}}{\widetilde{F}_k B} x_{k-1} + \frac{\tilde{g}_k}{\widetilde{F}_k B}\right)^2 \\ &\triangleq \widetilde{J}_{k-1}(x_{k-1}, c_{k-1}). \end{aligned} \qquad (29)$$

Assuming that $k-1 \notin \mathcal{A}$ (the case $k-1 \in \mathcal{A}$ will be dealt with later), there are two additive terms involving c_{k-1} namely

$$J_{k-1}(x_{k-1}, c_{k-1}) + \widetilde{J}_{k-1}(x_{k-1}, c_{k-1}) = J'_{k-1}(x_{k-1}, c_{k-1}) + \widetilde{J}_{k-1}(x_{k-1}) \tag{30}$$

where the expression in the RHS can be obtained making use of Lemma 3.1. By Lemma 3.1, in fact, we have

$$\begin{aligned} \varrho'_{k-1} &= \varrho_{k-1} + \tilde{\varrho}_k(\widetilde{F}_k B)^2, \quad F'_{k-1} = \frac{\varrho_{k-1}F_{k-1} - \tilde{\varrho}_k \widetilde{F}_k B \widetilde{F}_k \overline{A}}{\varrho'_{k-1}}, \\ g'_{k-1} &= \frac{\varrho_{k-1}g_{k-1} - \tilde{\varrho}_k \widetilde{F}_k B \tilde{g}_k}{\varrho'_{k-1}} \end{aligned} \tag{31}$$

$$\tilde{\varrho}_{k-1} = \frac{\varrho_{k-1}\tilde{\varrho}_k(\widetilde{F}_k B)^2}{\varrho'_{k-1}}, \quad \widetilde{F}_{k-1} = F_{k-1} + \frac{\widetilde{F}_k \overline{A}}{\widetilde{F}_k B}, \quad \tilde{g}_{k-1} = g_{k-1} + \frac{\tilde{g}_k}{\widetilde{F}_k B}. \tag{32}$$

Notice that: $c_{k-1} = F'_{k-1}x_{k-1} + g'_{k-1}$ gives the new optimal control law, while $\widetilde{J}_{k-1}(x_{k-1})$ is the penalty incurred by this choice if the same choice for c_0, c_1, ..., c_{k-2} is preserved. Clearly, one can proceed in the same way as from (29) to (30), for $i = k-1, \ldots, 1$ ($\forall i \notin \mathcal{A}$) and finally get the control law for $c_{k-1}, \ldots, c_0$ and the cost

$$\widetilde{J}_0(x_0) \triangleq \tilde{\varrho}_0(\widetilde{F}_0 x_0 + \tilde{g}_0)^2$$

which is the final penalty due to the inclusion of the new active constraint (16). Hence the minimum cost on $\mathcal{A}'$ is given by

$$V_{\mathcal{A}'}(x_0) = V_{\mathcal{A}}(x_0) + \widetilde{J}_0(x_0) = x_0^T P_{\mathcal{A}'} x_0 + 2x_0^T p_{\mathcal{A}'} + \varrho_{\mathcal{A}'}$$

from which

$$P_{\mathcal{A}'} = P_{\mathcal{A}} + \tilde{\varrho}_0 \widetilde{F}_0^T \widetilde{F}_0, \quad p_{\mathcal{A}'} = p_{\mathcal{A}} + \tilde{\varrho}_0 \widetilde{F}_0^T \tilde{g}_0, \quad \varrho_{\mathcal{A}'} = \varrho_{\mathcal{A}} + \tilde{\varrho}_0 \tilde{g}_0^2. \tag{33}$$

Thanks to Lemma 3.5, for each active constraint

$$c_i = -\overline{E}x_i + a_i - d, \qquad a_i \in \{-1, 1\} \tag{34}$$

one needs only to store an associated cost

$$J_i(x_i, c_i) = \varrho_i(c_i - F_i x_i - g_i)^2 \tag{35}$$

such that when (34) is removed from $\mathcal{A}$, the following incremental cost

$$\delta J(x_i, c_i) = J_i(x_i, c_i) - J_i(x_i, -\overline{E}x_i + a_i - d)$$

must be added to $J_{\mathcal{A}}(x_0, c)$. By Lemma 3.5, such equivalent costs $J_i(x_i, c_i)$, $i \in \mathcal{A}$, must be updated in the same way as if they were really present in $J_{\mathcal{A}}(x_0, c)$. Hence, the new cost $J'_i(x_i, c_i)$ is obtained from the old one $J_i(x_i, c_i)$ exactly as in (30), replacing $k-1$ by i, even if $i \in \mathcal{A}$. Conversely $\widetilde{J}_i(x_i)$ is obtained from $\widetilde{J}_i(x_i, c_i)$ taking into account that (34) is active, viz.

$$\text{if } i \in \mathcal{A} : \widetilde{J}_i(x_i) = \widetilde{J}(x_i, -\overline{E}x_i + a_i - d) \Longrightarrow$$

$$\Longrightarrow \begin{cases} \tilde{\varrho}_i &= \tilde{\varrho}_{i+1}(\widetilde{F}_{i+1}B)^2 \\ \widetilde{F}_i &= \dfrac{\widetilde{F}_{i+1}\overline{A}}{\widetilde{F}_{i+1}B} - \overline{E} \\ \tilde{g}_i &= \dfrac{\tilde{g}_{i+1}}{\widetilde{F}_{i+1}B} + a_i - d \end{cases} \tag{36}$$

The overall constraint inclusion algorithm is summarized in the Appendix (whenever $\pm$ or $\mp$ appears, the upper sign refers to inclusion).

3.2. Constraint deletion

Assume now that the constraint (16) must be removed from the active set, hence the objective is to pass from $J_{\mathcal{A}}(x_0, c)$ to $J_{\mathcal{A}'}(x_0, c)$ with $\mathcal{A}' = \mathcal{A} \backslash \{k\}$. The required procedure is almost the same as for inclusion, the only difference being that the costs $\widetilde{J}_i(x_i, c_i)$ in (29) must now be subtracted from, instead of being added to, $J_{\mathcal{A}}(x_0, c)$. Hence the main recursion (30) becomes

$$J_{k-1}(x_{k-1}, c_{k-1}) - \widetilde{J}_{k-1}(x_{k-1}, c_{k-1}) = J'_{k-1}(x_{k-1}, c_{k-1}) - \widetilde{J}_{k-1}(x_{k-1}) \tag{37}$$

and the corresponding formulas in (31) are the only ones modified, just by the replacement of the minus sign by the plus sign and viceversa. The algorithm for constraint deletion is therefore the same algorithm in the Appendix with a few sign replacements (whenever $\pm$ or $\mp$ appears, the lower sign refers to deletion).

4. Numerical example

The proposed algorithm, called *FQP (Fast QP)*, has been coded in MATLAB. To show its effectiveness, the following example concerning a third-order single-input system has been considered

$$A = \begin{bmatrix} 1 & 1.1752 & 0.5431 \\ 0 & 1.5431 & 1.1752 \\ 0 & 1.1752 & 1.5431 \end{bmatrix}, \quad B = \begin{bmatrix} 0.1752 \\ 0.5431 \\ 1.1752 \end{bmatrix},$$

$$Q = \begin{bmatrix} 1 & 0 & 0 \\ 0 & 0 & 0 \\ 0 & 0 & 0 \end{bmatrix}, \quad R = 0.01, \; E = [0, 0, 0]^T, \; d = 0.$$

The CLQ$_N$ problem (10) has been solved for an initial state $x_0 = [1, 1.5, -1.5]^T$ for which the optimal active set turns out to be $\mathcal{A} = \{2, 3, 4, 5\}$. The following table compares the number of flops (floating-point operations) required by FQP and by the standard QP routine of MATLAB, for various values of N.

As can be seen, there is a speedup of FQP over QP ranging from 3.7 for $N = 5$ to 233 for $N = 50$.

N	QP	FQP
5	13.4	3.6
10	108.4	16.9
15	323.3	20.4
20	723.4	23.8
30	2311.4	30.8
40	5340.9	37.7
50	10278.0	44.7

TABLE 1. Operation count in Kflops

5. Conclusions

A fast algorithm for the solution of constrained LQ control has been derived using the Active Set Method and Dynamic Programming. The algorithm provides a numerically and computationally appealing solution for on-line MPC design. Only the single-input case has been considered for simplicity; nevertheless the same approach can be applied, with appropriate modifications, to multi-input systems yielding the same computational benefits.

References

[1] J.A. Rossiter and B. Kouvaritakis, "Constrained stable generalized predictive control", *Proc. IEE Pt. D*, vol. 140, pp. 243–254, 1993.

[2] L. Chisci and J.A. Rossiter, "Iterative weighted least-squares approach to constrained LQ predictive control", *Proc. IEEE CCA'98*, Trieste, Italy, September 1998.

[3] L. Chisci, A. Marongiu, F. Schoen and G. Zappa, "Efficient algorithms for predictive control of systems with bounded inputs". In *Computer intensive methods in control and signal processing*, M. Karny & K. Warwick (eds.), pp. 93–104, Birkhäuser, Berlin, 1997.

[4] D. Chmielewski and V. Manousiouthakis, "On constrained infinite-time linear quadratic optimal control", *Systems and Control Letters*, vol. 29, pp. 121–129, 1996.

[5] J.A. Rossiter, B. Kouvaritakis and M.J. Rice, "A numerically robust state-space approach to stable predictive control strategies", *Automatica*, vol. 34, pp. 65–73, 1998.

[6] M.S. Bazaraa, H.D. Sherali and C.M. Shetty, *Nonlinear Programming: Theory and Algorithms, II edition*, John Wiley & Sons, New York, 1993.

[7] L. Chisci, A. Garulli and G. Zappa, "Fast parallel LQ regulator design for adaptive control", *Int. Journal of Adaptive Control and Signal Processing*, vol. 11, pp. 395–413, 1997.

Appendix – Constraint inclusion/deletion algorithm

Include/Delete the constraint $c_k = -\overline{E}x_k - d + \beta_k, \beta_k \in \{-1, 1\}$.

$$a_k := \begin{cases} \beta_k, & \text{for inclusion} \\ 0, & \text{for deletion} \end{cases}$$

$\tilde{\varrho}_k := \varrho_k, \quad \widetilde{F}_k := F_k + \overline{E}, \quad \tilde{g}_k := d + g_k - \beta_k$

for $i = k-1, k-2, \ldots, 1, 0$

if $\widetilde{F}_{i+1}B \neq 0$ then if $a_i = 0$ then begin

$$\tilde{\varrho}_i := \frac{\varrho_i \tilde{\varrho}_{i+1}(\widetilde{F}_{i+1}B)^2}{\varrho_i \pm \tilde{\varrho}_{i+1}(\widetilde{F}_{i+1}B)^2};$$

$$\widetilde{F}_i := F_i + \frac{\widetilde{F}_{i+1}\overline{A}}{\widetilde{F}_{i+1}B};$$

$$\tilde{g}_i := g_i + \frac{\tilde{g}_{i+1}}{\widetilde{F}_{i+1}B}$$

end

else begin

$$\tilde{\varrho}_i := \tilde{\varrho}_{i+1}(\widetilde{F}_{i+1}B)^2;$$

$$\widetilde{F}_i := \frac{\widetilde{F}_{i+1}\overline{A}}{\widetilde{F}_{i+1}B} - \overline{E};$$

$$\tilde{g}_i := \frac{\tilde{g}_{i+1}}{\widetilde{F}_{i+1}B} + a_i - d$$

end

else begin

$\tilde{\varrho}_i := \varrho_i;$

$\widetilde{F}_i := \widetilde{F}_{i+1}\overline{A};$

$\tilde{g}_i := \tilde{g}_{i+1}$

end

$$F_i := \frac{\varrho_i F_i \mp \tilde{\varrho}_{i+1}\widetilde{F}_{i+1}B\widetilde{F}_{i+1}\overline{A}}{\varrho_i \pm \tilde{\varrho}_{i+1}(\widetilde{F}_{i+1}B)^2}; \; g_i := \frac{\varrho_i g_i \mp \tilde{\varrho}_{i+1}\widetilde{F}_{i+1}B\tilde{g}_{i+1}}{\varrho_i \pm \tilde{\varrho}_{i+1}(\widetilde{F}_{i+1}B)^2};$$

$\varrho_i := \varrho_i \pm \tilde{\varrho}_{i+1}(\widetilde{F}_{i+1}B)^2;$

End loop i.

$P := P \pm \tilde{\varrho}_0 \widetilde{F}_0^T \widetilde{F}_0; \; p := p \pm \tilde{\varrho}_0 \widetilde{F}_0^T \tilde{g}_0; \; \varrho := \varrho \pm \tilde{\varrho}_0 \tilde{g}_0^2.$

[1] Dip. Sistemi e Informatica,
Università di Firenze,
Via di Santa Marta 3,
50139 Firenze, Italy.
E-mail: chisci,zappa@dsi.unifi.it

[2]Dept. Mathematical Sciences,
Loughborough University,
Leicestershire LE11 3TU, United Kingdom
E-mail: J.A.Rossiter@lboro.ac.uk

Progress in Systems and Control Theory, Vol. 26

Nonlinear Predictive Control Algorithms with Different Input Sequence Parameterizations Applied for the Quadratic Hammerstein and Volterra Models

Robert Haber, Ruth Bars and Orsolya Lengyel

Abstract. GPC (Generalized Predictive Control) developed by Clarke [1] is extended for different nonlinear input/output models. The nonlinear process models applied are the Hammerstein series model, the Volterra series model, the parametric generalized Hammerstein model and the parametric Volterra model. Extended horizon one-step-ahead and long-range optimal predictive control algorithms are derived. A quadratic cost function is minimized, which considers the quadratic deviations of the reference signal and the output signal predicted in a future point beyond the dead time and also punishes big control signal increments. The future process output is predicted in dependence of the control increment. Using a special assumption on the form of the input signal during the control horizon, the multi-dimensional optimization reduces to a one-dimensional problem. Hard input limitations can be considered, as well.

1. Introduction

Predictive control algorithms determine a series of the control signal minimizing the deviation between the reference signal and the process output signal in a given future horizon. According to the so called receding horizon strategy only the first control signal value is applied and in the next sampling period the computation procedure is repeated. Clarke [1] developed GPC (Generalized Predictive Control) for linear pulse-transfer function models. Considering no hard constraints the control signal can be calculated by an analytical expression. Further on, steady-state error is eliminated by using a model with control signal increments.

In most papers dealing with predictive control of nonlinear systems, the predicted output signal is computed by a sequential simulation of the non-predictive model equations till the end of the prediction horizon. The cost function is minimized numerically by selecting an optimal input signal sequence in the control horizon. Any constraints can be considered easily. There exists, however, also an other way, which is a direct extension of the linear GPC for the nonlinear case.

The predicted output signal can be calculated in the whole optimization horizon with the predictive models valid for the corresponding (different) prediction steps. Therefore, transformation equations for calculating the predictive nonlinear models are given for the generalized Hammerstein and the parametric Volterra models.

Using of predictive models is especially advantageous when a special input sequence parameterization is assumed, that means the future input increments in the control horizon are assumed to be any functions of the actual input signal (e.g. they are equal to each other). Then all predicted output signals in the prediction horizon and also the cost function become analytical functions purely of the actual control increment. Instead of using a multi-dimensional optimization the actual input increment can be computed – even in presence of hard input constraints – by a one-dimensional minimization. In the case of the generalized Hammerstein and parametric Volterra models with polynomial steady-state characteristics the cost function is a polynomial function of the actual control increment. The optimal input increment can be computed – in the absence of hard input constraints – by a root finding algorithm and by a straightforward selection of the roots. In the following single-input, single-output quadratic models are investigated.

As mentioned, the predictive form of the nonlinear process models is needed. The weighting function series models are predictive themselves. The predictive form of the parametric Hammerstein and Volterra models was derived in [2], and the results will be summarized here. Models with input signal increments are advantageous since the cost function contains the control increment and not the control signal itself. Incremental transformation of the non-parametric and parametric models was derived in [4]. An extension to the prediction equations is straightforward.

2. Nonlinear process models

In the following the input signal is denoted by u, the output signal by y, and the discrete time by k. The dead time relative to the sampling time is d. There are two types of process models being linear in the parameters:

- weighting function series models,
- parametric process models.

Only polynomial models with quadratic steady-state characteristics are considered.

2.1. Non-parametric process models

There are two nonlinear weighting function series:

- Hammerstein weighting function series:

$$y(k+d) = h_0 + \sum_{i=0}^{m} h_{1i} u(k-i) + \sum_{i=0}^{m} h_{2ii} u^2(k-i) \qquad (1)$$

- Volterra weighting function series:

$$y(k+d) = h_0 + \sum_{i=0}^{m} h_{1i} u(k-i) + \sum_{i=0}^{m}\sum_{j=i}^{m} h_{2ij} u(k-i) u(k-j) \tag{2}$$

In (1) and (2) h_0 is the constant term, h_{1i} and h_{2ij} are the parameters of the linear and quadratic terms, respectively. A sufficient modeling requires long memory m ($m > 10$), which results in more than 60 parameters with the Volterra series and the parameter estimation causes a big problem. Otherwise this model class is predictive itself, there is no need for a transformation to the predictive form.

2.2. Parametric process models

Two simple nonlinear parametric models will be considered:

- Parametric generalized Hammerstein model:

$$A(q^{-1}) y(k+d) = b_0 + B_1(q^{-1}) u(k) + B_2(q^{-1}) u^2(k) \tag{3}$$

- Parametric Volterra model:

$$A(q^{-1}) y(k+d) = b_0 + B_1(q^{-1}) u(k) + B_2(q_1^{-1}, q_2^{-1}) u^2(k) \tag{4}$$

The polynomials of the backward shift operator q^{-1} mean:

$$\begin{aligned} A(q^{-1}) &= 1 + a_1 q^{-1} + \cdots + a_{na} q^{-na} \\ B_1(q^{-1}) &= b_{10} + b_{11} q^{-1} + \cdots + b_{1nb1} q^{-n_b 1} \\ B_2(q^{-1}) &= b_{20} + b_{21} q^{-1} + \cdots + b_{2nb2} q^{-nb2} \\ B_2(q_1^{-1}, q_2^{-1}) u^2(k) &= \sum_{i=0}^{nb2}\sum_{j=i}^{nb2} b_{2ij} u(k-i) u(k-j) \end{aligned} \tag{5}$$

Both models are special classes of the parametric recursive polynomial NARMAX model. The parametric Volterra model (4) contains cross-product terms between differently delayed input components. Therefore, it approximates a general nonlinear dynamical process better than the generalized Hammerstein model. Assuming a dynamic behavior with the order $na = nb1 = nb2 = 2$, the parametric Volterra model can be modeled by 12 parameters.

3. Predictive nonlinear model equations

3.1. Non-incremental predictive equations

As the weighting function series models are predictive themselves, only the parametric models have to be transformed to the predictive form. As the parametric Volterra model includes the generalized Hammerstein one as a special case with

$$b_{2ij} = 0; \quad i \neq j; \quad b_{2i} = b_{2ii}; \qquad i = 0, 1, \ldots, nb2$$

only the case of the parametric Volterra model will be treated. Its predictive form for n_e steps ahead is given in [2]:

$$\hat{y}(k+d+n_e|k) = c_0^p + \alpha(q^{-1})y(k) + \beta_1(q^{-1})u(k+n_e) + \beta_2(q_1^{-1},q_2^{-1})u^2(k+n_e) \tag{6}$$

with

$$\begin{aligned} \alpha(q^{-1}) &= G(q^{-1}) \\ c_0^p &= F(1)b_0 \\ \beta_1(q^{-1}) &= F(q^{-1})B_1(q^{-1}) \\ \beta_2(q_1^{-1},q_2^{-1}) &= F(q^{-1})B_2(q_1^{-1},q_2^{-1}) \end{aligned} \tag{7}$$

The polynomials $F(q^{-1})$ and $G(q^{-1})$ are defined as

$$\begin{aligned} F(q^{-1}) &= 1 + f_1 q^{-1} + \cdots + f_{d+ne-1} q^{-(d+ne-1)} \\ G(q^{-1}) &= g_0 + g_1 q^{-1} + \cdots + g_{na-1} q^{-(na-1)} \end{aligned} \tag{8}$$

and are calculated by solving the following Diophantine equation

$$1 = F(q^{-1})A(q^{-1}) + q^{-(d+n_e)}G(q^{-1}) \tag{9}$$

3.2. Incremental predictive equations

The incremental form can be obtained from the non-incremental form by substituting the actual $u(k)$ and the future control signals $u(k+i); i > 0$ by a known previous control signal $u(k-1)$ and by the actual and future control increments until the actual time point k $\Delta u(k+j);\ j \geq 0$.

$$u(k+i) = u(k-1) + \sum_{j=0}^{i} \Delta u(k+j) \tag{10}$$

The resulting incremental form was given for non-predictive nonlinear models in [4]. For the predictive form of (6) similar expressions can be obtained.

The incremental predictive form includes terms depending on the past input, on the past and actual output signals, and on the actual and future control increments. The past terms and the actual measured output signal can not be changed. Therefore they are considered altogether as a constant term in the predictive equation (of course changing in each calculation step). The actual and future control increments appear in linear and quadratic form:

$$\begin{aligned} \hat{y}(k+d+n_e|k) &= p_0^{(n_e)} + P_1^{(n_e)}(q^{-1})\Delta u(k+n_e) + \\ &\quad + P_2^{(n_e)}(q_1^{-1},q_2^{-1})\Delta u^2(k+n_e) \end{aligned} \tag{11}$$

The coefficients can be calculated according to [4] with some modifications. The degree of the polynomials in (11) is given by

$$\deg(P_1(q^{-1})) = n_e; \quad \deg(P_2(q_1^{-1},q_2^{-1})) = [n_e, n_e].$$

The upper index n_e means that the parameters are functions of the prediction step, as well.

4. The control algorithm

The control signal is calculated by minimizing the following quadratic cost function:

$$J = \sum_{n_e=n_{e1}}^{n_{e2}} \gamma_{yn_e}[y_r(k+d+n_e) - \hat{y}(k+d+n_e|k)]^2 + \qquad (12)$$

$$+\sum_{j=0}^{n_u} \gamma_{u,j-1}\Delta u^2(k+j-1) \Rightarrow \min_{\Delta u}$$

Here y_r denotes the reference signal, $\hat{y}$ is the predicted value of the output signal on the basis of the system model using information available up to the actual time point k, while d denotes the discrete dead time. The tuning parameters of the control algorithm are:

- $n_{e2} - n_{e1}$ the prediction or optimization horizon,
- n_u the control horizon (the number of the supposed consecutive changes in the control signal),
- $\gamma_{yn_{e1}}, \ldots, \gamma_{yn_{e2}}$ weighting factors, generally equal to each other and being of value 1,
- $\gamma_{u_0}, \ldots, \gamma_{n_u-1}$ weighting factors, (denoted by γ_u if equal to each other).

The control increments $\Delta u(k), \ldots, \Delta u(k+n_u-1)$ have to be calculated. Only the first one is applied as input signal, and in the next time point the procedure is repeated (receding horizon strategy).

4.1. Solution of the control algorithm without input sequence parametrization

Multi-dimensional optimization is required to compute the optimal control signal. The control signal should be inside the control limits. This restriction can be considered during the optimization.

4.2. Solution of the control algorithm with input signal parametrization

The minimization of the cost function (12) can be simplified if only the actual control signal is searched with some assumptions for the future increments. In such cases the minimization problem becomes one-dimensional. Two sub-optimal control strategies are considered which simplify the calculations:

- Strategy 1: only one change is taken into account in the control signal in the actual time point k, and during the control horizon the control signal is considered constant

$$\Delta u(k) \neq 0; \Delta u(k+1) = \cdots = \Delta u(k+n_u-1) = 0. \qquad (13)$$

- Strategy 2: the changes in the control signal during the control horizon are considered equal to each other

$$\Delta u(k) = \Delta u(k+1) = \cdots = \Delta u(k+n_u-1). \qquad (14)$$

Strategy 2 is more cautious than Strategy 1. It is expected that it causes smoother transients than Strategy 1.

In sub-optimal cases assuming an input sequence parametrization only the actual control signal increment has to be searched. There are two different cases depending on the input constraints:

- numerical minimization under hard input constraints,
- analytical solution of the control equation without constraints.

Without considering hard input constratints the minimitzation of the fourth degree cost function (12) according to the control increment $\Delta u(k)$ leads to the solution of the third-degree one-dimensional equation:

$$k_0 + k_1\Delta u(k) + k_2\Delta u^2(k) + k_3\Delta u^3(k) = 0. \tag{15}$$

The coefficients of equation (15) depend on the strategy selected [3]:

$$\begin{aligned}
k_0 &= \sum_{n_e=n_{e1}}^{n_{e2}} \gamma_{yn_e}[p_0^{(n_e)} - y_r(k+d+n_e)][\sum_{i=r}^{n_e} p_{1i}^{(n_e)}], \\
k_1 &= 2\sum_{n_e=n_{e1}}^{n_{e2}} [p_0^{(n_e)} - y_r(k+d+n_e)][\sum_{i=r}^{n_e}\sum_{j=i}^{n_e} p_{2ij}^{(n_e)}] + \\
&+ \sum_{n_e=n_{e1}}^{n_{e2}} [\sum_{i=r}^{n_e} p_{1i}^{(n_e)}]^2 + \sum_{j=r}^{n_{e2}} \gamma_{uj}, \\
k_2 &= 3\sum_{n_e=n_{e1}}^{n_{e2}} [\sum_{i=r}^{n_e} p_{1i}^{(n_e)}][\sum_{i=r}^{n_e}\sum_{j=i}^{n_e} p_{2ij}^{(n_e)}], \\
k_3 &= 2\sum_{n_e=n_{e1}}^{n_{e2}} [\sum_{i=r}^{n_e}\sum_{j=i}^{n_e} p_{2ij}^{(n_e)}]^2.
\end{aligned} \tag{16}$$

The one-step-ahead case can be obtained from the long-range case with $n_{e1} = n_{e2} = n_e$. The parameter r is [3]

- $r = n_e$ for Strategy 1 (constant control signals in horizon) and
- $r = 0$ for Strategy 2 (constant control increments in horizon).

There are three solutions of the third degree polynomial equation. From among three real roots that one is chosen as the actual control increment for which the cost function is the less. The real root is chosen if conjugate complex pairs also exist. The control signal should be inside the control limits. The experiences have shown that the results of the suboptimal optimization with input sequence parametrization are almost as good as those of the optimal solution, if the tuning parameters are well selected. Because of less computation demand they seem to be more convenient for real time applications.

5. Simulation results

First some simulation results are shown for the control of the Hammerstein model. The cascade form of the Hammerstein model is illustrated in Fig.1 on the left. This form can be transformed for the generalized Hammerstein model. The parameters of the nonlinear static characteristics are: $c_0 = 2$; $c_1 = 1$; and $c_2 = 0.5$, respectively. Three different dynamic parts are considered, given by the continuous minimum phase transfer functions $e^{-10s}/(1+10s)$ (System 1) or $1/(1+5s)(1+10s)$ (System 2) or by the non-minimum-phase function $(1-7s)/(1+5s)(1+10s)$ (System 3). The unit step response of the linear part of System 3 is illustrated in Fig.1 on the right. The process is sampled with sampling time $h = 10$ with System 1 and with $h = 5$ with Systems 2 and 3. The common weighting factor of the control increments was in all simulations $\gamma_u = 0.01$.

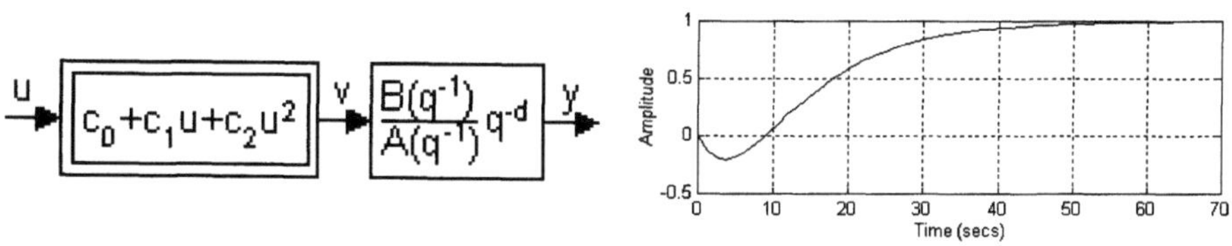

FIG. 1. Cascade Hammerstein model Non-minimum phase step response

First the performance of the long-range and the extended horizon one-step-ahead strategies is shown in the case of System 2 for Strategy 1. The output signals (Fig.2, left) and the control signals (Fig.2, right) are also given. Here $n_{e1} = 2$, while $n_{e2} = 2,\ 4\ \mathit{and}\ 6$, respectively. As is seen, the long-range responses are smoother.

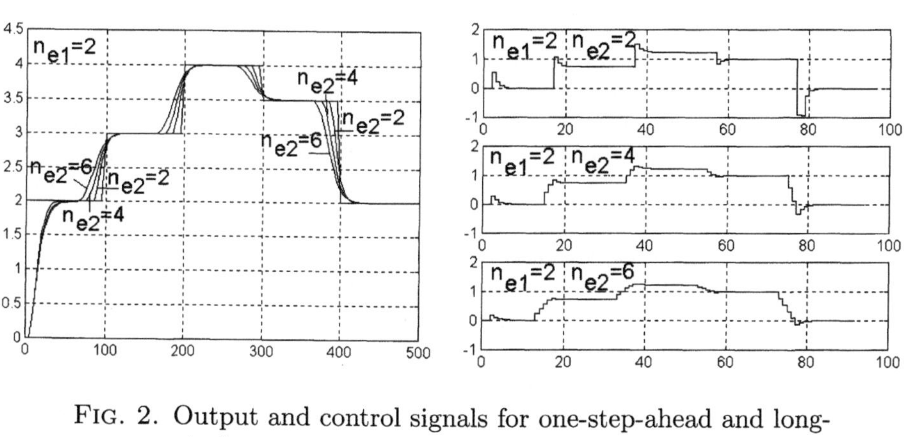

FIG. 2. Output and control signals for one-step-ahead and long-range control

The effect of the optimal and the sub-optimal strategies is compared for one-step-ahead extended horizon control for the case of System 1. Here $n_e = 2$. The output signals are given in Fig.3 (left) while the control signals are shown in Fig.3 (right). On the figures "1" and "2" denote the strategies with input sequence parametrization, respectively, and "3" indicates the optimal algorithm. It is seen that the suboptimal Strategy 2 and the optimal strategy give very close results. As the optimal solution is much more time consuming, for real-time applications the sub-optimal solutions with input sequence parametrization could be suggested.

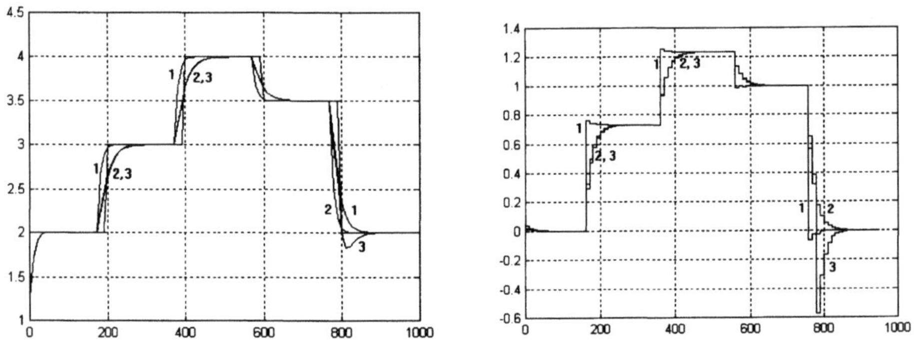

FIG. 3. Optimum and sub-optimal one-step-ahead extended horizon control

Extended horizon control could be applied for non-minimum-phase systems as well. System 3 is investigated now with one-step-ahead control according to Strategy 1 for $n_{e1} = n_{e2} = n_e = 2,\ 3\ and\ 8$, respectively (Fig.4, left). There is a peak first in the output signal, then it tracks the reference signal. This peak is less if the horizon extension is higher.

With long-range control ($n_{e1} = 1\ and\ n_{e2} = 8$) this peak almost disappears. Fig.4 (right) compares the output signals with the one-step-ahead case ($n_{e1} = n_{e2} = 8.$).

The effect of the control strategies and of the tuning parameters is similar for the parametric Volterra model, as well. The parameters of the plant can be assumed known or could be identified for example with the LS (Least Squares) algorithm. Here an example for the control of a Wiener model is shown with the Hammerstein and with the Volterra extended horizon algorithms. The Wiener model is given in Fig.5.

The static characteristics is as previously. The dynamic part is $0.5/(1-0.5q^{-1})$; $d = 2\ and\ \gamma_u = 0.01$. One-step-ahead control with Strategy 1 is applied with $n_e = 3$. The system was identified off-line in the form of the generalized Hammerstein and the Volterra models, respectively. Then the control algorithms were applied using also on-line identification further on. The output and the input signals are shown in Fig.6 (left) and Fig.6 (right). The Volterra approach seems to be better because there is no overshoot.

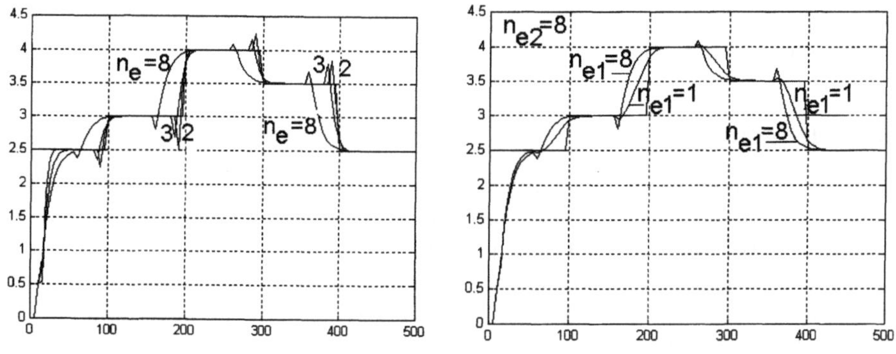

FIG. 4. One-step-ahead extended horizon versus long-range control of System 3

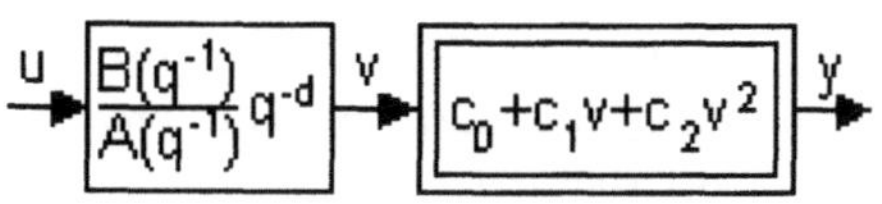

FIG. 5. The Wiener model

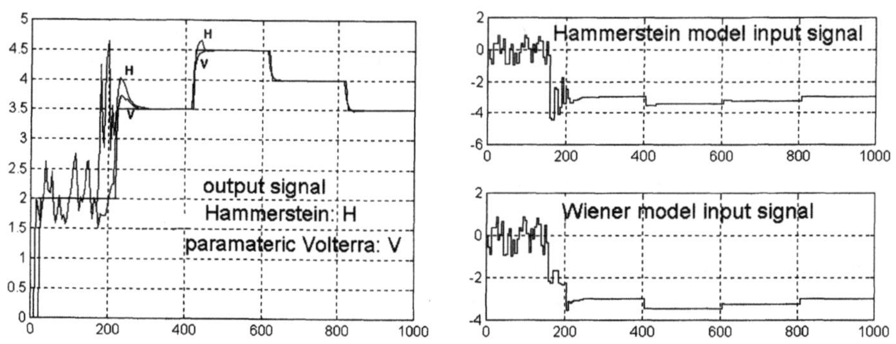

FIG. 6. Output and control signals for Wiener model control

6. Conclusion

GPC has been extended for different nonlinear input/output models. Extended horizon one-step-ahead and long-range optimal predictive control algorithms were derived. Using the predictive model equations with the input increments and assuming a special dependence between the input signals during the control horizon, the multi-dimensional optimization problem was reduced to a one-dimensional algorithm. The new algorithm can be tuned by several parameters and a control behavior similar to the optimal multi-dimensional solution can be achieved. It was shown that the parametric Volterra model is prior to the Hammerstein model if real processes are controlled.

Acknowledgements

The cooperation between the University of Applied Sciences Cologne and the Technical University of Budapest was supported by the Ministry of Science and Research of NRW (FRG) in the program "Förderung der Europafähigkeit der Hochschulen". The research work of the second author was also partly supported by the fund of the Hungarian Academy of Sciences for control research and partly by OTKA fund No. T015776. The supports are kindly acknowledged.

References

[1] D.W. Clarke, C. Mohtadi and P.S. Tuffs, *Generalized predictive control – Part I. The basic control algorithm,* Automatica., **23(2)** (1987), 137–148.

[2] R. Haber, *Predictive control of nonlinear dynamic processes,* Applied Mathematics and Computation, **70** (1995), 169–184.

[3] R. Haber, R. Bars and O. Lengyel, *Long-range predictive control of the parametric Hammerstein model,* IFAC NOLCOS'98, Enschede, The Netherlands, **Vol. 2.** (1998), 434–439.

[4] R. Haber, *Transformation of the absolute descriptions of linear and non-linear processes to their incremental forms,* International Journal of Systems Science, **23(6)** (1992), 935–955.

Robert Haber,
Department of Process Engineering,
Laboratory for Process Control,
University of Applied Sciences, Cologne
D-50679, Köln, Betzdorfer Str. 2, Germany
E-mail address: `robert.haber@fh-koeln.de`

Ruth Bars, Orsolya Lengyel,
Department of Automation,
Technical University of Budapest, Hungary,
H-1111, Budapest, Goldmann Gy. tér 3, Hungary
E-mail address: `bars@aut.bme.hu`

Progress in Systems and Control Theory, Vol. 26

Nonlinear Model Predictive Control Based on Stable Wiener and Hammerstein Models

Matthias Wellers, Heinrich Rake

Abstract. The presented Nonlinear Model Predictive Control (NMPC) scheme based on stable Wiener and Hammerstein models retains the characteristics of conventional linear Model Predictive Control (MPC) including capabilities to control stable nonlinear processes. The future process behavior is predicted by the q-th degree Wiener and Hammerstein model represented by Volterra-series. The optimal manipulated variable is calculated minimizing a quadratic cost function depending on the nonlinear predictor subject to input and output constraints.

The NMPC algorithm is applied to two case studies of chemical processes in comparison with other control schemes. A pH neutralization process, the basic structure of which fits to a Wiener model, and a highly nonlinear benchmark CSTR (continuously fed stirred tank reactor), are examined.

1. Introduction

Model Predictive Control (MPC), which is also called receding or moving horizon control, is an attractive feedback strategy, particularly for nonlinear real-world processes. Therefore, conventional MPC schemes have enjoyed wide-spread acceptance and success as an effective technique for dealing with control problems especially in the chemical/petrochemical industries.

In recent years research is focused on MPC using nonlinear models. But the derivation of these models can be very time consuming especially if the process is not well understood. Furthermore, the nonlinear control schemes that employ more realistic nonlinear process descriptions typically sacrifice the simplicity associated with linear techniques in order to achieve improved performance. Thus, the direct use of nonlinear models often leads to high order nonlinear optimization problems. Chen et al. give in [2] an overview of some nonlinear MPC schemes.

This paper presents a MPC algorithm based on stable Wiener and Hammerstein models of the degree q. Using the Volterra-series, which is a nonlinear extension of the linear impulse response model, the prediction is incorporated into a quadratic cost function. The optimization problem is solved subject to hard input and soft output constraints.

The nonlinear control algorithm is tested in two case studies. In the first case the nonlinear controller is applied to a pH neutralization process. The time invariant nonlinear titration curve within the process can be approximated in the relevant operating range by a 3rd degree polynomial. Several setpoint variations illustrate the performance of the predictive controller in relation to an adaptive PI-controller.

The second case study represents a highly nonlinear benchmark CSTR (continuously fed stirred tank reactor, [1]). The non-minimum-phase process is identified – using reduced Volterra-series ([3]) – as a 2nd degree Wiener model. The reduced Volterra-series stands for the compression of the unknown parameters by introducing orthonormal basis functions. Although the Wiener model is only a very simplified description of the process the optimal operating point could be reached by the NMPC. The benefits of the predictive controller are underlined by the comparison with a NMPC assuming exact modeling and measuring of all states at every time instant.

2. Nonlinear controller synthesis

2.1. Nonlinear model structure and prediction

Apart from recursive difference equations the Volterra-series is well known for representing time invariant nonlinear systems. For stable time-delayed systems the Volterra-series

$$\begin{aligned} y(k) = y_{00} &+ \sum_{i=1}^{m} g_1(i)u(k-i) \\ &+ \sum_{i_1=1}^{m} \sum_{i_2=i_1}^{m} g_2(i_1,i_2)u(k-i_1)u(k-i_2) + \cdots \\ &+ \sum_{i_1=1}^{m} \cdots \sum_{i_q=i_{q-1}}^{m} g_q(i_1,\ldots,i_q)u(k-i_1)\cdots u(k-i_q) \end{aligned} \tag{1}$$

can be truncated if the upper limit m is sufficiently high. If the series is considered as the nonlinear extension of the linear impulse response model then g_1 is the linear, g_2 the quadratic and g_q the q-th degree kernel. y_{00} is the value of y if all input signals are equal to zero. $y(k)$ represents the output and $u(k)$ the input signal at the sampling instant k. The Volterra-series is very suitable for modeling nonlinear processes, since it covers a large class of nonlinear time invariant systems like stable Wiener and Hammerstein models. Assuming the model structure shown in figure 1 the Volterra-series can be transformed into:

$$\begin{aligned} &y(k) \\ &= y_{00} + \underbrace{\sum_{i=1}^{m} g_l(i)u(k-i)}_{\text{linear}} + \underbrace{\sum_{n=2}^{q}\left(\sum_{i=1}^{m} g_{Wn}(i)u(k-i)\right)^{n}}_{\text{Wiener}} + \underbrace{\sum_{n=2}^{q}\left(\sum_{i=1}^{m} g_{Hn}(i)u^{n}(k-i)\right)}_{\text{Hammerstein}} \end{aligned} \tag{2}$$

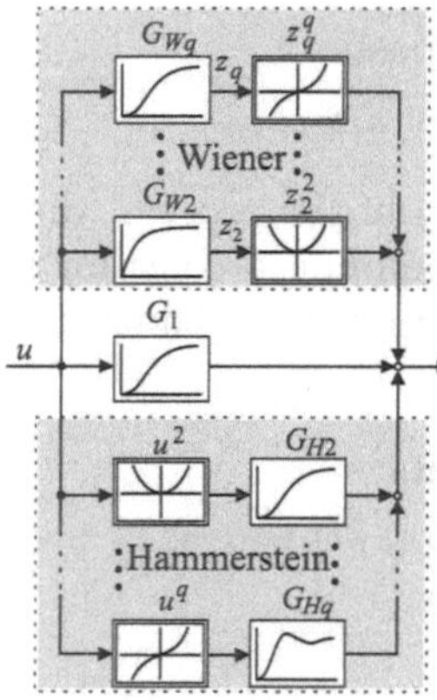

FIGURE 1. Wiener and Hammerstein model

where g_l, g_{Wn} and g_{Hn} indicate the impulse responses of the linear, Wiener and Hammerstein part of the model, respectively.

The description is used to estimate future output signals. In order to predict the output signal $\hat{y}(k+j\,|k)$ j sampling instants in advance at the actual time instant k the equation 2 is shifted j steps into the future. Therefore, the input signals

$$\boldsymbol{u}_{p_j} = [u(k-m+j), u(k-m+1+j), \cdots u(k-1)]^T \tag{3}$$

$$\boldsymbol{u}_{f_j} = [u(k\,|k), u(k+1\,|k), \cdots u(k+j-1\,|k)]^T \tag{4}$$

and each impulse response

$$\boldsymbol{g}_{np_j} = [g_n(m), g_n(m-1), g_n(m-2), \cdots, g_n(j+1)]^T \tag{5}$$

$$\boldsymbol{g}_{nf_j} = [g_n(j), g_n(j-1), g_n(j-2), \cdots, g_n(1)]^T \tag{6}$$

are divided into a future (index f) and past (index p) part. The predictor of the combined Wiener and Hammerstein model is built up by

$$\hat{y}(k+j\,|k) = y_{00} + \left[\boldsymbol{g}_{lp_j}^T, \boldsymbol{g}_{lf_j}^T\right] \cdot \begin{bmatrix} \boldsymbol{u}_{p_j} \\ \boldsymbol{u}_{f_j} \end{bmatrix} + \sum_{n=2}^{q} \left(\left[\boldsymbol{g}_{Wnp_j}^T, \boldsymbol{g}_{Wnf_j}^T\right] \cdot \begin{bmatrix} \boldsymbol{u}_{p_j} \\ \boldsymbol{u}_{f_j} \end{bmatrix} \right)^n + \sum_{n=2}^{q} \left[\boldsymbol{g}_{Hnp_j}^T, \boldsymbol{g}_{Hnf_j}^T\right] \cdot \begin{bmatrix} \boldsymbol{u}_{p_j} \\ \boldsymbol{u}_{f_j} \end{bmatrix}^{*n} . \tag{7}$$

The exponent $*n$ indicates that each element in the vector is raised to the power of n. Furthermore, in equation (7) y_{00} is a bias term the value of which

$$y_{00} = y(k) - \boldsymbol{g}_l^T \boldsymbol{u} - \sum_{n=2}^{q} \left(\boldsymbol{g}_{Wn}^T \boldsymbol{u}\right)^n - \sum_{n=2}^{q} \boldsymbol{g}_{Hn}^T (\boldsymbol{u})^{*n} \tag{8}$$

is fitted using input data of the past

$$\boldsymbol{u} = [u(k-m), u(k-m+1), \cdots u(k-1)]^T \tag{9}$$

and the actual output signal $y(k)$ of the process, thus the predictor is adjusted to the process at every instant. Assuming a constant mismatch between the model and the process as well as a constant disturbance the bias y_{00} is set constant over the prediction horizon.

2.2. Cost function and constraints

The resulting predictor of equation (7) is used in the cost function to calculate the performance objective

$$J(\boldsymbol{u}(\cdot\,|k)) = \sum_{j=N_1}^{N_2} (r(j+k\,|k) - \hat{y}(k+j\,|k))^2 + \lambda \sum_{j=0}^{N_u-1} (u(k+j\,|k) - u(k+j-1\,|k))^2 \tag{10}$$

where N_1 and N_2 determine the prediction horizon; $r(j\,|k)$ is the reference trajectory calculated at time k and λ weights the N_u steps of the future manipulated variable

$$\boldsymbol{u}(\cdot\,|k) = [u(k\,|k), \cdots, u(k+N_u - |k), u(k+N_u\,|k), \cdots u(k+N_2-1\,|k)]^T. \tag{11}$$

Thereby, the future manipulated variable is calculated subject to the input

$$u_{\min} \leq \quad u(k) \quad \leq u_{\max} \tag{12}$$

$$\Delta u_{\min} \leq \quad \Delta u(k) \quad \leq \Delta u_{\max} \tag{13}$$

and output constraints

$$y_{\min} \leq y(k) \leq y_{\max} \tag{14}$$

of the process. Where Δ denotes the difference operator. The constraints are implemented in the NMPC structure as hard input and soft output constraints. Therefore, the input constraints are transformed to a linear inequality restriction and the output constraints are integrated by a penalty function

$$J_{pe}(\boldsymbol{u}(\cdot\,|k)) = \sum_{j=1}^{N_2} \begin{cases} (y_{\max} - \hat{y}(k+j\,|k))^2 & \text{if} \quad \hat{y}(k+j\,|k) > y_{\max} \\ 0 & \text{if} \quad y_{\max} \geq \hat{y}(k+j\,|k) \geq y_{\min} \\ (y_{\min} - \hat{y}(k+j\,|k))^2 & \text{if} \quad y_{\min} > \hat{y}(k+j\,|k) \end{cases} \tag{15}$$

which is added to the performance objective in equation (10).

A possibility to achieve a large flexibility of the predicted input sequence $\boldsymbol{u}(\cdot\,|k)$ without calculating too many steps is partitioning the control horizon into several parts and assigning a constant value of the manipulated variable to each part. If every step of the manipulated variable is held constant over a period of T_u time increments the minimization algorithm has to optimize the signal

$$\boldsymbol{u}(\cdot\,|k) = [\underbrace{u_1, \cdots, u_1}_{T_u}, \underbrace{u_2, \cdots, u_2}_{T_u}, \cdots, \underbrace{u_{Nu}, \cdots, u_{Nu}}_{N_2-(N_u-1)T_u}]. \tag{16}$$

The control horizon partition T_u emphasizes the first step in comparison to the other ones. This idea of *blocking* ([4]) is used in the second case study.

3. Simulation

The nonlinear model based controller described above was applied to two processes. The first one, a pH neutralization process, is a mathematical model of a water fed laboratory plant. The second case study under consideration is a continuously fed stirred tank reactor with a cooling jacket. In order to show the possibilities of the NMPC based on Wiener and Hammerstein models it is compared to another controller scheme in every case study.

3.1. Case study I: pH neutralization process

The pH neutralization process shown in figure 2 is divided into two stages. Stage I is used to produce water of a constant pH value of 2.3 by adjusting acid flow. Stage II consists of the neutralization reactor and a measuring chamber. The fluid in stage II is mixed by stirring.

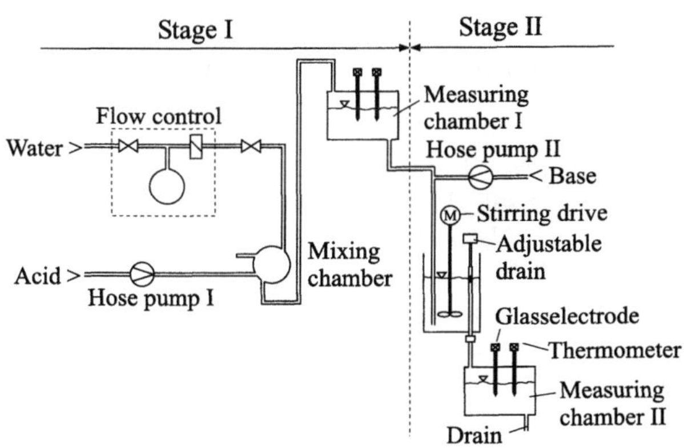

FIGURE 2. pH neutralization process

The objective of this application is to control the pH value of the second stage (pH_{II}). The manipulated voltage u of the hose pump II determines the basic solution flow $\dot{V}_B$ into stage II. Industrial applications of this case study can be found in wastewater neutralization and in many chemical processes especially with the need of varying pH during process run.

In the simulation the threeprotic orthophosphor acid H_3PO_4 and sodium hydroxide base $NaOH$ are used, thus the acid dissociates in three stages.

$$
\begin{aligned}
H_3PO_4 + H_2O &\leftrightarrow H_2PO_4^- + H_3O^+ \\
H_2PO_4^- + H_2O &\leftrightarrow HPO_4^{2-} + H_3O^+ \\
HPO_4^{2-} + H_2O &\leftrightarrow PO_4^{3-} + H_3O^+
\end{aligned}
\tag{17}
$$

The three stage dissociation results in a time invariant nonlinear titration curve including three buffering areas.

3.1.1. MODEL OF THE PROCESS. The basic structure of the pH neutralization process fits to a Wiener model. The linear dynamic elements represent the hose pump, the pipe and the mixing in stage II. They are approximated by a dead time of 10 sec and a PT_1 element whereas the titration curve can be substituted in the relevant operating range by a 3rd degree polynomial. The impulse response of the controller model and the time invariant nonlinear relation are shown in figure 3. The sampling time is set to $T = 2$ sec.

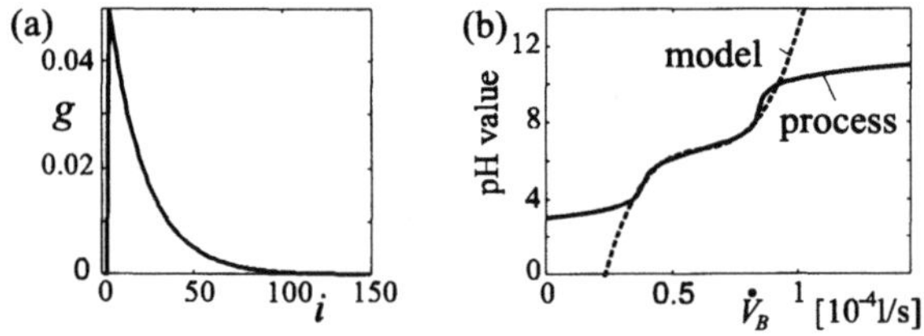

FIGURE 3. (a) Impulse response of the linear part and (b) titration curve of the Wiener model in comparison with the process

3.1.2. SIMULATION RESULTS. In order to implement the nonlinear controller several parameters have to be set. The values which are listed in table 1 were determined by simulations. N_1, N_2 and N_u may be selected within a relatively wide range. For the selection of the lower cost horizon N_1 it is not useful to take values which are smaller than the discrete dead time. The NMPC is compared with an adaptive PI-controller. Therefore, the transfer factor of the PI-controller is adapted by the controlled variable y of the process using an inverse titration curve.

TABLE 1. Control parameters

	control parameter	value
N_u	control horizon	1
N_1	lower cost horizon	5
N_2	upper cos horizon	10
λ	input variable wieghting factor	0

Figure 4 shows the closed-loop response of the pH neutralization process to step changes in the setpoint of the pH value of the second stage without disturbance. The first setpoint change takes place after 100 sec. Up to this time the NMPC is initialized measuring input and output signals of the process. During the simulation run it is shown that the correcting motion of NMPC is much more

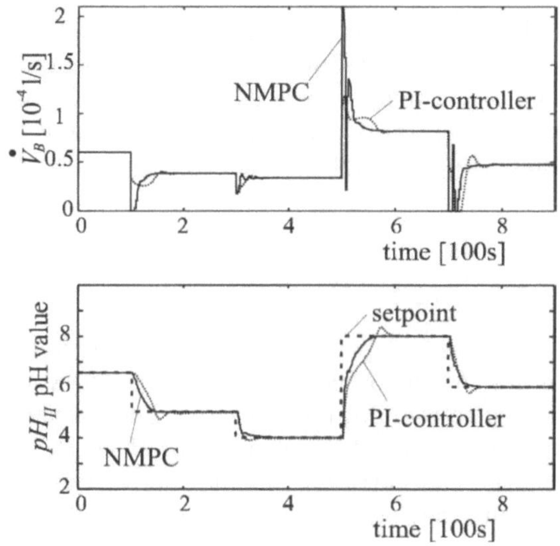

FIGURE 4. Setpoint tracking of the pH neutralization process

aggressive and uses the full allowed range of the input signal. Several times the volumetric flow $\dot{V}_B$ is set to zero. The PI-controller moves the process much more slowly but still with shooting over the setpoint.

The pH value of the process controlled by the NMPC reaches the setpoint without overshooting. In order to reduce the corrective motion of the manipulated variable it is possible to increase λ or to increase the upper cost horizon. But then the pH value would reach the setpoint after a longer period of time.

3.2. Case Study II: Continuously fed stirred tank reactor

The second case study considers a continuously fed stirred tank reactor (CSTR). A schematic diagram of the reactor with a cooling jacket in which cyclopentol is produced from cyclopentadiene by acid-catalyzed electrophylic hydration in aqueous solution is shown in figure 5. Further details on the derivation of this benchmark process and the chemical background can be found in Chen et al., [1] .

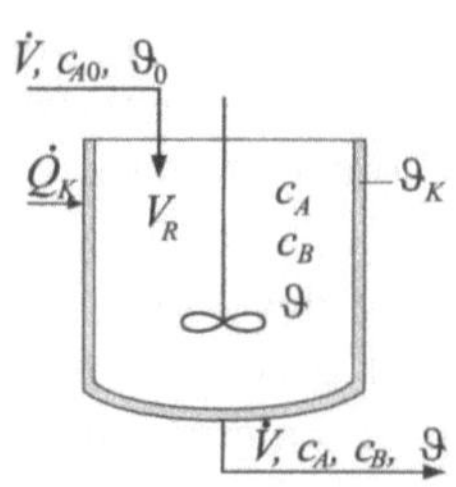

FIGURE 5. Continuously fed stirred tank reactor

The highly nonlinear behavior of the process is a result of the so-called van de Vusse reaction. The reactor is fed with diluted cyclopentadiene (substance A) of the concentration c_{A0} and temperature ϑ_0. In a thermal reaction cyclopentenol (substance B) of the concentration c_B is produced and reacts further on in an unwanted reaction to C. In parallel, the initial reactant A reacts to D which is not wanted either

$$\begin{array}{ccccc} A & \xrightarrow{k_1} & B & \xrightarrow{k_2} & C \\ 2A & \xrightarrow{k_3} & D & & \end{array} \qquad (18)$$

where the reaction velocities k_i depend on the temperature in the reactor ϑ according to the Arrhenius law. As the process is exogeneous, an external heat exchanger (energy flow $\dot{Q}_K$) is used to cool down the reactor. The concentration of substance B, c_B, and the temperature ϑ are the controlled variables. We assume ideal mixing of the contents, thus the concentrations in the center equal those at the outlet. Manipulated variables are the volumetric flow rate $\dot{V}$ and the energy flow rate $\dot{Q}_K$.

3.2.1. Model of the process. The aim is to operate the reactor at a point where optimal yield with respect to the product B is achieved (within a tolerance of $0.02\frac{\text{mol}}{\text{l}}$). At this operating point the transfer factor of the process changes its sign resulting into a very difficult control task. The operating point which is under consideration in this paper was taken from Chen et al., [1].

In order to control the concentration of the product B the volumetric flow rate $\dot{V}$, which is fed to the reactor, and the energy flow rate $\dot{Q}_K$ are manipulated. The heat removal $\dot{Q}_K$ is controlled by a proportional controller using the controlled variable ϑ. The command signal is held constant at 114^0 C and the P factor is set to $600\frac{\text{kJ}}{^0\text{C h}}$. Consequently, only the volumetric flow rate $\dot{V}$ (manipulated variable) and the concentration c_B (controlled variable) are taken into account for the nonlinear predictive controller.

After the implementation of the linear control circuit the CSTR was identified using reduced Volterra-series ([3]). The identified 2nd degree Wiener model consists of different impulse responses for the linear and quadratic path (see figure 6). Since the Wiener model is only a very simplified representation of the highly nonlinear benchmark process, the identified static and dynamic behavior can only be an inexact approximation of the process. This is pointed out in the comparison of the process with the modeled static characteristic. The sampling time for the identification and the predictive control was set to $T = 18$ sec.

3.2.2. Simulation results. The developed nonlinear predictive control algorithm based on the Wiener model described above is tested in this case study using two different sets of tuning parameters. In order to outline the capability of controlling nonlinear chemical processes using a simplified Volterra-series model the closed-loop responses are compared with an *all knowing NMPC.* This AKNMPC knows the exact model of the process as well as all states of the process

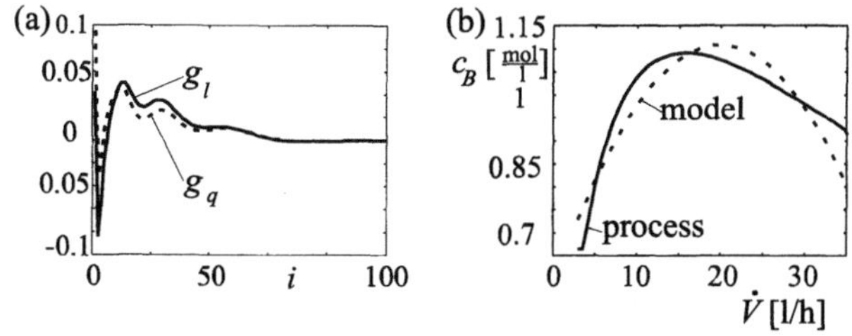

FIGURE 6. (a) impulse responses of the linear/quadratic part of the model and (b) static nonlinear curve of the model and the process

at every sampling instant. Thus the results of this controller are the best possible ones, but it will never be possible to create such a controller due to model plant mismatch and disturbances. The parameters of all controllers, the Volterra based NMPC and the AKNMPC, are listed in table 2. During the production the

TABLE 2. Control parameters

	control parameter	NMPC1	NMPC2	AKNMPC
N_u	control horizon	1	2	2
T_u	control horizon deviation	-	4	1
N_1	lower cost horizon	1	1	1
N_2	upper cos horizon	5	20	20
λ	input variable wieghting factor	0	0	0

setpoint $c_{B|S}$ of the controlled variable c_B can vary in the following range:

$$0.95\frac{\text{mol}}{\text{l}} \leq c_{B|S} \leq 1.09\frac{\text{mol}}{\text{l}} \tag{19}$$

The optimal yield with respect to the product B is achieved when the concentration c_B reaches $1.09\frac{\text{mol}}{\text{l}}$.

Figure 7 shows the closed-loop response of the CSTR to step changes in the setpoint of the maximal to the minimal and back to the maximal concentration of the product B. Both Volterra based NMPCs bring the output to the setpoint almost as quickly as the AKNMPC and keep it in the tolerated area of $0.02\frac{\text{mol}}{\text{l}}$. The NMPC1 shows a less damped behavior than NMPC2. The very small upper cost horizon of the first controller leads to fast movement of the process. The NMPC2 behaves a little bit slower but does not tend to become unstable because of the longer upper cost horizon.

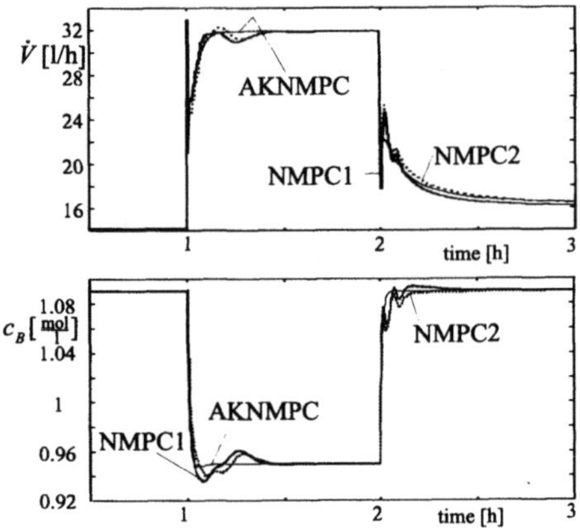

FIGURE 7. Setpoint tracking of the CSTR

4. Conclusions

Based on the Volterra-series a nonlinear predictive control algorithm is presented in this paper. For this algorithm a nonlinear predictor is developed including Wiener and Hammerstein models. In two case studies the developed NMPC is tested. The capability of controlling nonlinear chemical processes using 2nd and higher order Volterra-series is outlined. Further research is focused on the stability proof of the NMPC including input and output constraints.

References

[1] H. Chen, A. Kremling and F. Allgöwer, *Nonlinear Predictive Control of a benchmark CSTR,* In: *Proc. of the 3rd European Control Conference.* Rome, Italy, (1995), 3247–3252.

[2] H. Chen and F. Allgöwer, *Nonlinear Model Predictive Control Schemes with Guaranteed Stability,* In: *NATO Advanced Study Institute on Nonlinear Model Based Process Control* (R. Berber, Ed). Kluwer Academic Publishers. Dordrecht, The Netherlands. (in press).

[3] J. Kurth and H. Rake, *Identification of nonlinear systems with reduced Volterra-series,* In: *Proc. of the 10th IFAC Symposium on System Identification,* Kopenhagen, Denmark, **1**, (1994), 143–150.

[4] N.L. Ricker, *Use of Quadratic Programming for Constrained Internal Model Control,* Ind. Eng. Chem. Process Des. Devel., **24**(4), (1985), 925–936.

Institute of Automatic Control
RWTH Aachen
Steinbachstraße 54
D-52056 Aachen, Germany
E-mail address: `we@irt.rwth-aachen.de`

Part III

Applications of Nonlinear Predictive Control

Progress in Systems and Control Theory, Vol. 26

An Overview of Nonlinear Model Predictive Control Applications

S. Joe Qin and Thomas A. Badgwell

Abstract. This paper provides an overview of nonlinear model predictive control (NMPC) applications in industry, focusing primarily on recent applications reported by NMPC vendors. A brief summary of NMPC theory is presented to highlight issues pertinent to NMPC applications. Five industrial NMPC implementations are then discussed with reference to modeling, control, optimization, and implementation issues. Results from several industrial applications are presented to illustrate the benefits possible with NMPC technology. A discussion of future needs in NMPC theory and practice is provided to conclude the paper.

1. Introduction

The term Model Predictive Control (MPC) describes a class of computer control algorithms that control the future behavior of a plant through the use of an explicit process model. At each control interval the MPC algorithm computes an open-loop sequence of manipulated variable adjustments in order to optimize future plant behavior. The first input in the optimal sequence is injected into the plant, and the entire optimization is repeated at subsequent control intervals. MPC technology was originally developed for power plant and petroleum refinery applications, but can now be found in a wide variety of manufacturing environments including chemicals, food processing, automotive, aerospace, metallurgy, and pulp and paper. Theoretical and practical issues associated with MPC technology are summarized in several recent review articles. Qin and Badgwell (1997) present a brief history of MPC technology and a survey of industrial applications in [26]. Meadows and Rawlings summarize theoretical properties of MPC algorithms in [17]. Morari and Lee discuss the past, present, and future of MPC technology in [19]. A more recent overview of MPC theory development can be found in [6].

The success of MPC technology as a process control paradigm can be attributed to three important factors. First and foremost is the incorporation of an explicit process model into the control calculation. This allows the controller, in principle, to deal directly with all significant features of the process dynamics. Secondly the MPC algorithm considers plant behavior over a future horizon in time. This means that the effects of feedforward and feedback disturbances can be anticipated and removed, allowing the controller to drive the plant more closely along a desired future trajectory. Finally the MPC controller considers process input, state and output constraints directly in the control calculation. This means that constraint violations are far less likely, resulting in tighter control at the optimal constrained steady-state for the process. It is the inclusion of constraints that most clearly distinguishes MPC from other process control paradigms.

Though manufacturing processes are inherently nonlinear, the vast majority of MPC applications to date are based on linear dynamic models, the most common being step and impulse response models derived from the convolution integral. There are several potential reasons for this. Linear empirical models can be identified in a straightforward manner from process test data. In addition, most applications to date have been in refinery processing [26], where the goal is largely to maintain the process at a desired steady-state (regulator problem), rather than moving rapidly from one operating point to another (servo problem). A carefully identified linear model is sufficiently accurate in the neighborhood of a single operating point for such applications, especially if high quality feedback measurements are available. Finally, by using a linear model and a quadratic objective, the nominal MPC algorithm takes the form of a highly structured convex Quadratic Program (QP), for which reliable solution algorithms and software can easily be found [32]. This is important because the solution algorithm must converge reliably to the optimum in no more than a few tens of seconds to be useful in manufacturing applications. For these reasons, in many cases a linear model will provide the majority of the benefits possible with MPC technology.

Nevertheless, there are cases where nonlinear effects are significant enough to justify the use of NMPC technology. These include at least two broad categories of applications:

- Regulator control problems where the process is highly nonlinear and subject to large frequent disturbances (pH control, etc.)
- Servo control problems where the operating points change frequently and span a sufficiently wide range of nonlinear process dynamics (polymer manufacturing, ammonia synthesis, etc.).

It is interesting to note that some of the very first MPC papers describe ways to address nonlinear process behavior while still retaining a linear dynamic model in the control algorithm. Richalet et al. [29], for example, describe how nonlinear behavior due to load changes in a steam power plant application was handled by executing their Identification and Command (IDCOM) algorithm at a variable frequency. Prett and Gillette [25] describe applying a Dynamic Matrix Control (DMC) algorithm to control a fluid catalytic cracking unit. Model gains were obtained at each control iteration by perturbing a detailed nonlinear steady-state model.

While theoretical aspects of NMPC algorithms have been discussed quite effectively in several recent publications (see, for example, [15] and [17]), descriptions of industrial NMPC applications are much more difficult to find. A rare exception

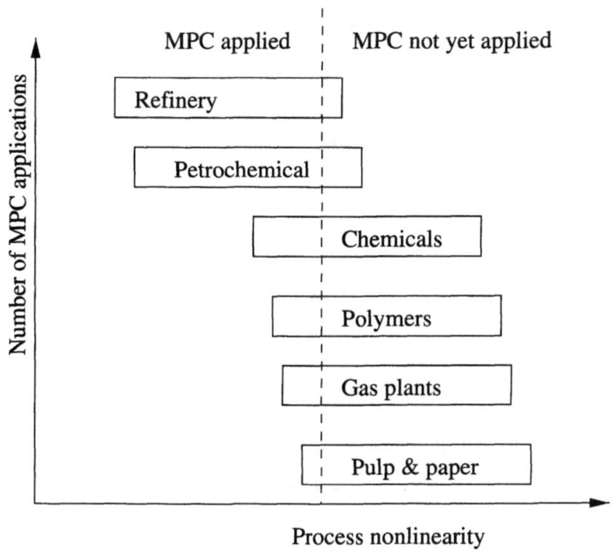

FIGURE 1. Distribution of MPC applications versus the degree of process nonlinearity.

can be found in the paper by Ogunnaike and Wright presented at the CPC-V conference [21]. This is probably due to the fact that industrial activity in NMPC applications has only begun to take off in the last few years.

In a previous survey of MPC technology [26], over 2200 commercial applications were discovered. However, almost all of these were implemented with linear models and were clustered in refinery and petrochemical processes. In preparing this paper the authors found a sizable number of NMPC applications in areas where MPC has not traditionally been applied. Figure 1 shows a rough distribution of the number of MPC applications versus the degree of process nonlinearity. MPC technology has not yet penetrated deeply into areas where process nonlinearities are strong and market demands require frequent changes in operating conditions. It is these areas that provide the greatest opportunity for NMPC applications.

The primary purpose of this paper is to provide a snapshot of the current state-of-the-art in NMPC applications. A brief summary of NMPC theory is presented to highlight what is known about closed-loop properties and to emphasize issues pertinent to NMPC applications. Then several industrial NMPC implementations are discussed in terms of modeling, control, optimization, and implementation issues. A few illustrative industrial applications are then discussed in detail. The paper concludes with a discussion of future needs and trends in NMPC theory and applications.

2. Theoretical foundations of NMPC

For the sake of discussion we first define a simplified NMPC algorithm. The calculations necessary for an implementable MPC algorithm are described in detail in [26]. For a more thorough discussion of theoretical issues pertaining to NMPC the reader is referred to [15].

Assume that the plant to be controlled can be described by the following discrete-time, nonlinear, state-space model:

$$\mathbf{x}_{k+1} = \mathbf{f}(\mathbf{x}_k, \mathbf{u}_k, \mathbf{v}_k, \mathbf{w}_k) \tag{1}$$

$$\mathbf{y}_k = \mathbf{g}(\mathbf{x}_k) + \xi_k \tag{2}$$

where $\mathbf{u}_k \in \Re^{m_u}$ is a vector of m_u process inputs or manipulated variables (MV's), $\mathbf{y}_k \in \Re^{m_y}$ is a vector of m_y process outputs or controlled variables (CV's), $\mathbf{x}_k \in \Re^n$ is a vector of n state variables, $\mathbf{v}_k \in \Re^{m_v}$ is a vector of m_v measured disturbance variables (DV's), $\mathbf{w}_k \in \Re^{m_w}$ is a vector of m_w unmeasured DV's or noise, and $\xi_k \in \Re^{m_y}$ is a vector of measurement noise.

The control problem to be solved is to compute a sequence of inputs $\{\mathbf{u}_k\}$ that will take the process from its current state $\mathbf{x}_k$ to a desired steady-state $\mathbf{x}_s$. The desired steady-state $(\mathbf{y}_s, \mathbf{x}_s, \mathbf{u}_s)$ is determined by a local steady-state optimization, which may be based on an economic objective. The optimal steady-state must be recalculated at each time step because disturbances entering the plant may change the location of the optimal operating point. Feedforward disturbances are removed by incorporating their effects into the model $\mathbf{f}$. Feedback disturbances are typically handled by assuming that a step disturbance has entered at the output and that it will remain constant for all future time. To accomplish this, a bias term that compares the current predicted output $\mathbf{y}_k$ to the current measured output $\mathbf{y}_k^m$ is computed:

$$\mathbf{b}_k = \mathbf{y}_k^m - \mathbf{y}_k \tag{3}$$

The bias $\mathbf{b}_k$ term is added to the model for use in subsequent predictions:

$$\mathbf{y}_{k+j} = \mathbf{g}(\mathbf{x}_{k+j}) + \mathbf{b}_k \tag{4}$$

The NMPC control algorithms described in this paper minimize the following dynamic objective:

$$J = \sum_{j=1}^{P} \|\mathbf{e}_{k+j}^y\|_{\mathbf{Q}_j}^q + \sum_{j=0}^{M-1} \|\Delta\mathbf{u}_{k+j}\|_{\mathbf{S}_j}^q + \sum_{j=0}^{M-1} \|\mathbf{e}_{k+j}^u\|_{\mathbf{R}_j}^q + \|\mathbf{s}\|_{\mathbf{T}}^q \tag{5}$$

subject to a model constraint:

$$\begin{aligned} \mathbf{x}_{k+j} &= \mathbf{f}(\mathbf{x}_{k+j-1}, \mathbf{u}_{k+j-1}) \quad \forall j = 1, P \\ \mathbf{y}_{k+j} &= \mathbf{g}(\mathbf{x}_{k+j}) + \mathbf{b}_k \quad \forall j = 1, P \end{aligned}$$

and subject to inequality constraints:

$$\begin{array}{ccccccl} \underline{\mathbf{y}}_j - \mathbf{s} & \leq & \mathbf{y}_{k+j} & \leq & \bar{\mathbf{y}}_j + \mathbf{s} & & \forall\, j = 1, P \\ \underline{\mathbf{u}} & \leq & \mathbf{u}_{k+j} & \leq & \bar{\mathbf{u}} & & \forall\, j = 0, M-1 \\ \Delta\underline{\mathbf{u}} & \leq & \Delta\mathbf{u}_{k+j} & \leq & \Delta\bar{\mathbf{u}} & & \forall\, j = 0, M-1 \\ \mathbf{s} & \geq & 0 & & & & \end{array}$$

The objective function in Eq. 5 involves three conflicting contributions. Future output behavior is controlled by penalizing deviations from the desired steady-state $\mathbf{y}_s$, defined as $\mathbf{e}^y_{k+j} \equiv \mathbf{y}_{k+j} - \mathbf{y}_s$, over a *prediction horizon* of length P. Future input deviations from the desired steady-state input $\mathbf{u}_s$ are controlled using input penalties defined as $\mathbf{e}^u_{k+j} \equiv \mathbf{u}_{k+j} - \mathbf{u}_s$, over a *control horizon* of length M. Rapid input changes are penalized with a separate term involving the moves $\Delta\mathbf{u}_{k+j}$. The size of the deviations is measured by a vector norm, usually either an L_1 or L_2 norm ($q = 1, 2$). The relative importance of the objective function contributions is controlled by setting the time dependent weight matrices $\mathbf{Q}_j$, $\mathbf{S}_j$, and $\mathbf{R}_j$; these are chosen to be positive definite. The last term in the objective is used to minimize the size of output constraint violations; the weight matrix $\mathbf{T}$ is also chosen to be positive definite. The solution is a set of M input adjustments:

$$\mathbf{u}^M = (\mathbf{u}_k,\ \mathbf{u}_{k+1}, \ldots, \mathbf{u}_{k+M-1}) \tag{6}$$

The first input $\mathbf{u}_k$ is injected into the plant and the calculation is repeated at the next sample time. The fact that one can solve the closed-loop control problem through a sequence of open-loop optimizations was recognized very early in the development of optimal control theory [12]. One can view the NMPC solution as a way of turning an intractable closed-loop computation into a sequence of tractable open-loop calculations [15].

In principle the NMPC method is limited to those problems for which a global optimal solution to the dynamic optimization can be found between one control execution and the next. With a linear model and a quadratic objective, the resulting optimization problem takes the form of a highly structured convex Quadratic Program (QP) for which there exists a unique optimal solution. Several reliable standard solution codes are available for this problem. Introduction of a nonlinear model leads, in the general case, to a loss of convexity. This means that it is much more difficult to find a solution, and once found, it cannot be guaranteed to be globally optimal. For both cases, recent research efforts are aimed at exploiting the structure to improve the efficiency and reliability of solution codes [32].

It was also recognized early in the development of optimal control theory that no matter how the control problem is solved, optimality does not necessarily imply closed-loop stability, even when the model represents the true plant perfectly [7]. Under certain conditions, however, this problem can be overcome through proper construction of the NMPC algorithm. The decoupling of optimality and closed-loop stability is an issue that is still not widely appreciated by industrial practitioners.

Recent research efforts on the problem of stability of NMPC with a perfect model (the so-called *nominal stability problem*) has produced three basic solutions.

The first solution, proposed by Keerthi and Gilbert [10], involves adding a terminal state constraint to the NMPC algorithm of the form:

$$\mathbf{x}_{k+P} = \mathbf{x}_s \tag{7}$$

With such a constraint enforced, the objective function for the controller (Eq. 5) becomes a Lyapunov function for the closed loop system, leading to nominal stability. Unfortunately such a constraint may be quite difficult to satisfy in real time; exact satisfaction requires an infinite number of iterations for the numerical solution code. This motivated Michalska and Mayne [18] to seek a less stringent stability requirement. Their main idea is to define a neighborhood W around the desired steady-state $\mathbf{x}_s$ within which the system can be steered to $\mathbf{x}_s$ by a constant linear feedback controller. They add to the NMPC algorithm a constraint of the form:

$$(\mathbf{x}_{k+P} - \mathbf{x}_s) \in W \tag{8}$$

If the current state $\mathbf{x}_k$ lies outside this region then the NMPC algorithm described above is solved with constraint 8. Once inside the region W the control switches to the previously determined constant linear feedback controller. Michalska and Mayne describe this as a *dual-mode* controller.

A third solution to the nominal stability problem, described by Meadows et al. [16], involves setting the prediction horizon and control horizons to infinity $P, M \to \infty$. For this case the objective function in Eq. 5 also serves as a suitable Lyapunov function, leading to nominal stability. They demonstrate that if the initial NMPC calculation has a feasible solution, then a feasible solution exists at each subsequent time step.

These theoretical results provide a foundation upon which to build an implementable NMPC controller. The challenge of the industrial practitioner is to take these ideas to the marketplace, which means that a number of additional practical issues must be confronted. Among other things, one must choose an appropriate model form, decide how best to identify or derive the model, and develop a reliable numerical solution method. The following section describes how five NMPC vendors have addressed these issues.

3. Industrial implementations of NMPC

In this section we describe the control algorithms used in several commercial NMPC products. Table 1 lists the products that we examined and the companies supplying them. This list is by no means exhaustive. However we believe that the technology sold by these companies is representative of the current state-of-the-art.

Table 2 provides information on the details of each algorithm, including the model types used, options at each step in the control calculation, and the optimization algorithm used to compute the solution. The control algorithm entries

TABLE 1. NMPC Companies and Product Names

Company	Product Name (Acronym)
Adersa	Predictive Functional Control (PFC)
Aspen Technology	Aspen Target
Continental Controls	Multivariable Control (MVC)
DOT Products	NOVA Nonlinear Controller (NOVA-NLC)
Pavilion Technologies	Process Perfecter

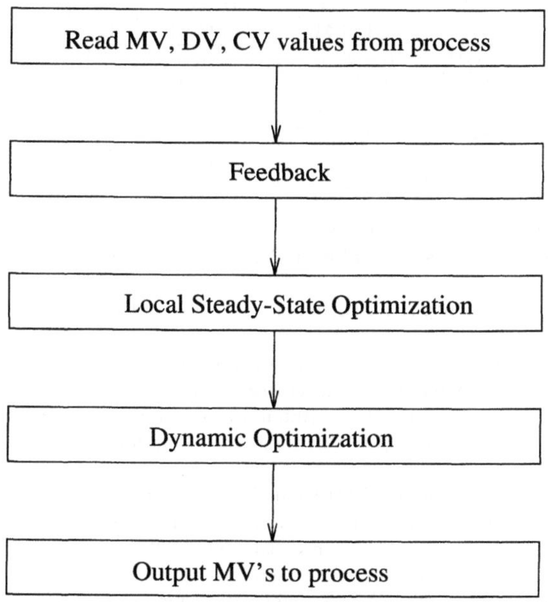

FIGURE 2. A general NMPC control calculation: MV – manipulated variable; DV – disturbance variable; CV – controlled variable.

correspond to the steps of the simplified NMPC control calculation illustrated in Figure 2. The following sub-sections describe these aspects in greater detail.

3.1. Models

The first issue encountered in NMPC implementation is the derivation of a dynamic nonlinear model suitable for model predictive control. In the general practice of linear MPC, the majority of dynamic models are derived from plant testing and system identification. For NMPC, however, the issue of plant testing and system identification becomes much more complicated. In this subsection we present process modeling methods used in the industrial practice of NMPC, which include system identification methods and first principles approaches.

TABLE 2. Comparison of Industrial NMPC Control Technology

Company	Adersa	Aspen Technology	Continental Controls	DOT Products	Pavilion Technologies
Product	PFC	Aspen Target	MVC	NOVA NLC	Process Perfecter
Model Forms[1]	NSS-FP S,I,U	NSS S,I,U	SNP-IO S	NSS-FP S,I	NNN-IO S,I,U
Feedback[2]	CD,ID	CD,ID,EKF	CD	CD,EKF	CD,ID
SS Opt Obj[3]	Q[I,O]	Q[I,O]	Q[I,O]	-	Q[I,O]
SS Opt Const[4]	IH,OH	IH,OH	IH,OS	-	IH,OH,OS
Dyn Opt Obj[5]	Q[I,O]	Q[I,O,M]	Q[I,O,M]	(Q,A)[I,O,M]	Q[I,O]
Dyn Opt Const[6]	IC,OH	IH,OS-l1	IH,OS	IH,OH,OS	IH,OS
Output Traj[7]	S,Z,RT	S,Z,RT	S,Z,RT	S,Z,RT	S,Z,TW
Output Horiz[8]	FH,CP	FH,CP	FH	FH	FH
Input Param[9]	BF,SM	MM	SM	MM	MM
Sol. Method[10]	NLS	MSN-QPKWIK	GRG GRG2	MCNLP Nova	GRG2

[1] Model Form: (IO) Input-Output, (FP) First-Principles, (NSS) Nonlinear State-Space, (NNN) Nonlinear Neural Net, (SNP) Static Nonlinear Polynomial, (S) Stable, (I) Integrating, (U) Unstable
[2] Feedback: (CD) Constant Output Disturbance, (ID) Integrating Output Disturbance, (EKF) Extended Kalman Filter
[3] Steady-State Optimization Objective: (Q) Quadratic, (I) Inputs, (O) Outputs
[4] Steady-State Optimization Constraints: (IH) Input Hard maximum, minimum, and rate of change constraints, (OH) Output Hard maximum and minimum constraints
[5] Dynamic Optimization Objective: (Q) Quadratic, (A) One norm, (I) Inputs, (O) Outputs, (M) Input Moves
[6] Dynamic Optimization Constraints: (IH) Input Hard maximum, minimum, and rate of change constraints, (IC) Input Clipped maximum, minimum, and rate of change constraints, (OH) Output Hard maximum and minimum constraints, (OS) Output Soft maximum and minimum constraints, (OS-l1) Output Soft constraints with l_1 exact penalty treatment [22].
[7] Output Trajectory: (S) Setpoint, (Z) Zone, (RT) Reference Trajectory, (TW) Trajectory Weighting
[8] Output Horizon: (FH) Finite Horizon, (CP) Coincidence Points
[9] Input Parameterization: (SM) Single Move, (MM) Multiple Move, (BF) Basis Functions
[10] Solution Method: (NLS) Nonlinear Least Squares, (MSN-QPKWIK) Multi-step Newton Method based on QPKWIK, (GRG/GRG2) Generalized Reduced Gradient, (MCNLP) Mixed Complementarity Nonlinear Program

3.1.1. STATE-SPACE MODELS. Because step response and impulse response models are non-parsimonic, a class of state-space model is adopted in the Aspen Target[1] product, which has a linear dynamic state equation and a nonlinear output relation:

$$\mathbf{x}_{k+1} = \mathbf{A}\mathbf{x}_k + \mathbf{B}_u\mathbf{u}_k + \mathbf{B}_v\mathbf{v}_k \quad (9)$$

$$\mathbf{y}_k = \mathbf{g}(\mathbf{x}_k) \quad (10)$$

[1]This product was formerly known as NeuCOP II.

where the MV's and DV's can have predetermined time-delays. More specifically, the output nonlinearity is modeled with a linear relation superimposed with a nonlinear neural network, that is,

$$\mathbf{g}(\mathbf{x}_k) = \mathbf{C}\mathbf{x}_k + \mathcal{N}(\mathbf{x}_k) \tag{11}$$

Since the state vector $\mathbf{x}$ is not necessarily limited to physical variables, this nonlinear model appears to be more general than measurement nonlinearity. For example, a Wiener model with a dynamic linear model followed by a static nonlinear mapping can be represented in this form. It is claimed that this type of nonlinear model can approximate any discrete time nonlinear processes with fading memory[30].

A difficult issue in nonlinear modeling is not the selection of a nonlinear relation, but rather the selection of a robust and reliable identification algorithm. The identification algorithm discussed in Zhao, et al. (1998) builds one model for each output separately. For a process having m_y output variables, overall m_y MISO sub-models are built. The following procedure is employed to identify each sub-model from process data.

1. Specify a rough time constant for each input-output pair, then a series of first order filters or a Laguerre model is constructed for each input (Zhao, et al. 1998; Sentoni, et al. 1998). The filter states for all inputs comprise the state vector $\mathbf{x}$.
2. A static linear model is built for each output $\{y_j, j = 1, 2, \ldots, m_y\}$ using the state vector $\mathbf{x}$ as inputs using partial least squares (PLS).
3. Model reduction is then performed on the input-state-output model identified in Steps 1 and 2 using principal component analysis and internal balancing to eliminate highly collinear state variables.
4. The reduced model is rearranged in a state-space model $(\mathbf{A}, \mathbf{B})$, which is used to generate the state sequence $\{\mathbf{x}_k, k = 1, 2, \ldots, K\}$. If the model converges, i.e., no further reduction in model order, go to the next step; otherwise, return to step 2.
5. A PLS model is built between the state vector $\mathbf{x}$ and the output y_j. The PLS model coefficients form the $\mathbf{C}$ matrix.
6. A neural network model is built between the PLS latent factors in the previous step and the PLS residual of the output y_j. This step generates the nonlinear static map $g_j(\mathbf{x})$. The use of the PLS latent factors instead of the state vectors is to improve the robustness of the neural network training and reduce the size of the neural network.

Besides the identification of the state-space model, a *model confidence index* (MCI) is also calculated on-line. If the MCI indicates that the neural network prediction is unreliable, the neural net nonlinear map is gradually turned off and the model calculation relies on the linear portion $\{\mathbf{A}, \mathbf{B}, \mathbf{C}\}$ only. Another feature of this modeling algorithm is the use of extended Kalman filters (EKF) to correct for model-plant mismatch and unmeasured disturbances (Zhao, et al. 1998). The EKF provides a bias and gain correction to the model on-line. This function replaces the constant output error feedback scheme typically employed in MPC practice.

A novel feature of the identification algorithm is that the dynamic model is built with filters and the filter states are used to predict the output variables. Due to the simplistic filter structure, each input variable has its own set of state variables, making the **A** matrix block-diagonal. This treatment assumes that each state variable is only affected by one input variable, i.e., the inputs are decoupled. For the typical case where input variables are coupled, the algorithm could generate state variables that are linearly dependent or collinear. In other words, the resulting state vector would not be a minimal realization. Nevertheless, the use of PLS algorithm makes the estimation of the **C** matrix well-conditioned. The iteration between the estimation of $\mathbf{A}, \mathbf{B}$ and **C** matrices will likely eliminate the initial error in estimating the process time constants.

Process nonlinearity is added to the model with concern for model validity using the model confidence index. When the model is used for extrapolation, only the linear portion of the model is used. The use of EKF for output error feedback is interesting; the benefit of this treatment is yet to be demonstrated.

3.1.2. INPUT-OUTPUT MODELS. The MVC algorithm and the Process Perfecter use input-output models. To simplify the system identification task, both products use a static nonlinear model superimposed upon a linear dynamic model.

Martin, et al.[13] describe the details of the Process Perfecter modeling approach. Their presentation is in single-input-single-output form, but the concept is applicable to multi-input-multi-output models. It is assumed that the process input and output can be decomposed into a steady-state portion which obeys a nonlinear static model and a deviation portion that follows a dynamic model. For any input u_k and output y_k, the deviation variables are calculated as follows,

$$\delta u_k = u_k - u_s \tag{12}$$

$$\delta y_k = y_k - y_s \tag{13}$$

where u_s and y_s are the steady-state values for the input and output, respectively, and follow a rather general nonlinear relation:

$$y_s = h_s(u_s) \tag{14}$$

The deviation variables follow a general linear dynamic relation:

$$\delta y_k = \sum_{i=1}^{n} a_i \delta y_{k-i} + b_i \delta u_{k-i} \tag{15}$$

The identification of the linear dynamic model is based on plant test data from pulse tests, while the nonlinear static model is a neural network built from historical data. It is believed that the historical data contain rich steady-state information and plant test is needed only for the dynamic sub-model.

The use of the composite model in the control step can be described as follows. Based on the desired output target y_s^d, a nonlinear optimization program calculates the best input and output values u_s^f and y_s^f using the nonlinear static model. During the dynamic controller calculation, the nonlinear static gain is approximated by a linear interpolation of the initial and final steady-state gains,

$$K_s(u_k) = K_s^i + \frac{K_s^f - K_s^i}{u_s^f - u_s^i}\delta u_k \tag{16}$$

where u_s^i and u_s^f are the current and the next steady-state values for the input, respectively, and

$$K_s^i = \frac{dy_s}{du_s}|_{u_s^i} \tag{17}$$

$$K_s^f = \frac{dy_s}{du_s}|_{u_s^f} \tag{18}$$

which are evaluated using the static nonlinear model. Bounds on K_s^i and K_s^f can be applied. Substituting the approximate gain Eq. 16 into the linear sub-model yields,

$$\delta y_k = \sum_{i=1}^{n} a_i \delta y_{k-i} + \bar{b}_i \delta u_{k-i} + g_i \delta u_{k-i}^2 \tag{19}$$

where

$$\bar{b}_i = \frac{b_i K_s^i (1 - \sum_{j=1}^n a_j)}{\sum_{j=1}^n b_j} \tag{20}$$

$$g_i = \frac{b_i (1 - \sum_{j=1}^n a_j)}{\sum_{j=1}^n b_j} \frac{K_s^f - K_s^i}{u_s^f - u_s^i} \tag{21}$$

The purpose of this approximation is to reduce computational complexity during the control calculation.

It can be seen that the steady-state target values are calculated from a nonlinear static model, whereas the dynamic control moves are calculated based on the quadratic model in Eq. 19. However, the quadratic model coefficients (i.e., the local gain) change from one control execution to the next, simply because they are rescaled to match the local gain of the static nonlinear model. This approximation strategy can be interpreted as a successive linearization at the initial and final states followed by a linear interpolation of the linearized gains. The interpolation strategy resembles gain-scheduling, but the overall model is different from gain scheduling because of the gain re-scaling. This model makes the assumption that the process dynamics remain linear over the entire range of operation. Asymmetric dynamics (e.g., different local time constants), as a result, cannot be represented by this model.

3.1.3. First principles models. Since empirical modeling approaches can be unreliable and require tremendous amount of experimental data, two of the vendors provide the option to use first principles models. These products usually ask the user to provide the first principles models with some kind of open equation editor, then the control algorithms can use the user-supplied models to calculate future control moves. The NOVA-NLC falls in this category. In both cases, model parameters must be estimated from plant data.

Hybrid modeling approaches that combine first principles knowledge with empirical modeling are also found in the commercial packages. The Process Perfecter uses a combination of first principles models in conjunction with empirical models[3]. The first principles models can be steady-state balance equations, a nonlinear function of physical variables that generates another physically meaning variable, such as production rate, or simply gain directions to validate empirical models.

3.2. Output feedback

In the face of unmeasured disturbances and model errors, some form of feedback is required to remove steady-state offset. As discussed earlier, the most common method for incorporating feedback into MPC algorithms involves comparing the measured and predicted process outputs [26]. The difference between the two is added to future output predictions to bias them in the direction of the measured output. This can be interpreted as assuming that an unmeasured step disturbance enters at the process output and remains constant for all future time. For the case of a linear model and no active constraints, Rawlings, et al. [28] have shown that this form of feedback leads to offset-free control. As can be seen in Table 2, all five NMPC algorithms described here provide the constant output feedback option.

When the process has a pure integrator, the constant output disturbance assumption will no longer lead to offset-free control. For this case it is common to assume that an integrating disturbance with a constraint ramp rate has entered at the output [26]. The PFC, Aspen Target, and Process Pefecter algorithms provide this feedback option.

It is well known from linear control theory that additional knowledge about unmeasured disturbances can be exploited to provide better feedback by designing a Kalman Filter [8]. Muske and Rawlings demonstrate how this can be accomplished in the context of MPC [20]. It is interesting to note that two of the NMPC algorithms in Table 2 provide options for output feedback based on a nonlinear generalization of the Kalman Filter known as the Extended Kalman Filter (EKF) [27]. Aspen Target provides an EKF to estimate both a bias and a feedback gain. NOVA-NLC uses an EKF to develop complete state and noise estimates.

3.3. Steady-state optimization

The PFC, Aspen Target, MVC, and Process Perfecter controllers split the control calculation into a local steady-state optimization followed by a dynamic optimization. Optimal steady-state targets are computed for each input and output; these are then passed to a dynamic optimization to compute the optimal input sequence required to move toward these targets. From Table 2 it can be seen that these calculations involve optimizing a quadratic objective that includes input and output contributions. The exception is the NOVA-NLC controller that performs the dynamic and steady-state optimizations simultaneously.

3.4. Dynamic optimization

At the dynamic optimization level, an MPC controller must compute a set of MV adjustments that will drive the process to the steady-state operating point without violating constraints. All of the algorithms described here use a form of the objective given in Eq. 5.

The PFC controller includes only the process input and output terms in the dynamic objective, and uses constant weight matrices ($Q_j = Q$, $R_j = R$, $S_j = 0$, $q = 2$). The Aspen Target and MVC products include all three terms with constant weights ($Q_j = Q$, $R_j = R$, $S_j = S$, $q = 2$). The NOVA-NLC product adds to this the option of one norms ($Q_j = Q$, $R_j = R$, $S_j = S$, $q = 1, 2$).

Instead of using a reference trajectory ($R_j = 0$, $S_j = 0$,$q = 2$), the Process Perfecter product uses a dynamic objective of the form, *trajectory weighting* that makes $\mathbf{Q}_j$ gradually increase over the horizon P. With this type of weighting, control errors at the beginning of the horizon are less important than those towards the end of the horizon, thus allowing a smoother control action.

3.5. Constraint formulations

There are basically two types of constraints used in industrial MPC technology; hard and soft [26]. Hard constraints are those which should never be violated. Soft constraints allow the possibility of a violation; the magnitude of the violation is generally subjected to a quadratic penalty in the objective function.

All of the NMPC algorithms described here allow hard input maximum, minimum, and rate of change constraints to be defined. These are generally defined so as to keep the lower level MV controllers in a controllable range, and to prevent violent movement of the MV's at any single control execution. The PFC algorithm also accommodates maximum and minimum input acceleration constraints which are useful in mechanical servo control applications.

The Aspen Target, MVC, NOVA-NLC, and Process Perfecter algorithms perform rigorous optimizations subject to the hard input constraints. The PFC algorithm, however, enforces input hard constraints only after performing an unconstrained optimization. This is accomplished by clipping input values that exceed the input constraints. It should be noted that this method does not, in general, result in an optimal solution in the sense of satisfying the Karush-Kuhn-Tucker (KKT) conditions for optimality.

All control products enforce output constraints as part of the dynamic optimization, as shown in Eq. 5. The Aspen Target, NOVA-NLC, and Process Perfecter products allow options for both hard and soft output constraints. The PFC product allows only hard output constraints, while the MVC product allows only soft output constraints. The exclusive use of hard output constraints is generally

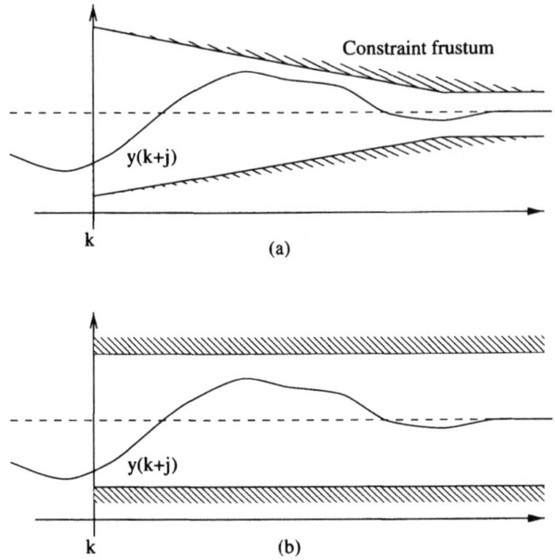

FIGURE 3. Soft constraint formulations: (a) a frustum; (b) a rectangular zone.

avoided in MPC technology because a disturbance can cause such a controller to lose feasibility.

The Process Perfecter product applies soft constraints by using a *frustum method*, as depicted in Figure 3(a). Compared to the typical zone formulation as shown in Figure 3(b), the frustum permits a larger control error in the beginning of the horizon than in the end, but no error is allowed outside the frustum. At the end of the horizon the frustum can have a non-zero zone, instead of merging to a single line, which is determined based on the accuracy of the process model to allow for model errors.

3.6. Output trajectories

Industrial MPC controllers use four basic options to specify future CV behavior; a setpoint, zone, reference trajectory or funnel [26]. All of the NMPC controllers described here provide the option to drive the CV's to a fixed setpoint, with deviations on both sides penalized in the objective function. In practice this type of specification is very aggressive and may lead to very large input adjustments, unless the controller is detuned in some fashion. This is particularly important when the model differs significantly from the true process. For this reason all of the controllers provide some way to detune the controller using either move suppression, a reference trajectory, or time-dependent weights.

All of the controllers also provide a CV zone control option, designed to keep the CV within a zone defined by upper and lower boundaries. A simple way to implement zone control is to define soft output constraints at the upper and lower boundaries.

The PFC, Aspen Target, MVC, and NOVA-NLC algorithms provide a CV reference trajectory option, in which the CV is required to follow a smooth path from its current value to the setpoint. Typically a first order path is defined using an operator entered closed loop time constant. In the limit of a zero time constant the reference trajectory reverts back to a pure setpoint; for this case, however, the controller would be sensitive to model mismatch unless some other strategy such as move suppression is also being used. In general, as the reference trajectory time constant increases, the controller is able to tolerate larger model mismatch.

3.7. Output horizon and input parameterization

Industrial MPC controllers generally evaluate future CV behavior over a finite set of future time intervals called the *prediction horizon* [26]. This finite output horizon formulation is used by all of the algorithms discussed in this paper. The length of the horizon P is a basic tuning parameter for these controllers, and is generally set long enough to capture the steady-state effects of all computed future MV moves. This is an approximation of the infinite horizon solution for closed loop stability discussed earlier, and may explain why none of the industrial NMPC algorithms considered here include a terminal state constraint.

The PFC and Aspen Target controllers allow the option to simplify the calculation by considering only a subset of future points called *coincidence points*, so named because the desired and predicted future outputs are required to coincide at these points. A separate set of coincidence points can be defined for each output, which is useful when one output responds quickly relative to another.

Industrial MPC controllers use three different methods to parameterize the MV profile; a single move, multiple moves, and basis functions [26]. The MVC product computes a single future input value; the PFC controller also provides this option. The Aspen Target, NOVA-NLC, and Process Perfecter controllers can compute a sequence of future moves spread over a finite *control horizon*. The length of the control horizon M is another basic tuning parameter for these controllers. Better control performance is obtained as M increases, at the expense of additional computation.

The PFC controller parameterizes the input function using a set of polynomial basis functions. This allows a relatively complex input profile to be specified over a large (potentially infinite) control horizon, using a small number of unknown parameters. This may provide an advantage when controlling nonlinear systems. Choosing the family of basis functions establishes many of the features of the computed input profile; this is one way to ensure a smooth input signal, for example. If a polynomial basis is chosen then the order can be selected so as to follow a polynomial setpoint signal with no lag. This feature is important for mechanical servo control applications.

3.8. Solution methods

The PFC controller performs an unconstrained optimization using a nonlinear least-squares algorithm. The solution can be computed very rapidly, allowing the controller to be used for short sample time applications such as missile tracking. Some performance loss may be expected, however, since input constraints are enforced by clipping.

The Aspen Target product uses a multi-step Newton-type algorithm developed by Oliveira and Biegler [22, 23], and makes use of analytical model derivatives. Due to the sparseness of the state space model in Aspen Target, the derivative computation is straightforward. The Newton algorithm makes use of the QPKWIK solver which has the advantage that intermediate solutions, although not optimal, are guaranteed feasible. This permits early termination of the optimization algorithm if the optimum is not found within the sampling time. Aspen Target uses the same QPKWIK engine for local steady-state optimization and the dynamic MV calculation.

The MVC and Process Perfecter products use a generalized reduced gradient (GRG) code called GRG2 developed by Lasdon and Warren [11]. The NOVA-NLC product uses the NOVA optimization package, a proprietary mixed complementarity nonlinear programming code developed by DOT Products.

3.9. Implementation issues

The following implementation issues must be considered in a commercial NMPC application.

1. *Modeling issues.* The first difficult issue is how to perform plant tests. To capture any nonlinearity in the process, extensive testing using a multi-level design is desired. This will make the testing period much longer than that of a linear plant test. To reduce the plant test period, the Process Perfecter uses historical data to build a nonlinear steady-state model and a rather short pulse test to build a linear local model. Alternatively, first principles models are used when they are available.
2. *Control and optimization issues.* Because of the use of a nonlinear model, the NMPC calculation usually involves a non-convex nonlinear program, for which the numerical solution is very challenging.
3. *Justification of the NMPC effort.* Because of the difficulties involved in NMPC implementations, the added benefit of applying NMPC has to be justified. To deal with this problem most products provide linear MPC as a back-up. In the case that NMPC is not needed or difficult to implement, linear MPC is implemented instead.
4. *Graphical user interface.* As the Windows NT operating system becomes widely acceptable in the process industry, most NMPC products provide an easy-to-use graphical user interface which takes advantage of Windows based features.

4. Summary of NMPC applications

The past three years have seen rapid progress in the development and application of industrial NMPC technology. Table 3 summarizes reported applications for the products considered in this paper. A total of 88 applications are reported in a wide range of application areas. The number of actual NMPC applications is likely to be significantly larger since only a handful of vendors were included in this survey, and only those applications known to the vendors were reported. It is possible that end-users have applied these products to other problems or have developed their own NMPC algorithms. Most of the applications reported here have taken place in the last three years, so the number of applications can be expected to grow rapidly in the near future.

TABLE 3. Summary of Reported NMPC Vendor Applications

Area	Adersa	Aspen Technology	Continental Controls	DOT Products	Pavilion Technologies	Total
Air and Gas			18			18
Chemicals	2		15		5	22
Food Processing					9	9
Polymers		1		5	15	20
Pulp & Paper					1	1
Refining					13	13
Utilities		5	2			3
Others	1		1			2
Total	3	6	36	5	43	88

While MPC applications are concentrated in refining [26], NMPC applications reported here cover a much broader range of application areas. Areas with the largest number of reported NMPC applications include chemicals, polymers, and air and gas processing.

Although not shown in the table, it has been observed that the size and scope of NMPC applications are typically much smaller than that of linear MPC applications [14]. This is likely due to the computational complexity of NMPC algorithms.

5. NMPC application examples

In this section we describe several successful applications of NMPC technology.

5.1. MVC: Application to an ammonia plant

Poe and Munsif [24] describe an application of the MVC product to control a plant producing 1450 tons/day of ammonia using natural gas and air as feedstocks. Key factors used to justify this application include:

- dynamic market supply and demand effects on natural gas price and product prices
- capacity and throughput limitations
- variations in gas feedstock rates, quality, and composition
- environmental limitations

The basic objective of the MVC controller is to maximize a profit function computed by subtracting natural gas feed and fuel gas costs from ammonia product, carbon dioxide, and steam export revenues. To achieve this the controller uses an overall economic optimization module to compute optimal steady-state targets for the plant, which are sent to seven separate dynamic controllers:

- hydraulic (steam and pressure balance) module
- primary reformer furnace temperature control module
- primary reformer riser temperature balance control module
- secondary reformer module
- shift/methanator module
- carbon dioxide removal module
- ammonia converter module

Poe and Munsif [24] describe the control modules in some detail; here we focus only on the ammonia converter module.

The ammonia converter is a standard Kellogg "quench converter" design, consisting of three nearly adiabatic catalyst beds, between which fresh feed is introduced to cool the reaction products. The converter control module manipulates the feed flow to the first bed as well as the quench flows to all three beds. This is done in order to maintain the three bed inlet temperatures at their optimal steady-state targets. Output constraints considered by the control include bed outlet temperatures and quench flow valve positions. Feedforward control is provided for changes in feed flowrate, temperature, and pressure, hydrogen/nitrogen ratio, and inert composition.

Figure 4 shows results for the second converter bed; results for the other beds were similar. The MVC controller was able to significantly reduce temperature variations at the bed inlet and outlet, allowing the average reaction temperature to be increased without violating the bed outlet constraint. Overall the plant's net fuel consumption was lowered by 1.8%, while the net production of ammonia increased by 0.7%.

5.2. Process Perfecter: Application to a polypropylene process

Demoro, et al. (1997) reported a successful application of NMPC to a polypropylene process using the Process Perfecter product. The polymer process is extremely nonlinear with complex reactions and involves frequent grade changes. Demoro, et al. (1997) also compare the results of a linear MPC with those of a NMPC on the same process, demonstrating the additional benefits of NMPC for this type of processes.

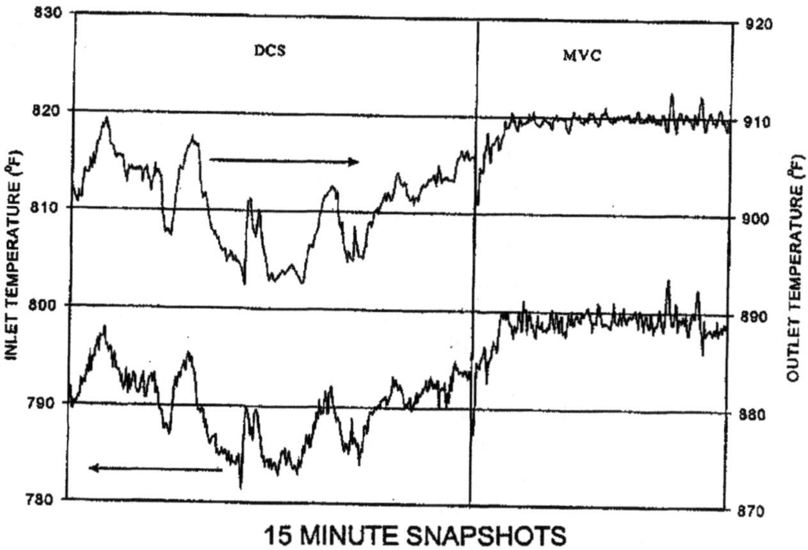

FIGURE 4. MVC Results: Ammonia converter bed 2 temperatures

The polypropylene process is a bulk liquid phase (slurry) process that carries out Ziegler-Natta polymerization in a loop reactor. The control and optimization objectives are:

- minimize product variability;
- minimize grade transition time; and
- maximize production rate.

The following three controlled variables were selected:

- production rate, which is calculated from a heat balance;
- solids in slurry, which is calculated from the density of the slurry; and
- melt flow, which is measured via lab analysis using melt flow rheometry about every 4 hours.

The three selected manipulated variables are:

- catalyst flow;
- feed flow of monomers; and
- modifier concentration, which is cascaded with the modifier flow controller.

The following two feedforward disturbance variables are used in the NMPC:

- inerts concentration in the monomer feed; and
- reactor temperature.

Because none of the CV's are directly measured on-line, soft sensors were built using either first principles or neural networks from process data. The reactor production rate was calculated with steady state mass and energy balance models. The amount of solids in slurry was calculated from the density. The melt flow rate was estimated using a neural network, with input variables being either original sensor measurements or transformed based on physical understanding of the process, for instance, production rate and component hold-up. Since several sensors are analytical measurements which are prone to gross errors, a sensor validation module screens the data before it is used in optimization and control.

The static nonlinear process model was then built from historical data using neural networks, which also estimated the melt flow rate from lab test data. This model was integrated with a dynamic linear submodel using the Process Perfecter software. To determine the benefits of using NMPC for this process, a linear model was also built for the process to configure a linear MPC.

Demoro et al. (1997) reported three control experiments on the polypropylene process. The first one used the linear model, the second used the nonlinear model, and third used the nonlinear model in conjunction with an upper level nonlinear optimizer. In each experiment the polymer melt flow rate was changed from 30 to 35. It was observed that the linear MPC was unable to accomplish the transition within 30 minutes, since the process is highly nonlinear and the controller was detuned to achieve stability. The nonlinear MPC algorithm was able to accomplish the transition very quickly. In the third experiment where an optimizer was used, production rate was maximized while grade transition was accomplished within the process constraints.

5.3. Aspen Target: Application to a pulverized coal fired boiler

Aspen Technology [4] reported an application of the Aspen Target to a pulverized coal fired boiler control. The objectives are to (i) improve boiler efficiency, (ii) reduce NOx emission, and (iii) reduce lose of ignition. The process consists of pulverizers for crushing the coal to improving firing, boilers, and a turbine. The coal burners are swirled type low-NOx burners with a boiler steam capacity of 650 tons/hour.

During coal combustion, moisture and oxygen are believed to dominate the formation of NOx and the model relation is nonlinear. Without detailed knowledge about the implementation, it is reported that the Aspen Target controller was able to reduce NOx emission by 15-25%, increase boiler efficiency by 0.1-0.3%, and decrease loss of ignition by 2%.

6. Future needs for NMPC technology development

Although many academic research results are available for NMPC, many issues that affect industrial practice are as yet unresolved:

1. *Modeling approaches.* There is no systematic approach for building nonlinear dynamic models for NMPC. In the case of empirical approaches, guidelines for plant tests are needed to build a reliable model.

2. *Optimization.* Speed and the assurance of a reliable solution in real-time are major limiting factors in existing applications.
3. *Output feedback.* Most current NMPC implementations use the traditional bias correction to the model prediction based on current measurements. While this approach is meaningful for linear MPC because of the principle of superposition, it is questionable how general this approach is to nonlinear processes. Nonlinear state estimation may provide an optimal approach to this issue.
4. *Justification of NMPC.* Criteria on where NMPC is needed are desirable but difficult to obtain. Benchmarks on the justification of NMPC are needed on a array of industrial processes. Unfortunately, only one such activity has been reported so far[3].
5. *Other issues.* Other issues that are applicable to linear MPC technology [26] should also be of the same level of concern for NMPC, if not more. These issues include multiple prioritized objective functions, determining controllable sub-processes, tuning, ill-conditioning, and fault tolerance.

7. Conclusions

We can draw the following conclusions.

- The past three years have seen rapid progress in the development and application of NMPC algorithms, with a total of 88 applications reported by the vendors included in this survey.
- The algorithms reported here differ in the simplifications used to generate a tractable control calculation; all of them, however, are based on adding a nonlinear model to a proven MPC formulation.
- None of the currently available NMPC algorithms includes the terminal state constraints or infinite prediction horizon required by control theory for nominal stability; instead they rely implicitly upon setting the prediction horizon long enough to effectively approximate an infinite horizon.
- The three most significant obstacles to NMPC applications are: nonlinear model development, state estimation, and rapid, reliable solution of the control algorithm in real time.
- Future needs for NMPC technology include development of a systematic approach for nonlinear model identification, nonlinear estimation methods, reliable numerical solution techniques, and better methods for justifying NMPC applications.

Acknowledgments

The authors would like to thank Jacques Richalet of Adersa, Hong Zhao of Aspen Technology, Humil Munsif, William Poe, and Ujjal Basu of Continental Controls, Mike Morshedi and Jeff Renfro of DOT Products, Jim Keeler, Steve Piche, and Doug Johnston of Pavilion Technologies, and Steve Treiber of Treiber Controls for their cooperation in providing information for this paper. The financial support of the National Science Foundation (grant CTS-9714950) and the Texas Advanced Technology Program (grant 003604-033) is gratefully acknowledged.

References

[1] P. Berkowitz and M. Papadopoulos. Multivariable process control method and apparatus. U.S. Patent, March 7 1995. No. 5396416.

[2] P. Berkowitz, M. Papadopoulos, L. Colwell, and M. Moran. Multivariable process control method and apparatus. U.S. Patent, Jan. 30 1996. No. 5488561.

[3] E. Demoro, C. Axelrud, D. Johnston, and G. Martin. Neural network modeling and control of polypropylene process. In *Society of Plastic Engineers Int'l Conference*, Houston, Texas, Feb. 23-26 1997.

[4] Aspen Technology, Inc. Pulverized coal fired boiler optimization and control. Aspen Technology Product Literature, 1998.

[5] Continental Controls, Inc. Production description: MVC 3.0. Houston, Texas, August 1995.

[6] G. De Nicolao, L. Magni and R. Scattolini. *Chapter of this Book*, 1998.

[7] R. E. Kalman. Contributions to the theory of optimal control. *Bull. Soc. Math. Mex.*, 5:102–119, 1960.

[8] R. E. Kalman and R. S. Bucy. New results in linear filtering and prediction theory. *Trans. ASME, J. Basic Engineeering*, pages 95–108, March 1961.

[9] J. Keeler, G. Martin, G. Boe, S. Piche, U. Mathur, and D. Johnston. The process perfecter: The next step in multivariable control and optimization. Technical report, Pavilion Technologies, Inc., 1996.

[10] S. S. Keerthi and E. G. Gilbert. Optimal infinite-horizon feedback laws for a general class of constrained discrete-time systems: Stability and moving-horizon approximations. *J. Optim. Theory Appl.*, 57(2):265–293, May 1988.

[11] L. S. Lasdon and A. D. Waren. *GRG2 User's Guide*. Department of Computer and Information Science, Cleveland State University, Cleveland, Ohio, 1986.

[12] E. B. Lee and L. Markus. *Foundations of Optimal Control Theory.* John Wiley and Sons, New York, 1967.

[13] G. Martin, G. Boe, J. Keeler, D. Timmer, and J. Havener. Method and apparatus for modeling dynamic and steady-state processes for prediction, control and optimization. US Patent, 1998.

[14] G. Martin and D. Johnston. Continuous model-based optimization. In *Process Optimization Conference*, Houston, Texas, March 24-26 1998. Gulf Publishing Co. and Hydrocarbon Processing.

[15] D. Q. Mayne. Nonlinear model predictive control: An assessment. In Jeffrey C. Kantor, Carlos E. Garcia, and Brice Carnahan, editors, *Fifth International Conference on Chemical Process Control*, pages 217–231. AIChE and CACHE, 1997.

[16] Edward S. Meadows, Michael A. Henson, John W. Eaton, and James B. Rawlings. Receding horizon control and discontinuous state feedback stabilization. *Int. J. Control*, pages 1217–1229, 1995.

[17] Edward S. Meadows and James B. Rawlings. Model Predictive Control. In Michael A. Henson and Dale E. Seborg, editors, *Nonlinear Process Control*, chapter 5, pages 233–310. Prentice Hall, 1997.

[18] Hanna Michalska and David Q. Mayne. Robust receding horizon control of constrained nonlinear systems. *IEEE Trans. Auto. Cont.*, 38(11):1623–1633, 1993.

[19] Manfred Morari and Jay H. Lee. Model Predictive Control: Past, Present and Future. In *Proceedings of* PSE/Escape '97, *Trondheim, Norway*, 1997.

[20] Kenneth R. Muske and James B. Rawlings. Model predictive control with linear models. *AIChE J.*, 39(2):262–287, 1993.

[21] B. A. Ogunnaike and R. A. Wright. Industrial applications of nonlinear control. In Jeffrey C. Kantor, Carlos E. Garcia, and Brice Carnahan, editors, *Fifth International Conference on Chemical Process Control*, pages 47–59. AIChE and CACHE, 1997.

[22] N.M.C. Oliveira and L.T. Biegler. Constraint handling and stability properties of model-predictive control. *AIChE J.*, 40:1138–1155, 1994.

[23] N.M.C. Oliveira and L.T. Biegler. An extension of Newton-type algorithms for nonlinear process control. *Automatica*, 31:281–286, 1995.

[24] William Poe and Himal Munsif. Benefits of advanced process control and economic optimization to petrochemical processes. Hydrocarbon Processing's Process Optimization Conference and Exhibition, March 1998.

[25] D. M. Prett and R. D. Gillette. Optimization and constrained multivariable control of a catalytic cracking unit. In *Proceedings of the Joint Automatic Control Conference*, 1980.

[26] S. Joe Qin and Thomas A. Badgwell. An Overview of Industrial Model Predictive Control Technology. In Jeffrey C. Kantor, Carlos E. Garcia, and Brice Carnahan, editors, *Fifth International Conference on Chemical Process Control*, pages 232–256. AIChE and CACHE, 1997.

[27] W. Fred Ramirez. *Process Control and Identification.* Academic Press, New York, New York, 1994.

[28] James B. Rawlings, Edward S. Meadows, and Ken Muske. Nonlinear model predictive control: a tutorial and survey. In *Proceedings of IFAC ADCHEM*, Japan, 1994.

[29] J. Richalet, A. Rault, J. L. Testud, and J. Papon. Model predictive heuristic control: Applications to industrial processes. *Automatica*, 14:413–428, 1978.

[30] G.B. Sentoni, J.P. Guiver, H. Zhao, and L.T. Biegler. State space nonlinear process modeling: identification and universality. Revised for AIChE Journal, March 1998.

[31] Vance J. VanDoren. Multivariable controllers enter the mainstream. *Control Engineering*, pages 107–112, March 1997.

[32] Steven J. Wright. Applying New Optimization Algorithms to Model Predictive Control. In Jeffrey C. Kantor, Carlos E. Garcia, and Brice Carnahan, editors, *Chemical*

Process Control: Assessment and New Directions for Research, AIChE Symposium Series 316, pages 147–155. AIChE and CACHE, 1997.

[33] H. Zhao, J.P. Guiver, and G.B. Sentoni. An identification approach to nonlinear state space model for industrial multivariable model predictive control. In *American Control Conference*, Philadelphia, PA, June 21-24 1998.

S. Joe Qin
Department of Chemical Engineering
The University of Texas at Austin
Austin, TX 78712
email: `qin@che.utexas.edu`
and
Thomas A. Badgwell
Chemical Engineering Department, MS-362
6100 Main Street
Rice University
Houston, TX 77005-1892
email: `tab@rice.edu`

Progress in Systems and Control Theory, Vol. 26

Multi-zone Control under Enterprise Optimization: Needs, Challenges and Requirements

Joseph Z. Lu

Abstract. In the past 15–20 years, MPC technology has penetrated a significant portion of the process industries and harvested large economic benefits. In the next 5–10 years, we will see widespread use of MPC in two ways: standalone and integrated. We will also see the emergence of enterprise optimization, i.e., the cross-functional integration and optimization of the supply chain, the production process and the distribution channel. What is not so clear is how MPC technology will be effectively integrated into an enterprise optimization scheme.

This paper takes a holistic view of process control and optimization in a complex plant. It describes the need, benefit and challenges of integrating MPC technology into enterprise optimization. It illustrates the interplay between MPC and enterprise optimization. Finally, it discusses where the new opportunities lie and what the main challenges are, focusing on advanced process control (APC) in general and NMPC in particular.

1. Motivation

MPC has been proven to be the most viable multivariable control solution in continuous process industries. It is becoming more and more popular in semi-batch and batch industries. More than 90% of industrial implementations of multivariable control solutions are in some form of MPC. Most industrial MPC products also include a proprietary economic optimization algorithm that is essential for guiding or even driving the process to deliver more profit. In industry, the integration of MPC and economic optimization is often referred to as advanced process control (APC). Since MPC plays such an important role in APC, the terms MPC and APC are often used interchangeably.

The estimated, and sometimes audited, payback time for APC applications (calculated often in transfer price[1]) ranges from a few weeks to 18 months, with most reported payback times falling between 3-12 months [4, 5, 9, 10 & 11]. In terms of economic benefit, MPC is one of the most significant enabling technologies for the process industries in recent years.

[1]Transfer price is an artificial price assigned to determine the benefit contribution of a process unit to the overall plant, under an assumed level of global coordination.

The extent of industrial APC application varies in different parts of the world and in different industries. Geographically, U.S. and Europe take the lead, while some parts of Asia and Africa follow three or more years behind. In different industries, oil refining takes the lead followed tightly by chemicals. In recent years pulp and paper, utility, batch, semi-conductor, mining, and ore processing industries have begun to catch up to speed.

As linear MPC becomes the standard technology in industry, nonlinear MPC is debuting both in academic and industrial settings [8]. One of the challenges for industrial researchers is to find large classes of practical NMPC problems on which the NMPC research can be motivated, and on which the research results can be analyzed, benchmarked, tested, and finally, implemented.

One such NMPC application class comes from a much greater scheme – cross-functional integration – which is an essential first step of enterprise optimization. Cross-functional integration is devised to capture the latent operating benefit in a complex production plant such as a refinery or a commodity chemicals plant. The latent benefit is captured by instituting global optimization and delivered by coordinating multiple APC applications. Cross-functional integration requires the process units to operate more responsively, not only to the setpoint changes, but also to the coordination instructions. Thus, it introduces some additional control requirements, some of which require that the controllers be designed with the consideration of enterprise optimization.

In the following sections, we will introduce the concept of cross-functional integration, describe the problems it is intended to solve, outline the industrial trend and the short term and long term goals, illustrate the new opportunities and challenges for NMPC, and present the new requirements on the underlying control solutions.

2. New trend–cross-functional integration

Traditionally, most MPC technologies are used for stabilizing operations, reducing variability, improving product qualities and optimizing unit production. Most benefits are achieved with a divide-and-conquer approach to a complex plant-wide production problem. In this approach, a large plant is divided into many process units, and APC is then applied on appropriate units. The divide-and-conquer approach reduces the complexity of the plant-wide problem, but each APC application can reach only its local optimum at best. In a complex plant, the sum of local optima can be significantly less than the potential global optimum. For example, the estimated latent benefit for a typical refinery is 2–10 times more than what APC by itself can capture [1].

One practical approach for delivering global optimization benefit is to add a coordination layer on top of all the MPC applications to achieve the global optimum. The coordination layer would usually cover multiple functions of the plant, such as operations, production scheduling, production planning and others.

2.1. Complexity of APC coordination

With the divide-and-conquer approach, transfer price is traditionally used in measuring the merit of APC application from a plant-wide view and in an appropriate time period. The transfer price of a product (or a feed) in a process unit often contains various assumptions. One common assumption is that the product produced in a unit will travel through the designed paths (or designed processes) with the designed fractions to reach the final designated markets. This assumption is not always valid due to a lack of dynamic coordination among the units.

This phenomenon is referred to as benefit erosion[2] in many publications such as in [1], but the fundamental cause is the lack of global coordination or optimization. Generally speaking, the more complex the production scheme, the more likely the benefit erosion will occur. Therefore, a complex plant, such as a refinery or a chemicals plant, presents a higher benefit potential for cross-functional integration.

The common approach to prevent benefit erosion is human coordination between different sections of the production processes. For example, after an APC application is implemented on a refinery's crude unit, the yield of the most valuable component often increases, while the yield of the least valuable component decreases. The scheduling group would detect the component inventory imbalances (which could cause tank level problems if not corrected in time) rippling through various parts of the refinery. The scheduling group would coordinate the affected parts of the refinery to "digest" the imbalances. The planning group would take feedback from both the scheduling group and the operations group, update its yield models to reflect the yield improvement, and rerun the plant-wide optimization to generate a new production plan and schedule.

However, if a new dynamic or steady-state bottleneck is encountered, or if an infeasible production schedule results, the scheduling group and the planning group would have to work together with the operations group to devise a new solution. The final solution may take a few adjustments or iterations. For complex cases, this process can take a few weeks or even months.

Only when all operations and activities in the refinery are coordinated together will the APC benefit erosion be minimized. Although the situation described in this refinery example may sound primitive, it is still one of the better cases. In reality, different parts of a refinery usually use different tools with different models on different platforms. Their jobs require them to look at the same problem with very different time horizons. All of those factors complicate the plant-wide coordination in real life.

This example demonstrates the complexity of plant-wide coordination, and also illustrates the vast latent benefits remaining even after successful applications of MPC across an entire plant. The cross-functional integration approach is believed to be a viable solution to capture these benefits. It is causing a paradigm shift for both the producers and the solution providers in the process industry.

[2]The benefit loss when the assumed level of global coordination is not reached.

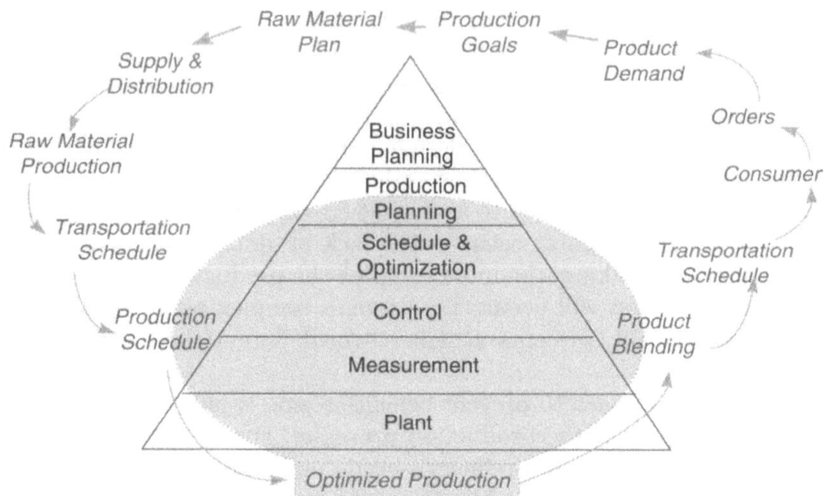

FIGURE 1. A refinery example of enterprise optimization.

It requires the existing infrastructures to be revamped, from the DCS hardware and supporting software systems to the applications and tools. In the following section, the global picture of cross-function integration will be discussed, and its MPC-relevant part will be discussed in detail.

2.2. The long and short term goals

Cross-functional integration takes a holistic approach to the plant-wide problem. The integration encompasses a large number of process units and operating activities. Considering the practical operability, it would be best built from the bottom up, integrating one layer at a time until finally reaching the level of enterprise optimization as depicted in Figure 1.

In Figure 1, the various layers in the pyramid describe the plant-wide automation solution structure and the decision-making hierarchy. Around the pyramid structure is the circle of supply chain, production planning and scheduling, process control, global optimization, and product distribution. As more layers are integrated into the cross-functional optimization, the integrated system will perform more tasks and make more decisions that are made heuristically and manually now. Many industrial researchers, such as Prett and Garcia [7], have long envisioned this. Now the vision is being further developed with design details of hardware systems, software structures, networking interfaces, and application structures.

As a long-term goal, enterprise optimization integrates all activities in the whole business process, from the supply chain to the production, and further to the distribution channel. In addition, risk management could also be included as a key technical differentiator, which is outside the scope of this paper. Parallel to the structural development for such a general-purpose complex system, some proof-of-concept projects have been piloted, and experiments were conducted on a number of different structures and reported in [2,3 &12]. The multiple layer structure described in Figure 1 is considered to provide flexibility in implementing and operating such a system. Moreover, each layer can be built in an appropriate level of abstraction and in a suitable length of time horizon. The lower layers capture more detailed information of various local process units in a shorter time horizon, whereas the higher layers capture more of the business essence of the plant in a longer time horizon.

As a short term goal, the cross-functional integration would include, in a refinery for example, raw material allocation, inventory management, production management, unit process control, real-time optimization, product blending, production planning and production scheduling. The instrumentation and real-time database are in the bottom layer, then the regulatory control, MPC, global/multi-unit optimization, and the production scheduling layer. Production planning is in the top layer.

More relevant to the control community, this integration will have a large impact on almost all control, optimization, and scheduling technologies currently employed in the industries. Advanced process control, primarily MPC and real-time optimization (RTO), will certainly be affected, particularly in terms of connectability, responsiveness and compatibility. This integration requires us to re-assess the MPC and RTO designs in terms of their online connectability and their dynamic integration. There is a need for designing the MPC and RTO together, or at least designing one with the consideration of working with the other dynamically. Furthermore, from the control perspective, global coordination requires MPC applications to perform in a much wider operating region and more responsively. This poses a new challenge to the MPC technology and introduces opportunities for NMPC.

2.3. Need of NMPC for multi-zone control

The majority of chemical process dynamics is, loosely speaking, slightly nonlinear and can be well modeled with a linearized model around a given operating point. With the application of CV/MV linearizing transformations, linear MPC can be extended to effectively handle a variety of simple nonlinear dynamics. Such practical examples include pH control, valve position control, high purity quality control, and differential pressure control. However, under cross-functional integration, a process would be required to promptly move its operating point over a significant span, or else be required to operate under very different conditions. The linear MPC can fall short in those instances.

One such category is migration control. As the operating point of the process migrates, the dynamics of some processes may change dramatically, even to the extent that the gain will change its sign. For example, some yields of the catalytic cracking unit in a refinery would change their sign when operated between under-cracking and over-cracking regions. In such a case, a linear MPC controller can fall short in that it may operate in either region with different models, but not maneuver between them smoothly, if at all.

Another category is transition control, which is a unique type of nonlinear control problem. During a transition, the process operating point typically moves much more quickly than usual, and the process typically responds more nonlinearly than usual. Such examples include crude switch in a refinery, and grade transition in many series production processes such as polymer and paper. The linear MPC may again fall short as in the first category. Furthermore, what makes the transition control unique is that the main control objective changes from minimizing the variation of controlled variables to minimizing the transition cost or time.

Both categories can be further abstracted into the multi-zone control, where the process needs to be regulated most of the time in each of the zones and to be occasionally maneuvered from one zone to another according to three performance criteria. The first criterion is for normal continuous operation in the first zone. The second criterion is for normal operation in the second zone. The third criterion is for the special requirements during the transition or the migration.

Although solving the multi-zone control problem alone has merit, solving it while coordinating with other parts of the plant will potentially provide much greater benefit. As enterprise optimization further advances, there is an increasing need for an NMPC tool that can solve multi-zone control problems. Likewise, there is an increasing need for an MPC formulation, linear and nonlinear, that takes the requirements of cross-functional optimization into account.

In the following section, we will discuss some industrial preferences and requirements on the NMPC solution with respect to the needs and requirements of cross-functional integration.

3. Preferences and requirements

3.1. Model structure

The model structure is preferred to have a linear backbone or linear sub-structure. The linear (or even nonlinear) portion of the model can be developed experimentally, if so desired. The nonlinear portion of the dynamics can be added as the need arises. This preference stems from the fact that, in most cases, one does not know a priori if he needs a nonlinear model or not.

The model structure is also preferred to be scaleable. Adding controlled variables (CVs) and manipulated variables (MVs) should not cause the existing model to be re-identified or re-generated. The user should easily be able to eliminate any input-output pair of the model. The preferred NMPC solution should not require the system to be square or all the MVs to be available all the time.

The ability to share model information in the cross-functional integration scheme is important. The model obtained in an APC project is often the most costly single item in the implementation. Sharing model information with the other layers of cross-functional integration can save a substantial amount of money. This model sharing can be a two-way process, bottom-up and top-down.

The preferred model-sharing scheme is bottom-up. It is much easier for engineers to find and correct modeling error on a unit-by-unit basis in the APC implementation than on a multi-unit or plant-wide basis in the cross-functional integration. The bottom-up model sharing enhances usability, as engineers can verify their models as they commission the MPC applications unit by unit.

The solution technique needs to be algorithmically robust. If multiple solutions exist, it is preferred to stick to a branch of solutions throughout. Or better yet, stick to the branch of solutions that requires only the minimum amount of control effect in some criterion.

3.2. Coordination port

This issue is actually relevant to both linear and nonlinear MPC. The MPC formulation needs to have an independent port for coordination. This requirement is essential for cross-functional integration. Simply sending a steady state target as setpoint to MPC is not adequate enough for cross-functional integration. The dynamic response required for coordination is often very different from that for setpoint change, and the performance criterion can be completely different from setpoint tracking or disturbance rejection. With a single port, it is difficult, if not impossible, to always satisfy both requirements.

A good coordination port implementation in MPC is one of the essential links in instituting the cross-functional optimization. The nature of the problem is illustrated in Figure 2 as a simplified example of three CVs and two MVs. The first CV has a setpoint, while the other two CVs both have high and low bounds. The MVs are not shown in the diagram. The coordination or optimization target is solved by an outside global optimizer. The controller needs to drive the system to the target in an independent response for coordination.

The task at hand is very similar to designing a two-degree-of-freedom controller for the coordination port, except that the input direction and the response requirement may vary from one transition to another.

Furthermore, the implementation of the coordination port is more than just solving a two-degree-of-freedom control problem. When the coordination port problem is not fully specified dynamically, as in many practical problems that we encounter, multiple solution paths to the destination will exist, as depicted in Figure 2. Treating this like a traditional control problem by specifying a desired response trajectory is not always suitable for two reasons. First, the CV error violation is usually much less important in the transition than in the normal operation. Second, except for trivial problems, one rarely knows a priori which transitional path would be financially optimal yet dynamically feasible.

An alternative solution is to solve for all equal solution paths in terms of the performance criterion (including the financial terms), and choose the one that requires the minimum MV movement in some criterion.

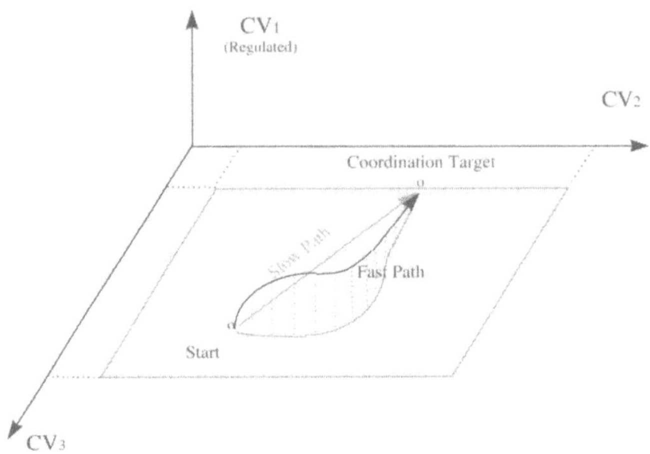

FIGURE 2. An example of different dynamic optimization paths.

In the linear MPC setting, Lu and Escarcega [6] described one such solution to the coordination port implementation. The original motivation of the work was to dynamically unify the local process (unit) optimization, regulatory control and constraint control. The straightforward extension to the coordination port was also discussed. A small-scale cross-functional integration of three RMPCT[3] applications was implemented in 1997 on a butane inventory management problem in a refinery. For this application, global coordination and optimization were required to maintain the balance of, and to distribute more profitably, each of the four C4 components (nC4, iC4, nC4= and iC4=) between the producing units and the consumption units. The coordination difficulty was that the main consumption unit took iC4 and C4= only at a strict 1-to-1 mole ratio, while the producing units did not adhere to any fixed mole ratios or production rates. The cross-functional RMPCT integration delivered significant benefits as reported in [11]. More recently, there have been three more successful cross-functional integration applications, though the benefit results have not yet been compiled. The largest covers five RMPCT applications (three crude distillation columns, two hydrotreaters, and their product pools).

[3]RMPCT, a Honeywell MPC product, stands for Robust Multivariable Predictive Control Technology.

3.3. Reliable fallback or shedding

Reliable fallback position is as important as any control performance improvement. In the real-life control settings, reliable fallback position is often a prerequisite of any additional control algorithm to achieve better performance.

Theoretically, a sophisticated NMPC controller, when applied on a highly nonlinear process that, presumably, linear MPC or PID cannot control, may not always have a sound fallback position when the NMPC controller must be turned off. When the model contains errors, the situation can get worse.

This raises a new problem: how to avoid driving a nonlinear process to such a state that the nonlinear controller may not have a sound fallback position to its downstream linear MPC or PID controller. Or, the reverse problem would be how to detect such a state and shed the nonlinear controller only when it is safe to do so.

3.4. Overrun protection or recovery

According to Moore's law, the computer speed to price ratio doubles about every 18 months. Therefore, the speed of the solution algorithm becomes less and less relevant. However, a controller will occasionally overrun[4] for several reasons. First, nonlinear optimization algorithms can occasionally take a much longer time to converge to a solution. An NMPC controller may employ such an optimization routine. Second, the computer users may schedule many other CPU intensive programs to run in about the same time the NMPC control is scheduled to run.

One solution is to terminate the algorithm prematurely and use a feasible intermediate result. This can be a double-edged sword, particularly when data I/O program is jammed in the same time. Alternatively, a recovery scheme can be more effective. A good scheme will detect the overrun, assess the potential effect, and try to catch up in the next scheduled interval.

4. Conclusion

If linear MPC is the main theme of the 80's and 90's in the process control industries, cross-functional integration and enterprise optimization will be the main theme of the next two decades. As we advance toward a higher level of computer integrated manufacturing, the problem definition and the scope of classical model predictive control will be expanded. Nonlinear MPC will play an important role in many areas, particularly in the area of multi-zone control. Under the cross-functional integration approach, MPC, linear or nonlinear, must be coordinated. One practical way of achieving the dynamic coordination is by designing a coordination port in the MPC/NMPC control strategies.

---

[4]A condition in which the controller does not complete its calculation within its allocated execution interval.

References

[1] Bodington, Edward, *Planning, Scheduling, and Control Integration in the Process Industries,* McGraw-Hill, 1995.

[2] Bain, Michael L., et. al, *Gasoline Blending with an Integrated On-Line Optimization, Scheduling, and Control System,* National Petroleum Refiners Association Computer Conference, Nov 15, 1993. New Orleans, LA.

[3] del Toro, Jorge L., *Computer-Integrated Manufacturing at the Monsanto Pensacola Plant,* AIChE, National Spring Meeting, April, 1991. Houston, TX.

[4] Hall, J. and Verne, T., *RCU Optimization in a DCS Gives Fast Payback,* Hydrocarbon Processing, October 1993, pp. 85–92.

[5] Hall, J. and Verne, T., *Advanced Controls Improve Operation of Lubes Plant Dewaxing Unit,* Oil and Gas Journal, July 24, 1995, pp. 25–28.

[6] Lu, Joseph and Escarcega J., *RMPCT: Robust MPC Technology Simplifies APC,* AIChE 1997 Spring Meeting, Houston.

[7] Prett, D., and Garcia, C., *Fundamental Process Control,* Butterworths, Boston. 1988.

[8] Qin, S. Joe, and Badgwell, Thomas A., *An Overview of Nonlinear Model Predictive Control Applications,* Workshop on NMPC-Assessment and Future Directions, June, 1998. Ascona, Switzerland.

[9] Smith, Fred B., *An FCCU Multivariable Predictive Control Application in a DCS,* The 48th Annual Symposium on Instrumentation for the Process Industries, Texas, January 1993.

[10] Sheehan, Brendan and Reid, Linton, *Robust Controls Optimize Productivity,* Chemical Engineering, August 1997.

[11] Verne, T. and Escarcega, J., *Multi-unit Refinery Optimization: Minimum Investment, Maximum Return,* The Second International Conference and Exhibition on Process Optimization, pp. 1–7, Houston, Texas, March, 1998.

[12] Watano, Takahiko, et. al, *Integration of Production, Planning, Operations and Engineering at Idemistu Petrochemical Company,* AIChE, National Spring Meeting, April, 1993, Houston, TX.

Hi-Spec Solutions
Honeywell IAC
16404 N. Black Canyon Highway
Phoenix, AZ 85023
E-mail address: Joseph.Lu@IAC.Honeywell.com

Progress in Systems and Control Theory, Vol. 26

Nonlinear Model Predictive Control of a Styrene Polymerization Reactor

Matthias Schley, Vinay Prasad, Louis P. Russo
and B. Wayne Bequette

Abstract. In this paper we apply a multivariable nonlinear quadratic dynamic matrix control (NL-QDMC) strategy to the control of a styrene polymerization reactor. NL-QDMC is an extension of a well-known technique for handling constrained processes based on linear models (QDMC) to nonlinear models. The NL-QDMC algorithm incorporates an Extended Kalman Filter (EKF) to handle state variable and parameter estimation; also included is an integrated white noise disturbance model. We consider (through simulations) the control of polymer properties such as number average molecular weight (NAMW) and polydispersity or branching. Temperature and volume measurements are available frequently while concentration and molecular weight distribution measurements are only available every 30 minutes with a delay of 30 minutes. There will always be both parametric and structural uncertainty in any model-based estimation and control scheme. We illustrate the effect of model structure uncertainty by using a simplified model for the control strategy, and a more complex model for the "plant" (incorporating the "gel effect", for example).

1. Motivation

In a workshop of this nature it is not necessary to provide the usual justification for nonlinear model predictive control, so we will dispense with such formalities. Also, the paper length limitations do not allow the typical lengthy literature review, so our review will focus primarily on our contributions and experience with nonlinear model predictive control. We assume that the reader understands the history and development of linear model predictive control techniques and is familiar with some of the basic approaches to nonlinear model predictive control.

This paper does not purport to make a new theoretical contribution (rigorous robustness results, etc.) to the field, but rather seeks to provide an example of the type of system where nonlinear model predictive control can have a significant performance impact over traditional control methods. Much of the nonlinear model predictive control literature has focused on problems such as control of systems with multiple steady-states, with temperature as the output variable. While these problems are interesting and certainly provide motivation for nonlinear control, we are interested in the control of variables more directly related to product quality or product properties.

Styrene polymerization reactor control is the topic of this paper. We consider the control of polymer properties such as number average molecular weight (NAMW) and polydispersity or branching. This requires a model/estimator that can estimate additional states, such as molecular weight moments, even when a limited number of measurements is available. If the molecular weight moments are accurately estimated, the NAMW can be readily inferred and controlled.

This paper has the following structure. In Section 2 we review related work in nonlinear model predictive control, state estimation and styrene polymerization control. The styrene reactor control problem is presented in detail in Section 3. Decentralized control is discussed in Section 4, followed by the EKF-based NL-QDMC strategy in Section 5. A summary and perspective on future research is presented in Section 6.

2. Background

2.1. Nonlinear model predictive control

MPC techniques involving the solution of the nonlinear equations for the control move optimization have been presented by a number of researchers [5], [11], [26]. Example applications have often focused on systems with output multiplicities [26], where processes can move from open-loop stable to open-loop unstable operating points, and input multiplicities [28], where the process gain can change sign (and right-half-plane zeros, causing "inverse response" behavior, can form) as the operating conditions change. In [28] we show the connection between right-half-plane zeros and input multiplicity, and in [29] use a numerical Lyapunov approach to develop regions of attraction; in both papers an isothermal CSTR with series-parallel reactions is used as an example system.

Garcia [6] presented one of the first extensions of linear model predictive control to include a nonlinear model. At time step k the nonlinear model was integrated from the state values at time step k-1, using the known input applied over the previous time step. The standard "additive disturbance term" was used to compensate for the difference between the measured states and the model predictions of those states. The future states were predicted by integrating the nonlinear model assuming no change in process input, then using the linear step response model (based on state values at the current time step) to optimize the set of future control moves using the standard quadratic dynamic matrix control (QDMC) procedure. This approach is henceforth known as the NL-QDMC (nonlinear quadratic dynamic matrix control) technique.

In [27] we compared the computation time of several different methods of solving the nonlinear model; we found that the NL-QDMC approach was significantly faster than methods solving the nonlinear model, and had virtually identical closed-loop performance. Gattu and Zafiriou [7] extended NL-QDMC to include a steady-state Kalman filter, which allows control of unstable systems. Lee and Ricker [13] noted some limitations to this approach and incorporated an extended Kalman filter for state and disturbance estimation.

A number of studies have involved measurement horizon, optimization-based methods for state and parameter estimation [11], [21], [25]. In [2] we show how multirate sampling can be included in the estimation horizon-based approach. In [17] we develop a two-level strategy for state and parameter estimation and show a significant computational savings compared to a single-level optimization-based approach.

A general review of nonlinear process control, including model predictive approaches, is presented in [1]. More recent nonlinear model predictive control tutorials are presented in [14] and [18].

2.2. Styrene polymerization

Synthetic polymer production exceeds one million metric tons per year worldwide, with a variety of polymers produced, ranging from high volume material at roughly one dollar per kilogram to high-value-added specialty materials at several thousand dollars per kilogram [12], [20]. The actual desired product property values are rarely measured accurately on-line, so the development of model-based estimation and control techniques is considered very important; for an example industrial application to a terpolymer process (3 monomers used to produce the polymer), see [16].

Most styrene polymers are produced by batch suspension or continuous mass polymerization processes. Polystyrene mass polymerization is homogeneous (reactions take place in a single phase viscous fluid) and often a combination of different reactors is used to meet product requirements. In general, CSTRs are more appropriate during the earlier polymerization stages where viscosities are relatively low. This corresponds to low conversions but high reaction rates. A single CSTR operating at such conversions may not be economical due to low space-time yields. A discussion of different reactors for the polymerization of styrene can be found in [9]. Only CSTRs are considered further in our paper.

Hidalgo and Brosilow [10] implement a low-order (4-state) nonlinear model predictive control strategy to regulate temperature in a styrene polymerization reactor. Their procedure is based on using two manipulated variables (cooling jacket and monomer flowrates) to control a single output variable (temperature). Doyle et al. [4] apply Volterra model-based MPC to a polymerization reaction in a CSTR where the controller is a linear controller augmented by an auxiliary loop of nonlinear corrections. Russo and Bequette [22] perform a complete operability analysis of the reactor studied in [10] and show the existence of infeasible operating regions where a reactor temperature cannot be maintained by any single input-single output control strategy. They also add three moment equations to understand the effect of process design on important polymer properties.

Gazi et al. [8] develop a systematic approach to verify the stability and performance of nonlinear closed-loop systems under parametric and model structure uncertainty, using the styrene polymerization example from [10]. They use a strategy where four manipulated variables are used to regulate three output variables.

3. Styrene polymerization problem description

This work will be focused on a single CSTR with a cooling jacket as shown in figure 1. Two different models will be used to describe the behavior of the reactor. A simplified model which assumes constant physical properties, constant volume and neglects the so-called "gel effect" will be used for the Kalman-filter based NL-QDMC strategy. A more detailed model which accounts for varying physical properties and includes the gel effect will be used to represent the "plant". This approach will demonstrate the ability of the model-based control strategy to handle model structure uncertainty.

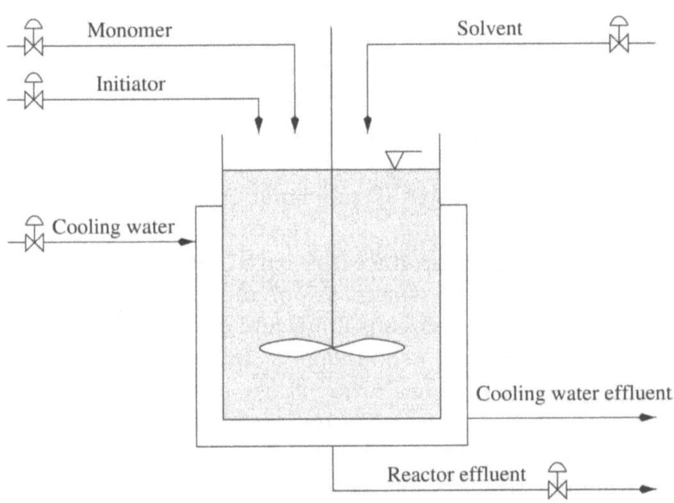

FIGURE 1. Solution polymerization of styrene in a CSTR

3.1. Simplified model

The simplified model used for estimation and control is an extension of one presented in [10] and [22], to include varying solvent concentration. The rather standard assumptions for a free radical polymerization are made, including the initiator (AIBN) decomposition, chain initiation, propagation and termination steps. The quasi-steady-state approximation and the long-chain hypothesis are made and the "gel effect" is neglected. The eight state variables are initiator concentration, monomer concentration, solvent concentration, reactor temperature, cooling jacket temperature, and the first three moments of the molecular weight distribution:

$$x^T = [C_I \quad C_M \quad C_S \quad T \quad T_c \quad \lambda_0 \quad \lambda_1 \quad \lambda_2]$$

The number M_n and weight M_w average molecular weights (NAMW and WAMW) are calculated by

$$M_n = \frac{\lambda_1}{\lambda_0} \tag{1}$$

$$M_w = \frac{\lambda_2}{\lambda_1} \tag{2}$$

The breadth of the MWD can be characterized by the polydispersity (M_w/M_n)

$$PD = \frac{\lambda_2 \lambda_0}{{\lambda_1}^2} \tag{3}$$

The manipulated inputs are the coolant, initiator, monomer and solvent flowrates:

$$u^T = [Q_c \quad Q_I \quad Q_M \quad Q_S]$$

3.2. Plant model

The complete plant model includes 9 state variable differential equations, parameters that vary with concentration and temperature (density, viscosity, heat transfer coefficient, etc.), and includes the "gel effect". A complete description of this model is provided by [23]. It is not presented here due to space limitations, and may be obtained by correspondence with the authors.

3.3. Operating conditions

This reactor has three steady-states; the selected operating point is the middle, unstable steady-state. Thermal initiation would be an important factor at the high temperature steady-state, while the conversion would be too low at the low temperature steady-state. Note that the use of a single manipulated variable, cooling jacket flowrate, will not be successful since the operating point is near an infeasible temperature region [22].

4. Decentralized control

Temperatures are sampled every 5 minutes whereas concentration measurements are analyzed by a laboratory and are only available every 30 minutes with a delay of 30 minutes, which leads to a multi-rate control scheme. Constant volume (perfect level control) is assumed so volume is not displayed in the following results. Temperature is regulated by a cascade controller. An upper constraint of 2 l/sec is implemented on the manipulated variables. Furthermore, anti-reset-windup is implemented on all control loops to assure that integrating action is suspended when a constraint is hit. Controller settings for the concentration loops were obtained via the linearization of the isothermal plant, using IMC-based PI-controller rules [24]. Extensive retuning was necessary to achieve desireable controller performance. The choice of the right pairing of controlled and manipulated variables is trivial for this plant and can be verified with a RGA analysis for the isothermal case.

Control loop responses will be directly compared with NL-QDMC results in the next section. A relatively high proportional gain for the cooling water loop was necessary in order to follow the demanded setpoint from the cascade controller. This can lead to "noisy" control action in the cooling water flow. Attempts to use smaller gains, though, lead to instability as the T_c-controller could not follow the demanded setpoint from the T-controller fast enough. Also, the choice of small integral times for smoother control action was not satisfying as it usually led to overshoot and ringing in the controlled variables. Tuning can therefore become a tough task when the plant is intended to be operated at different operating points. A gain scheduled controller could therefore be of value [3].

5. EKF-based NL-QDMC

The control of polymer molecular weight by a decentralized approach is a difficult task and may become impossible as the operating conditions, i.e. concentration setpoints, cause temperature or NAMW infeasibilies. This section presents results of the use of an extended Kalman filter based nonlinear model predictive control algorithm, which is an extended version of the approach presented by Lee and Ricker [13]. A summary of the EKF equations is provided in table 1. The interaction between the EKF and the controller is shown in figure 2.

A linear MPC approach has also been considered for this particular control problem. However, the polymerization CSTR shows highly nonlinear behavior, which results in the linear MPC controller not being able to provide good performance over a wide range of operating conditions. We present results for EKF-based NL-QDMC control for setpoint changes in polymer properties in a later section. The linear MPC scheme does not provide good control for similar setpoint changes, though it provides satisfactory control for disturbance rejection (regulatory control). We do not show linear MPC results for the sake of brevity.

5.1. Multirate EKF

As in the decentralized approach, temperatures and volume are sampled frequently (every 5 minutes), whereas the concentrations and the leading moments are obtained via laboratory analysis every 30 minutes with a delay of 30 minutes.

The original extended Kalman filter is augmented with an additional updating Kalman filter which only operates when laboratory data is available, thus providing a relatively easily implemented suboptimal solution to the multirate estimation problem. The important feature of this implementation is the conservation of the estimated data from the fast EKF tier. The structure of this multirate EKF is shown in figure 3. At the fast sample rate, few measurements y_k are available and only EKF_1 operates. As soon as laboratory measurements become available

TABLE 1. Kalman filter equations

Model Prediction step:

$$\begin{bmatrix} x_{k|k-1} \\ x^w_{k|k-1} \end{bmatrix} = \begin{bmatrix} F_{t_s}(x_{k-1|k-1}, u_{k-1}, C^w x^w_{k-1|k-1}) \\ A^w x^w_{k-1|k-1} \end{bmatrix} \quad (4)$$

$$\Sigma_{k|k-1} = \Phi_{k-1}\Sigma_{k-1|k-1}\Phi^T_{k-1} + \Gamma^w R^w (\Gamma^w)^T \quad (5)$$

Kalman filter gain and covariance:

$$L_k = \Sigma_{k|k-1}\Xi^T_k(\Xi_k\Sigma_{k|k-1}\Xi^T_k + R^v)^{-1} \quad (6)$$

$$\Sigma_{k|k} = (I - L_k\Xi_k)\Sigma_{k|k-1} \quad (7)$$

Measurement correction step:

$$\begin{bmatrix} x_{k|k} \\ x^w_{k|k} \end{bmatrix} = \begin{bmatrix} x_{k|k-1} \\ x^w_{k|k-1} \end{bmatrix} + L_k(y_k - \hat{y}_k) \quad (8)$$

the following state estimation step is performed by EKF^*_1 and EKF^*_2. EKF^*_2 uses the complete state measurement vector y^*_{k-d} to estimate the plant state $x^*_{k-d|k-d}$ for the time t_{k-d} when the laboratory measurement was taken. For the results presented in Section 5.3 the measurement vectors are

$$\begin{aligned} y_k &= [T \quad T_c]^T \\ y^*_k &= [T \quad T_c \quad C_I \quad C_M \quad C_S \quad \lambda_0 \quad \lambda_1 \quad \lambda_2]^T \end{aligned}$$

The new estimated state vector $x^*_{k-d|k-d}$ is then forwarded to EKF^*_1 where the current, "corrected" state vector $x^{corr}_{k|k}$ is evaluated in a recursive manner. For this purpose, past measurements and past control moves must be stored. It should be noted that EKF^*_1 has the same structure as EKF_1. Figure 4 shows the functioning of the updating algorithm. The delay of the measurements is denoted by d. The results from the updating algorithm are plotted as dotted lines. Note, that the corrected estimates $x^{corr}_{k|k}$ do not lie on the true states. This is due to the model based nature of the Kalman filter, which will, depending on the tuning, more or less "trust" the inferred *a priori* estimates.

5.2. Concentration and temperature control

The decentralized and EKF-based NL-QDMC approaches are compared in this section; controlled variables are the concentrations C_I, C_M and C_S as well as reactor temperature. All controlled variables are weighted equally, which is not necessarily advantageous. The response to a -5 K setpoint change in reactor temperature is shown in figure 5, for decentralized control, while figure 6 shows the NL-QDMC results (manipulated variable responses are not shown because of lack

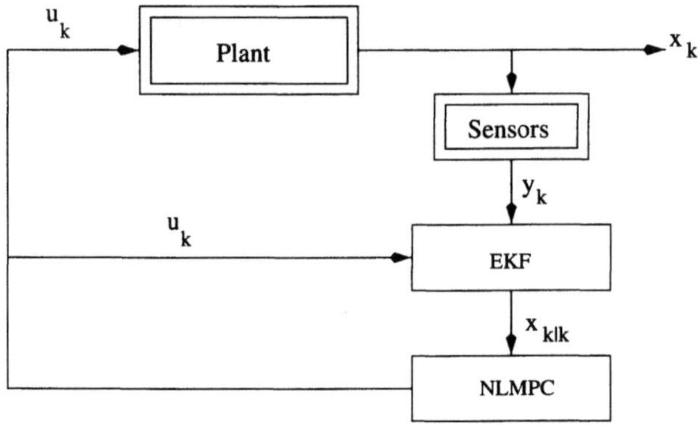

FIGURE 2. EKF and MPC interaction

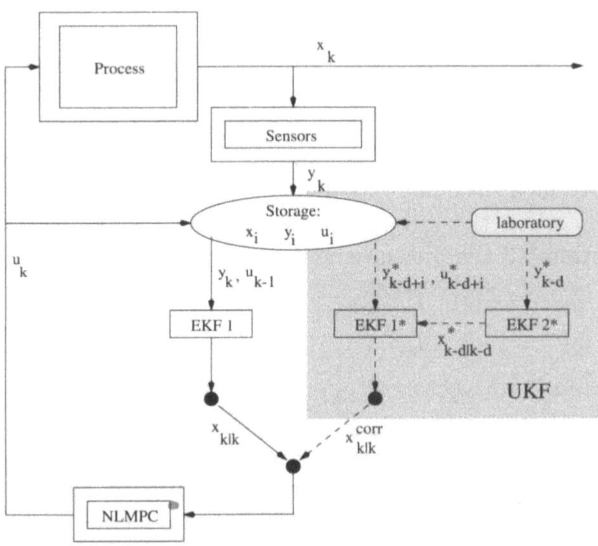

FIGURE 3. Multirate EKF

of space). One instantly observes the better performance of the NL-QDMC, as the controlled variables almost reach steady state after 7 hours. Also, reactor temperature gradually follows its setpoint trajectory showing little offset. Note, that the setpoint trajectories are modeled as second order transfer functions. This was shown to be advantageous for the controller performance as less smooth setpoint trajectories led to overshoot in the reactor temperature.

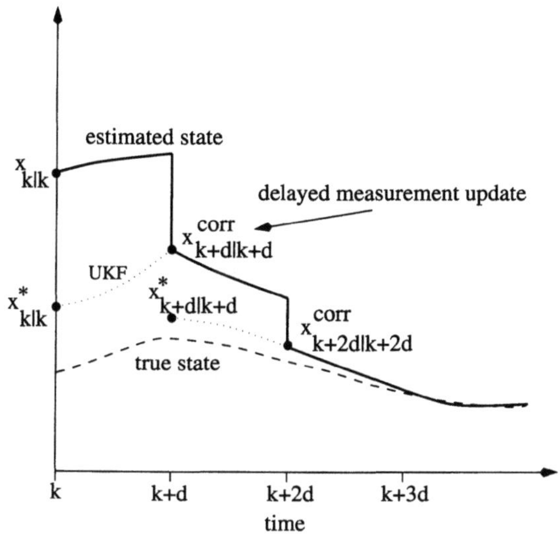

FIGURE 4. Kalman filter updating

The previous simulation results were based on control of concentrations (initiator, monomer and solvent) and temperature with a setpoint change in temperature. Similar improvement in performance of NL-QDMC over decentralized control can also be shown for disturbance rejection [23].

5.3. Polymer properties and temperature control

The previous simulation results were provided to directly compare decentralized control with EKF-based NL-QDMC. A primary motivation for model-based control, however, should be to obtain better control of variables more directly related to polymer product quality or properties. The following simulations are based on the control of molecular weight, polymer concentration, polydispersity and reactor temperature. It is assumed that temperature measurements are disturbed by normally distributed noise of standard deviation 0.1 K. Measurements of concentrations and leading moment are assumed to be corrupted with Gaussian noise of standard deviation 2% of the nominal value of the variable. Since the investigations of this work are based on a plant model, as opposed to real measurements, the following results will compare the "true" plant states to the model states. Therefore the effect of measurement noise will only be seen in the model states. The response to a setpoint change in the NAMW is shown in figures 7 and 8. Temperature and polymer concentration, i.e. λ_0, are kept constant. The prediction horizon is chosen to be 20 whereas an input horizon of only 1 is implemented. Figure 8 shows that it is possible to produce higher molecular weight polymer even if temperature and polymer concentration are kept constant. The molecular weight gradually follows the setpoint trajectory. Unfortunately, the decreasing solvent fraction and the increasing molecular weight lead to a higher reactor viscosity. Relatively small

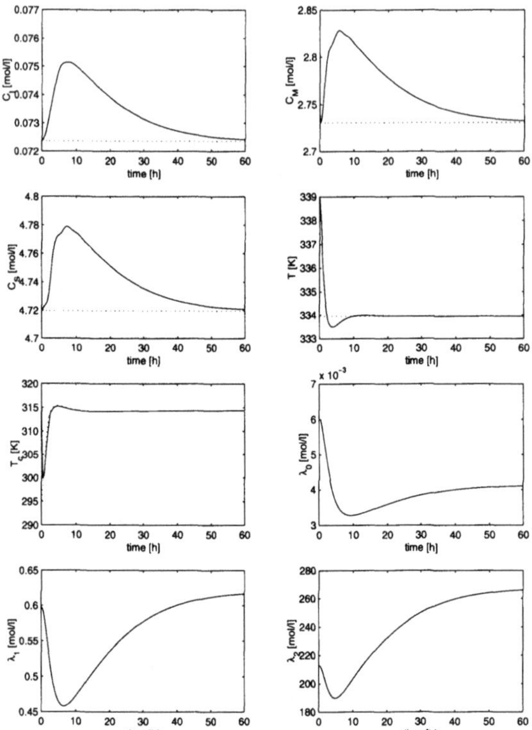

FIGURE 5. Plant states for -5 K setpoint change in T with decentralized control

changes in the setpoints of NAMW and λ_0 can lead to a high viscosity. Thus the choice of reasonable setpoint pairing for T, λ_0 and $NAMW$ can be a difficult task, when viscosity is not a controlled variable.

The most difficult task when controlling NAMW and λ_0, is the choice of a feasible temperature setpoint. Even though the previous simulation showed that a higher molecular weight is possible for the same temperature and polymer concentration, the reader should be aware that steady state infeasibilities make certain setpoint combinations infeasible. A controllability analysis may also fail to provide any useful information, as the selected controlled variables might be controllable at one operating point but not at the other. Also, only stable operating points can be analysed.

Figure 9 shows that the molecular weight distribution (MWD) broadens when the NAMW is increased. This is obviously undesirable, but inherent in the underlying reaction kinetics. As Ray [19] shows for a simplified case, this effect is indispensable when neglecting chain transfer reaction and making the long chain approximation. However, since the underlying plant model does not neglect those

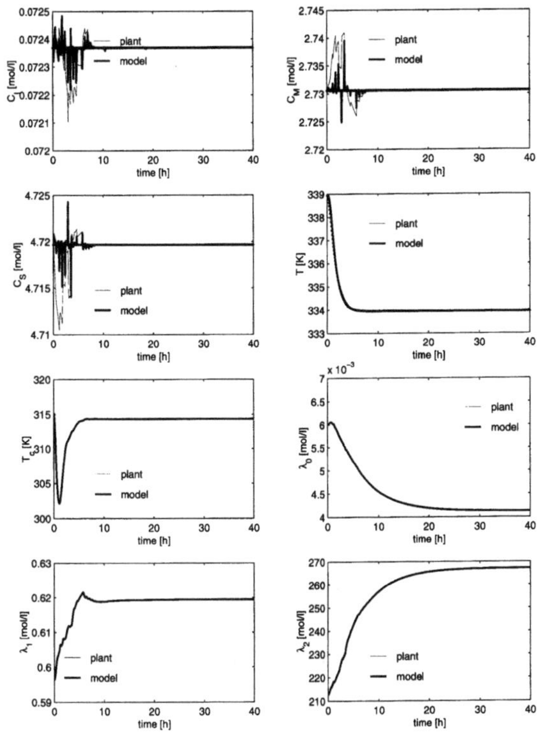

FIGURE 6. States for −5 K setpoint change in T with NL-QDMC

effects, the control of polydispersity, as a measure for the breadth of the distribution, can be controlled.

6. Summary and future work

The model-based results presented in this paper are extended from an EKF-based NL-QDMC approach presented by Lee and Ricker [13]. We have extended their approach to account for multirate measurements with significant analysis delay. We have directly compared EKF-based NL-QDMC with decentralized (MIMO PI) control for the regulation of concentrations and temperature. We have also shown how the EKF-based NL-QDMC approach can be used to regulate variables more directly related to polymer product properties (molecular weight and polydispersity). It was assumed that the moments of the molecular weight distribution could be determined every 30 minutes, with a 30 minute analysis delay. Additionally, the number average and weight average molecular weight could be frequently measured on-line with osmometry and light-scattering to improve the estimated values.

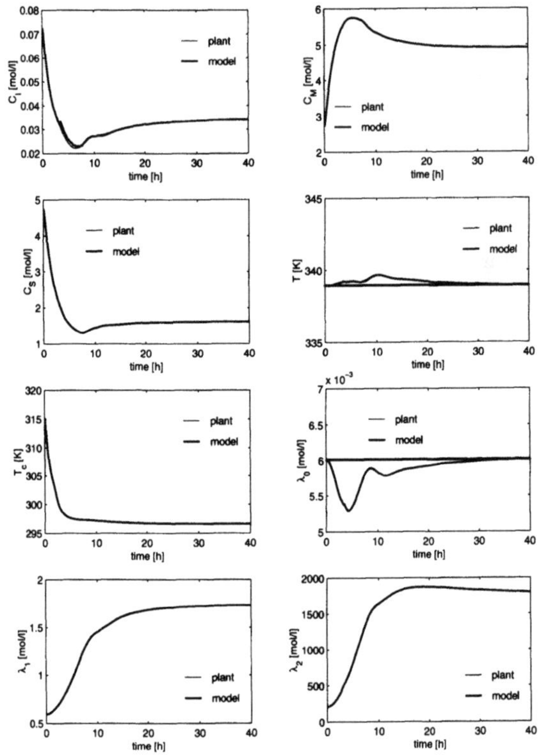

FIGURE 7. States for setpoint change in NAMW with NL-QDMC

A clear disadvantage to our approach is that a reasonably well-developed model (with appropriate parameter values) must be available or the model development could be quite time-consuming. As it becomes more and more important to produce product with consistent quality specifications, it will become easier to justify the time for model development. The EKF-based estimator will always require a deep understanding from the control system development and implementation team. Of course, it would be desirable to have other state estimation techniques that require less expertise for successful implementation. Another option is to use optimization-based estimation horizon methods such as those presented in [15], [17], [21] and [25].

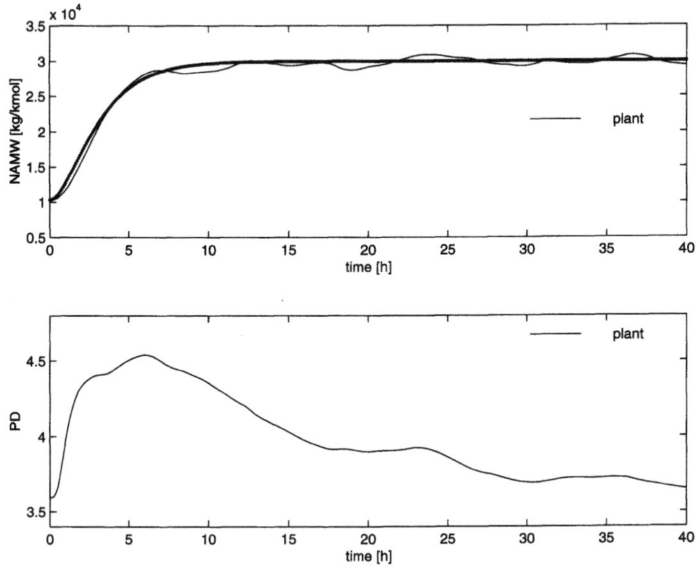

FIGURE 8. NAMW and PD for setpoint change in NAMW with NLMPC

References

[1] B.W. Bequette, *Nonlinear control of chemical processes: a review,* Ind. Eng. Chem. Res., **30** (1991a), 1391–1413.

[2] B.W. Bequette, *Nonlinear predictive control using multi-rate sampling,* Can. J. Chem. Eng., **69** (1991b), 136–143.

[3] B.W. Bequette, *Practical approaches to nonlinear control: a review of process applications,* in *Nonlinear Model-based Process Control,* R. Berber and C. Kravaris (eds.), NATO ASI Series, Vol. E., Kluwer, Dordrecht, **1998**, in press.

[4] F.J. Doyle, B.A. Ogunnaike and R.K. Pearson, *Nonlinear model-based control using second-order Volterra models,* Automatica, **31** (1995), 697–714.

[5] J.W. Eaton and J.B. Rawlings, *Feedback control of nonlinear processes using on-line optimization techniques,* Comp. Chem. Engng., **14** (1990), 469–479.

[6] C.E. Garcia, *Quadratic dynamic matrix control of nonlinear processes: an application to a batch reaction process,* AIChE Annual Meeting, San Francisco, CA **1984**.

[7] G. Gattu and E. Zafiriou, *Nonlinear quadratic dynamic matrix control with state estimation,* Ind. Eng. Chem. Res., **31** (1992), 1096–1104.

[8] E. Gazi, W.D. Seider and L.H. Ungar, *Verification of controllers in the presence of uncertainty: application to styrene polymerization,* Ind. Eng. Chem. Res., **35** (1996), 2277–2287.

[9] J.N. Henderson and T.C. Bouton, *Polymerization Reactors and Processes,* ACS Symposium Series, **104** (1979).

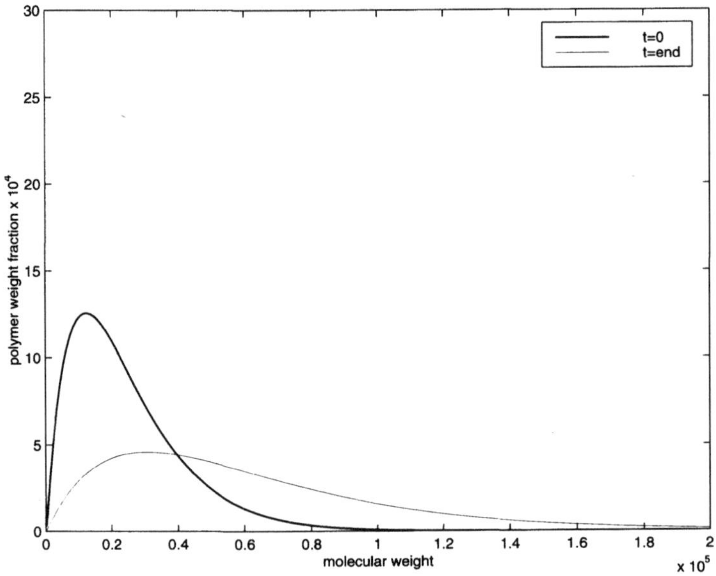

FIGURE 9. MWD for setpoint change in NAMW with NL-QDMC

[10] P.M. Hidalgo and C.B. Brosilow, *Nonlinear model predictive control of styrene polymerization at unstable operating points,* Comp. Chem. Eng., **14** (1990), 481–494.

[11] S. S. Jang, B. Joseph and H. Mukai, *Control of constrained multivariable nonlinear process using a two-phase approach,* Ind. Eng. Chem. Res., **26** (1987), 2106–2114.

[12] C. Kiparissides, *Polymerization reactor modeling: a review of recent developments and future directions,* Chem. Eng. Sci., **51** (1996), 1637–1659.

[13] J.H. Lee. and N.L. Ricker, *Extended Kalman filter based nonlinear model predictive control,* Ind. Eng. Chem. Res., **33** (1994), 1530–1541.

[14] E.S. Meadows and J.B. Rawlings, *Model predictive control,* Chapter 5 in *Nonlinear Process Control,* M.A. Henson and D.E. Seborg (eds.), Prentice Hall, Upper Saddle River, NJ (1997).

[15] K.R. Muske and T.F. Edgar, *Nonlinear state estimation,* Chapter 6 in *Nonlinear Process Control,* M.A. Henson and D.E. Seborg (eds.), Prentice Hall, Upper Saddle River, NJ (1997).

[16] B.A. Ogunnaike, *On-line modelling and predictive control of an industrial terpolymerization reactor,* Int. J. Cont., **59** (1994), 711–729.

[17] Y. Ramamurthi, P.B. Sistu and B.W. Bequette, *Control-relevant data reconcilation and parameter estimation,* Comp. Chem. Engng., **17** (1993), 41–59.

[18] J.B. Rawlings, E.S. Meadows and K.R. Muske, *Nonlinear model predictive control: a tutorial and survey,* in *Advanced Control of Chemical Processes,* ADCHEM '94, IFAC, Kyoto, Japan, **1994**, 203–214.

[19] W.H. Ray, *On the mathematical modeling of polymerization reactors*, J. Macromol. Sci.-Revs. Macromol. Chem., **C8** (1972) 1.

[20] W.H. Ray, *Modeling and control of polymerization reactors,* in Dynamics and Control of Chemical Reactors, Distillation Columns and Batch Processes (DYCORD+ '92), J.G. Balchen (ed.), Pergamon Press, Oxford (1993), 161–170.

[21] D. Robertson, J.H. Lee and J.B. Rawlings, *A moving horizon based approach for least squares estimation,* AIChE J., **42** (1994), 2209–2224.

[22] L.P. Russo and B.W. Bequette, *Operability of chemical reactors: multiplicity behaviour of a jacketed styrene polymerization reactor,* Chem. Eng. Sci., **53** (1998), 27–45.

[23] M. Schley, *Extended Kalman filter based nonlinear model predictive control of a styrene polymerisation reactor,* Diplomarbeit (Diploma thesis), University of Aachen, Aachen, Germany, May **1998**.

[24] D.E. Seborg, T.F. Edgar and D.A. Mellichamp, *Process Dynamics and Control,*, John Wiley and Sons (1989).

[25] P.B. Sistu and B.W. Bequette, *Process identification using nonlinear programming techniques,* Proceedings of the American Control Conference, San Diego, CA, **1990**, 1534–1539.

[26] P.B. Sistu and B.W. Bequette, *Nonlinear predictive control of uncertain processes: application to a CSTR,* AIChE J., **37** (1991), 1711–1723.

[27] P.B. Sistu, R.S. Gopinath and B.W. Bequette, *Computational issues in nonlinear predictive control,* Comp. Chem. Engng., **17** (1993), 361–366.

[28] P.B. Sistu and B.W. Bequette, *Model predictive control of processes with input multiplicities,* Chem. Eng. Sci., **50** (1995), 921–936.

[29] P.B. Sistu and B.W. Bequette, *Nonlinear model predictive control. Closed-loop stability analysis,* AIChE J., **42** (1996), 3388–3402.

Department of Chemical Engineering,
Rensselaer Polytechnic Institute,
Troy, NY 12180-3590,
USA
E-mail address: `bequette@rpi.edu`

Progress in Systems and Control Theory, Vol. 26

Nonlinear Multi-rate MPC with Large Scale Fundamental Models: Application to a Continuous Kamyr Digester

Francis J. Doyle III, Philip A. Wisnewski

Abstract. Given that many industries are relying more heavily on complex, fundamental process models, there exists a need to systematically incorporate these complex models into the controller design. In this work, a complex nonlinear fundamental model is used to develop an MPC controller and is applied to a nonlinear, fundamental, continuous digester "plant".

1. Introduction

The goal of this work is to investigate the use of model-based control for a large scale system, the continuous pulp digester. There are several attributes of the continuous digester that make this a challenging control problem: 1) the primary control variable, Kappa #, is infrequently measured and has long time delays associated with its measurement; 2) there are long time delays between the manipulated inputs and their observed effects on the Kappa #; 3) the digester exhibits nonlinear behavior, especially at lower Kappa #; and 4) the biological feedstock (wood chips) are subject to unmeasured variations in wood component concentrations, moisture and density and are a source of continuous variations in the Kappa #.

The control of the Kappa # in the continuous digester has been examined by many researchers. Lee and Datta [7] used extended Kalman filtering (EKF) to predict and control (with MPC) the Kappa # of a simulated batch digester. Using a controller model with the same structure as the plant model, their focus was on controller performance in the presence of initial state and model parameter uncertainty. For many of the cases examined, the EKF-MPC combination performed well. Amirthalingam and Lee [1] examined measurement selection for the continuous pulp digester using a least squares regression technique to select an optimal set of liquor measurements and then using subspace identification to identify a linear model and construct a multi-rate Kalman filter and model predictive controller. Michaelsen *et al.* [9] implemented an MPC algorithm using a mechanistic model, based on the Purdue model, with an optimal state estimator on a Kamyr digester in Sweden. Wisnewski and Doyle have successfully applied measurement and control structure selection, both based on robust control analysis tools, to the extended Purdue model and to a simplified, physically-based digester model [11].

In this work, a multi-rate sampled-data model predictive controller is applied to a two vessel continuous pulp digester (the extended Purdue model). The MPC controller incorporates the nonlinear, fundamental digester model for the output prediction given past inputs, and successively linearizes the fundamental model at each time step to obtain a linear model for the computation of future inputs. The nonlinear MPC control algorithm incorporates multi-rate sampled-data and a novel technique such that large scale linear model reduction can be performed on-line.

2. Continuous pulp digester case study

The continuous digester in this case study consists of two vessels, an impregnation vessel and a reaction vessel. The reaction vessel has one co-current and two counter-current cooking zones. The flow configuration, physical dimensions, and operating conditions are representative of an industrial continuous digester and a schematic of the case study is shown in Figure 1. The fundamental digester model developed in [14] will be used as the "plant" for simulation purposes. Only the pulping of the hardwoods are considered in this case study; thus, all physical parameters of the "plant" are those for hardwoods.

The model employed as the simulation plant (referred to as the plant) consists of 110 CSTRs in series; and the fundamental model employed in the MPC controller design (referred to as the controller model) consists of 50 CSTRs in series. To make a fair comparison in the evaluation of the three controllers in this study, reality is simulated by introducing uncertainty in several of the controller model's parameters; thus, the controller model *does not* equal the plant.

Uncertainty is incorporated by randomly deviating the activation energies, $E_{(1,2),i}$, in Equation (4) of [14], between $\pm 1\%$ from their nominal values. The high parametric sensitivity of the kinetic model to the activation energies yields high Kappa # variations for these deviations. The pre-exponential factors, $A_{(1,2),i}$ of Equation (5) in [14], are adjusted such that the steady-state outputs of the plant and model are in agreement. Table 1 gives the multipliers by which the kinetic parameters are adjusted. In addition to model parameter uncertainty, the fact that the plant and controller models differ in the number of CSTRs used to approximate the digester also introduces structural uncertainty between the models through the residence time behavior.

The reaction free parameter, e_f, in Equation (5) of [14], and the stoichiometric coefficients for EA consumption by hardwood given in Table (3) of [14], are tuned to reduce the steady-state differences in the Kappa # and secondary measurements between the plant and controller model. The final values for these parameters used in the controller model are $e_f(impregnation, cook, mcc, emcc) =$ [0.90, 0.915, 0.74, 0.74] and $[\beta_{OHL}, \beta_{OHC}] =$ [0.20, 0.48].

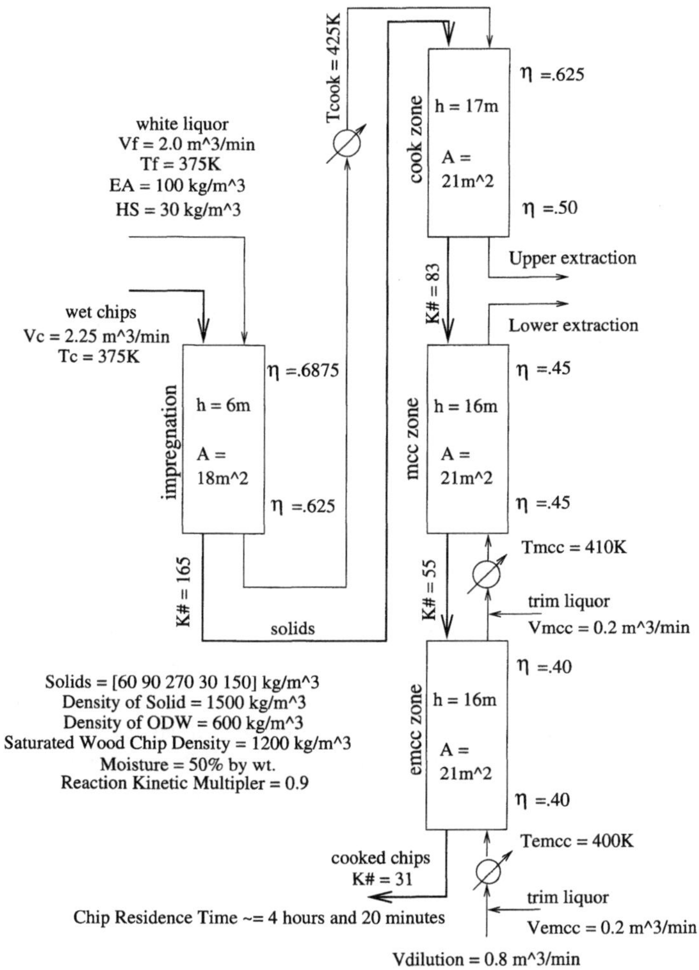

FIGURE 1. Flow configuration, digester dimensions, and operating conditions for the case study digester.

For this case study, there are three manipulated inputs: the white liquor flowrate to the impregnation vessel, V_f; the temperature of liquor entering the cooking zone, T_{cook}; and the trim flowrate to the mcc zone, V_{mcc}. Adjustment of the manipulated inputs are assumed to occur, if necessary, every 10 minutes. Liquor measurements at the bottom of cook zone (upper extraction flow) and the top of the mcc zone (lower extraction flow) are assumed possible. There are six secondary measurements assumed available: the EA concentration measured at the upper and lower extraction flow, EA(UE) and EA(LE), respectively; the dissolved lignin concentration measured at the upper and lower extraction flow, DL(UE)

TABLE 1. Factors multiplying the hardwood kinetic parameters to introduce uncertainty into the fundamental model.
High-reactivity lignin, s_1; low-reactivity lignin, s_2; cellulose, s_3; araboxylan, s_4; galactoglucomman, s_5.

i =	s_1	s_2	s_3	s_4	s_5
$A_{1,i}$	0.9766	1.2694	0.9822	1.0205	0.8287
$A_{2,i}$	0.9515	0.9359	1.0790	0.9178	1.1224
$E_{1,i}$	0.9972	1.0072	0.9982	1.0028	0.9911
$E_{2,i}$	0.9945	0.9939	1.0063	0.9921	1.0024

and DL(LE), respectively; and the free liquor temperature at the upper and lower extraction flow, T(UE) and T(LE), respectively. These secondary measurements are assumed to be corrupted with white noise, the variances of which are given in Table 2, and are available every 10 minutes with no measurement delay. It is assumed that there two measured disturbances, EA and HS concentrations in the white liquor, and five unmeasured disturbances, the five wood component concentrations. Note that as the composition of the wood chips change, the chip moisture and density also change through the porosity relationship given by Equation (8) of [14]. The disturbances are simulated as integrated white noise; the variances of these white noise signals used to generate the disturbances are also given in Table 2. A noisy measurement of the primary control variable, Kappa # at the bottom of the emcc zone, is assumed available every 30 minutes with a 20 minute measurement delay. The control objective of this case study is to minimize the variations in the Kappa # around its setpoint ($K\#_{sp} = 31.1$) in the presence of stochastic, measured and unmeasured disturbances.

TABLE 2. Variances of the white noise signals for simulation of measurement noise for the primary/secondary outputs and stochastic measured/unmeasured disturbances.

	Kappa #	EA(UE)	DL(UE)	T(UE)	EA(LE)	DL(LE)	T(LE)
$\sigma^2(noise)$	0.05	0.015	0.015	0.005	0.015	0.015	0.005
	EA	HS	ρ_{s_1}	ρ_{s_2}	ρ_{s_3}	ρ_{s_4}	ρ_{s_5}
$\sigma^2(dist.)$	0.0009	0.00008	0.0003	0.0007	0.007	0.0001	0.002

The model reduction technique, discussed in [13], requires that the linear model, derived from the linearization of the fundamental model, have a block diagonal structure. To obtain a block diagonal structure, the bulk flow term in Equation (9) of [14], $\dot{V}_b$, is removed from the modeling equations. This results in an inconsistency regarding the contributions of volume from reacting/diffusing

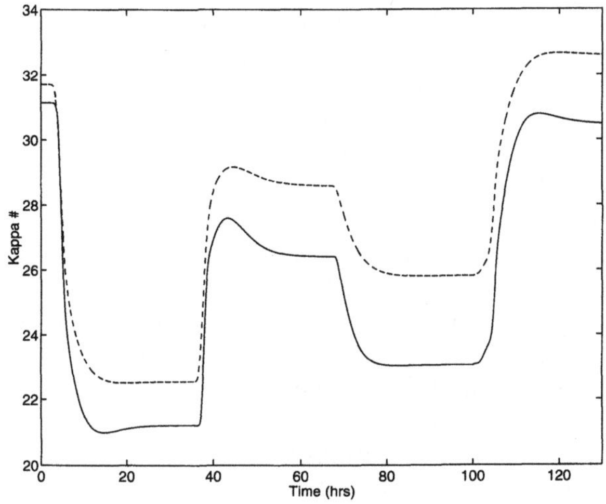

FIGURE 2. Kappa # responses for the plant (solid) and the model (dashed) to the series of steps in the manipulated inputs and measured disturbances.

mass in the entrapped and free liquor phases; however, removal of this term does not adversely effect the numerical solution of the modeling equations. However, two additional states for the water concentrations in the free and entrapped liquor are required in the modeling equations so that the mass and energy balances close. Comparisons between the plant and model are shown in Figure 2, where the dynamic response of Kappa # is compared for a series of steps in the manipulated inputs and measured disturbances. The series of steps are: a $+0.225\, m^3/min$ step in V_f at time zero; a $-5\, K$ step in T_{cook} at time 33.3 hours; a $+0.03\, m^3/min$ step in V_{mcc} at time 66.6 hours; and $+5\, kg/m^3$ and $+3\, kg/m^3$ steps in EA and HS concentration at time 100 hours. The model compares favorably to the plant, both in the process gains and, more importantly, the dynamic behavior, where higher-frequency model fidelity is needed for good closed-loop performance. Off-set between the plant and model exists due to plant/model mismatch; however, this off-set is small and would not be detected in noisy measurements of the primary/secondary outputs.

3. Nonlinear model predictive control design

For most control problems, controllers based on linear models perform adequately. However, chemical and biological processes are inherently nonlinear, and thus one would expect better closed-loop performance if nonlinear process models are employed in model-based controllers. However, the use of nonlinear process models can lead to nonlinear and/or nonconvex optimization problems. Though efficient nonlinear optimization methods exist (such as successive quadratic programming (SQP) and generalized reduced gradient methods) [3], nonlinear optimization may require a significant increase in computational complexity and cost, while, in general, not being able to guarantee a feasible, much less, optimal solution [10]. The various papers presented at this workshop (in this proceedings volume) suggest that new and more efficient methods may soon overcome some of these issues for large scale nonlinear systems.

García [5] is generally credited as the first to incorporate nonlinear models in model predictive control. He examined the use of successively linearizing a nonlinear plant model to compute future control moves. The advantage of this approach is that the nonlinear plant model is used to predict future outputs given past inputs, while the computation of the future input trajectory is still a quadratic programming problem, avoiding nonlinear programming. This algorithm requires that several on-line, numerical operations be performed: i) the linearization of the nonlinear process model, ii) the discretization of the linear model, and iii) the computation of a step response coefficient matrix. These on-line computations can be performed quickly and robustly for low-order nonlinear models, however, for large scale systems, these steps can be time consuming.

In this work, the inferential control of a two-vessel digester will be investigated using a multi-rate sampled-data model predictive controller. Full nonlinear optimization approaches to model predictive control, such as those described in [2, 3, 4, 8], are not attempted for the digester due to the high state dimensionality (> 1000 states) of the fundamental model, which could render the optimization problem infeasible within the given sampling time (10 minutes). The difficulty of using the high-order process models for the present design is the computational intensity required for generating the state estimation gain and future output prediction equations. In particular, two computationally intensive steps are the discretization of the high-order, linear model (derived from the nonlinear model) and the calculation of the Kalman filter gain.

A discrete-time model is needed to compute a step response coefficient matrix, a matrix which relates the future manipulated input moves to the future output predictions. Low-order, continuous-time systems can be quickly converted to discrete-time systems using standard computer algorithms. However, for high-order systems, this computation is very time consuming. For a one-thousand state model, the discretization takes approximately 3.5 hours on a Sun Enterprise 3000, 167 MHz. Calculating the steady-state Kalman filter gain is another time consuming step. For a one-thousand state model, calculation of the Kalman filter gain above takes between 10–12 hours on a Sun Enterprise 3000, 167 MHz.

For the derivation of a linear model predictive controller, these steps can be performed off-line, though tuning the Kalman filter may require several iterations. However, to incorporate the fundamental process model into a nonlinear model predictive controller, which utilizes successive linearization, the linearization and subsequent discretization of the high-order process model is required at each sampling time. This makes implementation impossible in that the discretization step cannot be performed within a short sampling time, such as *e.g.*, ≤ 10 minutes for the digester. Thus the order of the linear model must be reduced before the discretization step. Unfortunately, for large scale problems (>300 states), the current linear model reduction techniques are unstable as well as time consuming. Therefore, a novel approach to model reduction is incorporated into the controller such that a stable, low-order linear model can be obtained on-line.

3.1. MPC based on nonlinear fundamental model

In this section, a multi-rate data-sampling nonlinear model predictive controller is developed as a combination of the linear multi-rate MPC controller of [6] and the extended Kalman filter based nonlinear MPC controller of [7].

The fundamental process model is represented by the following continuous-time, nonlinear differential equation:

$$\begin{aligned} \dot{x} &= f(x, u, v, d) \\ y &= g(x) \end{aligned} \tag{1}$$

The discrete version the model is expressed as

$$\begin{aligned} x_k &= F_{t_s}(x_{k-1}, u_{k-1}, v_{k-1}, d_{k-1}) \\ y_k &= g(x_k) \end{aligned}$$

where $F_{t_s}(x_{k-1}, u_{k-1}, v_{k-1}, d_{k-1})$ denotes the terminal state vector obtained from integrating the fundamental model equations (Equation 1) one sampling time, t_s, from the initial state vector, x_{k-1}, holding the manipulated inputs (u_{k-1}), and the measured and unmeasured disturbances (v_{k-1}, d_{k-1}), constant.

The process model is augmented with the disturbance model, primary and secondary outputs, and the primary measurement delay to give:

$$\underbrace{\begin{bmatrix} x_k \\ x_k^w \\ y_k^c \\ y_k^{c_1} \\ y_k^{c_2} \\ \vdots \\ y_k^{c_\theta} \\ y_k^s \end{bmatrix}}_{\bar{X}_k} = \underbrace{\begin{bmatrix} F_{t_s}(x_{k-1}, u_{k-1}, v_{k-1}, C^w x_{k-1}^w) \\ A^w x_{k-1}^w \\ g^c(F_{t_s}(x_{k-1}, u_{k-1}, v_{k-1}, C^w x_{k-1}^w)) \\ y_{k-1}^c \\ y_{k-1}^{c_1} \\ \vdots \\ y_{k-1}^{c_{\theta-1}} \\ g^s(F_{t_s}(x_{k-1}, u_{k-1}, v_{k-1}, C^w x_{k-1}^w)) \end{bmatrix}}_{\Phi(x_{k-1}, u_{k-1}, v_{k-1}, x_{k-1}^w)} + \underbrace{\begin{bmatrix} 0 \\ B^w \\ 0 \\ 0 \\ 0 \\ \vdots \\ 0 \\ 0 \end{bmatrix}}_{\Gamma^w} w_{k-1}$$

$$\hat{y}_k = \Xi_k \bar{X}_k + \nu_k$$

The outputs, y_k, are corrupted with measurement noise, ν_k, which is assumed to be white noise with covariance R^ν.

First, model prediction is performed by integrating the nonlinear model one time step, using information at time step, $k-1$:

$$\bar{X}_{k|k-1} = \Phi(x_{k-1|k-1}, u_{k-1}, v_{k-1}, x_{k-1|k-1}^w) \tag{2}$$

The state vector is corrected with the current measurement vector:

$$\bar{X}_{k|k} = \bar{X}_{k|k-1} + K_f \left(\hat{y}_k^m - \Xi \bar{X}_{k|k-1}\right) \tag{3}$$

The Kalman filter gain, K_f, is computed from the nominal linear reduced model. Note, that since the full-order nonlinear model is being used in the prediction-correction step, the full-order steady-state Kalman filter gain is needed. For the 1050 state model, calculating the Kalman gain requires between 10 and 14 hours of computational time with a Sun Enterprise 3000 (167 MHz), which is a disadvantage to this approach in that iterative tuning of the Kalman filter gain can be burdensome.

In this development, the measured and unmeasured disturbances are assumed to remain constant over the future prediction horizon. It is possible to re-linearize the model equations at every sampling time into the future prediction; however, this is computationally intensive. A common assumption, that will be employed here, is that the linear model obtained at the current sampling time, k, is kept constant throughout the prediction horizon. The multi-step-ahead prediction equation is formulated such that the prediction equation is linear in the future *unknown* manipulated input moves.

$$\begin{bmatrix} x_{k+l|k} \\ x^w_{k+l|k} \\ y^c_{k+l|k} \end{bmatrix} \approx \begin{bmatrix} F_{l\cdot t_s}(x_{k|k}, u_{k-1}, v_k, C^w x^w_{k|k}) \\ (A^w)^l x^w_{k|k} \\ g(x_{k|k}) + C^c_k F_{l\cdot t_s}(x_{k|k}, u_{k-1}, v_k, C^w x^w_{k|k}) - C^c x_{k|k} \end{bmatrix}$$

$$+ \begin{bmatrix} A_k^{l-1} B_k^u & A_k^{l-2} B_k^u & \cdots & B_k^u \\ 0 & 0 & \cdots & 0 \\ C^c_k A_k^{l-1} B_k^u & C^c_k A_k^{l-2} B_k^u & \cdots & C^c_k B_k^u \end{bmatrix} \begin{bmatrix} u_k - u_{k-1} \\ u_{k+1} - u_{k-1} \\ \vdots \\ u_{k+l-1} - u_{k-1} \end{bmatrix}$$

$$\begin{aligned} A_k &= \left.\frac{\partial f(x,u,v,d)}{\partial x}\right|_{x=x_{k|k}, u=u_{k-1}, v=v_k, d=C^w x^w_{k|k}} \\ B_k^u &= \left.\frac{\partial f(x,u,v,d)}{\partial u}\right|_{x=x_{k|k}, u=u_{k-1}, v=v_k, d=C^w x^w_{k|k}} \\ C_k^c &= \left.\frac{\partial g(x)}{\partial x}\right|_{x=x_{k|k}} \end{aligned}$$

The superscript c denotes the primary output part of the output vector.

Substituting $\sum_{i=0}^{l} \Delta u_{k+i}$ in for $u_{k+l} - u_{k-1}$ (where $\Delta u_{k+i} = u_{k+i} - u_{k+i-1}$) in the above equation, the multi-step prediction equation for the primary control output, y^c, becomes:

$$\begin{bmatrix} y^c_{k+1|k} \\ y^c_{k+2|k} \\ \vdots \\ y_{k+p|k} \end{bmatrix} = \begin{bmatrix} I \\ I \\ \vdots \\ I \end{bmatrix} (g(x_{k|k}) - C^c x_{k|k}) + \tag{4}$$

$$\begin{bmatrix} C^c_k F_{t_s}(x_{k|k}, u_{k-1}, v_k, C^w x^w_{k|k}) \\ C^c_k F_{2 \cdot t_s}(x_{k|k}, u_{k-1}, v_k, C^w x^w_{k|k}) \\ \vdots \\ C^c_k F_{p \cdot t_s}(x_{k|k}, u_{k-1}, v_k, C^w x^w_{k|k}) \end{bmatrix} +$$

$$\begin{bmatrix} \tilde{C}^c_k \tilde{B}^u_k & 0 & \cdots & 0 \\ \tilde{C}^c_k (I + \tilde{A}_k) \tilde{B}^u_k & \tilde{C}^c_k \tilde{B}^u_k & \cdots & 0 \\ \vdots & \vdots & \ddots & \vdots \\ \tilde{C}^c_k \sum_{i=0}^{p-1} \tilde{A}^i_k \tilde{B}^u_k \cdots & \cdots & \tilde{C}^c_k \tilde{B}^u_k & \end{bmatrix} \begin{bmatrix} \Delta u_k \\ \Delta u_{k+1} \\ \vdots \\ \Delta u_{k+p-1} \end{bmatrix} \tag{5}$$

The terms denoted by $\tilde{}$ denote the corresponding elements of the reduced order linear model. This is expressed compactly as:

$$\begin{aligned} \mathcal{Y}_{k+1|k} &= \mathcal{I}(g(x_{k|k}) - C^c x_{k|k}) + \mathcal{S}^x_k(x_{k|k}, u_{k-1}, v_k, x^w_{k|k}) + \tilde{\mathcal{S}}^u_k \Delta \mathcal{U}_k \\ &= \mathcal{Y}^0_{k+1|k} + \tilde{\mathcal{S}}^u_k \Delta \mathcal{U}_k \end{aligned}$$

The term $\mathcal{Y}^0_{k+1|k}$ is computed by integrating the nonlinear model equations over the prediction horizon from the initial estimates provided by the Kalman filter.

The major advantage of the above formulation of nonlinear MPC is that the computation for the optimal control action is a quadratic, instead of a nonlinear, program.

$$\min_{\mathcal{U}_k} \|\Lambda^y(\mathcal{Y}_{k+1|k} - R_{k+1|k})\|_2^2 + \|\Lambda^u \Delta \mathcal{U}_k\|_2^2 \tag{6}$$

3.2. Results of MPC with nonlinear model

The closed-loop Kappa # response, for the controller using the nonlinear, fundamental process model, is examined for the set of stochastic composition disturbances, and is shown in Figure 3, along with the open-loop Kappa #. The disturbance model is used to estimate the five wood component concentrations and correlate the effects of the disturbances between the primary and secondary measurements. The controller tuning parameters are given in Table 3. Using the MPC controller based on the nonlinear, fundamental process model, the Kappa # deviates a maximum of 0.6 units (to 30.5) with a mean value of 31.1 and a variance of 0.04 for 500 hours of operation. Using a Sun Enterprise 3000 (167 MHz), the computation times for the various operations performed by the nonlinear MPC controller are as follows: model prediction/measurement correction, $\approx$ 0.5 seconds; linearization of the fundamental model given the current state vector, $\approx$ 12.5 seconds; reduction of the linear system, $\approx$ 4.5 seconds; discretization of the reduced linear model, ≈ 0.1 seconds; integration of the 1050 nonlinear model equations 900 minutes into the future, 55 seconds. In comparing the results of this study with two other linear designs (from [12]), one can see that there is little difference in the closed-loop performance between the MPC controllers based on a linear/reduced-order model and the full nonlinear model. Results of the three controllers are summarized in Table 4. In particular, it should be noted that the variances in Kappa # (noise-free measurements) for the MPC controllers using both the linear/reduced and nonlinear fundamental models are less than the variance of the Kappa # measurement noise, Table 2. The variances in the Kappa # (for noisy measurements) are close to variance of the measurement noise for Kappa #, which is as good as could be expected.

TABLE 3. Tuning parameters for the MPC controller utilizing the nonlinear, fundamental model.

α	$diag$[0.80 0.80 0.80 0.80 0.80]
p	90
q	4
Q	$diag$([0.0003 0.0007 0.0066 0.0001 0.002]
R	$diag$[0.05 0.015 0.015 0.005 0.015 0.015 0.005]
Λ^y	1
Λ^u	$diag$[5 0.3 50]

TABLE 4. Summary of closed-loop results. The first set of entries is for noise-free measurements, the second set of entries is for noisy measurements.

MPC Model	Max/Min Kappa #	Mean Kappa #	Standard Deviation (Variance)
Open-loop	47.5/30.9 48.1/30.4	40.0 40.0	4.06 (16.5) 4.07 (16.6)
Subspace Identified	35.4/28.6 35.8/28.3	31.3 31.3	1.36 (1.84) 1.38 (1.90)
Linear/ Reduced	31.7/30.8 32.2/30.3	31.2 31.2	0.16 (0.026) 0.27 (0.075)
Nonlinear Fundamental	31.6/30.5 32.1/30.0	31.1 31.1	0.20 (0.040) 0.30 (0.092)

The three MPC controllers are also compared for servo responses and rejection of deterministic, measured and unmeasured disturbances. The controllers were detuned so that the Kappa # responses to setpoint changes were not too aggressive. The move horizon, q, is set to 1 for each controller. Also, the manipulated

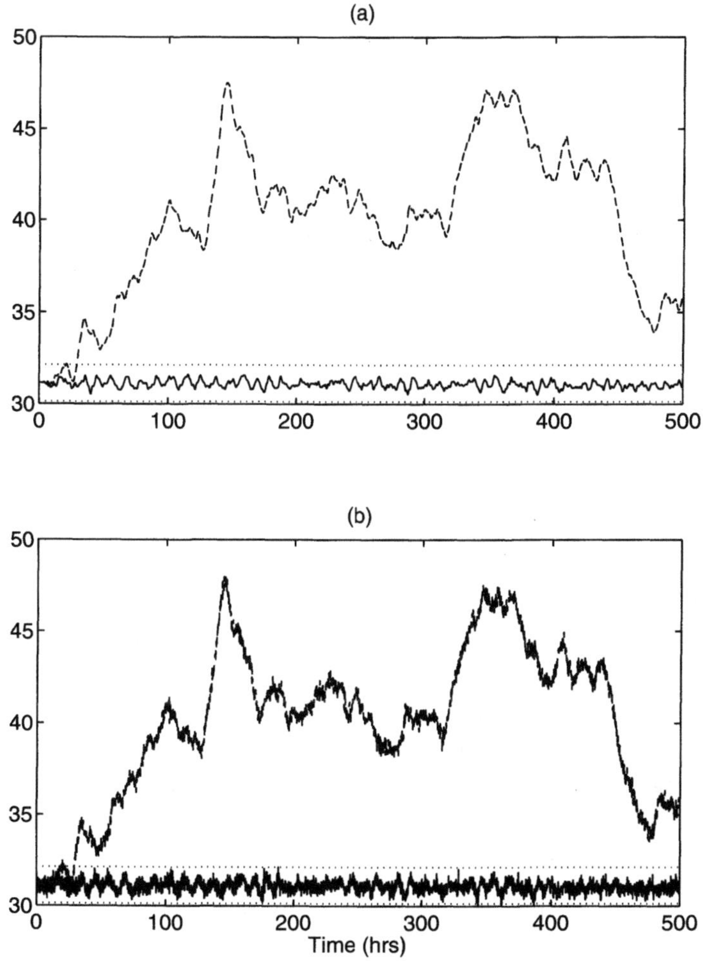

FIGURE 3. (a) Open- (dashed) and closed- (solid) loop Kappa # responses to stochastic disturbances in the wood and white liquor compositions using the nonlinear, multi-rate MPC controller. (b) Same as (a) with the addition of measurement noise. The dotted lines show the ± 1 bound around the setpoint.

input weighting matrix is the same for each controller. The closed-loop Kappa # responses of the three controllers to ± 10 setpoint changes are shown in Figure 4. The MPC controller utilizing the subspace identified model is sluggish to track the setpoint. The MPC controller utilizing the reduced linear model quickly tracks the -10 setpoint change, but has overshoot for the $+10$ setpoint change. The

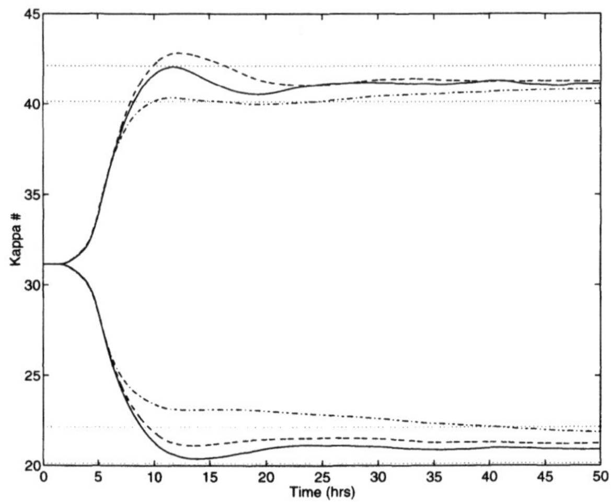

FIGURE 4. Kappa # responses to ±10 setpoint changes for the MPC controllers utilizing the nonlinear fundamental (solid), reduced linear (dashed), and subspace identified (dash-dot) models. The dotted lines represent a ±1 bound around the setpoints.

MPC controller utilizing the nonlinear process model gives a quick and symmetric response for both setpoint changes.

One observes similar trends in the closed-loop response to deterministic disturbance changes (both measured and unmeasured): the nonlinear model-based controller outperforms the linearized model-based controller which outperforms the empirical model-based controller [12].

4. Conclusions

The inferential, closed-loop performance of a nonlinear multi-rate, model predictive controller is demonstrated for stochastic, measured and unmeasured disturbances in the white liquor and wood component concentrations. The case study utilized for this comparison is a two-vessel digester, with one co-current and two counter-current cooking zones, and the physical dimensions and operating conditions are representative of an industrial digester. The controller utilizes the full nonlinear, fundamental process model in the prediction of future outputs for past inputs, but relies on successive linearization to obtain a linear model for computing optimal future control moves. A necessary component of this algorithm is an efficient method for reducing the high-order linear models (obtained from linearization) at each time step, so that these models can be discretized on-line, and used to generate the output prediction equation matrices. A modification to the model reduction algorithm is to use transformation matrices, determined at a nominal operating point, to quickly reduce the high-order linear model via matrix multiplication. This modification allows the 1050 state order model to be reduced to a 54 state order model in 4 seconds. However, even with the incorporation of the nonlinear process model in MPC, the closed-loop performance of the multi-rate nonlinear MPC controller was about the same as that for the linear MPC controller for the stochastic disturbances. However, for setpoint tracking and deterministic, measured and unmeasured disturbances, the nonlinear model predictive controller is superior. The nonlinear MPC controller tracks the setpoint changes faster than the MPC controller utilizing the subspace identified model, and provides a more symmetric response than the MPC controller with the reduced linear model. The nonlinear MPC controller also rejects the deterministic, measured and unmeasured disturbances faster than the other two controllers.

5. Acknowledgements

The authors gratefully acknowledge financial support from the Weyerhaeuser Company and the NSF (CTS 9257059). The authors also thank Dr. Ferhan Kayihan for his valuable input to this work.

References

[1] R. Amirthalingam and J. Lee. Subspace Identification Based Inferential Control of a Continuous Pulp Digester. *Comput. Chem. Engng.*, 21:S1143–S1148, 1997.

[2] B.W. Bequette. Nonlinear Predictive Control Using Multi-rate Sampling. *Can. J. Chem. Eng.*, (69):136–143, 1991.

[3] L. T. Biegler and J. B. Rawlings. Optimization approaches to nonlinear model predictive control. In Y. Arkun and W.H. Ray, editors, *Proceedings of the Fourth International Conference on Chemical Process Control*, pages 543–571, Padre Island, TX, 1991.

[4] D.B. Brengel and W.D. Seider. Multistep nonlinear predictive controller. *Ind. Eng. Chem. Res.*, 28:1812–1822, 1989.

[5] C. E. García. Quadratic dynamic matrix control of nonlinear processes. An application to a batch reaction process. *AIChE Annual Meeting*, 1984.

[6] J. H. Lee, M. S. Gelormino, and M. Morari. Model Predictive Control of Multi-Rate Sampled-Data Systems: A State-Space Approach. *Int. J. of Control*, 55(1):153–191, 1992.

[7] J.H. Lee and A.K. Datta. Nonlinear Inferential Control of Pulp Digesters. *AIChE J.*, 40(1):50–64, 1994.

[8] W. C. Li and L. T. Biegler. Process control strategies for constrained nonlinear systems. *Ind. Eng. Chem. Res.*, 27:1421–1433, 1988.

[9] R. Michaelsen, T. Christensen, G. Lunde, G. Lundman, and K. Johansson. Model Predictive Control of a Continuous Kamyr Digester at SCA-Nordliner, Munksund, Sweden. In *Proc. Control Systems '92 Conference*, pages 91–94, Whistler, B.C., Canada, 1992.

[10] M. Morari and J.H. Lee. Model Predictive Control: Past, Present and Future. In *PSE ESCAPE-7 Symposium*, Trondheim, Norway, 1997.

[11] P.A. Wisnewski and F.J. Doyle III. Control structure selection and model predictive control of the Weyerhaeuser digester problem. *J. Proc. Cont.*, 1998. In press.

[12] P.A. Wisnewski and F.J. Doyle III. Model-based predictive control studies for a continuous pulp digester. *IEEE Trans. Control Sys. Tech.*, 1998. Submitted.

[13] P.A. Wisnewski and F.J. Doyle III. Model reduction of large scale systems with application to a pulp digester. In *Proceedings of DYCOPS-5*, 1998.

[14] P.A. Wisnewski, F.J. Doyle III, and F. Kayihan. Fundamental continuous pulp digester model for simulation and control. *AIChE J.*, 43:3175–3192, 1997.

Department of Chemical Engineering
University of Delaware
Newark, DE 19716
USA
E-mail address: fdoyle@udel.edu

Progress in Systems and Control Theory, Vol. 26

Multivariable Nonlinear Control of Cement Mills

V. Wertz, L. Magni and G. Bastin

Abstract. In this paper, multivariable nonlinear control strategies for a cement grinding circuit are studied. It is shown why nonlinear control is requested in order to deal with typical unmeasured perturbations (hardness change of the clinker) and subsequently two alternative strategies are compared: a nonlinear predictive controller and a state feedback controller based on Lyapunov design. Merits and disadvantages of both approaches are discussed.

1. Introduction

The introduction of modern control methodologies in traditional process industries is in steady growth. Model based controllers usually allow to tackle multivariable control where classical *PID* controllers would only apply to single input single output feedback loops. In addition, recent nonlinear control techniques also provide additional improvements by allowing processes to be run over a larger operating range without controller retuning. This paper presents such an application of model based nonlinear multivariable control to a cement milling circuit.

Traditionally, the application of feedback control techniques to cement milling circuits is limited to monovariable classical *PI* control of the tailings flow rate (or of the circulating load) with either the feed flow rate or the classifier speed as control action. A typical example can be found in [1].

Recently, a multivariable *LQG* controller has been introduced to improve the performances of the milling circuit [3, 2]. However, this controller, whose design is based on a linear approximation of the process, is only effective in a limited range around the nominal operating conditions. On some occasions, it has been observed on real plants that important changes in operating conditions (due for example to changes of the hardness of the clinker) have driven the mill to a region where the controller cannot stabilize the plant.

The aim of this paper is to examine how nonlinear model based techniques can overcome this problem, called plugging. Hence, the paper starts with the presentation of a nonlinear model for the cement mill. One shows that the closed-loop with a classical *LQG* controller based on a linearization of this model indeed becomes unstable in the presence of large hardness changes. Then, two nonlinear

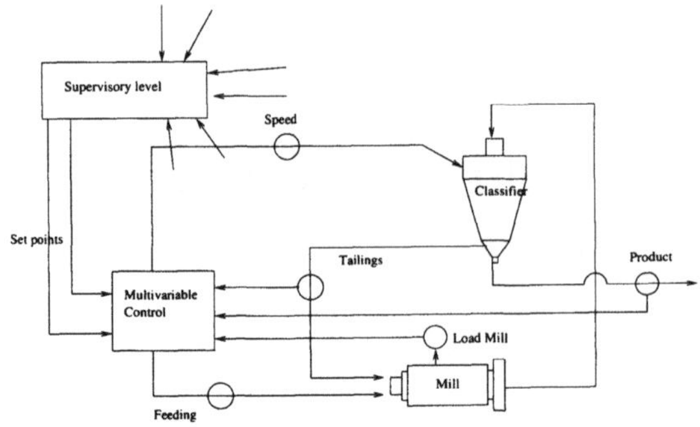

FIGURE 1. Milling circuit principle

control strategies are investigated. The first one is based on an algorithm for nonlinear receding horizon predictive control [5, 6, 7]. It will be demonstrated that the stability region is indeed enlarged, while keeping performances comparable to those obtained with LQG control.

The second control strategy aims at ensuring global stabilization. It is based on two PI controllers complemented with a nonlinear feedforward compensation.

The paper is organized as follows. The process and control strategy are described in Section 2. Then, the nonlinear model required to take into account the changes of hardness of input material is described in Section 3. The next section describes an *LQ* controller designed on the basis of this nonlinear model. This controller is similar (although not exactly identical) to the *LQ* controller that has been described in [2], and exhibits similar instabilities as those that have been occasionally observed on real plants. Section 5 describes the nonlinear predictive controller which has been designed and the simulation results which have been obtained. Section 6 then presents the globally stabilizing nonlinear controller. Discussion and conclusions follow in Section 7.

2. Process description and control strategy

The schematic lay-out of a cement milling circuit is shown in Figure 1.

The mill is fed with raw material (feed). After grinding, the material is introduced in a high-efficiency classifier and separated into two classes. The tailings (refused part) are fed back into the mill while the finished product (accepted part) exits the milling circuit. The classification of the material is driven by the rotational speed and by the air flow rate of the classifier.

In steady-state operation, the product flow rate is necessarily equal to the feed flow rate while the tailings flow rate and the load (i.e. the amount of material) in the mill may take any arbitrary constant values. The load in the mill depends on the input flow rate (fresh feed plus tailings flow rate) and on the output flow rate that depends in a nonlinear way on the load in the mill and on a very important and time-varying quantity: the hardness of the material. Sometimes this nonlinearity can also cause the instability of the system and the obstruction of the mill (a phenomenon called "plugging"), which then requires an interruption of the grinding process. These considerations motivated the introduction of a nonlinear model of the plant [4]. The load in the mill must be controlled at a well chosen level because too high a level of the load in the mill leads to the obstruction of the mill, while too low a circulating load contributes to fast wear of the mill internal equipment. A usual approach is to control the tailings flow rate by using the feed flow rate as control input. This control strategy is, however, not fully satisfactory since it indirectly induces a loss of the control of the product flow rate.

A correct fineness of the product is also very important [2]. The fineness, which is related to the particle size distribution, depends on the composition of the mill feed, but also on the rotational speed and on the air flow rate of the classifier. A natural control objective would therefore be to keep the fineness as close as possible to a desired value by acting, for example, on the rotational speed of the classifier. However, in the usual case where on-line fineness measurements are not available, an indirect control strategy is necessary.

The following can also be observed in practice:

- For a given feed composition, the fineness will be constant if the product flow rate and the load in the mill are kept constant.
- Similarly the energy consumption of the process (ratio Energy/Unit Product) will be fully under control if the product flow rate and the load in the mill are regulated at suitable set points.

This clearly implies that an overall control of both the fineness and the energy consumption can be achieved at a supervisory level in the control system, by manipulating and optimizing the set points of the tailings flow rate or of the product flow rate or of the load in the mill, using for instance off-line laboratory measurements of the fineness.

3. Nonlinear continuous time model

In this section the nonlinear model of the milling circuit which has been introduced in [4] is discussed. This model should describe the evolution of the load in the mill and it should be able to reproduce some (unstable) behaviour that has been observed with the LQ controller in the feedback loop. The model consists of three nonlinear differential equations explaining the evolution of three states.

$$\left\{ \begin{array}{rcl} T_f \dot{y}_f & = & -y_f + (1 - \alpha(z, v, d))\varphi(z, d) \\ \dot{z} & = & -\varphi(z, d) + u + y_r \\ T_r \dot{y}_r & = & -y_r + \alpha(z, v, d)\varphi(z, d) \end{array} \right. \quad (1)$$

with

$$\alpha(z, v, d) = \frac{\varphi^m v^n}{K_\alpha + \varphi^m v^n}$$

$$K_\alpha = (570)^m 170^n (\tfrac{570}{450} - 1)(Tons/hour)^m rpm^n$$

$$m = 0.8 \quad n = 4$$

$$\varphi(z,d) = \max\{0; (-dK_{\varphi 1}z^2 + K_{\varphi 2}z)\}$$

$$K_{\varphi 2} = 16.50(hour)^{-1} \quad K_{\varphi 1} = 0.1116(Tons * hour)^{-1}$$

$$T_f = 0.3 \text{ hour} \quad T_r = 0.01 \text{ hour} \quad d = 1$$

where y_f is the product flow rate (Tons/hour), z is the load in the mill (Tons), y_r is the tailings flow rate (Tons/hour), $\varphi(z,d)$ is the output flow rate of the mill (Tons/hour), u is the feed flow rate (Tons/hour), v is the classifier speed (rpm), and d represents the hardness of the material inside the mill with respect to the nominal one. The values of the various coefficients in this nonlinear model have been tuned in such a way that the model step responses fit with experimental step responses described in [3]. This constitutes a first validation of the model (see [11]).

From the second nonlinear differential equation and the definition of φ, it is possible to notice that there is a constraint on the maximum value of the circulating load (feed plus tailings flow rates). By an easy manipulation of the three nonlinear differential equations it follows that it is possible to choose independently only two of the three steady-state values, the third one being imposed from the model equations. Choosing setpoints for y_f and y_r would impose a given steady state value for φ which could be unachievable. Instead, it is suggested here to choose the load in the mill z and the value of the product flow rate y_f or of the tailings flow rate y_r as set-points.

Note however that a direct measurement of the load in the mill is not an easy task in practice (see discussion at the end of the paper). However, all the control strategies developed hereafter in the paper are state feedback control laws which assume full state measurements.

The main problem observed on industrial milling circuits, controlled by the "Linear Quadratic Multivariable Controller", is caused by large changes of the feed hardness. In some cases, the *LQ* controller is not able to control the mill after such large changes, plugging occurs in the mill and an interruption of the grinding process is necessary. So, our aim with the nonlinear model is to reproduce this behaviour at least qualitatively, when an *LQ* controller is applied. To see that the nonlinear model presented here is able to reproduce this practical observation, an *LQ* controller based on the linearized approximation of the nonlinear model is first considered and it is shown that simulations in closed-loop with the nonlinear model do exhibit instability when the hardness changes. Then, a nonlinear control law will be introduced to overcome this problem without loosing the good nominal performance of the *LQ* controller.

4. LQ control design

The nonlinear model (1) is first expanded with the following two sets of equations in order to ensure zero steady-state error and weighting of the control increments as well as of the control variables.

$$\left\{\begin{array}{lcl}\dot{u} & = & \delta_u \\ \dot{v} & = & \delta_v\end{array}\right. \tag{2}$$

$$\left\{\begin{array}{lcl}\dot{\sigma}_{yf} & = & y_f^* - y_f \\ \dot{\sigma}_z & = & z^* - z\end{array}\right. \tag{3}$$

With the model (1)-(3), zero steady-state error will be achieved provided asymptotic stability is obtained with the feedback controller. Note that parameter changes may modify the equilibrium values for some of the states, but (3) guarantees zero steady state error for the two selected outputs.

Rewrite the system (1)-(3) as

$$\dot{x} = f_c(x, U) \tag{4}$$

with $x = [y_f\ z\ y_r\ u\ v\ \sigma_{yf}\ \sigma_z]'$, $U = [\delta_u\ \delta_v]'$, $x^* = [y_f^*\ z^*\ y_r^*\ u^*\ v^*\ 0\ 0]'$ where x^* is such that $0 = f_c(x^*, 0)$. In order to investigate some properties of the linearized system, define

$$A(x^*) = \frac{\partial f_c(x,U)}{\partial x}\Big|_{x=x^*,U=0} \qquad B = \frac{\partial f_c(x,U)}{\partial U}\Big|_{x=x^*,U=0}$$

Then the system

$$\dot{x}_L = A(x^*)x_L + BU \tag{5}$$

with $x_L = x - x^*$ is the linearized system corresponding to system (4) around the equilibrium point $(x, U) = (x^*, 0)$. Note that the linearized system depends only on the values of v^* and z^*.

The corresponding state-space representation in discrete time is given by:

$$x_L(k+1) = A_D(x^*)x_L(k) + B_D(x^*)U(k) \tag{6}$$

where k is an integer index and $A_D(x^*)$ and $B_D(x^*)$ are obtained in a standard way from $A(x^*)$ and B with a sampling period $T_e = 1$ minute. The LQ-controller, based on the discrete time linear system state-space representation, has been designed using the Matlab Control System Toolbox. Recall that the design is based on the computation of the matrix gain $K(x^*)$ of the linear controller

$$U(k) = -K(x^*)x_L(k) \tag{7}$$

which minimizes the quadratic criterion:

$$J(x^*) = \sum_{i=0}^{\infty} x_L(t+i)'Qx_L(t+i) + U'(t+i)RU(t+i) \tag{8}$$

subject to (6), where Q and R are two symmetric positive definite matrices.

In Figures 2 and 3 (consider only the dotted lines first) two simulations obtained with $Q = I$ and $R = I * 10^{-3}$ are presented. This choice is similar to the one taken in [2], which has been applied on several industrial cement grinding circuits. Starting from a general equilibrium point x^*=[120 55 450 120 170 0 0]', the setpoint $[y_f^*\ z^*]'$=[120 75]' is chosen, corresponding to the maximum of the φ curve which is the optimum from an economical point of view, and then, the hardness d changes from its nominal value 1 to 1.25 in the first example (Figure 2) and from its nominal value 1 to 1.34 in the second example (Figure 3). This change is of course not known to the controller. For the LQ controller (dotted lines) it is clear that the nominal response to a set-point change in z is correct but that the rejection of a hardness change is poor in the first case while in the second case this perturbation causes instability of the closed loop (plugging). Indeed, right after the hardness change, the load in the mill increases steadily while the product and tailing flow rates both go to zero. After a while, the simulation becomes unrealistic (feed flow rate and classifier speed take negative values) which explains that the load in the mill does not continue to increase. These simulations show that the nonlinear model is able to represent the observed nonlinear behaviour of the plant when the hardness changes. Hence the nonlinear model is considered as a good model to synthesize a control law whose aim is to control the cement mill in a larger operating range.

5. Predictive control law

In this section the Nonlinear Receding Horizon (NRH) control law, first presented in [5] and then extended in [6] and [7], is used to control the cement mill (see also [11] for more details). This NRH control technique uses a suitable nonquadratic terminal state penalty retaining the computational advantages of finite-horizon optimization without terminal constraints and ensuring the closed loop stability under very mild conditions (only the stabilizability of the linearized system is required).

It will be seen in the sequel that the NRH regulator proposed here can be considered as a consistent nonlinear extension of the local linear control law. This observation justifies the use of this control law to improve the performance of an LQ controller when there are constraints or nonlinearities. The application of the NRH control law to the present problem goes as follows.

Consider the nonlinear discrete-time system

$$x(k+1) = f(x(k), U(k)), x(t) = \bar{x}, k \geq t \tag{9}$$

obtained from system (4) by discretization holding the manipulated variables U constant over the sampling interval. The state and the input vectors are subject to the constraints

$$x(k) \in \mathcal{X}, U(k) \in \mathcal{U}, k \geq t \tag{10}$$

where $\mathcal{X}$ and $\mathcal{U}$ are closed and bounded subsets of $\mathbb{R}^7$ and $\mathbb{R}^2$, respectively, containing x^* and the origin respectively as an interior point.

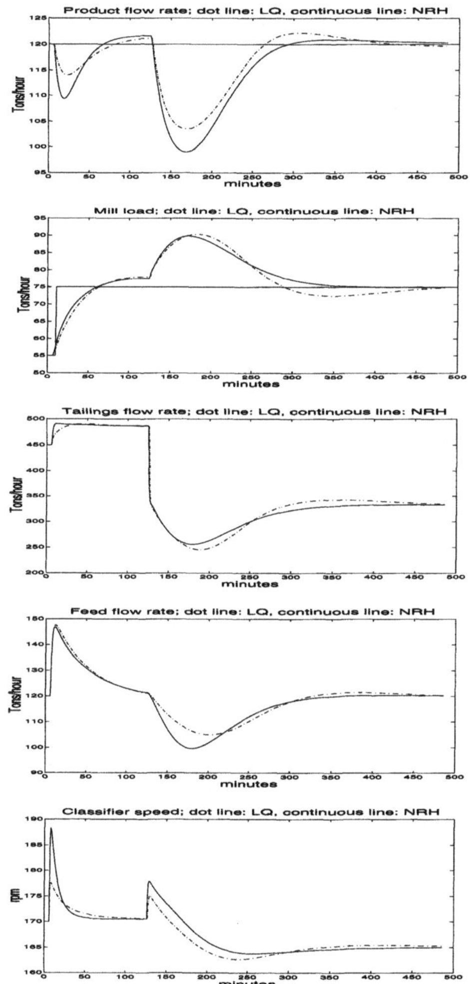

FIGURE 2. Step in z^* from 55 to 75 after 6 minutes and in d from 1 to 1.25 after 126 minutes

Define now the set:

$$\begin{aligned} X(K(x^*)) = \; \{ & \; \bar{x} \in \mathbb{R}^7 | lim_{k \to \infty} \|x(k)\| = x^*, with \quad \bar{x} = x(t), \\ & x(k) \in \mathcal{X}, \quad U(k) = K(x^*)x(k) \in \mathcal{U}, \quad k \geq t \}. \end{aligned}$$

In other words, $\bar{x}$ belongs to $X(K(x^*))$ if the application of the LQ control law (7) to the nonlinear system (9) generates an input sequence guaranteeing the satisfaction of the constraints (10) and driving the state to the equilibrium point x^* asymptotically. Then consider the following finite-horizon problem:

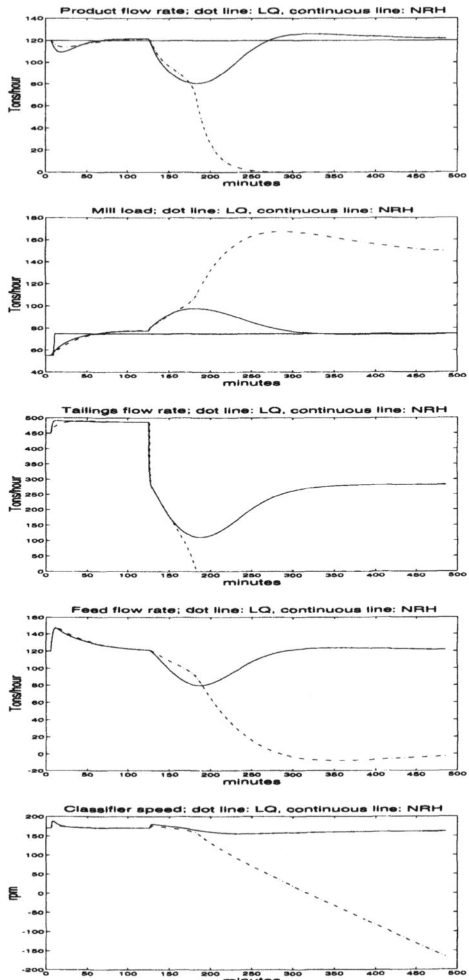

FIGURE 3. Step in z^* from 55 to 75 after 6 minutes and in d from 1 to 1.34 after 126 minutes

Finite-Horizon Optimal Control Problem (*FHOCP*). Minimize, with respect to $U_{t,t+N-1} := [U(t)U(t+1)\dots U(t+N-1)]$, the cost function

$$J(\bar{x}, U_{t,t+N-1}, N, x^*) = \sum_{i=0}^{N-1} \{(x(t+i)-x^*)'Q(x(t+i)-x^*) + U'(t+i)RU(t+i)\} + V_f(x(t+N), x^*) \tag{11}$$

subject to (9) with Q and R defined above. In (11) the nonquadratic terminal state penalty is defined as

$$V_f(x(t+N),x^*) = \sum_{i=N}^{\infty} \{(x(t+i)-x^*)'Q(x(t+i)-x^*)$$
$$+(x(t+i)-x^*)'K(x^*)'RK(x^*)(x(t+i)-x^*)\} \qquad (12)$$

In other words, $V_f(x(t+N),x^*)$ is the cost that is incurred over $[t+N,\infty]$ by applying the LQ control law (7) to the nonlinear system (9). Note that $V_f(x(t+N),x^*)$ is finite if $x(t+N)$ can be driven to x^* by the linear state feedback controller. In the following, given $K(x^*)$ such that $A_D(x^*)+B_D(x^*)K(x^*)$ is stable, a sequence $U_{t,t+N-1}$ is said *admissible* if when applied to (9) the following constraints are satisfied:

C1: $x(k) \in \mathcal{X}, u(k) \in \mathcal{U}, k = t,\dots,t+N-1$.

C2: $x(t+N) \in X(K(x^*))$.

Associated with (11), the following predictive NRH control strategy is introduced.

Predictive control law. Given the LQ time invariant control law, $K(x^*)$, define $\bar{x} = x(t)$ and find an admissible control sequence $U^o_{t,t+N-1}$ solving the *FHOCP*. Then, apply the control $U(t) = U^o(\bar{x})$, where $U^o(\bar{x})$ is the first column of the optimal sequence $U^o_{t,t+N-1}$ and solve the same problem again at time $t+1$ (receding horizon strategy).
In the following, $X_0(N,x^*)$ will denote the set of states $\bar{x}$ such that the NRH control $U^o(\bar{x})$ is computable (i.e. the set of admissible sequences is non empty). The stabilizing property of the predictive NRH controller is then stated in the following theorem.

Theorem 5.1. [5] *Assume that $A_D(x^*)+B_D(x^*)K(x^*)$ is stable. Then, if the NRH control law $U = U^o(x)$ is applied to the nonlinear system (9), the equilibrium point x^* is an asymptotically stable equilibrium point of the resulting closed-loop system with region of attraction $X_0(N,x^*)$.*

Remark 5.2. *Note that in view of the enlargement of the system with the integral action on the tracking error (3), the NRH control law guarantees zero steady-state error for the perturbed system if asymptotic stability is achieved.*

Remark 5.3. *As seen from the statement of the Theorem, the choice of the optimization horizon N is immaterial as far as local stability is concerned. However, increasing N is likely to enlarge the domain of attraction of the equilibrium point. This is an argument in favour of long optimization horizons, although computational constraints could pose a practical upper limit (the difficulty of solving the $FHOCP$ obviously increases with N).*

Further details on the computation of this nonlinear predictive control problem are described in [11].

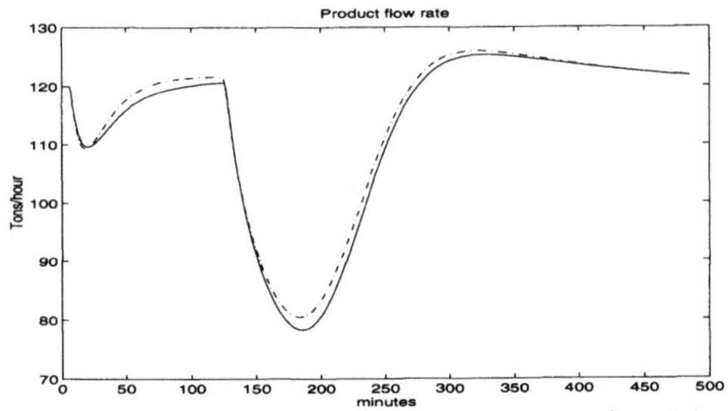

FIGURE 4. Control law synthesis with linear α (solid line) and with nonlinear $\alpha(z, v, d)$ (dotted line)

Results

Consider now the application of this nonlinear predictive control strategy to the cement mill model and its comparison with a classical LQ controller based on the linear approximation of the model. In Figures 2 and 3 the two simulations with the LQ controller are compared with the ones obtained using the NRH control law with the same Q and R and with $N = 2$ (continuous line). The hardness changes in these simulations are not known to the controllers. Note that the nominal performance, when the set-point in z changes, is very similar to the one obtained with the LQ controller. Similar conclusions can also be drawn when the hardness changes, as long as the LQ controller is still able to stabilize the plant. In that case (see Fig.2), the performances of both controllers are quite comparable. However, in Fig. 3 it is clear that the NRH control stabilizes the plant even when the LQ controller can not.

These simulations show that if the nonlinear model is a good representation of the nonlinear behaviour of the plant in the case of hardness changes inside the mill, then the nonlinear RH is able to solve one problem experienced with the LQ controller that is now in operation.

Remark 5.4. *One of the main problems of model-based algorithms is that it is not always possible to have a precise knowledge of all parameters of the model of the process to be controlled. Hence, it is important for the control algorithm to be robust with respect to some modelling uncertainties. As an indication of such robustness for the present algorithm, the following simulation has been performed. The process model is still the same nonlinear model (1) but the controller is computed on the basis of a simplified version where* $\alpha(z, v, d) = \alpha_1 + \alpha_2 v$. *In Figure 4 this simulation is reported under the same conditions as in Figure 3. It can be seen that the simplified model used in the computation of the control law does not alter the performance of the controller.*

Although the previous simulation indeed shows some robustness of the control law w.r.t. some modelling uncertainties, the control law still requires several parameters to be reasonably fitted to the actual plant. In the following section, another control strategy is presented which requires only qualitative knowledge of the process.

6. Global state feedback stabilization

The control strategy is developed based on model (1) with the classification function $\alpha(z, v, d)$ replaced by a simpler function $\alpha(v)$. Here, no precise parametrization of the grinding and classification functions $\varphi(z, d)$ and $\alpha(v)$ are required. Instead, it is only supposed that they are qualitatively described as follows: the grinding function $\varphi(z, d)$ is shown in Figure 5 for different values of d. It is a non monotonic function of the level of material z in the mill, reaching a maximum for some critical value of z. When z is too high, the grinding efficiency decreases and leads to the obstruction of the mill (plugging). On the other side, a too low value of z is also quite undesirable because it causes a fast wear of the balls.

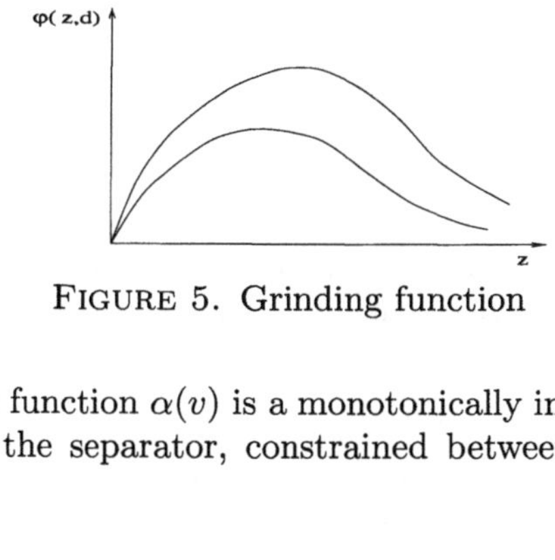

FIGURE 5. Grinding function

The classification function $\alpha(v)$ is a monotonically increasing function of the rotational speed v of the separator, constrained between 0 and 1 as shown in Figure 6.

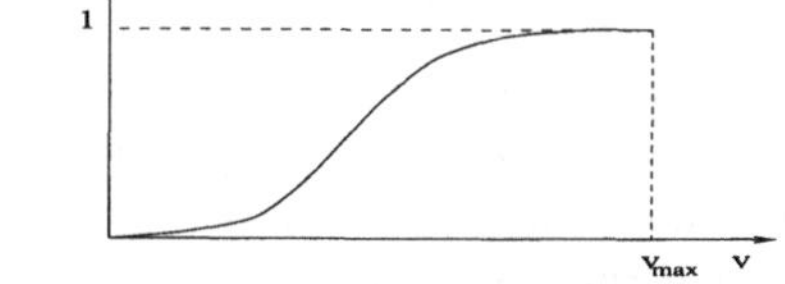

FIGURE 6. Classification function

Of paramount importance is the fact that with this modelling of the grinding and classification functions, the system (1) is positive (see e.g. [10]) in accordance with the physical reality :

$$\text{if } y_f(0) \geq 0, y_r(0) \geq 0, z(0) \geq 0, u(t) \geq 0 \text{ and } v(t) \geq 0 \text{ for all } t \geq 0 \text{ then}$$
$$y_f(t) \geq 0, y_r(t) \geq 0 \text{ and } z(t) \geq 0 \text{ for all } t \geq 0.$$

The global stabilizing controller is described in detail in [8, 9]. It is based on the following control laws:

$$\begin{aligned} u &= -y_r + k_1(z^* - z) + \theta \\ \dot{\theta} &= k_2(z^* - z) \\ v &= \eta \\ \dot{\eta} &= k_3(y_f^* - y_f). \end{aligned}$$

The first two equations describe a *PI* controller with feedforward compensation of the tailings flow rate which uses the feed flow rate to control the load in the mill. The last two equations describe an integral action on the deviation between product set point and actual flow rate, which acts on the classifier speed. Both control laws have been designed using Lyapunov design techniques, precisely in order to guarantee closed loop stability. The gains have been tuned by trial and error in simulation in order to achieve good performance.
As a result, the closed loop system can be rewritten as

$$\begin{aligned} T_f\dot{y}_f &= -y_f + (1 - \alpha(\eta))\varphi(z, d) \\ T_r\dot{y}_r &= -y_r + \alpha(\eta)\varphi(z, d) \\ \dot{z} &= -\varphi(z, d) + k_1(z^* - z) + \theta \\ \dot{\theta} &= k_2(z^* - z) \\ \dot{\eta} &= k_3(y_f^* - y_f). \end{aligned}$$

In order to take into account physical constraints on the control actions, saturations and anti-windup devices can be included in the control laws. Hence, one obtains

$$\begin{aligned} u &= m(\psi) \\ \psi &= -y_r + k_1(z^* - z) + \theta \\ \dot{\theta} &= k_2(z^* - z) + k_2(m(\psi) - \psi) \\ v &= l(\eta) \\ \dot{\eta} &= k_3(y_f^* - y_f) + k_3(l(\eta) - \eta) \end{aligned}$$

where $m(\psi) = sat_{[0,u_{\max}]}(\psi)$ and $l(\eta) = sat_{[0,v_{\max}]}(\eta)$.

A proof of global stability of the closed loop system with this controller can be found in [9]. In Figures 7 and 8, a simulation of this control law is shown, which compares quite favorably with the nonlinear predictive controller described in the previous section.

7. Discussion

The aim of this work has been to investigate further improvements to the multivariable *LQ* control law that was recently developed and succesfully implemented on several cement milling circuits. In particular practical observations had shown

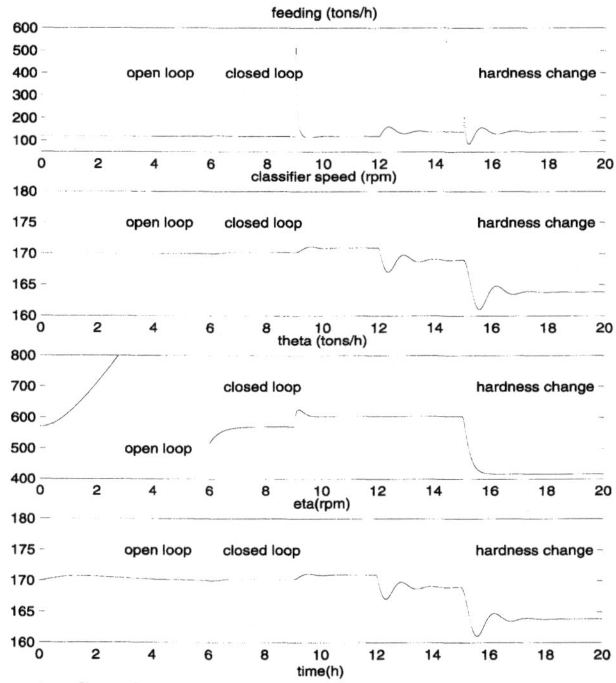

FIGURE 7. Simulation of the global state feedback controller - Process inputs

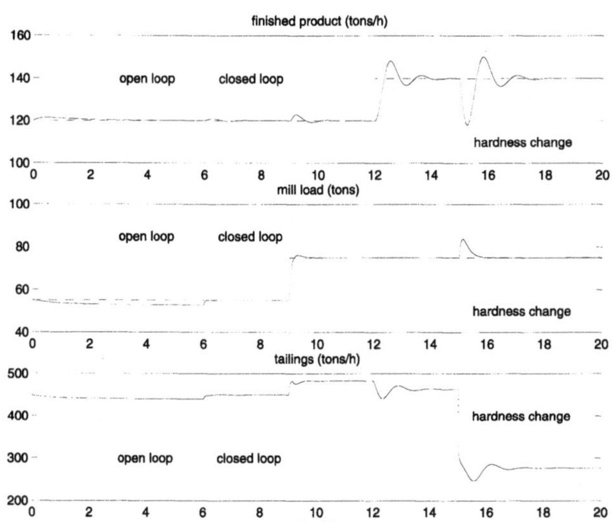

FIGURE 8. Simulation of the global state feedback controller - Process states

the appearance of mill plugging on some occasions, a phenomenon which the LQ controller could not prevent from happening. In this paper, a nonlinear model has been presented which specifically models the load in the mill, and subsequently two nonlinear control laws based on this model. The first one is based on predictive control ideas and preserves the main characteristics of the previous LQG controller but enlarges the operating region in which plugging can be avoided. The second one is only based on a qualitative description of the process and has global stabilizing properties. The fact that this controller only requires qualitative knowledge of the process for the analysis of the closed loop system should not conceal the difficulty of finding appropriate gains for the PI control laws. In order to achieve good performances, tuning of these gains has been necessary and if the real process is not available for several experiments, then an accurate model will be needed for an off-line tuning.

The real difficulty of the proposed controllers is the need for a measurement of the load of the mill. Various techniques can be implemented (some of them already exist on industrial milling circuits).

- The mill itself could be installed on a weighing device so that its total mass can be measured and the cement load deduced from it.
- On some mills, an "electronic ear" measures the noise of the mill, which is inversely proportional to the current load in the mill.

If precise measurements of the load cannot be obtained, then a nonlinear observer of this variable should be studied as an alternative to the presented scheme. This is left to future work.

Acknowledgments

This paper presents research results of the Belgiam Program on Interuniversity Attraction Poles, initiated by the Belgiam State, Prime Minister's Office for Science, Technology and Culture. The scientific responsability rests with its authors. The authors also wish to acknowledge partial support of the "S.A. Slegten" Company, Belgium. This paper has also been partially supported by MURST project "Model Identification, System Control, Signal Processing". Finally, the multivariable control structure mentioned in the paper is protected by a patent application (Patent nr. 93870040.8).

References

[1] Ciganek, C. and K. Kreysa, *Two-parameter control system for a cement grinding process*, Translation of Zement-kalk-gips, 202–206, 1991.

[2] V.Van Breusegen, L.Chen, G.Bastin and V.Wertz, V.Werbrouck, C. de Pierpont *An Industrial Application of Multivariable Linear Quadratic Control to a Cement Mill Circuit*, IEEE Transactions on Industry Application, vol.32, pp. 670–677, 1996.

[3] V.Van Breusegen, L.Chen, V.Werbrouck, G.Bastin and V.Wertz, *Multivariable linear quadratic control of a cement mill: an industrial application*, Control Eng. Practice, vol.2, pp. 605–611, 1994.

[4] B.de Haas, F. Jadot, G. Bastin and V. Wertz, Int. Research Report, CESAME, 1996.

[5] G. De Nicolao, L. Magni and R. Scattolini, *Stabilizing nonlinear receding horizon control via a nonquadratic terminal penalty*, IMACS Multiconference CESA '96, Lille France July 9–12, 1996.

[6] G. De Nicolao, L. Magni and R. Scattolini, *Stabilizing receding horizon control of Nonlinear Time-Varying Systems*, IEEE Trans on Autom. Contr., vol. 43, pp. 1030–1036, 1998.

[7] G. De Nicolao, L. Magni and R. Scattolini, *Stabilizing Predictive control of NARX models*, Automatica, Vol. 33, pp. 1691–1698, 1997.

[8] F. Jadot, G. Bastin, V. Wertz and L. Magni, *Preventing cement mills from plugging by robust state feedback*, Proc. Control 97, pp. 248–251, Sydney, Oct. 1997.

[9] F. Jadot, G. Bastin, V. Wertz and L. Magni, R. Sepulchre, F. Grognard *Robust global state feedback stabilization of cement mills* , Int. report, CESAME, 1998.

[10] Luenberger D.G., "Introduction to Dynamic Systems", Wiley, 1979.

[11] Magni L., G. Bastin, V. Wertz, *Multivariable nonlinear predictive control of cement mills*, IEEE Trans on Control Systems Technology, to appear.

CESAME
Catholic University of Louvain
avenue G. Lemaitre, 4
B-1348 Louvain-la-Neuve
Belgium
Tel : +32 10 47 23 80
Fax : +32 10 47 21 80
E-mail address: wertz@auto.ucl.ac.be

Progress in Systems and Control Theory, Vol. 26

Nonlinear Receding Horizon Control of Internal Combustion Engines

G. De Nicolao, L. Magni and R. Scattolini

Abstract. This paper presents the application of a Nonlinear Receding Horizon strategy to idle speed control of internal combustion engines. The controller synthesis is based on a Nonlinear Auto Regressive eXogenous model of the engine describing the dynamics of the pressure p inside the intake manifold and of the cranckshaft speed n as functions of the position of the idle-speed air actuation system α and of the spark advance ψ. The performance of the controller are tested against engine charges due to accessory torque loads. With reference to a commercial 1400 cm^3 engine model, comparisons between the LQ regulator presently used in large scale applications and the NRH approach are illustrated.

1. Introduction

The industrial interest in idle speed control is motivated by the fact that "on the average, vehicles consume about 30 percent of their fuel in city driving during idle" [12], [11]. Since the engine dynamics is highly nonlinear, the design of efficient idle speed controllers requires the availability of reliable nonlinear models, which must be obtained with a limited effort in order to reduce the development times and costs required to complete the control synthesis phase. Physical models of the engine in idle speed conditions usually contain a large number of unknown parameters to be estimated with expensive laboratory experiments. For these reasons, a black-box technique for the estimation of $NARX$ (Nonlinear AutoRegressive eXogenous) engine models has been presented in [7], [17] and [8]. These papers also illustrate the excellent results achieved by applying the proposed nonlinear identification method to three commercial engines, namely a 1200 cm^3, a 1400 cm^3 and a 1600 cm^3. The $NARX$ model describes the dynamics of the pressure inside the air intake manifold p, and of the crankshaft speed n, (measured in revolutions per minute (rpm)), as functions of the position of the idle speed additional air actuation system α and of the spark advance ψ. The models are usually identified in the crank-angle domain (θ) since it is commonly recognized that in the angle basis idle speed controllers are easier to design and calibrate see e.g. [3], [11]. The identification procedure is now currently being used by Magneti Marelli together with a control synthesis technique based on the Linear Quadratic (LQ) approach, see [8]. The use of a centralized LQ technique represents by itself a major improvement with respect to the standard schemes based on PI controllers, still widely applied in industrial applications.

Alternatively, the availability of a nonlinear model allows one to use more sophisticated strategies for the design of nonlinear controllers coping with the stringent requirements imposed by the specific control problem, see e.g. [14]. In this direction, the new approach proposed in this paper consists of using the Nonlinear Receding Horizon (NRH) technique presented in [4], [6], [5], [15] which can be viewed as a consistent extension of the LQ approach presently used in production.

The main problem with the application of the NRH approach is due to the computational time required to solve at any time instant (at any Top Dead Center) the FH optimization problem. This is obviously not possible with the present technology; however, since a reliable engine model is available, one can compute off-line the optimal NRH control law $\kappa^{RH}(x)$ for a (large) set of admissible values of the initial state $x = [p\ n]'$. Then, it is possible to approximate $\kappa^{RH}(x)$ with any suitable interpolation technique, for example by means of a Neural Network (NN). Finally, the approximating function so obtained is effectively implemented for on-line computations.

2. The $NARX$ model

The identification procedure presented in [7], [17], [8] allows one to estimate the parameter vector β of the $NARX$ model described by the coupled equations

$$\begin{cases} p(\theta) = f_p(p(\theta - i), n(\theta - i), \alpha(\theta - i), \psi(\theta - i), \beta), & i > 0 \\ n(\theta) = f_n(n(\theta - i), p(\theta - i), \alpha(\theta - i), \psi(\theta - i), \beta), & i > 0 \end{cases} \tag{1}$$

where the functions f_p and f_n are linear in β and their dependence on the state ($x = [p\ n]'$) and input ($u = [\psi\ \alpha]'$) vectors is determined during the estimation phase through a suitably modified stepwise regression algorithm (see e.g. [9]). Specifically, the available data are divided into two subsets which are used for identification and for validation respectively. Given alternative model structures, the identification data are first used to estimate β. Then, among the identified models, one chooses the one that minimizes a suitable performance index, computed on the validation data, according to an output error criterion.

2.1. Experimental results

In order to illustrate the performance of the identification method, some experimental results are here reported, which refer to a commercial 1400 cm^3 engine. The data were collected at 250 Hz rate on a car equipped with a development Electronic Control Unit (ECU) with serial link to an external development kit. Then they were transformed in the crank angle domain θ. The experiments were performed with warm engine and the clutch not coupled. Closed-loop idle speed control, air/fuel ratio control and knock control were disabled. The ECU was programmed in order to generate the pre-assigned stimuli (time-histories) on the

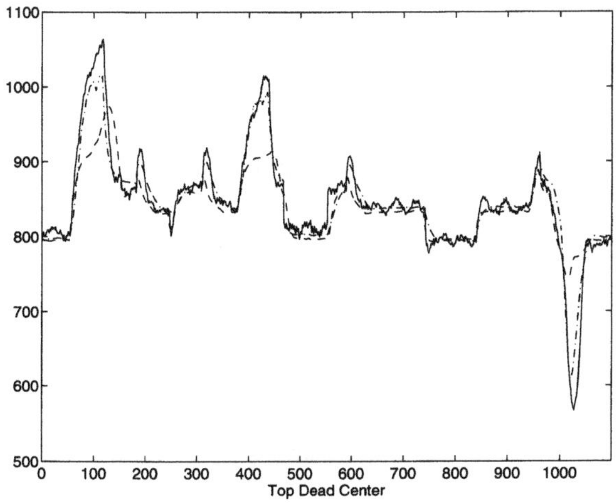

FIGURE 1. Crankshaft speed [rpm]: true (continuous line), simulation output of the *NARX* model (dash-dot) and simulation output of the linear model (dashed).

control variables α and ψ. Correspondingly, the transients of n and p were collected and the data were used to identify a *NARX* model with a maximum delay in the input and output variables equal to one. With reference to a (small) subset of the validation data, the transients of the measured values of n and those produced by the simulation of the *NARX* model are compared in Fig. 1. This figure also shows the transients of the output of a linear model identified with a standard *LS* algorithm. It is apparent that the nonlinear model describes satisfactorily the engine dynamics also for large deviations from the prescribed operating point, even at low ($500\,rpm$) speed values which can be easily reached due to sudden torque loads. On the contrary, the linear model is unable to represent the relevant plant dynamics outside a very limited region around the considered steady state conditions.

3. Linear control synthesis

The main controlled output at idle is the engine speed n, which in steady state conditions must reach a prescribed set-point value $\bar{n}$ (see [11]). The second controlled variable is the intake manifold pressure p, which usually is not required to maintain a prescribed value at idle. As for the inputs, the most important one is the total amount of air supplied to the engine by manipulating the position α of the air actuation system. The second control variable, namely the spark advance ψ, provides a faster actuation, although with limited authority. Then, ψ is usually retarded at idle, so as to guarantee the possibility to promptly react

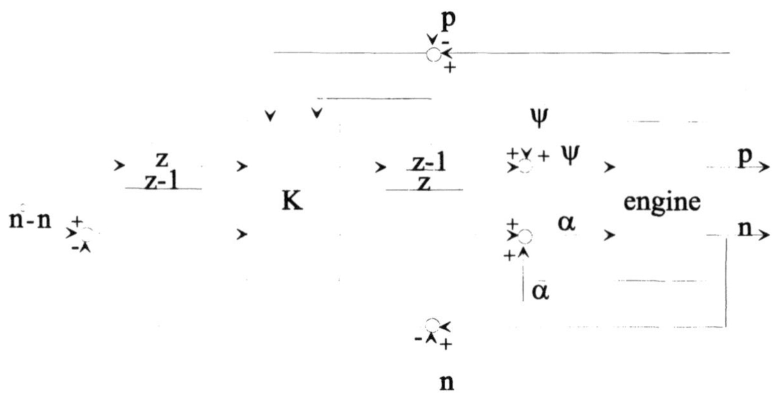

FIGURE 2. Control scheme

to sudden torque loads. For these reasons, a control scheme presently adopted in many commercial applications is the one proposed in [1] and reported in Fig. 2, where the matrix K must stabilize the ensemble of the linearization of the engine model around the nominal operating condition ($\bar{n}$, $\bar{p}$, $\bar{\alpha}$, $\bar{\psi}$), the integrator and the derivative term. The main peculiarities of this control strategy are: (a) being centralized, it automatically takes into account the couplings between input and output variables. (b) Steady state zero error regulation is guaranteed by the integral action on the crankshaft speed error. (c) Derivative action is forced on the spark advance variations so as to achieve the desired value $\bar{\psi}$ of ψ in steady state conditions. (d) The controller gain K can be computed by means of a standard LQ procedure for the enlarged system with state $x \in R^4$ composed by the linearized engine model, the integrator (with state γ) and the derivative term (with state η). In order to synthesize the matrix $K \in R^{2,4}$ from the $NARX$ model, the first step is to determine the steady state values $\bar{p}$ and $\bar{\alpha}$ corresponding to the required steady state conditions specified by $\bar{n}$ and $\bar{\psi}$. Then, the model is linearized and K is computed with the LQ technique for the enlarged system. Letting

$$\delta x(\theta) = \left[\begin{array}{cccc} p(\theta) - \bar{p} & n(\theta) - \bar{n} & \gamma & \eta \end{array} \right]'$$

be the state vector of the enlarged linearized system and

$$K = \left[\begin{array}{c} K_\psi \\ K_\alpha \end{array} \right], \quad K_\psi \in R^{1,4} \quad K_\alpha \in R^{1,4}$$

the computed LQ gain, the control algorithm finally generates the signals

$$\psi_d(\theta) = K_\psi \delta x(\theta) \tag{2}$$

$$\delta\alpha(\theta) = K_\alpha \delta x(\theta) \tag{3}$$

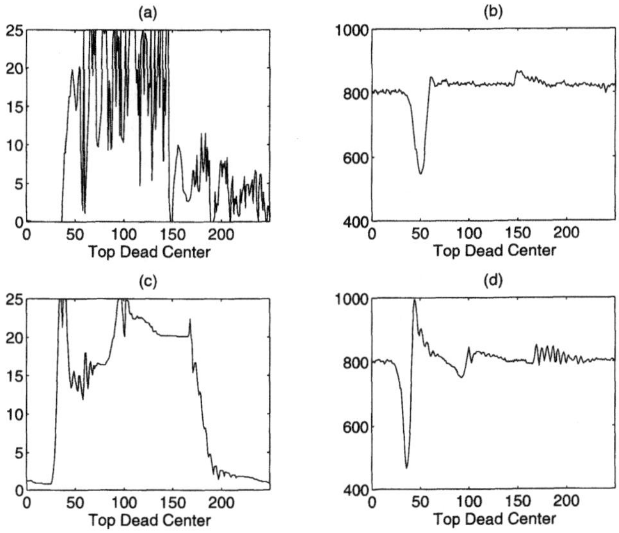

FIGURE 3. Linear controller synthesis based on the linearized *NARX* model: (a) Torque disturbance [Nm], (b) Crankshaft speed [rpm]; Linear controller synthesis based on the linear model: (c) Torque disturbance [Nm], (d) Crankshaft speed [rpm].

where $\psi_d(\theta)$ is the input of the derivative block in the scheme of Fig. 2. Correspondingly, the control variables are

$$\psi(\theta) = \bar{\psi} + \psi_d(\theta) - \psi_d(\theta - 1) \quad (4)$$

$$\alpha(\theta) = \bar{\alpha} + \delta\alpha(\theta) \quad (5)$$

Although the adopted synthesis method is based on a linearized model, the availability of a reliable nonlinear model of the engine is of crucial importance in the design phase. In fact, this allows one to test the performance of the controller in simulation, rather than through an heavily time consuming implementation on the *ECU* control unit of the car. This in turn means that the selection of the "optimal" weighting matrices of *LQ* control can be made with a simple yet effective sequential procedure, as confirmed by the results reported below (see also [8]).

3.1. Experimental results

The identified *NARX* model of the $1400\,cm^3$ engine has been linearized around the nominal operating point ($\bar{n} = 800\,rpm$, $\bar{\psi} = 5^0$). Then the controller gain matrix K has been synthesized and the overall controller structure of Fig. 2 has been implemented in the *ECU* of the car. Starting from idle speed conditions, the performance of the controller have been tested by charging the engine with a torque load disturbance obtained by forcing accessory loads. In Fig. 3(a) it is shown an estimate of the applied torque disturbance, which is roughly a square wave with maximum average value of 20 Nm. Correspondingly, the registered transient of n is reported in Fig. 3(b). Note that the maximum undershoot of n is $\cong 250\,rpm$, while the overshoots are negligeable. Note also that the variable n takes a quite long time to settle at its set-point value. However, the main goal of idle speed control is to reduce its peaks, while an error of $20 \div 30\,rpm$ is not detectable by the driver. In order to better appreciate the improvements which can be achieved with the availability of a reliable nonlinear engine model, the gain matrix K has also been computed with the LQ method applied to the identified linear system. The controller so obtained has also been implemented on the car and similar torque load disturbances have been imposed to the engine, as shown in Fig. 3(c). Correspondingly, the transient of n reported in Fig. 3(d) has been obtained. A comparison of Figs. 3(b) and 3(d) shows the benefits of having a reliable nonlinear engine model, even though the idle speed controller is designed in a linear setting.

4. Nonlinear receding horizon control

In this section, we briefly review the state feedback NRH control presented in [4], [6], [5], [15]. Consider the nonlinear discrete-time dynamic system

$$x(k+1) = f(x(k), u(k)), \qquad x(t) = \hat{x} \qquad k \geq t \tag{6}$$

where $x \in R^n$ is the state and $u \in R^m$ is the input. The functions $f(\cdot,\cdot)$ is a C^1 function of its arguments with $\bar{x} = f(\bar{x}, \bar{u})$. For system (6), we search for a NRH control law $u = \kappa^{RH}(x)$ which regulates the state to the equilibrium point $\bar{x}$, subject to the input and state constraints

$$x(k) \in X, \quad u(k) \in U, \qquad k \geq t \tag{7}$$

where X and U are closed subsets of R^n and R^m containing, respectively $\bar{x}$ and $\bar{u}$ as an interior point. In order to derive the NRH control law, first let

$$\delta x(k+1) = A\delta x(k) + B\delta u(k) \tag{8}$$

with $\delta x = x - \bar{x}$, $\delta u = u - \bar{u}$, be the linearization of system (6) around the equilibrium point $(x, u) = (\bar{x}, \bar{u})$, i.e.

$$A = \frac{\partial f}{\partial x}(\bar{x}, \bar{u}), \qquad B = \frac{\partial f}{\partial u}(\bar{x}, \bar{u})$$

Assuming that the pair $(A,\ B)$ is stabilizable, well known results of linear control theory state that it is possible to find a matrix K such that the eigenvalues of $(A+BK)$ are inside the unit circle in the complex plane. Hence K can be computed by means of standard synthesis methods for linear systems, for example with the LQ technique. Now, for a given stabilizing matrix K, at any time instant t let $\hat{x} = x(t)$ and minimize with respect to $u_{t,t+N-1} := [u(t) \quad u(t+1) \ldots u(t+N-1)]$,

$N \geq 1$, the cost function

$$\begin{aligned} & J(\hat{x}, u_{t,t+N-1}, N, \overline{x}, \overline{u}) = \\ = \ & \sum_{i=0}^{N-1} \{(x(t+i)-\bar{x})'Q(x(t+i)-\bar{x}) + (u(t+i)-\bar{u})'R(u(t+i)-\bar{u})\} + \\ & + V_f(x(t+N), \overline{x}) \end{aligned} \tag{9}$$

subject to (6) and (7), with $Q > 0$, $R > 0$ and the terminal state penalty V_f defined as

$$V_f(\hat{x}, \overline{x}) = \sum_{i=0}^{\infty} \{(x(t+i)-\bar{x})'(Q+K'RK)(x(t+i)-\bar{x})\}$$

where $x(t+i)$, $i \geq 0$ satisfies (6) with $u(k) = \bar{u} + K\delta x(k)$. The optimal control sequence $u^o_{t,t+N-1}$ solving the above optimization problem is termed *admissible* if, when applied to system (6)

$$x(k) \in X, \ u(k) \in U \ , t \leq k < t+N, \ x(t+N) \in X(K)$$

where $X(K)$ stands for a suitable exponential stability region (see [13], [6]) of the nonlinear closed-loop system composed by the nonlinear system (6) and the linear control law $\delta u(k) = K\delta x(k)$. In other words, $x(t) \in X(K)$ implies $x(k) \to \bar{x}$ (exponentially) and the fulfillment of the constraints (7) i.e. $x(k) \in X$, $\bar{u} + K\delta x(k) \in U$, $k \geq t$. Finally, the *state-feedback NRH control law* $u = \kappa^{RH}(x)$ is obtained by applying at any time instant t the control $u(t) = u^o(\hat{x})$ where $u^o(\hat{x})$ is the first column of $u^o_{t,t+N-1}$. Letting $X_0(N, K)$ be the set of states $\hat{x}$ such that any admissible control sequence $u^o_{t,t+N-1}$ exists, the following result holds:

Theorem 4.1. [6] *Assume that (A, B) is stabilizable and let K be such that the eigenvalues of $(A + BK)$ are inside the unit circle in the complex plane. Then, if the NRH control law $u = \kappa^{RH}(x)$ is applied to the nonlinear system (6), then $\bar{x}$ is an exponentially stable equilibrium point of the resulting closed-loop system having $X_0(N, K)$ as exponential stability region.*

A very practical procedure for stabilizing the nonlinear system (6) is to design first a linear control law $u = \kappa_L^{IH}(x) = \bar{u} + K(x - \bar{x})$ by minimizing an (Infinite Horizon) IH performance index subject to the linearized state dynamics (8). In this respect, well-established tools are available for the tuning of the weighting matrices Q and R in a standard LQ control problem so as to achieve the desired specifications for the linearized closed-loop system. Then, the same Q and R are used to implement the nonlinear RH controller. Under regularity assumptions, as $||x|| \to \bar{x}$, it turns out that $\kappa^{RH}(x) \to \kappa_L^{IH}(x) = \bar{u} + K(x - \bar{x})$. Moreover, $\left.\frac{\partial \kappa^{RH}(x)}{\partial x}\right|_{x=\bar{x}} = K$ so that the NRH control law can be regarded as a consistent nonlinear extension of the linear control law $u = \bar{u} + K(x - \bar{x})$. Once Q and R have been selected, the only free parameter is the optimization horizon N, which can be tuned to trade computational complexity (which grows with N) for performance ($\kappa^{RH}(x) \to \kappa^{IH}(x)$ as $N \to \infty$, where $\kappa^{IH}(x)$ is the unknown optimal control law for the IH nonlinear problem). Furthermore, as N grows, the stability region enlarges and $X_0(N,K) \to X^{IH}$, where X^{IH} is the region of attraction of the optimal IH nonlinear control law $\kappa^{IH}(x)$. It can also be proven that $X_0(N+1,K) \supseteq X_0(N,K) \supseteq X(K)\ \forall N > 0$, see [6].

4.1. Neural network implementation of the NRH control law

The RH approach requires to solve a nonlinear optimization problem on-line at any time instant. This is possible for "slow" systems, while a direct on-line implementation of NRH control is impossible for "fast" systems. In order to apply the NRH approach also with short sampling intervals, one can solve the optimization problem off-line for many different initial states x and store the computed sequences of control variables, i.e. the "realizations" of $\kappa^{RH}(x)$, in the computer memory. Then, in the real application, one has to select, or interpolate, the computed control sequence that best fits with the current state of the plant. Clearly, an excessive amount of computer memory may be required to store the closed-loop control law. Moreover, some interpolating procedure must be implemented in practice in order to effectively compute the control variable. An interesting way to solve these implementation problems is to resort to a functional approximation $\tilde{\kappa}^{RH}(x)$ of the optimal control law $\kappa^{RH}(x)$, see [16]. More specifically one has to search for a function $\tilde{\kappa}^{RH}(x,w)$, with a given structure, which depends on a vector of parameters w. The values of w must be optimized with respect to the approximation error

$$E(w) = \sum_{i=0}^{N_c} \left\| \kappa^{RH}(x_i) - \tilde{\kappa}^{RH}(x_i, w) \right\|$$

where x_i, $i = 1, \ldots, N_c$, are the states of the sequences in correspondence of which $\kappa^{RH}(x_i)$ has been computed off-line. The approximation strategy used in this work is based on multilayer feedforward neural networks [10]. In particular we assume that the approximating neural function $\tilde{\kappa}^{RH}(x,w)$ contains only one hidden layer with ν neural units (perceptrons) with a sigmoidal activation function $f_s(y) = (1+e^{-\beta y})^{-1}$ and that the output layer is composed of linear activation units (see [2] for a successfull application of this approach to control a laboratory mechanical system).

5. NRH idle speed control

The design procedure described in the previous section has been used in the idle speed control problem to improve the performances obtained with the LQ control law. In view of the limited authority of the spark advance only the position α of the air actuation system has been computed according to the NRH approach of Section 4. On the contrary, the spark advance has been still determined with the linear control law (2), (4). For this reason, the non linear system under control (6) is formed by the ensemble of the $NARX$ engine model, the integrator on the

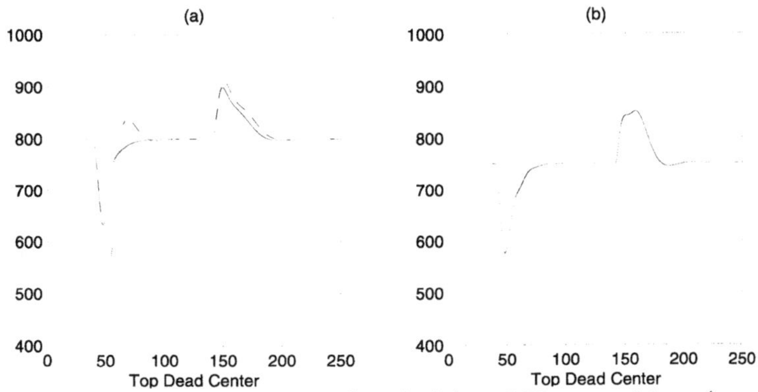

FIGURE 4. Crankshaft speed [rpm]: (a) NRH control law (continuous line) and LQ control law (dashed) with $\bar{n} = 800$; (b) NRH control law with $\bar{n} = 750$.

crankshaft speed error, the derivative term (4) and the linear control law (2). This system has again the state x=[n p γ η]' and α as only input. Assuming the steady state values $\bar{x}$ =[$\bar{n}$ $\bar{p}$ 0 0]', $\bar{\alpha}$ required in idle speed conditions, the NRH control law $\alpha(\theta) = \kappa^{RH}(x(\theta))$ can be computed for the enlarged non linear model starting from many different values of the state $x(\theta)$ and the corresponding sequences can be used to train the neural net. The overall synthesis procedure can be summarized according to the following steps:

Off-line tasks:

(a) Identify the $NARX$ model (1) (see [7], [8], [17])
(b) Design the LQ gain K for the control scheme of Fig. 2 (see [8])
(c) Synthetize the NRH control law for the additional air actuation system α starting from different values of the state $x(\theta)$ (see [6]).
(d) Interpolate the nonlinear control law by means of a neural net (see [16]).

On-line tasks:

(a) Compute the spark advance with the LQ control law.
(b) Compute the opening air actuation command with the neural net.

This procedure has been applied with an optimization horizon $N = 5$ and the sequences so obtained have been used to train a NN with 30 perceptrons in the hidden layer, which could be used for on-line computations even with the technology currently adopted in commercial applications. In order to test the performance of the neural controller trained with the NRH control law, many simulations have been performed with the $NARX$ model. In particular, a torque load disturbance similar to the real one reported in Figs 3(a) and 3(c) has been imposed. Correspondingly, the results illustrated in Fig. 4(a) have been obtained. Specifically Fig. 4(a) shows 2 different transients: the first one has obtained with the LQ control law (Section 3), the second one refers to the NRH control law. The transient produced by the trained neural net is not reported here because it is indistinguishable from the one produced by the optimal control sequence. Two facts are apparent: (a) The NN is an efficient interpolator of $\kappa^{RH}(x)$; (b) the $(NN) - NRH$ control law reduces the undershoot due to the disturbance more than 50 rpm. This in practice means a much higher driveability quality and the possibility to reduce the steady-state value $\bar{n}$, with considerable energy and economic saving. To better appreciate this point, the NRH synthesis procedures previously described has been repeated for a new steady-state condition with $\bar{n} = 750$ rpm. Then, the same disturbance previously considered has been applied to the engine simulator and the results reported in Fig. 4(b) have been obtained.

6. Conclusions

The feasibility study reported in this paper has clearly shown that if a reliable nonlinear engine model in idle speed conditions is available, the use of a NRH control law can lead to significant improvements. Moreover, the implementation problems can be overcome by means of suitable interpolation techniques. An extensive experimental phase must now be made to validate the $NN - NRH$ control law on real engines.

References

[1] M. Abate, B. Barmish, C. Murillo Sanchez, and R. Tempo. Application of some new tools to robust stability of spark ignition engine: a case study. *IEEE Trans. on Control Systems Technology*, pages 22–30, 1994.

[2] L. Cavagnari, L. Magni, and R. Scattolini. Neural network implementation of a nonlinear receding-horizon controller. In *Conference on Control Applications*, 1998.

[3] Y. Chin and F. Coast. Engine dynamics: time-based versus crank-angle based. *SAE Paper 860412*, 1986.

[4] G. De Nicolao, L. Magni, and R. Scattolini. Stabilizing nonlinear receding horizon control via a nonquadratic terminal penalty. In *IMACS Multiconference CESA '96*, pages Vol. 1, 185–187, 1996.

[5] G. De Nicolao, L. Magni, and R. Scattolini. Stabilizing predictive control of nonlinear *ARX* models. *Automatica*, 33:1691–1697, 1997.

[6] G. De Nicolao, L. Magni, and R. Scattolini. Stabilizing receding-horizon control of nonlinear time-varying systems. *IEEE Trans. on Automatic Control*, 1998. (To appear.)

[7] G. De Nicolao, O. Moro, L. Poggio, C. Rossi, R. Scattolini, and C. Siviero. Identification of nonlinear ic engine models for idle speed control. In *European Control Conference*, 1997.

[8] G. De Nicolao, C. Rossi, R. Scattolini, and M. Suffritti. An automatic synthesis procedure of idle speed linear quadratic regulators through identification experiments. In *IFAC Workshop Advances in Automotive Control*, 1998.

[9] N. Draper and H. Smith. *Applied Regression Analysis*. John Wiley & Sons, New York, 1981.

[10] K. Hornik, M. Stinchombe, and H. White. Multilayer feedforward networks are universal approximators. *Neural Networks*, 2:359–366, 1989.

[11] D. Hrovat and J. Sun. Models and control methodologies for IC engine idle speed control design. *Control Engineering Practice*, pages 1093–1100, 1997.

[12] R. Jurgen. *Automotive Electronics Handbook*. McGraw-Hill, New York, 1995.

[13] S. Keerthi and E. Gilbert. Optimal, infinite-horizon feedback laws for a general class of constrained discrete-time systems. *J. Optimiz. Th. Appl.*, 57:265–293, 1988.

[14] L. Kjergaard, S. Nielsen, T. Vesterholm, and E. Hendricks. Advanced nonlinear engine idle speed control systems. *SAE Paper n.940974*, 1994.

[15] L. Magni and R. Sepulchre. Stability margins of nonlinear receding horizon control via inverse optimality. *System & Control Letters*, 32:241–245, 1997.

[16] T. Parisini and R. Zoppoli. A receding-horizon regulator for nonlinear systems and a neural approximation. *Automatica*, 31:1443–1451, 1995.

[17] R. Scattolini, G. De Nicolao, M. Cittadini, C. Rossi, and C. Siviero. Modelling internal combustion engines via NARX identification techniques. In *IEEE Mediterranean Control Conference, Cyprus*, 1997.

Dipartimento di Informatica e Sistemistica
Università degli Studi di Pavia,
Via Ferrata 1
27100 Pavia
Italy
tel: +39-382-505362
fax: +39-382-505373/525638
E-mail address: {denicolao,magni,scatto}@conpro.unipv.it

Progress in Systems and Control Theory, Vol. 26

Performance and Computational Implementation of Nonlinear Model Predictive Control on a Submarine

Gordon J. Sutton and Robert R. Bitmead

Abstract. We consider a particular model predictive control (MPC) problem involving a nonlinear constrained submarine in order to explore the practicalities of implementation. The controller is designed using a discretised nonlinear model and is simulated on the continuous time submarine system. This controller comprises on-line constrained finite horizon optimisations, based on gradient methods, accompanied by an Extended Kalman Filter state estimator. Input and state constraints are implemented and the optimisation convergence rate is found to depend considerably on the cost function tuning.

1. Introduction

Model Predictive Control (MPC) is a control strategy which uses substantial in-loop computing power to move from a computational plant model to the implementation of a control law through on-line optimisation. This optimisation proceeds via receding-horizon, open-loop optimisations coupled with measurement feedback via a state estimator. The attractions of MPC are the immediate applicability of the receding-horizon control law (RHCL) to nonlinear and constrained systems which typically arise in applications, and the simplicity and intuition of the design variables. Our observations are that the state estimation side often is overlooked.

Our aim in this paper is to explore aspects of the implementation, design and theory of MPC via a specific computer example of a bottom-tracking submarine operating in an unknown environment. Earlier works have sought to recognise linear, unconstrained MPC (or Generalised Predictive Control(GPC)) as LQG control [1], identifying the generation of future output predictions with optimal state estimation as the RHCL with state estimate feedback. These observations are helpful in developing these GPC ideas in a nonlinear context. More recent works in robust, nonlinear, constrained MPC have concentrated on methods using full-state availability for the RHCL [4, 5, 10]. We are seeking to identify directions to proceed via a specific computational example. Accordingly, it does not provide mathematical guarantees but gives some valuable insights.

In [6, 7] a submarine example was used to demonstrate the performance of LQG control on a nonlinear, but linearisable, plant. This was compared and contrasted with the performance of an LQ RHCL with an Extended Kalman Filter (EKF). Limitations of the fully linear MPC (=GPC=LQG) control law were exposed as deriving initially from the inadequate state estimation. With the EKF, the performance and stability attraction region were improved. The authors [9] developed this comparison further to include a nonlinear, constrained RHCL coupled with an EKF. Predictably, performance improved dramatically and global capture became possible.

Here our aim is to explore in more detail the implementation of MPC on this constrained, nonlinear submarine example. Particularly, a gradient-based iterative optimisation is developed for calculating successive plant input sequences as part of the open-loop RHCL. This is embellished to accommodate input constraints by projection methods and state constraints via penalty methods. One of the features noted is that for cheap-control problems (which represent a large bulk of applications interest), the early elements in the open-loop control sequence are adjusted first and so are amenable for implementation in the RHCL after only a few iterations. Some theoretical explanations are developed.

The submarine example is a continuous-time, constrained, nonlinear system which, for the purposes of control design alone, we discretise. The simulations are performed using the discrete control on the continuous system. The submarine "lives" in a two dimensional world and experiences no external disturbances such as underwater currents. The physical submarine inspiring this model is a four meter long, one meter diameter, un(hu)manned yellow submarine depicted in Figure 1. It is driven by a single rear propeller and is controlled by a pair of rear wings each one meter wide, moving in tandem. A sonar system is in place which delivers the distance between the sea floor and the centre of the submarine, perpendicular to the sea floor. The propeller could be utilised as a control input, but for our purposes it will be held at a constant rotational speed. The control objective is to follow the sea bed at a height of fifty meters with no prior knowledge of the sea bed. The physical purpose is to survey the sea floor for the laying of communication cables. A third dimension and currents could be easily incorporated into the model by an extension in similar mathematical style, but the two dimensional system is a sufficiently complicated and revealing test for the theory. The system states are the submarine's forward, perpendicular and rotational velocities in a local submarine frame, the relative angle of the submarine to the sea floor, the distance from the submarine to 50 metres above the sea floor and the sea floor's angular velocity. The state estimates, in the order mentioned, are denoted $(\hat{u}, \hat{w}, \hat{q}, \hat{\theta}, \hat{k}, \hat{x}_f)$, the wing angle is denoted β and the sonar measurement of $\hat{k}$ is denoted y.

The submarine has a "non-minimum phase" property in the sense that it first moves the wrong way to achieve a step change. MPC is able to plan beyond this "non-minimum phase" horizon. The submarine is also non-holonomic, with acceleration not available perpendicular to the main axis, preventing a feedback linearisation control approach. The system is nonlinear but smooth, with continuous derivatives, and has an obvious control constraint that the wing angle be less than $\pi/2$ radians. Thus the submarine is a good example on which to explore MPC. It falls into the class of systems, whose robust stability was studied theoretically by the authors [8], composed of nonlinear state feedback laws with observers.

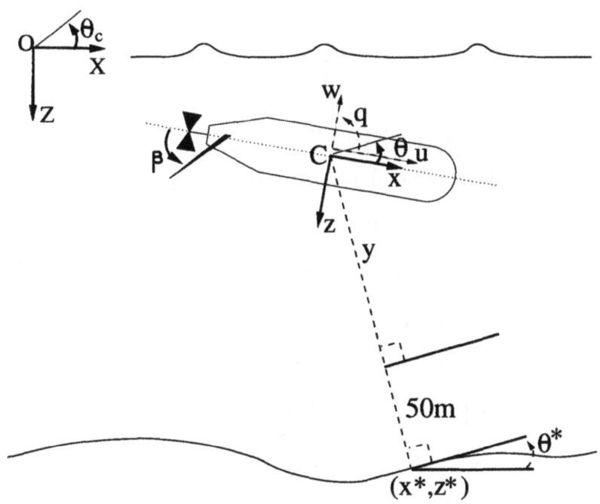

FIGURE 1. Autonomous Underwater Vehicle

The equations of motion of the submarine follow and were first derived by Aristide Santos for his doctoral thesis [7]. Some minor changes have been made.

$$M_I \begin{pmatrix} \dot{u}(t) \\ \dot{w}(t) \\ \dot{q}(t) \end{pmatrix} = mq(t) \begin{pmatrix} 0 & -1 & 0 \\ 1 & 0 & 0 \\ 0 & 0 & 0 \end{pmatrix} \begin{pmatrix} u(t) \\ w(t) \\ q(t) \end{pmatrix} + u(t) D_h \begin{pmatrix} u(t) \\ w(t) \\ q(t) \end{pmatrix} + \Gamma_g(t) + U_{cw}(t) + U_{cp}(t) \quad (1)$$

$\Gamma_g(t)$ is a buoyancy term and is zero since the submarine is neutrally buoyant. U_{cw} and U_{cp} are the forces and moments caused by the wing and propeller rotating and are given in (2)–(7) with the time dependence suppressed for notational clarity.

$$\varepsilon \triangleq \sin^{-1}\left(\frac{w}{\sqrt{u^2+w^2}}\right) \quad (2)$$

$$J_p \triangleq \frac{u}{|\nu D_p|} \quad (3)$$

$$C_{xsw} \triangleq -\left(C_{xow} + c_w \frac{C_{zow}^2(\varepsilon+\beta)^2}{2\pi b_w}\right) \tag{4}$$

$$C_{zsw} \triangleq -\left(C_{zow}(\varepsilon+\beta) + 2.1(\varepsilon+\beta)^2\right) \tag{5}$$

$$U_{cw} = \begin{pmatrix} 0.5\rho S_w V^2(C_{zsw}\sin(\beta) + C_{xsw}\cos(\beta)) \\ 0.5\rho S_w V^2(C_{zsw}\cos(\beta) - C_{xsw}\sin(\beta)) \\ -U_{cw}(2)(0.2c_w\cos(\beta) + d_{aw}) - U_{cw}(1)(0.2c_w\sin(\beta)) \end{pmatrix} \tag{6}$$

$$U_{cp} = \begin{pmatrix} \rho|\nu|\nu D_p^4(C_{t0p} + C_{t1p}J_p + C_{t2p}J_p^2 + C_{t3p}J_p^3) \\ -\rho|\nu|D_p^3 w C_n \\ \rho|\nu|D_p^3 w C_n D_{ap} \end{pmatrix} \tag{7}$$

The model parameters can be found in [7].

The rate of change of the absolute angle of the sea floor, $\dot{\theta}^\star(t)$, at $x^\star(t)$, is modelled as the output of a filtered white noise process, and is given by

$$\dot{x}_f(t) = A_f x_f(t) + B_f \xi(t), \quad \theta^\star(t) = -C_f x_f(t). \tag{8}$$

Expressing the submarine's relative angle to the sea floor, in terms of the absolute angles of the submarine, $\theta_c(t)$, and the sea floor, $\theta^\star(t)$, then differentiating gives

$$\dot{\theta}(t) = \dot{\theta}_c(t) - \dot{\theta}^\star(t) \tag{9}$$

$$= q(t) + Cx_f(t). \tag{10}$$

The rate of change of $y(t)$ in the absence of measurement noise is denoted $k(t)$. We will assume the measurements have additive white Gaussian noise, $\eta(t)$, giving

$$\dot{k}(t) = u(t)\sin(\theta(t)) - w(t)\cos(\theta(t)), \quad y(t) = k(t) + \eta(t). \tag{11}$$

Combining (1-11) gives the submarine's state equations and output $y(t)$.

2. Model predictive controller construction

Predictive control combines state estimation with receding-horizon control optimisation. An EKF is used for the state estimation and the indirect method of Lagrange Multipliers, often used for algebraic optimisations, is used for the open-loop input optimisation of the dynamic system at each step of the RHCL. The forward velocity, u, is measured exactly for use in the EKF and the EKF gain is held constant to obviate the need to compute the Riccati equation iteratively.

2.1. Open-loop optimisation

The approach taken is to transform the state update equation into a constraint of the discrete system and then to use the indirect method of Lagrange Multipliers to optimise the constrained cost function. The Lagrange Multipliers are time varying and are known as "costates" [2]. This enables us to find the gradient of the cost function with respect to the inputs alone, given an initial state.

Consider the discrete time system,

$$x_{i+1} = f(x_i, u_i), \quad x_0 \text{ given}, \tag{12}$$

and a cost function we wish to minimise

$$J = \phi(x_N) + \sum_{i=0}^{N-1} L(x_i, u_i), \tag{13}$$

which is typically a quadratic function of the form

$$J = \frac{1}{2} x_N' P_0 x_N + \frac{1}{2} \sum_{i=0}^{N-1} x_i' Q x_i + u_i' R u_i. \tag{14}$$

We rewrite J adding a sequence of value 0 by including the vector Lagrange multiplier sequence $\{\lambda_i : i = 1, \ldots, N\}$.

$$J = \phi(x_N) + \sum_{i=0}^{N-1} L(x_i, u_i) + \lambda_{i+1}' [f(x_i, u_i) - x_{i+1}]. \tag{15}$$

By defining a scalar sequence H_i, known as the Hamiltonian, as

$$H_i = L(x_i, u_i) + \lambda_{i+1}' f(x_i, u_i), \tag{16}$$

J can be written as

$$J = \phi(x_N) + \lambda_N' x_N + \sum_{i=1}^{N-1} [H_i - \lambda_i' x_i] + H_0. \tag{17}$$

We wish to examine differential changes in J due to differential changes in u_i. We have

$$\begin{aligned} dJ &= \left[\frac{\partial \phi}{\partial x_N} - \lambda_N'\right] dx_N + \sum_{i=1}^{N-1} \left\{ \left[\frac{\partial H_i}{\partial x_i} - \lambda_i'\right] dx_i \right. \\ &\quad \left. + \frac{\partial H_i}{\partial u_i} du_i \right\} + \frac{\partial H_0}{\partial x_0} dx_0 + \frac{\partial H_0}{\partial u_0} du_0. \end{aligned} \tag{18}$$

Since the vector multiplier sequence $\{\lambda_i\}$ is multiplying zero terms, they can be arbitrarily chosen. They are chosen so as to simplify (18) and are defined by

$$\lambda_N' = \frac{\partial \phi}{\partial x_N}, \quad \lambda_i' = \frac{\partial H_i}{\partial x_i} = \frac{\partial L_i}{\partial x_i} + \lambda_{i+1}' \frac{\partial f_i}{\partial x_i}. \tag{19}$$

As a consequence of this choice, dJ simplifies to

$$dJ = \sum_{i=0}^{N-1} \frac{\partial H_i}{\partial u_i} du_i + \lambda_0' dx_0 \tag{20}$$

and λ_i takes on a direct interpretation. λ_0 is the gradient of J with respect to x_0 while holding u_i constant. If we think of x_i as the initial state and J_i as the cost to go from state i using the remaining control sequence $(u_i, \ldots, u_{N-1})$ then λ_i is the gradient of J_i with respect to x_i while holding the remaining u_i constant. Also $\frac{\partial H_i}{\partial u_i}$ is the gradient of J with respect to u_i while holding x_0 and $u_{j,j\neq i}$ constant. Thus, if x_o is constant, an N dimensional search direction from the vector $(u_o, u_1, \ldots, u_{N-1})$ can be obtained which will lead to a lower cost. At a local minimum of J, we can conclude that $\frac{\partial H_i}{\partial u_i} = 0$.

So the optimisation process is:

- solve (12) forwards in time using a candidate input sequence $(u_0 \ldots u_{N-1})$ to find $\{x_i\}$,
- find λ_N using (19) and iterate backwards in time to find $\{\lambda_i\}$
- use each $\{x_i, u_i, \lambda_{i+1}\}$ set to calculate $\frac{\partial H_i}{\partial u_i}$,
- update $\{u_i\}$ by moving a distance along the gradient vector $(\frac{\partial H_0}{\partial u_0} \ldots \frac{\partial H_{N-1}}{\partial u_{N-1}})$ from $(u_0 \ldots u_{N-1})$, reducing the distance moved if required to ensure J decreases.
- repeat the process with the new input sequence to the desired accuracy.

This optimisation is not in itself novel, but our investigation here is to seek a computationally feasible optimisation method suited to NMPC. From the initial state $x(0)$ the input sequence can be initialised as the N-step finite horizon LQ optimal control. At each subsequent sampling time the input sequence is initialised as $(u_2, \ldots, u_N, u_N)$ from the previous optimisation. We can thus theoretically find a local minimum of J. There are still the implementation issues, besides designing J, of how to incorporate input and state constraints.

Input constraints can be implemented by projecting the particular u_i onto the constraint set. That is, if the search step takes a particular u_i beyond the constraint set, the u_i becomes the value at which the constraint is crossed. State constraints are computationally expensive to implement strictly in this framework, requiring the piecing together of constrained and unconstrained segments. They may be accommodated approximately, with little computational expense, by penalty functions[3], in which the performance objective is modified by the inclusion of a function which increases rapidly near and outside the constraint set.

2.2. Starting and stopping the optimisation

Initialising each open-loop optimisation with $(u_1, \ldots, u_N, u_N)$ has advantages of computation, continuity and maintaining feasibility. Since we only require the first input from each optimisation, there is the issue of how accurate the optimisation needs to be. The optimal inputs are dependent, so the whole sequence needs to be optimised to some extent. At the same time, there would be wasted computation on fine tuning the later inputs, since the estimated state for the start of the next optimisation will almost certainly not be that predicted by the model due to modelling errors, sensor errors, and unpredictable disturbances.

2.3. Effect of tuning R on the optimisation

We now consider the effect of R from (14) on the fine tuning end of the optimisation. From (20) we have $\frac{\partial H_i}{\partial u_i}$ equivalent to $\frac{\partial J}{\partial u_i}$, so the search direction is equivalent to $(\frac{\partial J}{\partial u_0}, \ldots, \frac{\partial J}{\partial u_{N-1}})$. If $\{|\frac{\partial J}{\partial u_i}|\}$ is a decreasing sequence then, with a fixed search distance, the early inputs will govern a larger portion of this distance and be adjusted faster. It also implies that u_1 is the most influential control value to the cost and hence will be calculated more accurately more quickly. Since for NMPC we only want the first input from each optimisation we could thus stop the optimisation before the later inputs are fine tuned and still obtain a reasonable approximation to the first input.

To obtain an understanding of how $\{\frac{\partial J}{\partial u_i}\}$ behaves around the optimal control sequence, we first recall that at the optimum each $\frac{\partial J}{\partial u_i} = 0$. A Taylor series expansion gives an approximation to $\frac{\partial J}{\partial u_i}$ around the optimal control $\{u_i^\star\}$ as

$$\begin{aligned}\frac{\partial J}{\partial u_i}(u^\star + \delta u) &= \frac{\partial J}{\partial u_i}(u^\star) + \sum_{j=1}^{N} \frac{\partial}{\partial u_j}\left(\frac{\partial J}{\partial u_i}\right)(u^\star)\delta u_j + O(\delta u^2) \\ &\approx \sum_{j=1}^{N} \frac{\partial}{\partial u_j}\left(\frac{\partial J}{\partial u_i}\right)(u^\star)\delta u_j. \end{aligned} \tag{21}$$

We note that convexity requires the matrix $\left\{\frac{\partial}{\partial u_j}\left(\frac{\partial J}{\partial u_i}\right)(u^\star)\right\}$ to be positive definite, so $\frac{\partial^2 J}{\partial u_i^2}(u^\star)$ are the dominating weights of the search direction in (21), which we wish to have decreasing. If a gradient descent optimisation algorithm is used we can pictorially see why it would be beneficial to have $\{\frac{\partial^2 J}{\partial u_i^2}\}$ decreasing with i around the optimum. Figure 2a depicts how along the steeper axis, with higher $\frac{\partial^2 J}{\partial u_i^2}$, u_i will be approximately minimised quickly and along the shallow axis u_i will settle more gradually in a standard gradient optimisation. The ovals are level curves of the cost criterion. A larger than ideal step size has been used to demonstrate how on the steeper axis, the input can oscillate while still reducing the cost after each step. With smaller steps the oscillations cease. Figures 2b and 2c compare some paths for evenly and unevenly weighted second derivatives around the optimum using small steps.

We now consider the linear case to demonstrate this. Take the system

$$x_{k+1} = Fx_k + Gu_k, \quad x_0 \text{ given}, \tag{22}$$

and the associated cost function (14) such that $J = J(x_0, u_0, \ldots, u_{N-1})$. We write x_j explicitly in terms of independent variables;

$$x_j = F^j x_0 + \sum_{i=0}^{j-1} F^{j-i-1} G u_i. \tag{23}$$

Differentiating and combining (14) and (23) gives

$$\frac{\partial J}{\partial u_i} = u_i' R + x_N' P_0 F^{N-i-1} G + \sum_{j=i+1}^{N-1} x_j' Q F^{j-i-1} G. \tag{24}$$

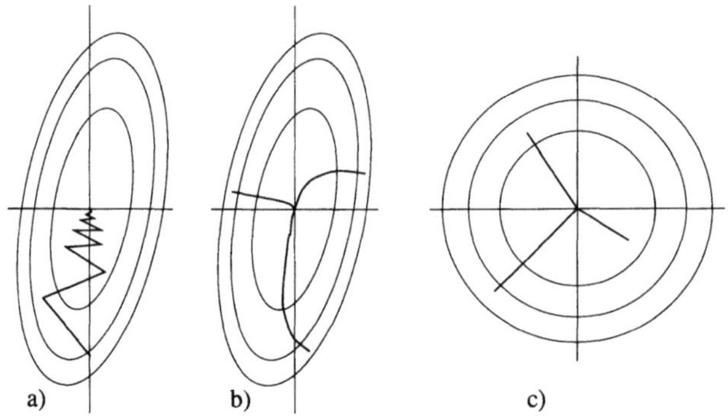

FIGURE 2. Effect of variations in $\frac{\partial^2 J}{\partial u_i^2}$ on the gradient descent path. (a) oscillations due to large step size, (b) paths with small step size and (c) paths when $\frac{\partial^2 J}{\partial u_1^2} = \frac{\partial^2 J}{\partial u_2^2}$.

Differentiating again we obtain

$$\frac{\partial^2 J}{\partial u_i^2} = R + G'F'^{N-i-1} P_0 F^{N-i-1} G + \sum_{j=i+1}^{N-1} G'F'^{j-i-1} Q F^{j-i-1} G. \tag{25}$$

The difference between consecutive second derivative values is thus

$$\frac{\partial^2 J}{\partial u_i^2} - \frac{\partial^2 J}{\partial u_{i+1}^2} = G'F'^{N-i-2}[Q - P_0 + F'P_0 F]F^{N-i-2}G. \tag{26}$$

With appropriate selection of P_0, such as $P_0 = Q$, analysis of (25) and (26) indicates that for large R (which might not be of practical interest), the $\frac{\partial^2 J}{\partial u_i^2}$ are roughly constant over i while for small R the early terms dominate. The effect of this, for small R, is to distend the performance function J in these earlier directions leading to two competing effects: (i) that the optimisation is ill-conditioned and hence supports only slow adjustments in the gradient direction, and (ii) that the first elements of the optimal control vector are found more accurately first. For receding-horizon control, where we are predominantly interested in the first control, the latter effect mitigates the former. This analysis in the linear case should be extensible to the nonlinear case, although we have yet to do this.

2.4. Specifics for the submarine

A strict input constraint of $|\beta| \leq 0.5$ radians is enforced due to the drag and excessive spiralling effect that a large β can cause. The choice of control input is dependent on the forward velocity and so, although the Extended Kalman Filter currently does not observe it, the forward velocity is assumed known.

We also wish the submarine to head in the positive OX direction (see Figure 1). This becomes an issue during the descent, where between the transient rotation and the state estimation errors the vessel can descend with the vessel angle,

θ, greater than $\pi/2$ radians and sometimes cause the submarine to head in the negative OX direction upon reaching the desired depth. This is resolved using the state constraint $|\hat{\theta}_k| < \pi/2$. To implement the constraint approximately we add a $\hat{\theta}_k$ dependent quadratic term to the cost function. This gives the cost function as

$$J = \frac{1}{2} P_0 \hat{y}_N^2 + \frac{1}{2} \sum_{k=0}^{N-1} \hat{y}_k^2 + R\beta_k^2 + S(\hat{\theta}_k)(|\hat{\theta}_k| - \theta_0)^2, \tag{27}$$

with the input constraint: $|\beta| \leq 0.5$rad, where $S(\hat{\theta}_k)$ is zero for $\hat{\theta}_k$ inside $(-\theta_0, \theta_0)$ and is S_c.signum$(\hat{\theta}_k)$ for $\hat{\theta}_k$ outside $(-\theta_0, \theta_0)$.

3. Model predictive control experiences

Unconstrained MPC works well with this example except for the need to prevent the submarine facing backwards at the conclusion of the descent. Figure 3 depicts the behaviour of MPC with the (artificially severe) control limit of $|\beta| < 0.1$rad. The upper picture illustrates the control input as a function of time, while the lower depicts the submarine trajectory in two dimensions.

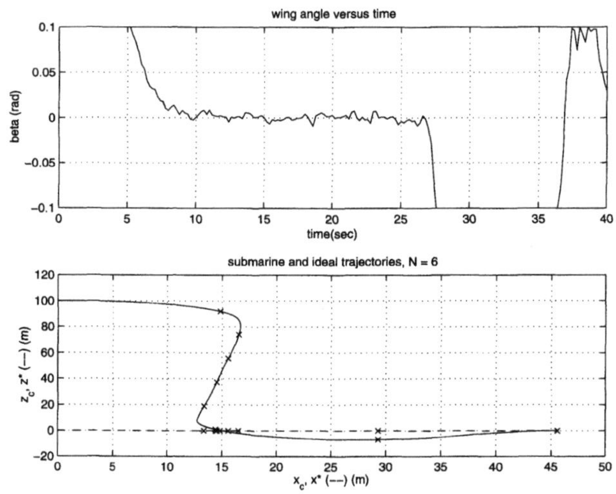

FIGURE 3. Wing angle and trajectory with input constraint $|\beta| \leq 0.1$rad. $R = 10$, $P_0 = 1$, $S_c = 0$, $T = 0.2$sec, $N = 20$. x: 5sec time points.

To achieve an almost vertical, yet forward, descent we implement a state constraint approximately. S_c has been set to 50 and θ_0 to 1.5rad. Figure 4 presents one trajectory without and one with the constraint $|\hat{\theta}_k| < \pi/2$rad. On occasions, $|\theta|$ does exceed $\pi/2$rad but only by a fraction of a degree. The important feature is ensuring the submarine is heading the correct way near the desired depth.

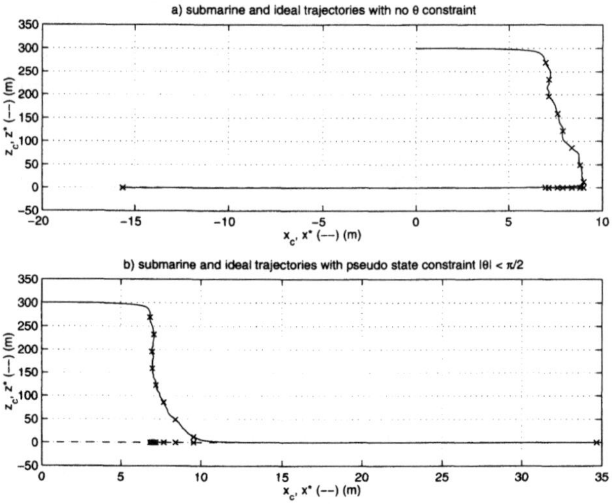

FIGURE 4. Trajectories without and with constraint $|\hat{\theta}_k| < \pi/2$. $R = 10$, $S_c = 350$, $P_0 = 1$, $\theta_0 = 1.5$rad, $T = 0.2$sec, $N = 30$. x: 10sec time points.

The simulations presented have used flat sea floors. We have looked at the more realistic problem of tracking an evolving sea floor, with very satisfactory results. These will be reported elsewhere.

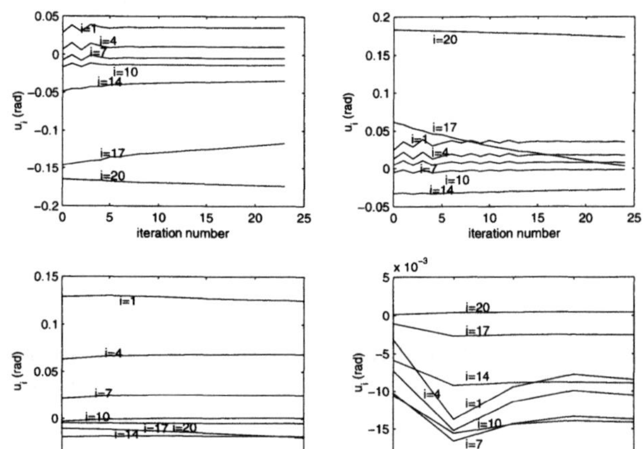

FIGURE 5. Evolution of $\{u_i\}$ during optimisation, for submarine going into descent(left) and coming out of descent (right). $R = 1$ above and $R = 700$ below.

Figure 5 shows some typical iterative optimisations of $\{u_i\}$ with $R = 1$ above and $R = 700$ below. Some of the inputs are removed for clarity. The left hand pair has the submarine turning into a descent so the inputs mainly decrease with i. The right hand pair shows the input evolution as the submarine is leaving its stable descent which results in u_0 not being the largest input. In both cases, the smaller R results in the early u_is being found first, indicated by the flat tails of their curves. For the smaller R it can take up to 25 steps to find the optimal input sequence, though u_0 is usually closely found before 8 steps. The larger R leads to a more even fine tuning across i and hence early u_is are no longer found first. There are however far fewer steps. The chattering effect due to ill conditioning for small R and large step sizes can be seen on the early inputs with small R. The overall behaviour matches that expected by extrapolating the linear theory to the nonlinear case.

Experience with selecting the input from an approximate optimisation indicates the behaviour predicted above, where the control term used in the RHCL is computed accurately early in the optimisation, produces an almost identical trajectory to when the optimisations are allowed to proceed to an accurate solution.

4. Conclusion

We consider the implementation of MPC on a submarine example. The open-loop input optimisations required for the receding-horizon control law are evaluated using a gradient-descent search, based on the indirect method of Lagrange Multipliers. The effect of tuning the quadratic input cost weight R on the optimisation and hence the appropriateness of this method for MPC is addressed. The use of penalty functions to implement state constraints approximately is also presented.

References

[1] R. R. Bitmead, M. Gevers and V. Wertz. *Adaptive Optimal Control, The Thinking Man's GPC*, Prentice Hall, 1990.

[2] A. E. Bryson Jr, Y-C. Ho, *Applied Optimal Control: Optimization, Estimation and Control*, Ginn and Co., 1969.

[3] D. G. Luenberger. Introduction to Linear and Nonlinear Programming, Addison-Wesley, 1973.

[4] D. Q. Mayne and H. Michalska. "Receding Horizon Control of Nonlinear Systems", *IEEE Trans. Automat. Contr.*, vol. 35, no. 7, pp. 814–824, July 1990.

[5] H. Michalska and D. Q. Mayne. "Robust Receding Horizon Control of Constrained Nonlinear Systems", *IEEE Trans. Automat. Contr.*, vol. 38, no. 11, pp. 1623–1632, November 1993.

[6] A. Santos and R. R. Bitmead, "Nonlinear Control for an Autonomous Underwater Vehicle (AUV) Preserving Linear Design Capabilities", 34th *IEEE CDC Proceedings*, pp. 3817–3822, Dec 1995.

[7] A. Santos, PhD Thesis, L'École Nationale Supérieure des Mines de Paris, 1995.

[8] G.J. Sutton and R.R. Bitmead, "Robust stability theorems for nonlinear predictive control," *36th IEEE Conf. Decision and Control,* San Diego CA, USA, 1997.

[9] G. J. Sutton and R. R. Bitmead, "Experiences with Model Predictive Control applied to a nonlinear Constrained Submarine", 37th *IEEE CDC,* Tampa, FL, USA, 1998.

[10] T. H. Yang and E. Polak, "Moving Horizon Control of Nonlinear Systems with Input Saturation, Disturbances, and Plant Uncertainty", *Inter. Journal of Contr.*, Vol.58. No.4, pp. 875–903, 1993.

Departments of Engineering and Systems Engineering
Australian National University
Canberra
ACT, 0200
Australia
E-mail address: gordons@syseng.anu.edu.au, bob@syseng.anu.edu.au